Series: Marine Science

Series Editor: Andrés H. Arias

Volume 2

Marine Ecology

An Ecosystemic View of Anthropogenic Impacts

Editors

Juan M. Molina

Instituto Argentino de Oceanografía
Universidad Nacional del Sur
CONICET, IADO
Bahía Blanca, Argentina
and
Departamento de Biología
Bioquímica y Farmacia
Universidad Nacional del Sur (UNS)
Bahía Blanca, Argentina

Gabriela E. Blasina

Instituto Argentino de Oceanografía
Universidad Nacional del Sur
CONICET, IADO
Bahía Blanca, Argentina
and
Departamento de Biología
Bioquímica y Farmacia
Universidad Nacional del Sur (UNS)
Bahía Blanca, Argentina

CRC Press
Taylor & Francis Group
Boca Raton London New York

CRC Press is an imprint of the
Taylor & Francis Group, an **informa** business
A SCIENCE PUBLISHERS BOOK

Cover credit: The picture is the coast of Monte Hermoso, Argentina. Reproduced by kind courtesy of Dr. María Emilia Croce.

First edition published 2025
by CRC Press
2385 NW Executive Center Drive, Suite 320, Boca Raton FL 33431

and by CRC Press
4 Park Square, Milton Park, Abingdon, Oxon, OX14 4RN

CRC Press is an imprint of Taylor & Francis Group, LLC

Library of Congress Cataloging-in-Publication Data (applied for)

ISBN: 978-0-367-45660-3 (hbk)
ISBN: 978-1-003-02461-3 (ebk)

DOI: 10.1201/9781003024613

Typeset in Times New Roman
by Prime Publishing Services

Preface to the Series

Covering more than 70% of the Earth, the oceans regulate our environment including the climate, water purification, carbon storage and primary production. In addition, they also provide provisioning and cultural services, including harvestable goods (fish, shellfish and seaweed), mangroves timber, algae, minerals, recreation, beauty and plenty of cultural benefits. At present, the human population is reaching 8000 million while about half of the planet's suitable land has already been cultivated for crops, exacerbating climate change, biodiversity loss and compromising other ecosystems services. In this scenario, our oceans, which hold 97 per cent of the Earth's water, emerge as the immediate supply answer, not only in the food production sector, but also as a source of energy, biomaterials and even biochemicals/pharmaceuticals.

In these times, when global attention is focusing on oceans and their role in present and future human development, we are humbled to deliver this Marine Book Series through Taylor & Francis, aiming to cover all aspects of oceans and marine research. The present series will provide state-of-the-art coverage of important topics in marine biology and biodiversity, marine ecology, marine biotechnology, marine bioproducts, marine bioprocessing and marine chemistry. We believe this series will be of interest to specialists and nonspecialists alike as well as to scientists and managers engaged in the study of marine biosphere, addressing natural phenomena occurring at coasts, wetlands, estuaries, continental slopes, continental rises, abyssal regions, mid-ocean ridges, along with neritic and pelagic waters of the ocean.

The Marine Series Collection starting point "Marine Biology: A Functional Approach to the Oceans and their Organisms", deliver contributions from an impressive group of international scientists bringing a comprehensive analysis which range from the ocean fundamentals to the coupling and functionalities of different ecosystems—from primary producers to the oceanic megafauna. The second pillar of the series introduce us into the intricate world of marine ecosystems with "Marine Ecology: An Ecosystemic View of Anthropogenic Impacts." This comprehensive volume synthesizes cutting-edge research from leading experts, revealing how human activities shape the structure and function of these vital environments. Each chapter tackles pressing challenges of the Anthropocene, including climate change, fisheries, and chemical contamination, while emphasizing vulnerable ecosystems and innovative mitigation strategies. Discover the delicate balance of sandy beaches, rocky shores, and deep-sea habitats, and learn about the tools necessary for preserving biodiversity. With a focus on practical solutions, this volume will equip readers with knowledge on taxonomic identification and physiological responses of marine organisms. Ideal for students, researchers, and environmental advocates alike, this essential resource aims to inspire action towards a sustainable future for our oceans. Join the movement to protect marine life and ensure that the rich diversity of our planet endures for generations to come!

Dr. Andrés H. Arias

Preface

Marine Ecology: An Ecosystemic View of Anthropogenic Impacts is a synthesis of the current knowledge about the structure and functioning of marine ecosystems, and their responses to the main anthropogenic impacts they are subjected to. Each chapter is spearheaded by experts in their respective fields, and contains much of the results of their own research on the subject, which has given a hands-on approach to each of them. In this book, the reader will find information on the major challenges of the Anthropocene, posed by humanity's current lifestyle, to the biota of the planet, a point of view from specifically vulnerable ecosystems, and a section outlining the tools that humankind has developed to mitigate the damage to, protect, and even restore the marine biodiversity. It is the hope of the editors and authors of this book that the perspectives outlined here will help future generations with the increasingly difficult task of ensuring that the wonderful diversity of life does not vanish.

The first section of the book addresses man-made challenges imposed on ecosystems from the optic of the challenge itself. These challenges are discussed in the light of recent research, and are accompanied by specific results produced by each group of experts on the subject. The major impacts on the biota are outlined, and mitigation strategies are discussed. Climate change, one of the major overarching phenomena, is discussed in the first chapter. Chapter 2 discusses the direct and indirect effects that fisheries at different scales have produced on the marine ecosystems that sustain them. In Chapter 3 the authors explore the issues of large-scale aquaculture, and the sustainability of small-scale alternatives. Chemical contamination is a major threat to marine life, and different aspects of this challenge are treated in Chapter 4. The section closes up with Chapter 5, dedicated to sea littering and the patterns of its dispersal based on recent findings.

The second section is dedicated to particular ecosystems, their major biological components and interactions, and how human impacts disrupt the natural dynamics that occur in them. The first three chapters in this section provide a detailed description of sandy beaches, rocky shores and estuarine environments. The chapters discuss environmental complexity and nuances of each marine habitat, the threats they are faced with, and strategies crucial to ensuring the long-term health and preservation of the biodiversity that they host. The section closes in Chapter 9, which is dedicated to deep sea ecology, a vast, once-thought barren, ecosystem, that is nothing of the sort.

The last section discusses tools and strategies to tackle these aforementioned issues. In order to understand, predict and restore the effects of natural or anthropogenic factors on the structure and function of ecosystems, it is essential to possess knowledge on the species inhabiting a particular environment. In this sense, accurate taxonomic identification of biological species is the first step. The Chapter 10 analyzes current molecular techniques as a tool, complementary to phenotypic description, for the correct identification of known species and the discovery of cryptic biodiversity. Physiology links the individual with its environment in a mechanistic fashion. How organisms respond physiologically to fluctuating environmental conditions, and what tools has science developed to answer that question is discussed in Chapter 11. We humans are inexorably intertwined with the health of marine environments. Given the unprecedented level of marine ecosystem's degradation, governance of these environments must effectively ensure a sustainable and equitable use of it. This topic is developed and exemplified in

Chapter 12 of the book, which advocates the adoption of a holistic view of marine ecosystems, emphasizing their complex and dynamic nature. The last chapter of this book looks at how we can revert the damage and degradation inflicted upon the marine ecosystems discussed throughout the previous chapters. The initiatives and strategies to restore damaged environments are discussed and explored in this final chapter.

Contents

Section I

Major Human Impacts on Marine Ecosystems

CHAPTER 1

Climate Change Impacts on Coastal Biological Communities

Henrique N. Cabral,[1,*] *Diana Madeira,*[2] *Mário Diniz,*[3,4] *Carolina Madeira,*[4,5] *Ana Luísa Maulvault,*[3,4,6] *Elorri Arevalo,*[1,7] *Noelle M. Lucey*[8] and *Sam Dupont*[9,10]

Main climate change drivers for coastal communities

Global climate change and associated extreme weather events are a major threat to marine biodiversity worldwide, impacting ecosystem services, ocean-dependent livelihoods, and economies (Intergovernmental Panel on Climate Change (IPCC) 2023a, Smith et al. 2021, Talukder et al. 2022). The continuously increasing anthropogenic greenhouse gas emissions have induced physical and chemical changes in the ocean (Intergovernmental Panel on Climate Change (IPCC) 2023a). Nevertheless, the attribution of these changes to specific drivers has been difficult as they often interact and overlap at different spatial, temporal, and organisational levels, among other interpretation problems (e.g., Abram et al. 2019, Gissi et al. 2021).

Global mean sea surface temperature has increased nearly 1°C from the beginning of the 20th century and an additional increase of 0.89°C to 2.89°C is predicted for the 2081–2100 period, according to IPCC's scenarios SSP1-2.6 or SSP5-8.5, respectively (Cooley et al. 2022). Climate velocity, i.e., the speed and direction at which isotherms move, has also been changing dramatically, especially

[1] INRAE, UR EABX, 50 Avenue de Verdun, 33610 Cestas, France.

[2] ECOMARE – Laboratory for Innovation and Sustainability of Marine Biological Resources, CESAM – Centre for Environmental and Marine Studies, Department of Biology, University of Aveiro, Estrada do Porto de Pesca, 3830-565 Gafanha da Nazaré, Aveiro, Portugal.

[3] UCIBIO – Applied Molecular Biosciences Unit, Department of Chemistry, NOVA School of Science and Technology, NOVA University Lisbon, 2819-516 Caparica, Portugal.

[4] Associate Laboratory i4HB - Institute for Health and Bioeconomy, NOVA School of Science and Technology, NOVA University Lisbon, 2819-516 Caparica, Portugal.

[5] UCIBIO – Applied Molecular Biosciences Unit, Department of Life Sciences, NOVA School of Science and Technology, NOVA University Lisbon, 2819-516 Caparica, Portugal.

[6] IPMA I.P. – Portuguese Institute for the Sea and Atmosphere, Division of Aquaculture, Upgrading and Bioprospection, Av. Dr. Alfredo Magalhães Ramalho, 6, 1495-165 Algés, Portugal.

[7] AZTI Marine Research, Basque Research and Technology Alliance (BRTA), Txatxarramendi Ugartea z/g, 48395 Sukarrieta, Spain.

[8] High Meadows Environmental Institute, Princeton University, Princeton, New Jersey 08544, USA.

[9] Department of Biological and Environmental Sciences, University of Gothenburg, 45178 Fiskebäckskil, Sweden.

[10] IAEA, Marine Environment Laboratories, Radioecology Laboratory, 4 Quai Antoine 1er, 98000 Monaco.

* Corresponding author: henrique.cabral@inrae.fr

in the poles and tropical areas, as well as in the deep ocean (Brito-Morales et al. 2020, Burrows et al. 2011, Cooley et al. 2022). Periods of extreme seawater temperature relative to the long-term mean values that persist for days to months, known as *marine heatwaves* (e.g., (Fox-Kemper et al. 2021, Hobday et al. 2016, Smale et al. 2019)), became more frequent in the last decades (almost the double of events compared to the 1980s, and more than 20-fold when considering only the largest and most impactful ones (Laufkötter et al. 2020)). According to IPCC predictions, marine heatwaves may increase 2–15 times (according to mild or extreme scenarios) in 2081–2100 compared to 1995–2014 (Cooley et al. 2022, Fox-Kemper et al. 2021, Frölicher et al. 2018). A relative 0.20 m sea level rise since the early 20th century has been affecting most of the coastal areas worldwide; however, an additional rise of 0.2–0.30 m is foreseen in the most pessimistic IPCC scenarios (SSP5-8.5) (Cooley et al. 2022, Fox-Kemper et al. 2021).

Evidence of other physical changes, including ocean circulation, stratification, upwelling, and sea ice patterns, are sparser and often reported at a regional level; still, their impacts may ripple to wider scales. Ocean currents drive biogeochemical processes and biological connectivity in marine ecosystems and thus changes in their patterns may affect ecosystem functioning drastically. Ocean stratification in the surface layers is expected to be accentuated in the future, while predictions regarding coastal upwelling are less consensual and probably differ according to geographical area (Cooley et al. 2022).

The most important climate-induced chemical changes in the ocean include a decrease in water pH and oxygen concentration. The uptake of carbon dioxide (CO_2) by the ocean leads to the formation of carbonic acid (H_2CO_3), which in turn dissociates into bicarbonate (HCO_3^-) and hydrogen (H^+) ions. The latter decrease water pH, thereby causing *ocean acidification*. Some secondary effects of this acidification are the decrease in the concentration of carbonate ions (CO_3^{2-}), affecting organisms with calcium carbonate shells and skeletons (e.g., Doney et al. 2009). Direct measurements since the 1980s have outlined a consistent decrease in ocean surface pH at a rate of 0.016–0.020 per decade, and a decrease has also been reported for deeper waters (Canadell et al. 2021). The decrease by the end of the 21st century is expected to be of 0.08–0.37 (according to the different scenarios) compared to reference values for the 1995–2014 period (Canadell et al. 2021). The decrease in oxygen concentration results from ocean warming (that diminishes water oxygen saturation capacity) and an increase in oxygen consumption, ocean stratification, and ventilation changes (Intergovernmental Panel on Climate Change 2019, Keeling et al. 2010). The decline in oxygen concentration was mainly noticed in the upper layers of the ocean, with a global mean decrease of 0.5% to 3.3% over 1970–2010; an expansion in oxygen minimum zones was also reported. Projections for 2081–2100 emphasize this decline in oxygen concentration to historically unprecedented levels over the 21st century but present a low agreement regarding oxygen minimum zones (Canadell et al. 2021, Kwiatkowski et al. 2020).

The assessment of the impacts of these climate-induced changes in marine ecosystems is extremely difficult due to several reasons. As most ecosystems, coastal systems are complex and highly dynamic, and for a wide range of regions, knowledge on these aspects is still extremely scarce. Climate drivers act in combination and the resulting effects may be difficult to predict. Furthermore, climate-induced pressures lead to changes at different biological organisation levels, and spatial and temporal scales, with a multitude of possible interactions that severely constrain assessments. A large volume of studies has been conducted in the past few decades to evaluate the impacts of climate-induced pressures in biological organisms, populations, and communities. However, major knowledge gaps exist due to most studies focusing on a single or few species and/or on a single or few drivers (Boyd et al. 2018). Efforts to tackle more complex contexts have increased in recent years, basically exploring capabilities in experimental setups and modelling approaches. According to Boyd and co-authors, most studies have addressed *acclimation* of organisms but very few refer *adaptation* (Fig. 1), which is critical to estimate changes and build predictions in a wider time frame (Boyd et al. 2018).

In this chapter, we review the existing knowledge regarding climate change impacts on coastal organisms, from molecular to ecosystem level. We present evidence and examples of documented impacts and highlight the main gaps and future research perspectives.

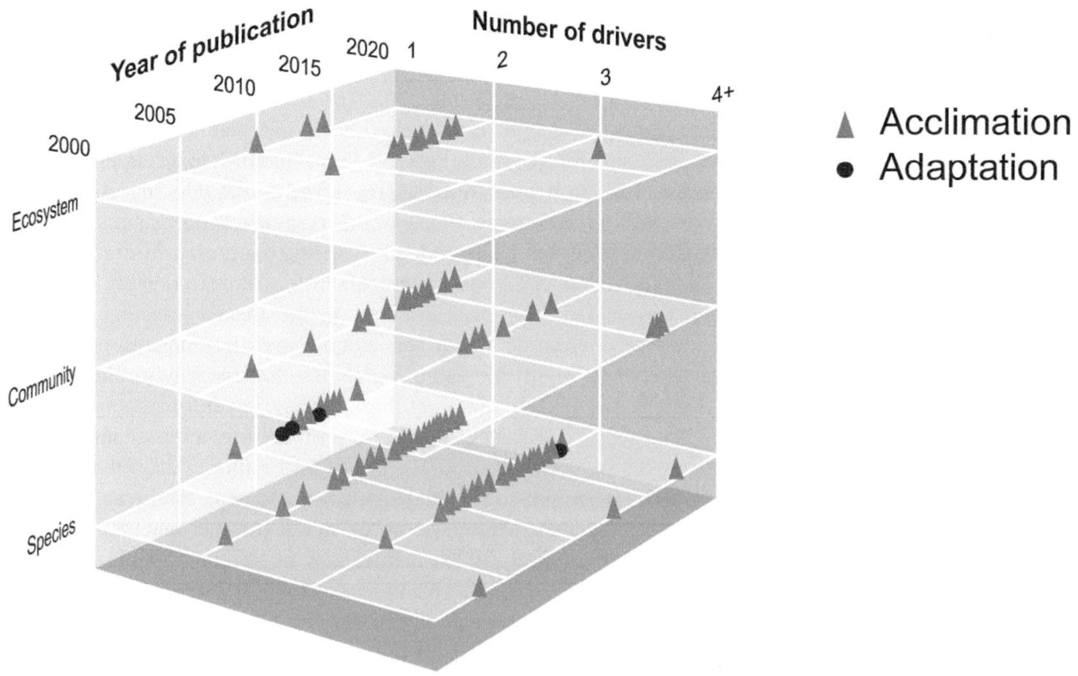

Figure 1. Trends in research from 2000 to 2020 according to a survey of 171 studies (adapted from Boyd et al. 2018). Studies using multiple drivers (generally 2 or 3) at the species level and mostly focusing on acclimation rather than adaptation were dominant over the analysed period.

Effects at the molecular and cellular levels

Cellular physiology under stress

The capacity of marine organisms to deal with novel or changing ocean conditions is, at a first instance, dependent on the regulation of their biochemistry and physiology. Changes in the environment may lead to stress to the inhabiting communities. Many definitions of stress have been proposed, the most common classifying it as a threat to the homeostasis of the organism. For the purpose of this chapter, we will adopt a general definition based on systems biology, according to which "stress occurs when a biological control system detects a failure to control a fitness-critical variable, i.e., with the potential to significantly impact organismal survival or reproductive success" (Del Giudice et al. 2018). Stressful environmental shifts are detected by organisms through their sensory systems, resulting in the activation of signal transduction pathways and eliciting a cellular response to the detected environmental shift. Despite phylogenetic diversity, most organisms display a core of evolutionary conserved molecular mechanisms to respond to cellular stresses (Kültz 2003). For example, when the homeostatic set-points of the cell are disrupted by environmental changes, the cellular stress response (CSR) is activated (Kültz 2020, Somero 2020). The timing and intensity of this response is often dependent on stress levels, signalled by the extent of macromolecular damage in the cell (Somero 2020). The CSR can be expressed via several pathways and has four central functions, which can be gradually activated as needed. These are: (i) to maintain and restore the integrity of macromolecules (for example, repair protein and DNA damage); (ii) to modulate and reallocate metabolic energy towards homeostasis and cellular repair processes; (iii) to regulate the cell cycle via the activation of cell cycle checkpoints, resulting in the regulation of cellular proliferation; and (iv) to induce apoptosis when repair mechanisms are not capable of preventing extensive cellular damage (Kültz 2020, Somero 2020). The CSR, or some of its components, may be activated in both marine vertebrates and invertebrates upon exposure to various environmental factors including cold or heat (e.g., Beemelmanns et al. 2021a, Beemelmanns et al. 2021b, Madeira et al. 2020a, 2014, Sleight et al. 2018, Veilleux et al. 2018, Yan et al. 2022), hypercapnia (e.g., (Mittermayer et al. 2019, Pimentel et al. 2015, Tomanek 2011)), hypoxia (e.g., Beemelmanns et al. 2021a, Beemelmanns et al. 2021b, Xin et al. 2022), pollution (e.g., (López-Pedrouso et al. 2022)), hypo- or hyper-salinity (e.g., (Evans and Kültz 2020)), ammonia toxicity (e.g., (Lu et al. 2022)), disease (e.g., (Rungrassamee et al. 2010)), biotoxins (associated with harmful algal blooms) (e.g., (Cao et al. 2018)), and several combinations of these factors (e.g., Beemelmanns et al. 2021a, Beemelmanns et al. 2021b, Huth and Place 2016, Pimentel et al. 2015, Rathburn et al. 2013, Sampaio et al. 2018), highlighting the non-specificity of this response towards stressor types.

When environmental shifts associated to global changes lead to stress, disruption of the structural integrity of cells and damage to their different components and macromolecules is likely to occur. This may be associated with direct physical or chemical impacts on the macromolecular structure (e.g., denaturation) or indirect impacts of environmental drivers elicited by an enhanced production of reactive oxygen species (ROS), which are highly unstable molecules that react with and damage cellular components (Kültz 2020, Somero 2020). To counteract these effects, several pathways may be up-regulated including heat shock response, proteolysis, cytoskeletal remodelling, antioxidant defence, and immune function (Bernal et al. 2020, DeSalvo et al. 2010, Evans and Kültz 2020, Ivanina et al. 2016, Madeira et al. 2017). For example, environmental changes such as elevated temperature, ocean acidification, and biotoxins strongly affect cytoskeletal proteins of many marine species, including molluscs (Johnson and Hofmann 2017, Martins et al. 2020, Tomanek et al. 2011), fish (Araújo et al. 2018, Jayasundara et al. 2015, Madeira et al. 2017), corals (Kaniewska et al. 2012, Petrou et al. 2021), anemones (Oakley et al. 2017), and crustaceans (Madeira et al. 2020a). Several authors have suggested that the cytoskeleton is a major target of stress-induced ROS production, triggering the activation of cytoprotective pathways such as the heat shock response (e.g., (Tomanek 2011)). Evidence thus suggests that environmental changes can induce cytoskeleton destabilization and further remodelling. Cytoskeletal adjustments during stress may be essential to maintain cellular functions such as cell motility, signalling, gene transcription, organelle movement, and ATP turnover rates (Garland et al. 2015, Madeira et al. 2017). Major macromolecules such as proteins, DNA, RNA, or lipids are also sensitive to damage under stressful environmental conditions. For instance, temperature is known to affect the integrity of the protein pool of marine ectothermic animals, as these cannot regulate body temperature. Increased thermal load can destabilize protein structure and lead to denaturation and cytotoxic aggregations in the cell cytosol and organelles (e.g., endoplasmic reticulum and mitochondria) (Lushchak 2011, Somero 2020). In addition, by increasing organism metabolic rates, elevated temperature (and other stressors including salinity shifts and ocean acidification) can enhance the production of ROS in mitochondria leading to oxidative stress, i.e., to enhanced ROS concentration in the cell (González et al. 2015, Lushchak 2011, Rivera-Ingraham et al. 2016). Increased production of ROS often results in damages to proteins (e.g., carbonylation), DNA (e.g., strand breaks) and lipids (e.g., peroxidation) (Somero 2020). These damages ultimately lead to impaired enzyme activity, genome instability, and membrane injury, potentially inducing cell apoptosis if the damage is too vast (Zhang et al. 2022). Regarding salinity shifts, changes in inorganic ion contents of body fluids can also lead to macromolecular damage because excess cations can interact with DNA and RNA (Evans and Kültz 2020). Both salinity shifts and ocean acidification can induce changes in extracellular and intracellular pH, with impacts on protein interactions and enzyme function (Evans and Kültz 2020, Finley and Chau 1991, Venn et al. 2013). This highlights the relevance of efficient osmotic and acid-base regulation to maintain performance under osmotic and pH stresses via control of organic osmolytes and ion transport (Araújo et al. 2018, Kaniewska et al. 2012, Kültz 2020).

Under damaging cellular conditions, chaperones such as heat shock proteins (HSPs) (e.g., Hsp70, Hsp90), which mount the heat shock response, and the ubiquitin-proteasome system may play a key role in the maintenance of protein folding/re-folding or prevent protein aggregations in the cytosol by fuelling proteolysis and promoting cell recovery following stress (Finley and Chau 1991, Maxwell et al. 2021, Whitley et al. 1999, Zhang et al. 2022). Thus, both chaperones and the ubiquitin-proteasome system play an important role in the quality control of protein pools in the cell. These mechanisms are known to display a selective value, as they contribute to the success of species across environmental gradients (Hofmann 2005) showing effects on life-history and fitness related traits (e.g., development, stress tolerance, longevity, fecundity) (Basu et al. 2002, Chen et al. 2018, Rutherford 2003, Sørensen et al. 2003). Recent advances in the study of HSP networks have accumulated evidence of a beneficial role of HSP expression in response to changing environmental conditions and patterns of local adaptation, suggesting HSPs play a role in adaptive responses to environmental change (Chen et al. 2018). Interestingly, further organelle-specific unfolded protein responses (UPRs) carried out by chaperones to address endoplasmic reticulum and mitochondria proteotoxic stress have also been pointed out as a primary mechanism in species acclimation to a changing ocean (e.g., corals (Aguilar et al. 2019, Dimos et al. 2019, Ruiz-Jones and Palumbi 2017)). In fact, a recent review has highlighted that endoplasmic reticulum stress management and up-regulation of the UPR is a universal mechanism shaping how animals respond to environmental stress and how they adapt to changing environments, mediating both inter- and intraspecific variation in life history traits (Yap et al. 2021).

In addition to carrying out protein quality control in the cell, HSPs and ubiquitin are both also involved in key cell signalling events related with redox sensing and DNA repair mechanisms. For example, Hsp70 functions as a redox sensor via ROS-induced post-translational modifications (oxidation of cysteine thiols), which change Hsp70's interactions with other proteins and co-chaperones, transferring redox status signals to the cell (Zhang et al. 2022). Moreover, ROS can further induce the expression of Hsp70, which in turn can help promote the expression of genes encoding for antioxidant enzymes such as superoxide dismutase, catalase, and glutathione-related enzymes (peroxidase and S-transferase) that convert the ROS superoxide ($O_2^{\cdot -}$), hydrogen peroxide (H_2O_2), and hydroxyl radical ($\cdot OH$), as well as lipid peroxides into less toxic products (Zhang et al. 2022). Non-enzymatic antioxidants are also crucial in oxidative stress management, including vitamins, carotenoids, and the reduced glutathione (GSH) pool (Chainy et al. 2016, González et al. 2015). For example, GSH functions as an electron donor to free radicals, but once it fulfils this function it becomes oxidized (GSSG) and needs to be reconverted to its reduced form to keep its antioxidant function. This is only possible via reducing equivalents, namely NADPH, which is produced via the pentose phosphate pathway (PPP). Thus, in cells under oxidative stress, an up-regulation of the PPP is frequently observed. This has been reported for corals (Axworthy et al. 2022, Hernández-Elizárraga et al. 2019), anemones (Tomanek 2011), fish (Araújo et al. 2018, Tomanek 2014), and molluscs (Madeira, Vinagre et al. 2016, Tomanek 2014). Moreover, many marine organisms display significant increases in the antioxidant enzyme network (either in terms of abundance or activity) upon exposure to various stresses (Dias et al. 2019, 2020, Khan et al. 2021, Lee, Kang et al. 2020, Lesser 2006, Madeira et al. 2018, 2016, Pimentel et al. 2015, Zheng et al. 2022). Still, extreme stress may result in the impairment of the antioxidant defence system, possibly associated to enzyme denaturation (Madeira, Costa et al. 2016). The ability of organisms to keep redox homeostasis is crucial as disruptions to cellular redox status have effects on cell proliferation, gene expression, immune responses, inflammatory processes, metabolic homeostasis, and cell death (Lesser 2006, Li et al. 2016, Song and Choi 2021, Wang et al. 2013, Zhang et al. 2022). In fact, redox status is known to set the environmental tolerance limits of metazoans and mediate adaptive responses to stressors, given the significant crosstalk between redox status and other cellular functions (Costantini 2019, Dilly et al. 2012).

The involvement of HSPs in DNA repair mechanisms has also been demonstrated, with these proteins modulating DNA repair pathways to cope with genomic instability during stress (Niu et al. 2006, Sottile and Nadin 2018). In particular, HSPs and heat shock transcription factors can interact with multiple DNA sensor and repair proteins (e.g., p53-binding protein, apurinic/apyrimidinic endonuclease 1), as well as cell cycle checkpoint kinases (e.g., CHK1), activating repair pathways and controlling cell-cycle progression to avoid the replication of damaged DNA (Sottile and Nadin 2018). Moreover, HSPs can contribute to nuclear transport of DNA repair proteins or be recruited to repair foci (Niu et al. 2006, Sottile and Nadin 2018). This interplay between HSPs and DNA repair pathways has been evidenced in vertebrates, warranting further studies in invertebrate groups. However, as most cell repair mechanisms are conserved across life kingdoms, we expect HSPs to be a central hub in the cell repair network of all taxa. Mounting evidence has also highlighted a role of ubiquitin in interlinked cellular processes such as DNA repair pathways (Boutou et al. 2015, Ciechanover 1998, He et al. 2016), transcription regulation (Conaway et al. 2002), activation of cell cycle checkpoints to facilitate DNA repair, and apoptosis (He et al. 2016). For example, DNA damage sensors are also regulated by ubiquitin chains, ultimately resulting in the recruitment of DNA repair proteins to the damage sites (He et al. 2016). Thus, both HSPs and ubiquitin ultimately affect cell fate decisions under stress, depending on the extent of cellular damage imposed by environmental changes.

The activation of cellular repair pathways and regulation of cell cycle progression/arrest during stressful periods imply an energy cost to the cell and to the organism. Several studies have shown that environmental stress leads to the depletion or re-arrangement of the energy budget of organisms, with potential implications in energy homeostasis and organism-level processes such as acclimation capacity, tolerance ranges, and survival (Leung et al. 2017, Madeira et al. 2021, Pörtner 2012, Sokolova et al. 2012). Thus, the bioenergetic framework of an organism has practical implications on its capacity to deal with environmental changes (Sokolova 2013, Sokolova et al. 2012). At the cellular level, cell cycle arrest helps saving energy during stress but repair pathways are ATP-demanding processes (Somero 2020) that require the reallocation of energy reserves towards stress management, homeostasis, and damage control. Thus, the available energy is used to sustain critical cellular and organismal functions, determining the physiological tolerance of individuals to stressful events (Calosi et al. 2013, Kühnhold et al. 2017, Somero 2020). Energy can be stored in three forms—lipids (triacylglycerol), carbohydrates (glycogen), and proteins—that can be mobilized to meet cellular energy demands. Upon stress, glycogen is easily broken down into glucose units to fuel glycolysis (Goh and Lai 2014) and thus produce pyruvate. Pyruvate can then be converted into acetyl-co-enzyme A (CoA), a central metabolite in the cell, which is used in the Krebs cycle (or Tricarboxylic Acid Cycle, TCA), occurring in the mitochondria, to produce ATP and its reducing equivalents NADH and $FADH_2$ (Alberts et al. 2017). These reducing equivalents are then directed to the mitochondrial electron transport chain (ETS) to produce ATP via oxidative phosphorylation. This process is critical to provide energy to the cell and keep basal metabolism. Recent evidence has also shown that levels of TCA intermediates play a major role in the control of cell function and fate under different contexts, regulating chromatin modifications, DNA methylation, hypoxic responses, and immunity (Martínez-Reyes and Chandel 2020).

When glycogen stores are depleted, lipid reserves, which yield large amounts of energy and can be used for long-term energy storage (Sibly et al. 2013), can be mobilized to produce fatty acids to fuel β-oxidation in the mitochondria. To do this, fatty acids are converted to fatty acyl-CoA, which then undergoes successive oxidations to produce the central metabolites acetyl-CoA, NADH, and $FADH_2$ (Alberts et al. 2017). Alternatively, some organisms can also use their protein pool as an energy supply, although this should not be the favoured route given that protein catabolism yields less energy than glycogen or lipid catabolism, and protein is needed for somatic growth (Kühnhold et al. 2017, Sibly et al. 2013). Still, when needed, proteins can be broken down into amino acids, which are then converted to TCA intermediates. For example, environmental changes such as elevated temperature, salinity shifts, ocean acidification, and combinations of these stressors have been shown to deplete the energy reserves of various marine species including sea cucumbers (e.g., carbohydrates during thermal stress (Kühnhold et al. 2017)), shrimps (e.g., protein contents during thermal stress (Madeira et al. 2019)), oysters (e.g., glycogen and lipids during acidification and/or low salinity (Dickinson et al. 2012)), polychaetes (e.g., lipids during combined thermal and hypo-saline stress (Madeira et al. 2021)), fish (e.g., liver and muscle glycogen during ocean acidification (Ruiz-Jarabo et al. 2021)), and corals (e.g., lipids, proteins, or carbohydrates in some species exposed to thermal stress (Grottoli et al. 2017, Rodrigues and Grottoli 2007, Schoepf et al. 2013)). This indicates that, upon stress, many organisms must metabolize their energy reserves to fuel ATP production and meet the increased cellular energy demands. Accordingly, an up-regulation of glycolysis has also been observed in marine organisms (including fish, crustaceans, bivalves, and corals) exposed to thermal stress (Leggat et al. 2011, Madeira et al. 2017, Song et al. 2019), ocean acidification (Araújo et al. 2018, Cao et al. 2018, Noisette et al. 2021, Ruiz-Jarabo et al. 2021), and de-oxygenation/hypoxia (Gracey et al. 2001, Li, Sun et al. 2019). However, especially during hypoxic stress, metabolic depression may mitigate energy demands (Hermes-Lima et al. 2015). According to stress type or level, the pyruvate resulting from glycolysis can then be targeted to feed the aerobic metabolism via the TCA cycle and ETS or the anaerobic metabolism via fermentation, with the production of lactic acid. Although TCA has been shown to increase in several marine species exposed to environmental stress (Leggat et al. 2011, Madeira et al. 2017, Noisette et al. 2021, Song et al. 2019), some organisms shift

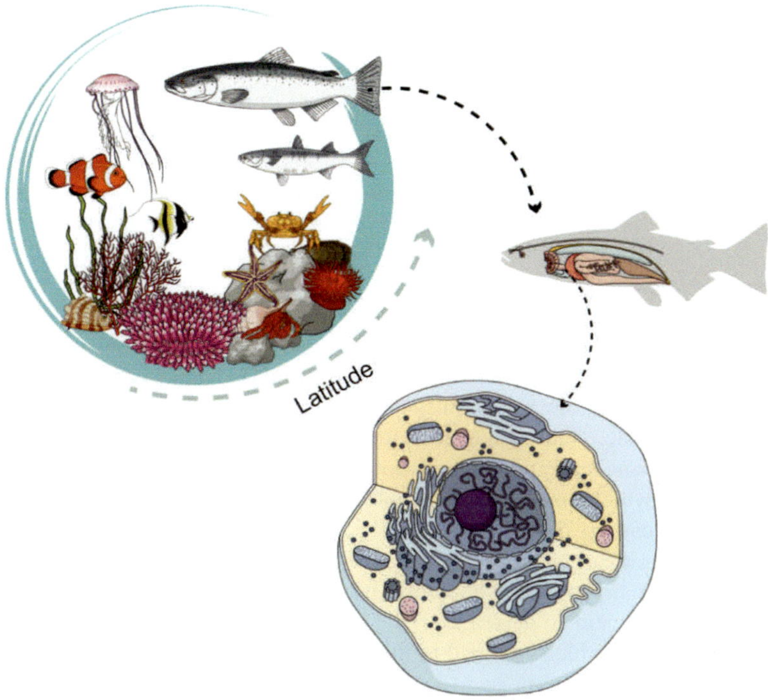

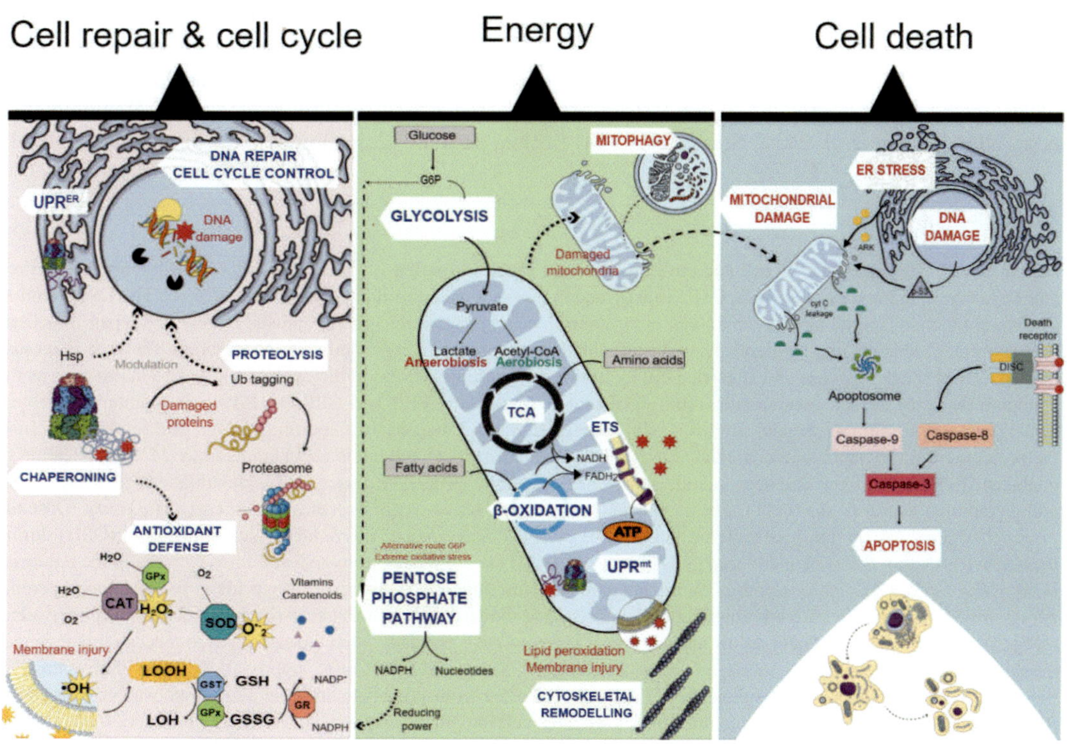

Figure 2 contd. ...

from aerobic to anaerobic pathways to rapidly sustain energetic needs (Anestis et al. 2007, Beemelmanns et al. 2021, Cao et al. 2018, Liao et al. 2018). However, long-term reliance on anaerobic pathways for energy production may not be sustainable, as accumulation of lactic acid can result in metabolic acidosis (Kieffer et al. 1994, van Raaij et al. 1996). Still, in the case of exposure to hypoxia, energy supply via anaerobic metabolism is extremely relevant, given oxygen limiting conditions (Alderdice et al. 2021). Additionally, the remodelling of bioenergetic pathways towards anaerobic metabolism may be favoured if environmental changes lead to extensive mitochondrial damage and ROS production; in this case, mitophagy may be used as a strategy to temporarily reduce ROS levels, as observed in corals (Alderdice et al. 2021).

When energy homeostasis is impaired, cellular signalling is disrupted and cellular damage reaches unsustainable levels, apoptosis may be triggered. Apoptotic networks are conserved in metazoans (Ketelut-Carneiro and Fitzgerald 2022, Kvitt et al. 2016) and may be initiated through two major pathways: (i) the extrinsic death receptor pathway via tumour necrosis factor (TNF) or Fas ligand (FasL) signalling, with subsequent binding of signalling molecules to cell-surface receptors, activation of the death-inducing signalling complex (DISC), and further recruitment of initiator caspase-8; and (ii) the intrinsic mitochondrial pathway via permeabilization of the mitochondrial membrane controlled by Bcl-2 and Bax proteins, further leakage of cytochrome C from the mitochondria into the cytosol, apoptosome formation, and activation of initiator caspase-9 (Cheng et al. 2015, Krumschnabel and Podrabsky 2009, Takle and Andersen 2007). The initiator caspase-8 and caspase-9 further activate caspase-3, an effector of cell death that inactivates DNA repair, supresses apoptosis inhibitors, dismantles the cytoskeleton, and cleaves cellular components (Takle and Andersen 2007). The mitochondrial pathway can also be activated by (i) excessive DNA damage, which can lead to the activation of p-53, ultimately resulting in the release of cytochrome C from mitochondria, and (ii) persistent endoplasmic reticulum stress that relays stress signals to the mitochondria via apoptosis-signal-regulating kinase (ARK), which further triggers mitogen-activated protein kinase (MAPK) and c-Jun N-terminal kinase (JNK), setting the apoptotic cascade in motion via leakage of cytochrome C from the mitochondria (Cheng et al. 2015, Sano and Reed 2013, Szegezdi et al. 2006). Several studies in marine organisms have shown that pro-apoptotic signals and apoptosis can be triggered upon exposure to high temperature (e.g., Beemelmanns et al. 2021, Cheng et al. 2015, Sleadd et al. 2014, Zhao et al. 2022, Rahman and Rahman 2021, Sleight

---

...Figure 2 contd.

Figure 2. Conceptual model of cellular physiology under stress in marine organisms. A set of evolutionary conserved pathways (the cellular stress response, CSR) is activated in marine organisms under stressful conditions. The CSR involves four main functions including cell repair, cell cycle control, energy regulation, and apoptosis. During cell repair, functions such as protein chaperoning, proteolysis, and antioxidant defence are up-regulated. The chaperoning function is usually termed as heat shock response and is carried out by heat shock proteins (HSPs) in the cytosol and in organelles such as the endoplasmic reticulum and mitochondria (unfolded protein response, UPR^{ER} and UPR^{mt}). HSPs help maintaining protein folding and stabilization. If proteins are irreversibly damaged, they are tagged by ubiquitin (Ub) and sent for degradation in the proteasome. HSPs and Ub also mediate DNA repair mechanisms and cell cycle checkpoints to avoid the replication of damaged DNA. Antioxidant defences include superoxide dismutase (SOD), catalase (CAT), glutathione peroxidase (GPx), and glutathione-S-transferase (GST). These enzymes transform reactive oxygen species (ROS; e.g., $O_2^{\cdot-}$, H_2O_2, ·OH) and lipid peroxides (LOOH) into less toxic products. GPx and GST make use of the reduced glutathione pool (GSH) during their activity. Once GSH is oxidized into GSSG, it needs to be converted back to its reduced form by glutathione reductase (GR), using NADPH. Regarding energy regulation, cells may use glucose, proteins, or fatty acids as fuel to support aerobic energy production via the tricarboxylic acid cycle (TCA) and oxidative phosphorylation (in the electron transport system, ETS). Anaerobic energy production may also be favoured (pyruvate → lactate) when the cell needs to meet energy demands quickly, when mitochondria are damaged, or under oxygen limiting conditions (e.g., hypoxia). When oxidative stress in the cell is extreme, glucose-6-phosphate (G6P) may be redirected from glycolysis into the pentose phosphate pathway to produce reducing power (NADPH) and nucleotides. To keep cell integrity, homeostasis, and energy production upon stress, cytoskeleton remodelling is usually enhanced. However, when cellular damage is too vast, apoptosis is induced via the death receptor pathway or the mitochondrial pathway. The former involves the binding of death signalling molecules to cell-surface receptors, assembly of the death-inducing signalling complex (DISC), and recruitment of caspase-8. The latter involves leakage of cytochrome C (cyt C) from mitochondria, apoptosome formation, and activation of caspase-9. Both caspases then activate caspase-3, and apoptosis is initiated.

et al. 2018, Zou et al. 2021, Kvitt et al. 2016, Pernice et al. 2011, Richier et al. 2006, Mayor et al. 2012, Johnstone et al. 2019), exposure to hypoxia (e.g., Xin et al. 2022, Wang et al. 2022, Clark et al. 2013), and ocean acidification (e.g., de Souza et al. 2014, Wang et al. 2016, 2021). Interestingly, a two-phase apoptotic response has been described for some metazoans (e.g., corals), including an acute pro-apoptotic phase in response to stress, followed by an activation of anti-apoptotic mechanisms and cellular defences, resulting in the termination of apoptosis and acclimation to the new environmental conditions (Kvitt et al. 2016). Moreover, some hypoxic tolerant species have developed the ability to delay or suppress apoptosis (Falfushynska et al. 2020).

In summary, cellular responses to stress are mounted through a complex interacting network of molecules and pathways that can be activated according to damage extension and energy budgets of the cell (Somero 2020) (Fig. 2). Despite the evolutionary conservation of such responses across the animal kingdom, significant variation has been reported in the marine realm. In the next section, we explore the sources of variation in cellular physiology under stress.

Sources of variation leading to different cellular physiology under stress

The cellular physiology of a given species under stress may be affected by different factors, such as ontogenetic stage, stressor temporal dynamics, local adaptation, multiple stressors, and species mobility (Fig. 3). Species-specific traits (e.g., eurytherm vs stenotherm, body size and morphology, genome size, mobility) can determine differences in tolerance limits and molecular responses of organisms to environmental changes (Leiva et al. 2019, Madeira et al. 2019). Significant diversity has been found in molecular responses to warming in invertebrates (Clark et al. 2017, Collins et al. 2021, Thibault et al. 2020) and fish (Bernal et al. 2020), related to species life-history traits, geographical distribution, and commonness/rareness. In particular, eurythermal, common, and broadly distributed species are less responsive to temperature changes, have wide tolerance ranges, and can better sustain activity in variable environments when compared to stenothermal, rare, and narrow distributed species (Bernal et al. 2020, Dahlke et al. 2020, Thibault et al. 2020). For example, while many marine species display an induction of heat shock response when exposed to warming, Antarctic fish often lack this response (Hofmann et al. 2000). In these fish, evolution in constant near-freezing water temperatures resulted in the constitutive expression of inducible HSPs (Place and Hofmann 2005). Interestingly, this pattern also seems to occur in warm-adapted species living in variable environments. These species display high constitutive levels of HSP instead of induced expression upon stress, a strategy known as constitutive frontloading (Barshis et al. 2013) or preparative defence strategy (Dong et al. 2008. Madeira et al. 2012). In addition, closely related species distributed in different latitudes (temperate vs tropical vs polar) can also differ in their molecular responses to environmental changes. For example, the tropical shrimp *Lysmata amboinensis* has lower levels of Hsp70 and higher increases in lactate and lipid peroxidation upon exposure to warming, when compared to the temperate *Lysmata seticaudata* (Rosa et al. 2014). For many fish and some invertebrates, stress responses seem to be tissue-specific with the distinct tissues responding differently to environmental stressors. Stress biomarkers (antioxidant enzymes, lipid peroxidation, Hsp70, and ubiquitin) analysed in several tissues of a common tide pool fish (*Lipophrys trigloides*) showed higher concentrations of all biomarkers in brain followed by gills, visceral mass, and muscle (Vinagre et al. 2021) and several other examples are documented in the literature. Another source of variation in exposure and response to environmental stressors is life-cycle stage. Typically, early and mature adult stages are the most vulnerable to environmental variations (Dahlke et al. 2020, Pandori and Sorte 2019, Pankhurst and Munday 2011). For example, in some benthic marine invertebrates, larvae and juveniles can cope with the projected near-future levels of warming but do not tolerate the projected far-future conditions, which have a stronger negative impact on larvae, namely decreased size and survival (Byrne and Przeslawski 2013). Moreover, increased temperature and decreased salinity, combined or alone, compromised early life-cycle stages of the ascidians *Styela plicata* and *Microcosmus squamiger* (Pineda et al. 2012). Because these early life-cycle stages are not yet fully developed, their defence mechanisms, including those related to oxidative stress, energetic budget, and metabolic pathways, may be less effective, failing to neutralize the effects of environmental stresses. A recent review of the thermal

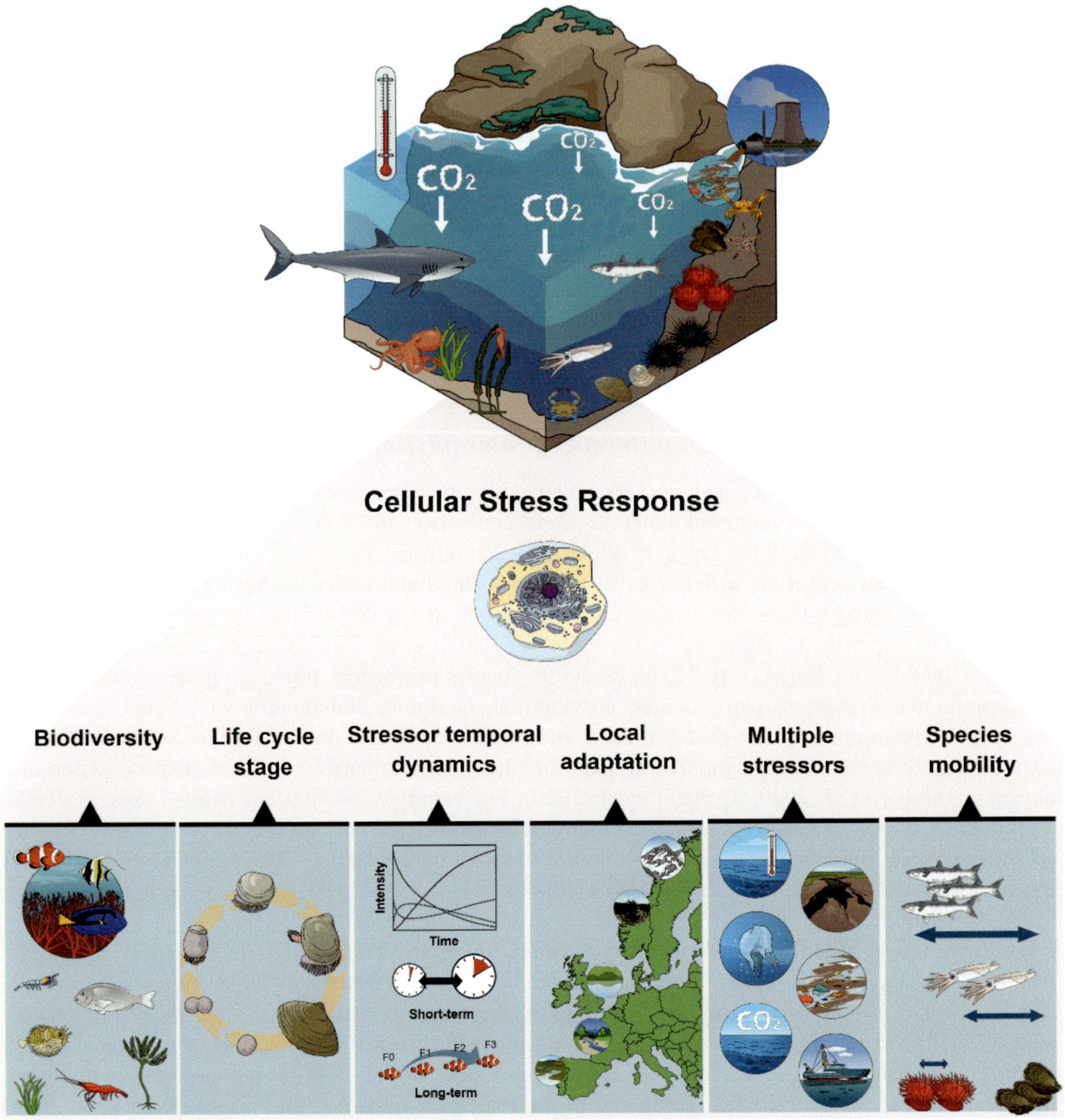

Figure 3. Sources of variation leading to different cellular physiology under stress.

tolerance of 694 marine and freshwater species distributed worldwide (Dahlke et al. 2020) disclosed that spawning adults and embryos have narrower tolerance ranges when compared to other life stages. However, an increased sensitivity of early life-cycle stages is not universal across all taxonomic groups (Kroeker et al. 2013).

Field and laboratory experiments have revealed that many marine organisms exhibit local physiological and molecular profiles, indicating that local adaptation is a powerful force even in high-dispersal marine species (Palumbi et al. 2019, Vargas et al. 2022, 2017). Adaptation to local conditions can also be related to molecular plasticity, reflecting the organism's ability to cope with environmental changes. For example, sea bream (*Sparus aurata*) juveniles, which inhabit variable estuarine areas, display significant molecular plasticity when exposed to warming, as opposed to sea bream larvae, which inhabit more stable ocean areas, and lack the capacity to acclimate to increased temperatures (Madeira et al. 2017, 2020a).

The co-occurrence of different environmental drivers on different locations can also impact how organisms respond and adapt to environmental changes. Numerous studies focused on the effects of single or combined stressors (e.g., warming, acidification, hypoxia) on marine organisms' cellular physiology (e.g., metabolic pathways, antioxidant system, protein damage) and their capability to adapt to environmental changes (Araújo et al. 2018, Madeira, Costa et al. 2016, Verlecar et al. 2007, Xu et al. 2021). However, other factors may also contribute to potentiate the effects of climate-related stressors on marine ecosystems (Madeira et al. 2021, Maulvault et al. 2019, Maulvault et al. 2018a). For instance, a study on the Antarctic key fish species *Trematomus bernacchii* exposed to temperature increase, pH reduction, and Cadmium (Cd) contamination showed modulation of the oxidative stress response to Cd by multiple stressors, as revealed by glutathione increase and genotoxicity under exposure to the combined stressors, suggesting that *T. bernacchii* is vulnerable to the predicted global change scenarios (Giuliani et al. 2021). Furthermore, meagre (*Argyrosomus regius*) exposed to a multi-stressor environment (warming, acidification, and venlafaxine) showed a strong response after exposure to these multi-stressors via water, as revealed by significant increases in lipid peroxidation, Hsp70, and total ubiquitin (Maulvault et al. 2019, Maulvault et al. 2018a). Overall, combined environmental stressors can complexify marine species responses to climate change via antagonistic, additive, or synergetic effects (Bautista and Crespel 2021).

When exposed to environmental changes such as sea warming, acidification, or hypoxia, many marine organisms, including fish, can move and escape towards new habitats, similar to those they are adapted to. However, sessile organisms or small-ranging animals have limited dispersal capacity and thus are more vulnerable to environmental changes (Poloczanska et al. 2016, Sunday et al. 2015). Accordingly, a study on cytoprotective proteins across marine taxa has shown that sessile species display higher levels (> two-fold) of CSR proteins and oxidative damage when compared to mobile species (Madeira et al. 2019). According to Habary et al. (2017), acclimatation can confer plasticity in some performance traits but ultimately, it does not determine the survival and distribution of mobile stenothermal species under global warming, following the biogeographical principle of 'adapt, move, or die'. This means that marine organisms can either move, acclimate to environmental changes by physiological adjustment, or adapt via genotype alteration that enables changes in the species' biology (Beaugrand and Kirby 2018, Donelson et al. 2019). While acclimation is an important driving force for phenotypic plasticity, it can be achieved at the cost of unfavourable energetic trade-offs for the organism or not be fast enough to keep pace with the rate of environmental shifts (Habary et al. 2017). For highly mobile species, climate change is leading to changes in their distribution, but they may also experience range contractions when environmental changes surpass their thermal niche, which is likely to affect their persistence (see Section 4).

The current responsiveness of many populations of marine organisms to climate change can be insufficient to cope with the predicted environmental shifts, and thus imposes challenges and resilience over generations. To be able to cope with continuous environmental changes, marine organisms must therefore adapt across generations through genetic inheritance (Bautista and Crespel 2021). Concomitantly, they may also display transgenerational plasticity via epigenetic mechanisms (Donelson et al. 2018). In fact, a transgenerational response is usually defined as a phenotypic alteration in progeny in response to environmental stress experienced by the progenitors and can influence progeny response to the same type of stress (Griffith and Gobler 2017, Ross et al. 2016). Veilleux et al. (2015) studied the molecular processes of transgenerational acclimation to a future warming ocean in adult tropical reef fish (*Acanthochromis polyacanthus*) and found that metabolic genes were the most up-regulated across generations, suggesting changes in energy production for keeping performance at increased temperatures. Immune- and stress-responsive genes were also up-regulated across generations, suggesting that a new complement of genes may arise to allow the second generation to cope with temperature increases. Interestingly, HSP genes were unresponsive, which suggests that short-term thermal stress may not be an adequate indicator for long-term adaptive capacity. A transgenerational study in the sea urchin *Strongylocentrotus intermedius* exposed to high temperature showed differentially expressed genes associated with neurotransmitter catabolic process, carboxypeptidase activity, metallopeptidase activity, and calcium ion binding (Shi et al. 2020). On the contrary, transgenerational exposure of the bivalves *Mercenaria mercenaria* and *Argopecten irradians* to ocean acidification resulted in progeny more vulnerable to experimental conditions (thermal stress, food limitation, lower pH) than their parents, suggesting absence of transgenerational acclimation to ocean acidification (Griffith and Gobler 2017). Negative transgenerational responses were also reported in amphipods, with increased oxidative stress in the progeny of parents exposed to ocean acidification (Lopes et al. 2019). This may indicate that transgenerational responses are also species dependent, and organisms respond in different ways to the predicted future environmental changes. In fact, in their recent review, Donelson et al. (2018) revealed that 22% of the analysed taxa showed negative transgenerational responses, 37% showed no transgenerational response, and 41% showed positive transgenerational responses to global change drivers.

Effects at the individual level

Climate-induced effects observed at the molecular level cascade to whole-body level through internal information pathways that set the functional responses of marine organisms. Such developmental, physiological, morphological, behavioural, reproductive (Wilson and Barker 2021), and other responses determine organisms' active tolerance to climate change stressors (Pörtner 2012), govern their performance, and establish the link between changes in molecular processes and those at population and community levels. Organisms therefore function as systems that have evolved to sense, respond, and catalyse acclimation and adaptation processes to the external environment, therefore being living sentinels of climate change effects.

Climate change effects at the whole-body level start to arise as organisms try to balance all their complex functional properties to persist and thrive in the environment, while tracking the different, intertwined patchworks of local climate patterns and velocities (Ashworth 2017, Pinsky et al. 2013). If acclimatization processes at the molecular level (i) promote a new steady homeostatic state, (ii) are inefficient to maintain internal homeostasis or, (iii) are too energetically costly for the organism, specific climate change signatures will become dominant at the individual phenotypic level, especially with the emergence and propagation of multiple drivers across ecosystems (Henson et al. 2017). The term 'climate syndrome' was coined by (Pörtner 2012) to describe the combined effects of stressors that change in parallel, both in seawater and body fluids of marine animals, i.e., ocean warming (temperature), ocean acidification (hypercapnia), and ocean deoxygenation (hypoxaemia). It became evident that when combined, environmental factors could strengthen or weaken each other's response signals observed at the individual level, including altered cardiocirculatory and metabolic patterns, as well as disturbances in growth, condition, and behaviour, and that this was crucial to define climate change sensitivity thresholds of marine organisms (see Fig. 4 for a summary of individual-level traits affected by climate change stressors).

Although individual-level responses to climate change are not uniform due to the sources of variation described previously, it is still possible to project outcomes at the whole-body level. Table A1 (Annex 1) provides a list of recent studies from the past 5 years (representative but not exhaustive) investigating different individual-level traits in response to climate change, encompassing different trait categories, single and multiple stressors, gradual and extreme climate events, animal taxa as well as oceans and climatic regions. A large fraction of studies (i) employed experimental approaches (whereas field studies are scarcer), (ii) focused on demersal fish or benthic macroinvertebrates (note that only animal studies were included), (iii) focus simultaneously on two-to-four individual-level trait categories and one or two climate stressors, (iv) are from temperate or subtropical regions, and (v) span all life stages.

Depending on the direction of change displayed by different traits when organisms are exposed to multiple dimensions of climate change, the consequence can either be the multiplication of threats or opportunities to individuals (Garcia et al. 2014). Below, we provide a summary of the main individual-level changes observed in marine animals exposed to climate change and increasing climate variability in the ocean.

Physiology

Environmental stress has generally led to changes in inter-related whole-body physiological parameters such as respiration, metabolic, and heart rates, as well as feeding, absorption, and excretion rates, as these

Table A1. Different research approaches to study individual-level effects of climate change stressors on marine organisms.

Type of study	Animal group	Species	Life-stage	Climate variables (S/M[1])		Climate zone	Ocean	Habitat type	Individual-level variables	Reference
Meta-analysis	Fish	694 species	Embryos, Larvae, Adults, Spawners	Single	Warming	All	Global	All	Physiology	(Dahlke et al. 2020)
Modelling	Zooplankton	*Paracyclopina nana*	Adults, Nauplii	Single	Exposure to UV-B radiation	NA	NA	NA	Reproduction, Longevity	(Dur et al. 2021)
Modelling	Bivalves	*Ruditapes philippinarum*	NA	Single	Marine Heatwave	Temperate, Subtropical	Pacific, Atlantic	Benthic	Physiology, Longevity	(Bertolini and Pastres 2021)
Review	Fish, Invertebrates, Zooplankton	NA	NA	Single	Warming	All	Global	All	Physiology, Reproduction, Longevity	(Assan et al. 2020)
Review	Invertebrates	NA	NA	Single	Acidification	Global	Global	NA	Physiology, Reproduction, Development, Morphology, Behaviour, Longevity	(Lee et al. 2020)
Review	Fish	NA	NA	Single	Hypoxia	All	Global	All	Reproduction, Behaviour	(Saha et al. 2022)
Review	Fish	NA	NA	Single	Warming, Salinity changes, Hypoxia, Acidification	All	Global	All	Reproduction	(Servili et al. 2020)

[1] Single/Multiple

Table A1 contd. ...

...Table A1 contd.

Type of study	Animal group	Species	Life-stage	Climate variables (S/M[1])		Climate zone	Ocean	Habitat type	Individual-level variables	Reference
Review	Marine mammals	18 species	Pups, Juveniles, Adults	Multiple	Warming-Acidification-Freshwater Introduction-Ice Melting-Altered Water Circulation	Polar, Temperate, Subtropical	Atlantic/Pacific	Pelagic	Reproduction, Longevity	(Gulland et al. 2022)
Field/Imagery	Sponges	NA	Adults	Single	Marine Heatwave	Temperate	*Southern*	Benthic	Longevity	(Perkins et al. 2022)
Experimental	Sponges	*Crella incrustans*	Adults	Single	Marine Heatwave	Temperate, Subtropical	Indian, Pacific	Benthic	Physiology, Morphology, Reproduction, Longevity	(Strano et al. 2022)
Experimental	Corals	*Astroides calycularis*	Adults	Single	Acidification	Temperate	Atlantic/ Mediterranean	Benthic	Reproduction	(Marchini et al. 2021)
Experimental	Trematodes	*Himasthla* sp.	Cercaria	Multiple	Warming-Acidification	Temperate	Pacific	Pelagic	Reproduction, Longevity	(Franzova et al. 2019)
Experimental	Gastropods	*Austrocochlea concamerata*	Adults	Multiple	Warming-Acidification	Tropical	Pacific	Benthic	Physiology, Morphology	(Leung et al. 2020)
Experimental	Gastropods	*Trochus sacellum* and *Astralium haematragum*	Adults	Single	Marine Heatwave	Temperate-Tropical	Pacific	Benthic	Physiology, Longevity	(Hemraj et al. 2020)
Experimental	Gastropods	*Turbo undulatus*	Adults	Multiple	Warming-Acidification	Tropical	Pacific	Benthic	Physiology, Longevity	(Leung et al. 2021)
Experimental	Bivalves	*Anomalocardia flexuosa*	Adults	Single	Marine Heatwave	Subtropical, Tropical	Atlantic	Benthic	Physiology, Longevity	(Carneiro et al. 2020)
Experimental	Bivalves	*Mytilus coruscus*	Adults	Multiple	Acidification-Hypoxia	Subtropical	Pacific	Benthic	Physiology, Longevity	(Shang et al. 2020)
Experimental	Bivalves	*Ostrea lurida*	Adults, Larvae	Single	Warming	Temperate	Pacific	Benthic	Reproduction, Development, Longevity	(Spencer et al. 2021)
Experimental	Bivalves	*Saccostrea glomerata*	Adults	Multiple	Warming, Acidification	Subtropical-Tropical	Pacific, Indic	Benthic	Longevity	(Scanes et al. 2021)

Experimental	Bivalves	*Scrobicularia plana*	Adults	Single	Marine Heatwave	Temperate	Atlantic	Benthic	Kinetics, Physiology, Longevity	(Amorim et al. 2020)
Experimental	Crustaceans	*Artemia franciscana*	Cysts, Nauplii, Adults	Single	Acidification	Temperate-tropical	Global	Benthic	Physiology, Reproduction, Longevity	(Thangal et al. 2021)
Experimental	Crustaceans	*Marsupenaeus japonicus*	Adults	Single	Hypoxia	Tropical, Subtropical	Atlantic, Indian, Pacific	Benthic	Morphology, Longevity	(Wang et al. 2022)
Experimental	Zooplankton	*Temora turbinata*	Adults	Single	Hypoxia	Tropical, Subtropical	Global	Pelagic	Physiology, Behaviour, Longevity	(He et al. 2021)
Experimental	Amphipods	*Ampelisca brevicornis*	Juveniles, Adults	Single	Acidification	Subtropical	Atlantic, Pacific	Benthic	Physiology, Behaviour, Reproduction, Longevity	(Bhuiyan et al. 2022)
Experimental	Amphipods	*Gammarus locusta*	Adults	Single	Acidification	Temperate	Atlantic	Benthic	Physiology, Behaviour, Reproduction, Longevity	(Borges et al. 2018)
Experimental	Isopods	*Cirolana harfordi*	Adults, Offspring	Single	Warming	Temperate	Pacific	Benthic	Reproduction, Development, Longevity	(Gavel et al. 2022)
Experimental	Echinoderms	*Heliocidaris erythrogramma*	Adults	Single	Marine Heatwave	Temperate	Indian	Benthic	Physiology, Longevity	(Minuti et al. 2021)
Experimental	Echinoderms	*Lytechinus variegatus*	Adults	Single	Warming	Temperate	Atlantic	Benthic	Health, Longevity	(Brothers et al. 2018)
Experimental	Echinoderms	*Strongylocentrotus intermedius*	Adults	Single	Acidification	Temperate	Pacific	Benthic	Physiology, Morphology, Reproduction, Longevity	(Cui et al. 2022)
Experimental	Fish	*Argyrosomus regius*	Juveniles	Multiple	Warming-Acidification	Temperate	Atlantic	Benthopelagic	Morphology, Behaviour	(Maulvault et al. 2018)
Experimental	Fish	*Dicentrachus labrax*	Juveniles	Multiple	Warming-Acidification	Temperate	Atlantic	Demersal	Physiology	(Cominassi et al. 2020)

Table A1 contd. ...

...Table A1 contd.

Type of study	Animal group	Species	Life-stage	Climate variables (S/M[1])		Climate zone	Ocean	Habitat type	Individual-level variables	Reference
Experimental	Fish	*Dicentrachus labrax*	Adults	Single	Exposure to UV-B radiation	Temperate	Atlantic	Demersal	Morphology, Behaviour, Longevity	(Alves et al. 2021)
Experimental	Fish	*Diplodus sargus*	Juveniles	Single	Warming	Temperate	Atlantic	Demersal	Physiology, Morphology, Longevity	(Anacleto et al. 2018)
Experimental	Fish	*Fundulus heteroclitus*	Adults	Single	Hypoxia	Temperate	Atlantic	Benthopelagic	Physiology, Longevity	(Borowiec and Scott 2021)
Experimental	Fish	*Harpagifer antarcticus*	Adults	Single	Warming	Polar	Antarctic	Demersal	Physiology, Longevity	(Saravia et al. 2021)
Experimental	Fish	*Labroides dimidiatus*	Adults	Single	Acidification	Tropical	Indian	Demersal	Behaviour	(Paula et al. 2019)
Experimental	Fish	*Larimichthys crocea*	Adults	Single	Hypoxia	Temperate	Pacific	Benthopelagic	Longevity	(Ding et al. 2020)
Experimental	Fish	*Menidia beryllina* and *Cyprinodon variegatus*	Adults	Multiple	Warming, Acidification, Food Supply	Subtropical	Atlantic	Pelagic-Neritic	Kinetics, Longevity	(Gobler et al. 2018)
Experimental	Fish	*Oncorhynchus mykiss*	Juveniles	Single	Hypoxia	Temperate	Atlantic/ Pacific/ Arctic	Benthopelagic	Physiology, Longevity	(Aksakal and Ekinci 2021)
Experimental	Fish	*Salmo salar*	Adults	Multiple	Warming-Hypoxia	Temperate	Atlantic	Benthopelagic	Longevity	(Zanuzzo et al. 2020)
Experimental	Fish	*Solea senegalensis*	Embryos, Larvae	Single	Exposure to UV-B radiation	Temperate	Atlantic	Benthic	Physiology, Morphology, Development, Longevity, Behaviour	(Araújo et al. 2021)
Experimental	Fish	*Symphodus ocellatus*	Adults	Single	Acidification	Temperate	Atlantic	Reef-associated	Reproduction	(Nissling and Larsson 2018)

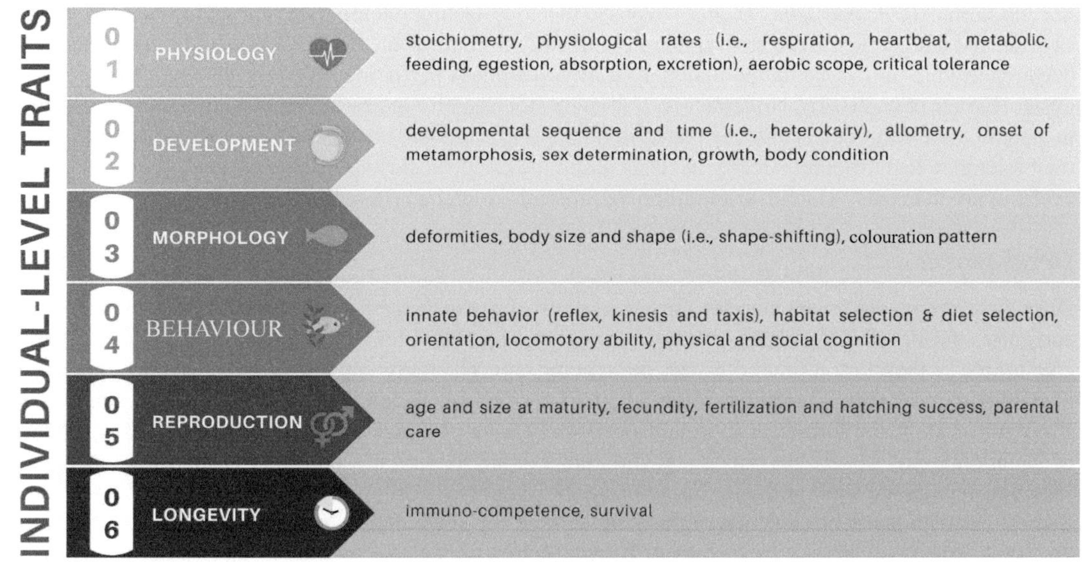

Figure 4. Individual-level traits altered by climate change stressors in marine organisms. Traits were grouped according to different, interlinked categories: (1) physiology, (2) development, (3) morphology, (4) behaviour, (5) reproduction, and (6) longevity.

are mechanisms used by organisms to adjust aerobic scope and tolerance ranges. These changes depend not only on stress intensity but also on individual kinetics and dynamic energy budget (DEB theory, see (van der Meer 2019)). For instance, global warming caused over 20% increase in physiological rates of ectothermic organisms, such as arthropods, molluscs, and fish, in the past couple of decades (Seebacher et al. 2015), leading to an overall higher caloric demand when organisms are exposed to unfavourable thermal conditions. Hypoxia, on the other hand, is known to reduce the maximum metabolic rate and metabolic scope of ectothermic animals below a critical oxygen saturation level (Thomas et al. 2019), whereas acidification's impact on physiological rates is comparatively smaller, at least in adult stages. Nevertheless, negative or positive impacts of acidification in physiology may be disproportionally higher in calcifying organisms or photosynthetic organisms, respectively (Nikinmaa and Anttila 2020, Thomas et al. 2022).

In the long run, the changed dynamics between energy supply and use in marine organisms exposed to multiple stressors, caused by the interplay between higher metabolic requirements, altered diet selection, changed feeding, and changed nutrient absorption rates (depending on food availability and nutritional quality), will lead to the disruption of whole organism stoichiometry (Liang et al. 2021). This means that response strategies used by organisms depend on: (i) the compensatory mechanisms employed to maintain the metabolic rate and/or aerobic scope (usually in gradual scenarios of climate change), which can lead to increases in physiological rates such as ventilation, heartbeat, and mobility; or (ii) the energy-saving mechanisms employed (usually during very intense stress or extreme climate change events), which can lead to metabolic depression, lowering the remaining physiological rates. Although metabolic rates are known to scale with temperature and body mass (see the metabolic theory of ecology (MTE) and the universal temperature dependence (UTD) of metabolism (Brown et al. 2004, Clarke 2004, Clarke and Fraser 2004)), organisms' energy budget can therefore change their sensitivity and ability to respond. This is most evident when stressors are combined. For instance, metabolic depression was observed in *Mytilus edulis* exposed to high temperature and low pH (Lesser 2016), and in the tropical marine snail *Turritella bacilum* when exposed to warming and hypoxia (Marshall and McQuaid 2020). On the contrary, in *Mytilus galloprovinciallis*, the combined exposure to hyposalinity and heat shock increased the mussels' metabolic rate (Vasquez et al. 2022). Moreover, many recent studies have reported that exposure to one stressor may reduce tolerance to a different, co-occurring stressor. This was observed in coho salmon (*Oncorhynchus kisutch*) exposed to a long-lasting photoperiod and acute warming, which experienced a reduction in the critical thermal maximum (Hines et al. 2019). Ocean acidification has also been reported to narrow acute thermal and salinity tolerances in marine invertebrates (e.g., Sydney rock oyster (Parker et al. 2017)). Nonetheless, if climate stochasticity is low and environmental stress occurs in a predictable way (i.e., is autocorrelated in time), organisms may be able to maintain or even improve their tolerance to a different stressor as observed in the caridean shrimp *Palaemon varians* acclimated to cyclic hypoxia events, which showed improved thermal tolerance (Peruzza et al. 2021).

Development

During ontogeny, organisms display critical periods of sensitivity to environmental cues, especially in early life, during which phenotypic switches in development occur in response to specific stressors. Cue timing is therefore a crucial aspect influencing trait plasticity and the development of alternative phenotypes under climate change. These alternative phenotypes can be reversible or irreversible upon stress removal. In the former case, organisms usually have prolonged sensitive periods and display full acclimation ability to new environmental conditions throughout their lifetime, whereas in the latter case, sensitive periods are short and time-limited, and the phenotype resulting from a developmental switch is permanent, with latent or carry-over effects in the next life stage. For example, (Pechenik 2006) revised the latent effects of larval experiences of marine invertebrates and concluded that larvae exposure to stressors such as UV-radiation, salinity changes, and reduced food concentration may have permanent effects in the next life stages, such as lower growth and survival of juveniles, and reduced fecundity of reproductive adults (observed in bryozoans, gastropods, bivalves, polychaetes, echinoderms, and crustaceans). Moreover, the timing and sequence of developmental processes (e.g., diapause, hatching, or metamorphosis) can be altered in a form of developmental plasticity known as heterokairy (see (Rundle and Spicer 2016)). Climate-related stress has generally led to faster development in individuals from species of lower trophic levels, regardless of development type (i.e., direct-transitions between life stages occur in a gradual manner-, or indirect-transition from one specific life stage to another undergoes metamorphic changes). In species with bi-phasic life cycles, which is the case in most marine invertebrates, the onset of metamorphosis may be delayed. As a result, the number of larval stages during development (and associated mortality risk) may increase, and larvae continue to grow until they (i) find suitable environmental conditions (while avoiding environmental extremes), and (ii) build up the energy reserves necessary to initiate metamorphosis (Lowe et al. 2021). For instance, decreased intermoult periods are known to occur in individuals from species with non-continuous growth (such as observed in the lesser blue crab *Callinectes similis* exposed to elevated temperature (Kuhn and Darnell 2019)). This effect was also observed when warming was combined with acidification in the shrimp *Palaemon elegans,* which displayed a greater number of moults due to the shortening of moulting cycles (Maia et al. 2022). Some studies have also reported the inability of megalopae larvae to moult into the first juvenile stage or their prolonged transition into post-larval stages upon altered environmental conditions (e.g., *Menipe mercenaria* zoea stage-V when exposed to acidification (Gravinese et al. 2018) and *Acanthaster* cf. *solaris* starfish larvae after parental exposure to acidification (Hue et al. 2022)). However, in individuals from species of higher trophic levels, developmental rate may take an alternative response route, as many of these species face additional climate-related challenges (e.g., starvation), which lead to slower development. In all cases, climate change-induced perturbations in development across the ontogeny of marine organisms (see an example in Fig. 5) usually lead to reduced overall individual growth, decreased immuno-competence, higher prevalence of disease, lower body condition, and size reductions in the order of 4% to 22% (Tian and Benton 2020). This pattern is consistent across taxa and environments, as evidenced by the prolific literature reporting. For instance, in a recent review of Actinopterygii fish from 30 orders comprising data from 1976 to 2018, the majority of species displayed significant negative growth changes at both global and local scales in response to climate stressors including temperature, precipitation, salinity, oxygen, and UV (Huang et al. 2021).

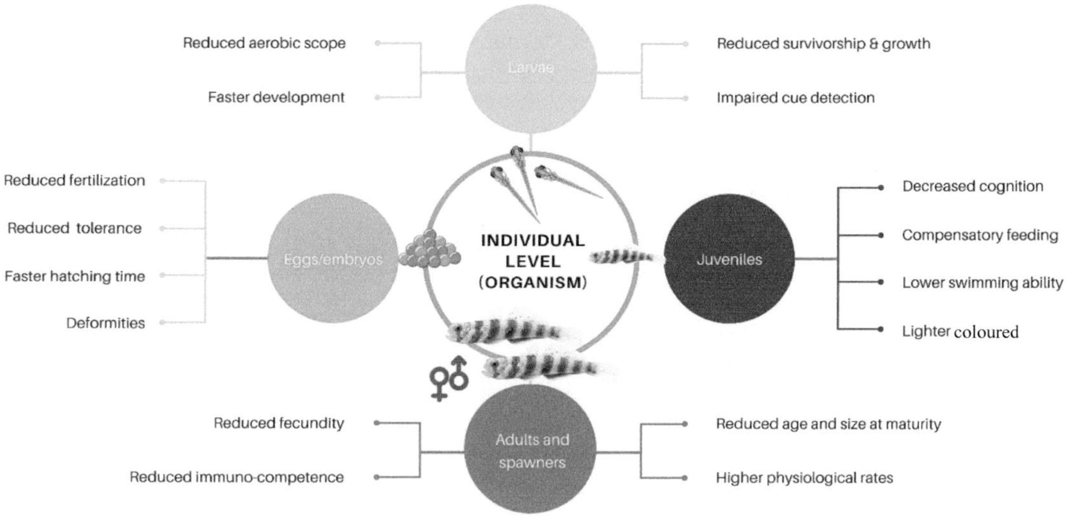

Figure 5. Example of climate-change induced alterations at the individual level across ontogeny in a marine fish (eggs/embryos, larvae, juveniles, adults, and spawners) combining multiple trait-categories.

Morphology

As organisms progress through and interact with different habitats and climate stressors along their life, drastic changes in individuals' morphology may occur in response to environmental fluctuations. These changes include deformities, altered sex, and variation in coloration patterns. Research from the past decade has shown that climate change stressors can induce abnormal cell divisions with later occurrence of malformed cell masses (e.g., corals exposed to high temperatures (Gleason and Hofmann 2011)). The induced morphological deformities can range from mild to lethal, if functions such as swimming are impaired (e.g., shell detachment in gastropods' veliger larvae (Doxa et al. 2021); spine, pelvic fin, and maxilla deformities in fish (Lee et al. 2017), among others). In many ectothermic animals, sex is determined by the temperature experienced during development (e.g., in marine turtles and many fish species). Therefore, temperature fluctuations occurring during development, such as marine heatwaves or other stochastic events, can lead to altered sex, which in turn affects individual environmental tolerance because females and males commonly differ in tolerance limits, with females generally displaying lower tolerance (e.g., the ditch shrimp *Palaemon varians* (Missionário et al. 2022)), depending on fecundity levels and parental investment in the offspring. Interestingly, climate-induced sex changes can also occur in organisms with chromosomal as well as polygenic sex determination, meaning that not even individuals from species with genetically-determined sex are protected against the (ir)reversible masculinization or feminization effects of climate stressors (see Valdivieso et al. 2022). Lastly, the effects of climate change on coloration patterns are relatively understudied but current knowledge suggests that melanic or dark morphs are particularly climate sensitive. In fact, a recent study has argued that due to the role of coloration in ectothermic animals' thermoregulation, as proposed by the thermal melanism hypothesis, lighter colours will likely be selected in the ocean's future stressful environments (Delhey et al. 2020). This will reduce the risk of overheating and provide an adaptive advantage to lighter-coloured morphs, contradicting the equally recent claim that darker animals would be favoured in warm and humid environments, according to Gobler's ecogeographic rule (Tian and Benton 2020). Nevertheless, both these claims require further experimental testing before final conclusions can be drawn.

Behaviour

Organisms' repertoire of behaviours are susceptible to being modified by climate change: (i) through changes in individuals' perception of their environment (e.g., via sensorial systems that affect organisms' ability to detect and respond to external cues or the activation of specific but already existing neural networks in novel environmental conditions), fatigue or habituation related to exposure time or frequency, climate-induced changes in resource availability (food items or suitable habitat), and interspecies encounters (predator abundance, parasite outbreaks, social interactions, etc.); or (ii) through changes in other individual-level traits (physiology, development, morphology) that affect cognitive or locomotory behaviours. Amongst the most conspicuous behavioural changes observed in animals under climate-related stress, the following can be highlighted:

i. Innate behaviour responses (reflex, kinesis, and taxis – changes in movements in response to an environmental cue). These include stationary behaviour, as observed in the fish *Dicentrarchus labrax* when exposed to UV-B in closed cages (Alves et al. 2021), or increased directional movement away from the stressor to avoid exposure, as observed in females of the copepod *Temora turbinata* in response to the hypoxic bottom water layer in a stratification column (He et al. 2021);
ii. Enhanced compensatory feeding, as observed in Southern long-fin goby (*Favonigobius lateralis*) exposed to acidification (Rodriguez-Dominguez et al. 2019), or its inhibition upon exposure to multi-stressors, as observed in the estuarine amphipod *Ampelisca brevicornis* exposed to ocean warming and acidification (Bhuiyan et al. 2022); impaired food detection, as observed in the Antarctic amphipod *Gondogeneia antarctica* subjected to low salinity combined with low pH water conditions (Park et al. 2020); and altered diet selection and foraging strategy, such as shifts from quality to quantity, increased or decreased bite/attack rate, and boldness, as recorded in estuarine fishes exposed to warming (O'Connor and Booth 2021);
iii. Changes in the locomotive ability of mobile animals, namely in swimming or crawling distance or the ability to sustain velocity. Generally, locomotion rates may increase if metabolic and respiration rates increase under stress as a compensatory mechanism (e.g., the crab *Portunus pelagicus* doubled the number of dislocations prior to moulting when exposed to 36°C vs 20°C (Azra et al. 2018)) but under chronic exposure/severe stress, animals may be negatively affected and display longer righting response time, shorter overall distances travelled, reduced velocity, and shorter time to fatigue (e.g., as observed in the sea urchin *Paracentrotus lividus* exposed to acidification (Asnicar et al. 2021); decreased swimming endurance was also observed in elasmobranchs exposed to acidification and warming (Lauder and Di Santo 2015)). Coupled to locomotor components, non-locomotor components affecting mobility capacity are also affected, including increased/reduced responsiveness to a stimulus, impaired directionality/orientation, and increased latency (i.e., response time) (e.g., as shown in the planktivorous coral reef fish *Pomacentrus wardi* exposed to warming and acidification (Allan et al. 2017));
iv. Decreased physical and social cognitive performance, as this type of goal-oriented behaviour relies essentially on perceived predictability of the environment, which can be significantly affected by climate instability and/or increased variability. Negative impacts in learning, memory, and decision making can therefore result in longer shelter seeking times or higher competition for shelter (e.g., as observed in intertidal invertebrates exposed to multiple stresses (Scapini et al. 2019)), longer times to find and capture preys (e.g., as observed in the reef fish *Patagonotothen cornucola* exposed to UV stress (Valiñas and Walter Helbling 2016)), decreased habitat exploration and changes in habitat preference (e.g., bearded seals shift habitat usage upon extended periods of reduced ice cover (Gulland et al. 2022)), slower or reduced harm-avoidance behaviours (e.g., as observed in schools of juvenile *Chromis viridis* exposed to high partial pressure of carbon dioxide (pCO_2) in the presence of a predator (Cattano et al. 2019)), and changes in reproductive behaviour (discussed in the next section), which are all common patterns of altered animal behaviour in a multiple climate change stressor environment.

Reproduction

Reproduction comprises a panoply of interconnected and interdependent biochemical, physiological, and behavioural processes (e.g., signalling, hormone synthesis, gametes production and release, breeding, hatching, embryogenesis, offspring survival) with high energetic costs and that require specific environmental cues to be triggered (Cushing 1969, Pankhurst and Porter 2003). When it comes to electing "actors with leading roles", photoperiod coupled to temperature fluctuation are undoubtedly the two environmental cues that influence organisms' reproduction the most (e.g., maturity/spawning/hatching timing, sex differentiation), especially in higher latitude regions where day length and temperature signal strongly vary among seasons (Pankhurst and Porter 2003). Warming tends to increase metabolism, accelerating organisms' growth, maturation, spermatogenesis/oogenesis and thus changes the timing of crucial reproductive events, like spermiation/ovulation and spawning (Assan et al. 2020). Other effects at the reproductive level have also been attributed to warming but the inconsistency of the findings reported in recent literature shows that their extent and nature will vary according to organisms' physiological tolerance (i.e., thermal range and the capacity to acclimate and adapt to sub-optimal conditions), period of occurrence, and the amplitude/severity of temperature increase (Assan et al. 2020). Hence, the observed effects elicited by warming at the individual level across different taxa include: (i) shifts in event timings such as spawning, maturity, and sex determination/differentiation (e.g., (Asch et al. 2019, Pauly and Liang 2022, Poloczanska et al. 2016)); (ii) improvements due to increased metabolic rates and/or acclimatory processes (e.g., (Gavel et al. 2022, Spencer et al. 2021)); or, in contrast, (iii) reproductive inhibitions (Franzova et al. 2019, Gulland et al. 2022) that most likely arise from changes in steroid hormone synthesis, metabolization, and excretion rates (Servili et al. 2020).

Compared to warming, the impacts of ocean acidification, salinity changes, deoxygenation, or increased UV radiation, either acting alone or combined with other stressors, on individuals' reproductive success are by far less well documented. Still, recent studies have shown that hypercapnia can modulate various behavioural cues crucial to a successful reproduction, such as nest building, parental care, social/group interactions, agonistic behaviour towards sexual competitors, mate tracking, and attraction of females for pair spawning (e.g., Borges et al. 2018, Maulvault et al. 2018b, Maulvault et al. 2018a, Milazzo et al. 2013) due to neurological and chemosensory impairments (Nilsson et al. 2012). Apart from the behavioural impairments, another negative aspect related with acidification is the fact that organisms exposed to lower environmental pH levels are obliged to prioritize the reallocation of energy towards the reestablishment of the internal acid-base and redox balance at the expense of reproduction (Bhuiyan et al. 2022, Boulais et al. 2017, Marchini et al. 2021, Thangal et al. 2021). The same principle is also applicable to salinity changes and deprived oxygen levels that imply great metabolic adjustments and energetic trade-offs to maintain homeostasis, meaning that less resources are therefore available to all the aforementioned steps of species reproduction, and to ensure gametes' quality (Servili et al. 2020). Hence, lower salinities affect the duration, speed, and number of motile spermatozoa (Nissling and Larsson 2018), as well as reproductive behaviours (e.g., nest building in sand gobies (Lehtonen et al. 2016)), and decrease fertilization, gamete development and viability, and hatching rate (Allen et al. 2017). Similarly, hypoxia was found to disrupt the overall functioning of the brain-pituitary-gonad axis, reducing gonadal development and gamete production and quality in both females and males, as well as fertilization, hatching success, and larval survival (Saha et al. 2022, Servili et al. 2020).

As for the effects of UV-B radiation, even though the present literature clearly evidences the need to strengthen research on marine organisms' reproductive traits upon chronic exposure to this stressor, as well as on potential interactions with concomitant variables (e.g., warming), a recent individual-based modelling study on the copepod *Paracyclopina nana* has revealed decreased reproductive success (lower egg and offspring production) (Dur et al. 2021).

Longevity

Recent literature indubitably shows a panoply of adverse effects at multiple biological complexity levels elicited by climate change-related stressors, all of which likely compromise marine organisms' survival. However, organisms may have the opportunity to overcome such environmental changes at the individual level by performing internal adjustments (via plasticity or genetic/transgenerational changes) that will provide them physiological tolerance to some extent (Lee, Jeong et al. 2020, Leung et al. 2021, Paula et al. 2019). Whether environmental changes are elicited by a single stressor or multiple ones can also be decisive to individuals' survival, depending on the effects triggered by the combination of several stressors (Deutsch et al. 2015, Franzova et al. 2019). For instance, in the study conducted by (Gobler et al. 2018), acidification significantly affected the survival of two estuarine fish species (*Menidia beryllina* and *Cyprinodon variegatus*) in combination with warmer temperatures but not when acting alone. Similarly, (Franzova et al. 2019) observed that the survival time of *Himasthla* sp. cercaria exposed to different temperatures considerably (and erratically) varied according to the surrounding pCO_2 conditions.

In both wild and aquaculture settings, one of the aspects that defines marine species' longevity the most is their ability to resist disease. Hence, hostile abiotic conditions due to climate change greatly challenge the survival and ecological success of these species because, as previously discussed, they can trigger profound, negative, and cascading physiological effects that can culminate in impaired animal metabolism and immunocompetence (Makrinos and Bowden 2016). Indeed, in recent years, several reports have associated increased occurrence of disease outbreaks of different ethology (bacterial, viral, parasitical) and sudden mass mortality (especially of marine invertebrates such as molluscs, echinoderms, and cnidarians (Assan et al. 2020, Burge et al. 2014, Sharma and Ravindran 2020)) with climatic events, particularly warming and heatwaves (e.g., Gulland et al. 2022, Montánchez and Kaberdin 2020, Tracy et al. 2019). This link is attributed to the fact that climate change plays a two-sided role in marine species' longevity: on one hand, it compromises their physiological status and immune responses; on the other hand, it may favour the growth, dispersion, distribution, and virulence of pathogenic organisms that thrive under hostile abiotic conditions (Abirami et al. 2021, Assan et al. 2020, Makrinos and Bowden 2016, Marcogliese 2008). In the study of (Amorim et al. 2020), *Scrobicularia plana* specimens exposed to different marine heatwave events scenarios revealed significantly lower total haemocyte counts and phagocytic activity than control specimens, with bivalves being most affected by longer heatwaves with shorter recovery periods. Alves et al. (2020) reported a significant impairment of *Sparus aurata* humoral immunity upon UVB chronic exposure, even at low and moderate doses. Contrasting the aforementioned studies (Ivanina et al. 2016), observed that acidification on its own had little effect on cellular and humoral immune traits of *Mercenaria mercenaria* and *Crassostrea virginica*. In addition to changes in innate and adaptive immunity mechanisms, climate change-related stressors can also shape the structure and alter the functionality of marine biota's microbiome (the first line of defence against infection), increasing species' susceptibility to disease, as pointed out in recent reports (Abirami et al. 2021, Brothers et al. 2018, Makrinos and Bowden 2016, Scanes et al. 2021). Still, the notorious lack of empirical and field studies specifically addressing this topic points out the need to strengthen research in this direction (including the various steps involved in marine organisms' immune responses and using innovative "omic" tools, such as metagenomics analysis), to more accurately predict the impacts of climate change at the individual level.

Overall, climate change effects at the individual level considering all the multiple response categories and their associated plasticity and adaptation are crucial aspects affecting inter-individual variation that will define population, species, and community trajectories regionally and globally. These aspects will be discussed in the next section.

Effects at population, community, and ecosystem level

Species distribution

Changes in species distribution have been one of the most well-documented effects of global changes, with implications at population, community, and ecosystem levels. A large volume of literature has analysed the distribution and abundance of marine organisms and provided evidence of species distribution shifts, as well as local extinctions and/or expansions (species new records) related to global changes. Perry et al. (2005) studied the demersal fish assemblage in the North Sea, based on surveys conducted from 1977 until 2001. They concluded that the centre of the distribution of 15 out of 36 species shifted northward (corresponding to distances of 48 to 403 km) because of the ~ 1°C increase in water temperature during this period. Several studies on fish species tackled other co-occurring pressures (e.g., fishing), as changes in distribution trends can result from changes in fishing effort and practices as well as from climate changes; however, such studies have generally evidenced the preponderance of climate change in species distribution shifts. Pinsky et al. (2013) hypothesised and demonstrated that local differences in climate velocity may better explain heterogeneity in species shifts based on 40 years (1968 to 2011) of data collected in scientific surveys of fish and invertebrates (360 species or species groups) from the continental shelves of North America.

Cheung et al. (2009) reviewed the literature on the effect of climate change on the distribution of a wide diversity of marine organisms (e.g., fish, benthic invertebrates, zooplankton) worldwide. Using a dynamic bioclimatic envelope model to evaluate changes in the distribution of more than one thousand marine species, the authors concluded that a dramatic species turnover of over 60% of the present biodiversity is foreseen for 2050. Moreover, they projected numerous local extinctions for tropical and sub-polar areas, as well as more pronounced species invasions in the Arctic and Southern Ocean. Globally, boreal species have expanded their distributions northward into the Arctic because high-latitude regions have warmed (Wassmann et al. 2011). The Barents Sea, for example, is a shallow marginal sea of the Arctic Ocean which shares many commonalities with the Chukchi Sea (Hunt et al. 2013). Rapid warming in the Barents Sea has changed the spatial distribution of fish as subarctic species expand northward on the shelf (Fossheim et al. 2015), further extending their distributions into the Arctic Basin with the increased northward transport of warm subarctic waters (Snoeijs-Leijonmalm et al. 2022). There is recent evidence of major shifts in the distribution of adult fishes in the Bering Sea, where boreal species such as pollock and Pacific cod have expanded their range northwards (Eisner et al. 2020, Spies et al. 2020, Stevenson and Lauth 2019). Similarly, the southern limits of Arctic species such as Arctic cod have shifted further north, with boreal species extending to these limits (Baker 2021, Marsh and Mueter 2020).

In general, the centres of fish distribution ranges are moving poleward, which roughly corresponds to an expansion of the distribution of tropical and temperate species and a contraction of boreal and polar species (Cheung et al. 2009, Perry et al. 2005, Pinsky et al. 2020, Rijnsdorp et al. 2009). Pinsky et al. (2020) proposed a theoretical framework to explain these processes acting on species range shifts and outlined that expansions and extirpations are favoured or constrained by both biological traits and abiotic factors, and that the resulting dynamics may be quite complex. Intuitively, narrow ecological niches and low dispersal potential will accelerate trailing-edge extirpations and slow leading-edge range expansions, while broad ecological niches, high mobility, and high local adaptation ability will have the opposite effect (i.e., slow range extirpations and accelerate expansions). Nonetheless, range expansion and settlement of fish species in new areas may be rather complex and driven by both biotic (e.g., food and species interactions such as predation and competition) and abiotic conditions (e.g., temperature and habitat conditions) encountered in the new range (Feary et al. 2014, Figueira and Booth 2010).

At the community level, these changes in species distribution impact assemblage structure and ecosystem functioning. Overall, changes in patterns of species richness may disrupt marine biodiversity and ecosystems, and impact commercial fisheries (Cheung et al. 2008, Roessig et al. 2004, Worm et al. 2006). With warming, range shifts often occur on the equatorward range boundary while invasions occur on the poleward range boundary, leading to the tropicalization of marine communities, i.e., an increase in the proportion of tropical species in higher latitude communities (Pinsky et al. 2020). An example of warming-driven range shifts has been observed on the western coast of Australia following a series of marine heatwaves starting in 2011. The events caused a major loss of coastal temperate kelp forest, constricting its range by 100 km. With the lost kelp forest, there was a community-wide shift toward warm-water species. The kelp-dominated community was replaced by persistent seaweed turfs, tropical fish, and tropical invertebrate communities. Moreover, there were changes in seaweed and fish abundances, with cool-water species decreasing and warm-water species increasing in abundance, and some of the species in these rearranged communities were first recorded at these sites. Likewise, some of the species previously documented at these sites are now completely absent (Wernberg et al. 2016). However, changes in species composition of marine communities may be more complex and subtle than a global tropicalization trend. McLean et al. (2021) examined fish communities of the Northern hemisphere and proposed the existence of four processes: tropicalization, increasing warm-affinity species; deborealization, decreasing cold-affinity species; borealization, increasing cold-affinity species; and detropicalization, decreasing warm-affinity species (Fig. 6). These authors reinforced the trend of a tropicalization in many regions but also highlighted that in more than 30% of the cases this was due to deborealization (i.e., decreases in cold-affinity species). Tropicalization was more pronounced in areas that were initially warmer, experienced greater warming, or were deeper. Moreover, species tendencies to increase or decrease resulted from a more diverse synergy of factors than climate change alone, and these interactions should be explored (McLean et al. 2021).

Ocean warming may also affect the vertical distribution of marine species, although this subject is poorly documented. Distribution shifts to deeper waters may be an alternative to a poleward shift as a response to ocean warming (Dulvy et al. 2008). These authors found a deepening of the North Sea demersal fish community as a response to increased (1.6°C) ocean temperatures from 1980 to 2004, at a rate of ~ 3.6 m per decade.

Temperature and ocean warming are not the only global stressors causing changes in species distributions, and therefore community composition. Ocean acidification has been driving changes in the populations' distribution ranges of calcifying species. For example, on the west coast of the United States, upwelling-driven ocean acidification has been shown to limit the vertical distribution of small, planktonic calcifying pteropod populations, which were more likely to be found in waters above the aragonite saturation horizon (Bednaršek and Ohman 2015). As pteropods are an important prey source for many marine fish, their loss in the California Current System, even if temporary, is likely to result in changes in predator-prey interactions with serious implications for the local productive marine fisheries. Calcifying prey species and their loss in marine communities due to acidification is expected to be most severe in polar regions, and particularly regarding changes in pH, with negative consequences for fisheries in Alaska (Mathis et al. 2015). Impacts of ocean acidification on marine community composition are by no means exclusively polar. There are several temperate rocky seagrass and coral reef communities worldwide that harbour unique volcanic carbon dioxide vent systems. These are natural analogues mimicking potential future community composition changes as a consequence of ocean acidification (Fabricius et al. 2011, Foo et al. 2018, Hall-Spencer et al. 2008). In these model systems, there is widespread agreement that benthic calcifiers (including marine invertebrates, corals, and coralline algae) will all decrease in their respective communities. As these diverse calcifying populations offer many different functions within their respective communities, including providing substrate for settlement and habitat structure, future changes in response to ocean acidification are projected to lead to major community modifications (Doney et al. 2009, Fabry et al. 2008, Guinotte and Fabry 2008, Nagelkerken and Munday 2016).

Ocean deoxygenation is also driving changes in species distributions and changes in community compositions, both in pelagic and coastal areas. Mediated primarily by organismal tolerance to low oxygen, large pelagic fish and shark populations (i.e., marlins, sailfish, tuna, and sharks) have experienced

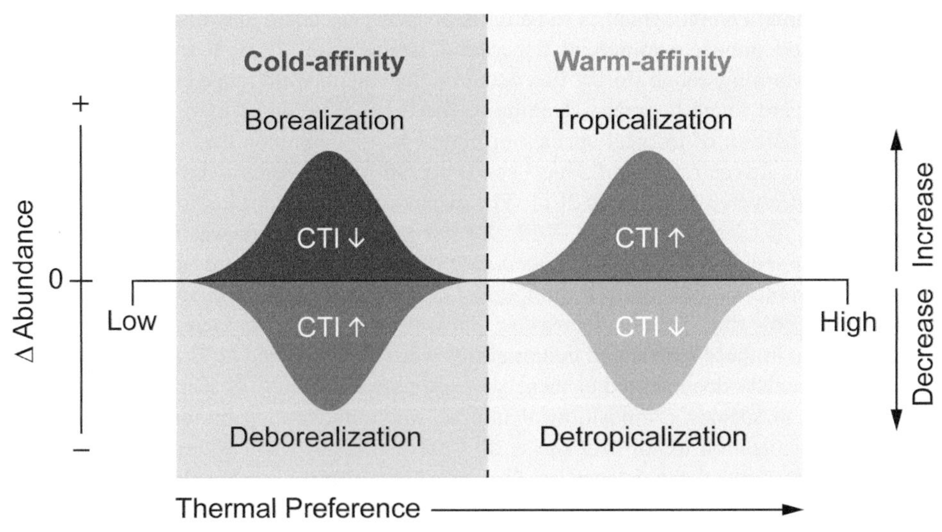

Figure 6. Graphical representation of the four underlying processes contributing to changes in marine communities related to thermal affinities of species (CTI – Community Thermal Index) (adapted from McLean et al. 2021).

vertical habitat compression as deeper, low oxygen waters drove them into shallower waters (Prince and Goodyear 2006, 2007, Vedor et al. 2021). Such habitat compression increases their vulnerability to human fishing pressures and is expected to result in changing trophic interactions as top-down pressure is lessened (apex predators absent). Through species-specific metabolic limitations, the combined effects of deoxygenation and warming have also proved valuable in understanding and projecting changes in biogeography (Deutsch et al. 2015, Howard et al. 2020).

Phenology, trophic interactions, and food-web dynamics

Species have evolved complex behavioural and life histories to match their phenology to predictable climate patterns to maximize fitness (Ji et al. 2010). Widespread shifts in phenology due to changes in climate have occurred over the last 50 years in a broad range of taxonomic groups from plankton to top predators, with spring phenology of all marine species advanced by 4.4 (SE = 0.7) days per decade on average and summer events occurring 4.4 (SE = 1.1) days per decade earlier (Poloczanska et al. 2014, 2013). This advancement of warmer weather, as well as extreme events, has impacted the timing of migrations, with species migrating either earlier in the spring or later in the fall (e.g., juvenile salmon migrate earlier in the spring when temperatures are cooler; Munsch et al. 2019). It has also altered the timing of spawning or reproductive events with earlier or unsynchronized spawning (e.g., corals in the Red Sea have lost their reproductive synchrony reducing the probability for successful fertilization and new recruits; Shlesinger and Loya 2019). Lastly, the timing of a populations' peak abundance within a community has been documented to change (e.g., phytoplankton and zooplankton bloom later or earlier in the year; Edwards and Richardson 2004). These mismatches cause predator and prey resources to become misaligned, leading to trophic mismatches, or trophic asynchrony (Sydeman et al. 2015), which is important in the restructuring of marine food webs (Edwards and Richardson 2004). For example, trophic coupling between an Arcto-boreal pelagic marine fish population and its zooplankton prey is decoupled by warming when the zooplankton prey population peaks over a month before the fish are ready to consume it. In the Arctic, desynchronization between fish and their zooplankton prey frequently occurs leading to declining fish populations. Compared to Arctic populations, temperate fish populations are more likely to track changes in the timing of their prey, thus avoiding population decline. This indicates that the effects of asynchrony are more detrimental for 'specialist' communities inhabiting higher latitudes (Durant et al. 2019).

Global changes lead to changes in the abundance, distribution, and phenology of species (Durant et al. 2019, Parmesan 2006, Van der Putten et al. 2010), as demonstrated in the previous sections. The resulting community restructuring can cause the loss of interspecific interactions as species disappear or move, or it can cause new interactions to emerge with the addition of new co-occurring species (Galasso et al. 2015, Gilman et al. 2010). Additionally, the strength of pre-existing interactions between species may be changed (Alexander et al. 2016, Milazzo et al. 2013, Wernberg et al. 2013). All of these interactions involve changing prey-predator relationships, and can span from primary producers, living at the base of the food chain and allowing energy and biomass to be transferred to consumers, to top predators (Doney et al. 2012). Because this flux of matter is not linear, with frequent omnivory, feeding loops, intra-guild predation, and cannibalism, changing interactions can be difficult to predict (Polis and Strong 1996, Thompson et al. 2007).

Global changes alter food web organisation and dynamics (Hoegh-Guldberg and Bruno 2010, O'Connor et al. 2009) through shifts in the number of trophic groups and links that connect species at the top of the food chain to those at its base (Ullah et al. 2018). For example, predators regulate the populations on which they prey in such a way that the disappearance of predators usually leads to a proliferation of prey populations (Bonaviri et al. 2017). One species may directly affect the abundance of another species but there may also exist indirect consequences on the abundance of other species, known as cascading effects. Additionally, the trophic sensitivity hypothesis establishes that predators are more sensitive to environmental changes than their prey (Petchey et al. 1999, Vasseur and McCann 2005, Voigt et al. 2003). Predators are more mobile than prey and respond to environmental stressors via emigration rather than by evolving tolerance to environmental stress (Menge and Sutherland 1987). Conversely, prey (particularly sessile species) may have limited behavioural mechanisms to cope with stress and may evolve greater tolerance via biochemical adaptation (e.g., Petes et al. 2008).

Energy demands of predators occur simultaneously with the peak availability of prey to maximise predator recruitment. However, as discussed in the previous section, mismatches between food requirements and food availability may be altered by changes in species phenology, with consequences in recruitment, growth, and survival (Bewick et al. 2016, Cushing 1969, 1990, Durant et al. 2007). For example, the breeding period of the Atlantic puffin (*Fratercula arctica*) is synchronized with the peak of herring abundance, which is its main food items (Durant et al. 2003, 2005). The spatial and temporal synchronization between prey availability and predator requirements increase the survival rate of the young, while a mismatch would produce a weaker cohort.

Global changes are also expected to deeply disrupt the co-evolutionary game between predator and prey by desynchronising, at an unknown rate, the higher trophic levels. Predators may have to adapt quickly to the phenological changes of their prey to persist (Durant et al. 2019). This is especially true for ectothermic organisms, such as fish and their prey. Facing the scarcity of their preferred prey, predators can also adapt by shifting their diet on new prey. The eastern Australian salmon (*Arripis trutta*) is an abundant pelagic fish species distributed in south-eastern Australia (Hughes et al. 2013). The diet of *A. trutta* was historically dominated by krill (*Nyctiphanes australis*) during the 1960s and is now mainly composed of small schooling zooplanktivorous fish, such as Australian sardine (*Sardinops sagax*), jack mackerel (*Trachurus declivis*), sandy sprat (*Hyperlophus vittatus*), blue mackerel (*Scomber australasicus*), and Australian anchovy (*Engraulis australis*). Global climate-related warming has increased the intensity of the East Australian Current (EAC) over recent decades (Ridgway 2007), which has resulted in changes to the structure of communities (e.g., Hallegraeff 2010, McLeod et al. 2012). The decline of *N. australis* led to a switch to alternative and more abundant prey (Bulman et al. 2001, Cury 2000).

Prey-predator interactions can be modified when temperature induces a spatial mismatch between species. From the Aleutian Islands in Alaska to the Channel Islands in California, the sunflower starfish, *Pycnopodia helianthoides*, plays a key role in shaping benthic communities by preying upon the purple sea urchin, *Strongylocentrotus purpuratus* (Bonaviri et al. 2017). When annual temperature was below 14°C, *P. heliantoides* was present in high abundances. In contrast, *S. purpuratus* abundance was highest when annual temperature was above 14°C and this was probably due to an absence of *P. heliantoides*, which regulated the sea urchin population. Recent temperature increases resulted in significant *P. heliantoides* mortality, which led to increased sea urchin abundance (Schultz et al. 2016). This proliferation of efficient thermophilic grazers (such as sea urchins) can decimate kelp forests and turn into impoverished sea urchin barrens (Ling et al. 2015). Warmer water temperatures are also affecting the behaviour and physiology of individuals (Hoffmann and Sgrò 2011). Physiological responses to temperature commonly follow a dome-shaped relationship where a maximum is reached as rates increase with temperature but responses decrease rapidly thereafter if temperatures exceeds the thermal optimum (i.e., the thermal window; Munday et al. 2009, Pörtner and Knust 2007). Temperature affects movement (Domenici et al. 2019), foraging efficiency (Miller et al. 2014), attack and ingestion rates (Englund et al. 2011), or prey escape responses (Allan et al. 2015, Domenici 2010). When temperatures rise, predators have higher energy demands (Smolinský and Gvoždík 2014). However, their feeding success depends on the response of the prey involved. For example, increased predation rates were observed in coral reef fish (dottyback, Serranidae) due to a wider tolerance to high temperatures than the prey (damselfish, Pomacentridae) (Allan et al. 2015). In contrast, reduced predation pressure on mosquitofish at high temperatures was due to its escape speed being greater than those of the predator (Australian bass) attack speed (Grigaltchik et al. 2012). These examples indicate temperature can affect swimming performance and escape response of prey (e.g., responsiveness, escape latency, reaction distance, and movement directionality away from the threat) as well as the attack rate and attack speed of the predator. This can also be influenced by predator hunger level and their motivation to forage (Allan et al. 2015, Domenici et al. 2019). In addition, a change in temperature also modifies environmental properties such as water viscosity, which can also affect the movement of individuals and prey-predator interactions.

Ocean acidification alters the sensory and nervous systems of organisms (Domenici et al. 2019), which may cause the failure of prey to detect the odour of predators. Fish, and particularly fish larvae, exposed to high CO_2 levels exhibit changes in their behavioural performances leading to increased predation risk and decreased foraging effort (Cattano et al. 2018). For example, hunting behaviours of the venomous marbled cone snail *Conus marmoreus* are impaired when exposed to elevated CO_2. This species inhabits coral reefs and is generally an ambush or stealth hunter of other gastropods (Watson et al. 2017). Laboratory experiments demonstrated that the activity rate of *C. marmoreus* individuals increased while their motivation to feed was reduced under high CO_2. Finally, the ability to detect or capture prey was reduced. Elevated CO_2 also affected the antipredator behaviour of its prey, the humpbacked conch *Gibberulus gibberulus gibbosus* (Watson et al. 2014).

Carbon dioxide enrichment boosts primary production (Brown et al. 2001, Connell and Russell 2010, Doubleday et al. 2019). However, the effect on herbivores is not yet clear as it depends on the quality of the primary producers. In some ecosystems, acidification boosts the nutritional quality of turf algae by increasing the relative nitrogen content (Leung et al. 2018). Consequently, high quality algae leads to higher biomass of herbivores (Connell et al. 2017, Vizzini et al. 2017). In contrast, some studies also showed that an increase in primary production negatively impacts herbivores due to a decline in plant protein (Facey et al. 2014). An increase in primary production may also result in a proliferation of cyanobacteria (Ullah et al. 2018). Unpalatable cyanobacteria produce potent allelochemicals that deter feeding and are difficult to control by grazers and can potentially divert productivity into alternate food web pathways, suppressing the flow of energy to the second trophic level (Dillon et al. 2010). There is still little work that considers simultaneously warmer water temperatures and higher CO_2 concentrations. In south-eastern Australia, Ullah et al. (2021) demonstrated that warming combined with acidification caused a strong decline in the biomass of invertebrate prey, particularly herbivorous invertebrates and lower trophic level omnivores. Indeed, warmer temperatures increased metabolic demands of predators and consumption rates by carnivores. However, taken separately, the warmer temperature induced a small decline in invertebrate biomass while acidification largely increased the invertebrate biomass. This highlights the complexity as well as the need to account for all the factors affected by global change simultaneously to understand future consumer-resource interactions.

The use of trophic models provides an understanding of how communities evolve according to global changes. Durant et al. (2019) used a predator-prey model accounting for the full life cycle of fish and zooplankton. The authors showed that fish populations in Arcto-boreal biomes would decline in abundance by 2060 due to a decrease in the temporal overlap between fish and prey. Fish populations in temperate biomes are projected to remain stable. Both the predator and prey advance timing, leading to maintenance of the temporal overlap. Match-mismatch dynamics become progressively more important toward higher latitudes, as the period available for phytoplankton and zooplankton production decreases with latitude due to increasing restrictions in light and nutrients. Moreover, Arcto-boreal biomes are composed of relatively few species thus giving each species a pivotal role. In other words, small, fast-growing species progressively dominated the ecosystems in historical records and this trend is expected to persist under the pressure of global changes (Hattab et al. 2016, Lotze et al. 2019, Maureaud et al. 2017). The biomass and production of all fish seem likely to decrease globally (Petrik et al. 2020) while biomass transfers will be drastically faster at the scale of the whole food web and within trophic classes (du Pontavice et al. 2020). Thus, temperature-induced changes in flow kinetic and trophic transfer efficiency may contribute independently and cumulatively to the decline in consumer biomass (du Pontavice et al. 2021). Body condition of top marine predators, such as the harp seal (*Pagophilus groenlandicus*), appear to decline due to poorer nutritional quality of prey fish (Griffith et al. 2019). The pelagic community should become increasingly dominated by energy (lipid, fatty acid)-poorer zooplankton species. However, an increase of the total amount and faster turnover of zooplankton might counteract the reduction of more lipid-rich species. The loss of top predators and trophic cascades may result in widespread trophic downgrading of ecosystems (Heupel et al. 2014) and may negatively affect the robustness of food webs to perturbation.

Connectivity

In coastal seas, movements of organisms between patchy habitats affect the structure and function of an ecosystem (Caldwell and Gergel 2013, Fullerton et al. 2010, Muir and Williams 2012, Perez et al.

2019). Highly mobile organisms transfer nutrients and energy from one location to another (Hyndes et al. 2014, Williams et al. 2018). Connectivity could then influence population dynamics and other ecological processes such as predation, competition, and recruitment (Olds et al. 2018, Perry et al. 2018). Species that are mobile in at least one stage of their life cycle are spatially organised in metapopulations (Cowen and Sponaugle 2009). The geographically separated subpopulations that constitute metapopulations could exchange individuals from source (or "origin") to sink (or "destination") subpopulations. Population connectivity plays a fundamental role in local, metapopulation, and community dynamics, genetic diversity, and population resilience (Botsford et al. 2001, Hastings and Harrison 1994). Pattern and magnitude of demographic exchange among populations is a crucial issue to understand the evolutionary and ecological processes of speciation, adaptation, and persistence. The degree of connectivity among habitat patches determines how subpopulations replenish themselves, whether they are demographically "open" or "closed," and whether they will persist in the face of climate change (White et al. 2019). Climate change, intensified by anthropogenic disturbances, has destroyed a large range of natural habitats, limiting the ability of species to migrate and blocking connectivity (Driver and Hoeinghaus 2016, Geist and Hawkins 2016, Horn et al. 2020, Kaymak et al. 2018).

Connectivity may be classified into two distinct categories: functional and structural connectivity. Functional connectivity refers the suite of biological or behavioural responses of organisms to the environment (Kindlmann and Burel 2008, Turgeon et al. 2010). It describes the effect of ocean conditions on the dispersal potential of organisms via physiological mechanisms, including their activity and performance (e.g., time available for dispersal and survival probability while dispersing; Gerber et al. 2014) (cf. Section 3). Structural connectivity reflects the spatial structure of the environment and its influence on the likelihood of individuals from one habitat reaching another habitat (Kindlmann and Burel 2008). It includes habitat size and quality, as well as the number and proximity of habitat patches. Climate change affects functional connectivity by directly affecting the biology of dispersing organisms and structural connectivity by modifying the spatial organisation of seascapes.

Most studies of connectivity between subpopulations in the marine environment have focused on larval stages, with benthic invertebrates being the most studied group (Gerber et al. 2014). Over 70% of marine species have a bipartite life cycle in which sessile or sedentary adults produce planktonic propagules (Bradbury and Snelgrove 2001, Levin 2006). These species exist in metapopulations, with subpopulations of benthic adults occupying habitat patches that are connected only by the movements of larvae (Cowen and Sponaugle 2009). The eggs, larvae, or spores of benthic adult fish, invertebrates, or seaweeds may travel tens or hundreds of kilometres before settling and transitioning to juvenile life stages (Kinlan and Gaines 2003, Siegel et al. 2003).

Larvae disperse from the "origin" population where they are spawned to the "destination" population where they settle to the benthos across the three-dimensional ocean. Most marine larvae drift passively, due to their small size and weak swimming ability, and are oceanographically advected by ambient currents. Ocean currents are chaotic and turbulent, and therefore larval transport is random and unpredictable from year to year (Siegel et al. 2008). Larvae from different origins encounter different conditions during oceanic transport, as well as in their settlement location. The conditions experienced affect the individual survival and the relative contribution of each origin population to the new cohort (Hamilton et al. 2008). Projected changes in ocean circulation in response to global warming have the potential to affect patterns of connectivity worldwide (Munday et al. 2009, van Gennip et al. 2017). Warmer temperatures accelerate egg and larval development, leading to reduced drifting period (Hirst and López-Urrutia 2006, O'Connor et al. 2007) and dispersal distances (Ayata et al. 2010, Huret et al. 2010, Lett et al. 2010), advanced settlement (Munday et al. 2009), changes in transport processes and settlement location (Huret et al. 2010, Munday et al. 2009, Tracey et al. 2012). Figueiredo et al. (2022) estimated that a 2°C increase in water temperature will greatly affect the survival and settlement competence of larvae of a coral species belonging to the Great Barrier Reef, *Acropora millepora*. Larval dispersal distance will be reduced by 6.6%, which will reduce the number of reached reefs. The proportion of larvae produced by a reef that settles back onto that reef (local retention) will increase by 19.5%. Consequently, reefs will become more isolated and more dependent on locally produced larvae for replenishment after disturbances. However, the effect of warmer temperatures on long-term larval survival and settlement, and how such changes may affect connectivity, remains largely unexplored.

Climate change is likely affecting larval sensitivity to environmental conditions (O'Connor et al. 2007). The effect of ocean acidification on larval behaviour and their sensory systems has been studied experimentally on tropical coral reef fish (Gerber et al. 2014). Habitat selection can be modified because environmental cues are not perceived in the same way by larvae and adults. Hence, the larvae of orange clownfish (*Amphiprion percula*) were attracted to stimuli that led to suboptimal locations for settlement (Munday et al. 2009). In addition, climate change influences patterns of larval production, including the redistribution of populations, as well as adult growth and the relative allocation of energy to gonadal and somatic production (Gerber et al. 2014), which could have implications for connectivity patterns (as discussed in Section 3).

Connectivity also plays an important role in gene flow between subpopulations. Coleman et al. (2017) studied kelp populations (*Ecklonia radiate*) along the coast of New South Wales, Australia. These authors detected an increase in poleward connectivity, while equatorward connectivity decreased. Kelp populations already underwent climate-mediated range retractions (Smale and Wernberg 2013, Vergés et al. 2014) driven by their low genetic diversity conferring an inability to respond to thermal stress (Coleman et al. 2011, Wernberg et al. 2016). The recovery of lower latitude populations via recolonization from genetically diverse propagules from higher latitudes should be hindered under future climate scenarios, resulting in continued poleward shifts in kelp distribution and extirpation of the equatorward populations. Decreased connectivity may induce subpopulations to be adapted to local environmental conditions, as environmental changes are likely to vary from one location to another (Young et al. 2018), thus reducing genetic diversity and evolutionary capacity. In a climate change context, adaptation may also favour individuals in areas that experience higher natural variability (Morikawa and Palumbi 2019) as the effects of climate change are often experienced through changes in the intensity and frequency of extreme environmental conditions (Malhi et al. 2020, Tittensor et al. 2019).

To fully understand species' connectivity patterns and potential climate change impacts, studies must consider the vulnerability and movement patterns of all life stages (Álvarez-Romero et al. 2018, Moffitt et al. 2009). Some species show very clear ontogenetic shifts or seasonal migrations between habitats (Andrews et al. 2009, Pittman et al. 2014). Staveley et al. (2019) studied the spatial and temporal movement patterns of juvenile and sub-adult Atlantic cod (*Gadus morhua*) in the Swedish Skagerrak region. Atlantic cod are physiologically adapted to tolerate high thermal variation (Metcalfe et al. 2012, Righton et al. 2010), particularly during juvenile stages (Bjornsson 2001). Then, they utilize a range of marine environments (e.g., seagrass beds and gravel areas) and display important seasonal changes in connectivity. A drastic decline in movement of juvenile and sub-adult cod occurred as the water temperature decreased, resulting in less connectivity within the fjord system during wintertime. Despite a decrease in connectivity during cold seasons, the study also highlighted areas where movement activity occurred throughout the entire study period according to both habitat availability and habitat choice. This example demonstrates that organisms depend on a range of distinct habitats and ecosystems, and link those habitat and ecosystem units into a complex mosaic (Sheaves 2009).

Marine population connectivity studies have focused mostly on larval dispersal (Andrello et al. 2015, Magris et al. 2016, Treml et al. 2008), with less emphasis on the adult stage (Friesen et al. 2019). It is now recognised that movement of adults is crucial for the persistence of populations, particularly for teleost and elasmobranch fish (Bryan-Brown et al. 2017, Frisk et al. 2014). Movement of adults may even contribute more significantly to genetic connectivity than classical larval-mediated processes (Frisk et al. 2010, 2008). For example, reproductive migrations over great distances to optimal spawning/pupping areas (e.g., as observed in menhaden, *Brevoortia tyrannus*, or bluefish, *Pomatomus saltatrix*) demonstrate how the degree of connectivity is determined by the behavioural ecology of adults (Friesen et al. 2021). Both species are common along the north-eastern US continental shelf, and their spawning occurs in coastal waters with the resulting offspring recruiting to inshore habitat. Adults are dispersed during winter but the mixing of spawners during spring sets the conditions for panmictic genetic populations (i.e., individuals originating from a single spawning area; Lynch et al. 2010). Environmental cues that synchronise mass movements of spawners may be altered by climate change, which may modify connectivity patterns (both the timing and intensity of migration). Species movements are limited by the suitability and distribution of the surrounding habitat (Irlandi and Crawford 1997, Stevenson-Holt et al. 2014). Habitat suitability is mainly determined by temperature, pH, and/or dissolved oxygen preferences of species (Deutsch et al. 2015) and all these factors are modified by climate change. As individuals track preferred temperatures on relatively short time scales (Li, Hollowed et al. 2019), this could have significant implications for connectivity patterns via adult movement, as well as via larval dispersal if spawning or larval release occur in different areas.

Regime shifts and no-analogue ecosystems

An ecological tipping point occurs when an ecosystem can no longer cope with one or more environmental changes, and the ecosystem suddenly shifts from one state to another. This shift, often referred to as a regime shift, can lead to a qualitatively different and persistent ecosystem structure and function. Tipping points and regime shifts have occurred, and are expected to occur, in Polar regions, coral reefs, and in oceanic, deep sea, and many coastal temperate systems (Heinze et al. 2021). One of the best-studied examples of a regime shift is the synchronous increase of dinoflagellates, reduction of diatoms, and decrease in the size of copepods that occurred in the North Sea because of warming seawater (Alheit et al. 2005). These changes triggered an increase in the demersal fish populations; however, continuously high fishing pressure coupled with additional changes in the plankton community led to a collapse of the entire demersal assemblage, particularly Atlantic cod (Beaugrand et al. 2003, Lynam et al. 2017, Sguotti et al. 2023).

Both temperature and oxygen-driven mass bleaching events have drastically modified tropical coral reef ecosystems by reducing coral cover on the Great Barrier Reef in Australia and the Caribbean (Hughes et al. 2019, Johnson et al. 2021, Johnston et al. 2018). Ocean acidification is also expected to cause regime shifts as the current day CO_2 vent analogues have shown a marked decline in calcifier-dominated benthos leading to habitat and biodiversity loss for many associated/reliant species (Kleypas and Yates 2009). Regime shifts have recently been attributed to an increased frequency of extreme climatic events. In the Australian kelp forest example described above, the kelp forest distribution shifts due to heatwaves caused a regime shift (Wernberg et al. 2016). This was prompted by a lack of time between extreme climate events that prevented kelp recovery, as well as by a doubling in the intensity and frequency of heatwaves (Cai et al. 2015). It is now expected that the range shifts of the remaining kelp forests on the western coast of Australia will eventually be extirpated as they move along the southern Australian coastline until there is no more coastal habitat to move into. This has negative implications for thousands of endemic species supported by the kelp forests of Australia's Southern Ocean, as well as serious economic implications related to fisheries loss. Like Australia's kelp forest loss, kelp habitat contractions with similar ecosystem regime shifts are occurring worldwide (Krumhansl et al. 2016, Kumagai et al. 2018, Ling 2008).

There are numerous examples of multiple global and local stressors, overexploitation, and extreme events causing regime shifts. However, assessments of the functioning and structure of the resultant ecosystem state is often overlooked. This is particularly true when identifying the recovery ability or a state of hysteresis (Blöcker et al. 2023) if the ecosystem is unable to return to its previous regime, even if the conditions that caused the shift are removed (e.g., after a system experiences a marine heatwave, temperatures return to ambient but the ecosystem is not able to shift back to its previous state). These processes are typically dynamic and non-linear, but there are a few emerging trends. For example, the more the species are simultaneously redistributed within an ecosystem, the greater the additive negative ecological impacts, and reduced likelihood of recovery (Marzloff et al. 2016). Improved methods for detecting, predicting, and managing ecosystem regime shifts is particularly important in the context of global change (de Young et al. 2008).

Assessment, monitoring, and future research

As outlined throughout this chapter, significant progress in documenting the impact of climate change and other global changes, such as ocean acidification and deoxygenation on marine species and ecosystems, has been made over the last two decades. This is illustrated in the latest IPCC report, which compiled thousands of scientific articles in three reports collectively comprising more than 10,000 pages (Intergovernmental Panel on Climate Change (IPCC) 2023b). These reports conclude with the highest level of scientific certainty that human-driven global changes are altering the ocean's key physicochemical parameters with observable consequences for marine species and ecosystems.

In the face of such over-whelming evidence, global changes and their impacts on marine life attracted attention from both the scientific and policy communities. The ocean (SDG 14), climate change (SDG 13), and other environmental drivers such as ocean acidification (SDG 14.3) are central in the Sustainable Development Goals' (SDGs) framework adopted by all United Nations Member States in 2015, providing a blueprint for peace and prosperity for people and the planet, now and into the future (United Nations: Department of Economic and Social Affairs 2022). The urgency to act to address and minimize the impact of global changes on the ocean led to the launch in 2021 of the United Nations Decade of Ocean Science for Sustainable Development (UN Ocean Decade 2021–2030; https://oceandecade.org/). Through this initiative, the global ocean science community is invited to work with ocean users and managers to co-design the plan for the next ten years in ocean science and technology to deliver the ocean we need for the future we want. This led to an increased recognition by the scientific community to work in close collaboration with ocean users and prioritize science toward action rather than a traditional slow and continuous increase in knowledge. For example, the project Ocean Acidification Research for Sustainability (OARS) was recently endorsed by the UN Ocean Decade with the central outcome to co-design and implement research and observation strategies in collaboration with data/information producers and end-users by 2025 (Dobson et al. 2022).

A focus on solutions to tackle global marine changes will provide a priority roadmap for future science. This roadmap can only be locally relevant as many factors have to be accounted for, including the specificities of the local marine species, ecosystems, and associated services, the unique array of local and global environmental challenges, the feasibility and availability within a reasonable timescale of the proposed actions, and other parameters limiting the implementation of these actions (e.g., scientific gap, lack of capacity, cultural features). Gattuso et al. (2015) identified four different categories of actions to tackle marine climate change and ocean acidification: (i) mitigation, aiming at reducing CO_2 emissions or removing it from the atmosphere; (ii) protection of existing ecosystems, for example by increasing the surface of marine protected areas (MPAs) to maintain or increase existing biodiversity and promote resilience; (iii) habitat restoration; and, (iv) societal adaptation to anticipate and minimize the effects of climate change on society and ecosystem services. The development and implementation of these different actions will require some advance in technology, science (e.g., new theoretical frameworks, development of best practices, collection of new data, synthesis of existing information), communication, and societal changes at the local and global scales (Jagers et al. 2020). Scientifically, such complex environmental issues can only be solved through a combination of approaches (observation, experimentation, and modelling; Dupont et al. 2021) and methods (from molecules to ecosystem; e.g., Rodríguez-Martínez et al. 2022). This chapter presents many scientific gaps that limit our understanding of the impacts global changes are imposing on marine ecosystems but this is no excuse for inaction. Future work must prioritize science that will lead to real actions and implementation of solutions that preserve marine biodiversity and ecosystem functioning.

Synthesis and theoretical frameworks

There is overwhelming evidence on global environmental changes and their impacts on marine species and ecosystems (Intergovernmental Panel on Climate Change (IPCC) 2023a). However, these data are not always adapted or summarized in a way that would allow informing ocean users on the best course of action to minimize and address these impacts. The data need is highly dependent on the scale (e.g., local vs global) and category of solution (e.g., mitigation, adaptation), and in many cases new data will be required to fully develop and implement solutions. However, in face of the global ocean crisis emergency, the community should take advantage of the existing data and information to move forward and avoid delaying actions. One of the main future challenges will be to take advantage of existing data and information to test new hypotheses, improve existing theoretical frameworks, develop and implement solutions, and evaluate gaps while identifying the factors responsible for these information gaps (e.g., absence of data sharing, poor communication between communities, lack of capacity or best-practices).

Many solutions are depending on our forecasting abilities for future changes under different scenarios. Results from experimental work and from *in situ* observing efforts are not always well integrated into synthesis and modelling efforts. Ocean observation and data and information delivery systems are often focused on one or a handful of physical and biogeochemical parameters. Moreover, they do not include biology and under-estimate the complexity of bridging physicochemical and biological changes. Projections rely mainly on proxy variables like temperature, oxygen, and salinity, and offer simplistic thresholds to speculate about the status and trends of biodiversity and ecosystem services. However, the impacts of global environmental changes on ecosystems are complex and dependent on a plethora of conditions. There is a need to consider factors such as adaptation to local chemical variability, evolutionary processes, ecological interactions, and the modulating role of other environmental drivers or stressors.

Some of these issues can be resolved with the existing data. For example, Vargas et al. (2022, 2017) took advantage of the bibliographic and biological responses' databases developed by the Ocean Acidification International Coordination Centre and tested the hypothesis that an organism's sensitivity to pH was driven by the present range of natural variability through local adaptation. Hypothesis-driven synthesis and meta-analysis exercises can significantly improve our ability to bridge physicochemical changes, biological responses, and our forecasting ability.

Multiple stressors, ecology, evolution, and best practices

One of the ten United Nations Ocean Decade Challenges is to "understand the effects of multiple stressors on ocean ecosystems". However, research on global ocean changes usually focus on single drivers, species, and short-term experiments (Riebesell and Gattuso 2015). Consequently, the current state of knowledge tends to under-estimate the complexity of the biological response to environmental changes leading to an over- or under-estimation of the real impacts. This is partially a consequence of practical constrains (e.g., funds, time, facilities) but also a lack of understanding of the best practices in the respected fields of ecology, evolution, and multiple stressors. Future research and synthesis efforts will have to embrace the complexity of marine ecosystems.

This will require the development and sharing of best practices. For example, the Scientific Committee on Oceanic Research working group for "Changing Ocean Biological Systems: How will biota respond to a changing ocean?" promotes new methods for assessing the effects of multiple environmental drivers on marine species and ecosystems (Boyd et al. 2019). Developing solutions for multiple stressors requires the ability to identify the main environmental drivers of biological response at each location as well as their sources (Beusen et al. 2022). Understanding the consequences of multiple environmental drivers on a given species or ecosystem is highly complex and cannot be solved by classic multiple stressors experiments (Boyd et al. 2018). A scientific strategy combining single stressor mechanistic experiments, multiple stressors experiments, field observations, and modelling and integrating existing theoretical frameworks is needed to resolve each driver's mode of action and potential interactions (Beusen et al. 2022). Similar challenges and needs for sophisticated strategies and experimental designs combining a wide range of approaches are required to address other levels of complexities (e.g., (Gaylord et al. 2015) for ecology; (Calosi et al. 2016) for evolution). In face of the global ocean change emergency, and to avoid delaying the needed actions, a fit-for-purpose strategy should be developed to reach the right level of complexity in our understanding.

Observation and monitoring

Marine environmental *in situ* observations are a prerequisite for ecosystem-based management and essential to ensure that sustainable management targets are met. Global ocean changes include the so-called 'slow onset events', such as ocean warming, acidification, deoxygenation, sea level rise, and glacial retreat, as well as an increase of extreme weather and biogeochemical anomalous conditions, such as marine heatwaves. Over the last decades, significant progresses have been made in the development of best-practices and organization of observing networks for physicochemical parameters. For example, the Global Ocean Acidification Observing Network is now gathering more than 900 scientists from 105 countries and continues to expand and enhance geographic extent and availability of coastal and open ocean observing data (Tilbrook et al. 2019). However, many challenges remain. A strong communication across disciplines will allow the development of a fit-for-purpose observing system and collecting physicochemical and biological data with the required spatiotemporal resolution, which will need innovative approaches across disciplines (e.g., Widdicombe et al. 2022).

Acknowledgments

C.M. acknowledges "Fundação Para a Ciência e Tecnologia (FCT, I.P.)" for funding projects ExtremeOceans (PTDC/BIA-BMA/1494/2020) and AQUATROPHYS (2021.09718.CBM) and Biodiversa+ for funding project EASMO (BioDivClim452). FCT, I.P. is also acknowledged for funding this work under the scope of the strategic projects i4HB (LA/P/0140/2020), UCIBIO (UIDB/04378/2020; UIDP/04378/2020), and for the individual research fellowship CEECIND/01526/2018 awarded to C.M. We acknowledge FCT I.P. (Portuguese Foundation for Science and Technology) for the researcher contracts to D.M. (CEECIND/00153/2022, DOI: https://doi.org/10.54499/2022.00153.CEECIND/CP1720/CT0016) and C.M. (CEECIND/01526/2018, DOI: https://doi.org/10.54499/CEECIND/01526/2018/CP1564/CT0008) and for funding projects ExtremeOceans (PTDC/BIA-BMA/1494/2020, DOI: https://doi.org/10.54499/PTDC/BIA-BMA/1494/2020), Aqua-CLIMADAPT (PTDC/CTA-AMB/0592/2021, DOI: https://doi.org/10.54499/PTDC/CTA-AMB/0592/2021) and AQUATROPHYS (2021.09718.CBM). We also acknowledge Biodiversa+ for funding project EASMO (BioDivClim452). FCT I.P is also acknowledged for the financial support to i4HB (LA/P/0140/2020), UCIBIO (UIBD/04378/2020; UIDP/04378/2020), and CESAM (UIDP/50017/2020+UIDB/50017/2020+LA/P/0094/2020). The IAEA is grateful to the Government of the Principality of Monaco for the support provided to its Marine Environment Laboratories.

References

Abirami, B., Radhakrishnan, M., Kumaran, S. and Wilson, A. (2021). Impacts of global warming on marine microbial communities. The Science of the Total Environment 791(147905): 147905. https://doi.org/10.1016/j.scitotenv.2021.147905.

Abram, N., Gattuso, J.-P., Prakash, A., Cheng, L., Chidichimo, M. P., Crate, S. et al. (2019). Framing and context of the report. pp. 73–130. *In*: Pörtner, H.-O., Roberts, D.C., Masson-Delmotte, V., Zhai, P., Tignor, M., Poloczanska, E. et al. (eds.). The Ocean and Cryosphere in a Changing Climate: Special Report of the Intergovernmental Panel on Climate Change. Cambridge University Press. https://doi.org/10.1017/9781009157964.003.

Aguilar, C., Raina, J.-B., Fôret, S., Hayward, D. C., Lapeyre, B., Bourne, D. G. et al. (2019). Transcriptomic analysis reveals protein homeostasis breakdown in the coral Acropora millepora during hypo-saline stress. BMC Genomics 20(1): 148. https://doi.org/10.1186/s12864-019-5527-2.

Aksakal, E. and Ekinci, D. (2021). Effects of hypoxia and hyperoxia on growth parameters and transcription levels of growth, immune system and stress related genes in rainbow trout. Comparative Biochemistry and Physiology. Part A, Molecular & Integrative Physiology 262(111060): 111060. https://doi.org/10.1016/j.cbpa.2021.111060.

Alberts, B., Johnson, A., Lewis, J., Morgan, D., Raff, M., Roberts, K. et al. (2017). Molecular Biology of the Cell (J. Wilson and T. Hunt, Eds.). W.W. Norton and Company. https://doi.org/10.1201/9781315735368.

Alderdice, R., Suggett, D. J., Cárdenas, A., Hughes, D. J., Kühl, M., Pernice, M. et al. (2021). Divergent expression of hypoxia response systems under deoxygenation in reef-forming corals aligns with bleaching susceptibility. Global Change Biology 27(2): 312–326. https://doi.org/10.1111/gcb.15436.

Alexander, J. M., Diez, J. M., Hart, S. P. and Levine, J. M. (2016). When climate reshuffles competitors: A call for experimental macroecology. Trends in Ecology and Evolution 31(11): 831–841. https://doi.org/10.1016/j.tree.2016.08.003.

Alheit, J., Möllmann, C., Dutz, J., Kornilovs, G., Loewe, P., Mohrholz, V. et al. (2005). Synchronous ecological regime shifts in the central Baltic and the North Sea in the late 1980s. ICES Journal of Marine Science: Journal Du Conseil 62(7); 1205–1215. https://doi.org/10.1016/j.icesjms.2005.04.024.

Allan, B. J. M., Domenici, P., Munday, P. L. and McCormick, M. I. (2015). Feeling the heat: the effect of acute temperature changes on predator–prey interactions in coral reef fish. Conservation Physiology 3(1): cov011. https://doi.org/10.1093/conphys/cov011.

Allan, B. J. M., Domenici, P., Watson, S. A., Munday, P. L. and McCormick, M. I. (2017). Warming has a greater effect than elevated CO_2 on predator–prey interactions in coral reef fish. Proceedings. Biological Sciences 284(1857): 20170784. https://doi.org/10.1098/rspb.2017.0784.

Allen, J., Schrage, K., Foo, S., Watson, S.-A. and Byrne, M. (2017). The effects of salinity and pH on fertilization, early development, and hatching in the crown-of-thorns seastar. Diversity 9(1): 13. https://doi.org/10.3390/d9010013.

Álvarez-Romero, J. G., Munguía-Vega, A., Beger, M., del Mar Mancha-Cisneros, M., Suárez-Castillo, A. N., Gurney, G. G. et al. (2018). Designing connected marine reserves in the face of global warming. Global Change Biology 24(2). https://doi.org/10.1111/gcb.13989.

Alves, R. N., Mahamed, A. H., Alarcon, J. F., Al Suwailem, A. and Agustí, S. (2020). Adverse effects of ultraviolet radiation on growth, behavior, skin condition, physiology, and immune function in gilthead seabream (Sparus aurata). Frontiers in Marine Science 7. https://doi.org/10.3389/fmars.2020.00306.

Alves, R. N., Justo, M. S. S., Laranja, J. L. Q., Alarcon, J. F., Al Suwailem, A. and Agustí, S. (2021). Exposure to natural ultraviolet B radiation levels has adverse effects on growth, behavior, physiology, and innate immune response in juvenile European seabass (Dicentrarchus labrax). Aquaculture (Amsterdam, Netherlands) 533(736215): 736215. https://doi.org/10.1016/j.aquaculture.2020.736215.

Amorim, V. E., Gonçalves, O., Capela, R., Fernández-Boo, S., Oliveira, M., Dolbeth, M. et al. (2020). Immunological and oxidative stress responses of the bivalve Scrobicularia plana to distinct patterns of heatwaves. Fish and Shellfish Immunology 106: 1067–1077. https://doi.org/10.1016/j.fsi.2020.09.024.

Anacleto, P., Figueiredo, C., Baptista, M., Maulvault, A. L., Camacho, C., Pousão-Ferreira, P. et al. (2018). Fish energy budget under ocean warming and flame retardant exposure. Environmental Research 164: 186–196. https://doi.org/10.1016/j.envres.2018.02.023.

Andrello, M., Mouillot, D., Somot, S., Thuiller, W. and Manel, S. (2015). Additive effects of climate change on connectivity between marine protected areas and larval supply to fished areas. Diversity and Distributions 21(2): 139–150. https://doi.org/10.1111/ddi.12250.

Andrews, K. S., Williams, G. D., Farrer, D., Tolimieri, N., Harvey, C. J., Bargmann, G. et al. (2009). Diel activity patterns of sixgill sharks, Hexanchus griseus: the ups and downs of an apex predator. Animal Behaviour 78(2): 525–536. https://doi.org/10.1016/j.anbehav.2009.05.027.

Anestis, A., Lazou, A., Pörtner, H. O. and Michaelidis, B. (2007). Behavioral, metabolic, and molecular stress responses of marine bivalve Mytilus galloprovincialis during long-term acclimation at increasing ambient temperature. American Journal of Physiology. Regulatory, Integrative and Comparative Physiology 293(2): R911–21. https://doi.org/10.1152/ajpregu.00124.2007.

Araújo, J. E., Madeira, D., Vitorino, R., Repolho, T., Rosa, R. and Diniz, M. (2018). Negative synergistic impacts of ocean warming and acidification on the survival and proteome of the commercial sea bream, Sparus aurata. Journal of Sea Research 139: 50–61. https://doi.org/10.1016/j.seares.2018.06.011.

Araújo, M. J., Quintaneiro, C., Soares, A. M. V. M. and Monteiro, M. S. (2021). Effects of ultraviolet radiation to Solea senegalensis during early development. The Science of the Total Environment 764(142899): 142899. https://doi.org/10.1016/j.scitotenv.2020.142899.

Asch, R. G., Stock, C. A. and Sarmiento, J. L. (2019). Climate change impacts on mismatches between phytoplankton blooms and fish spawning phenology. Global Change Biology 25(8): 2544–2559. https://doi.org/10.1111/gcb.14650.

Ashworth, J. (2017). Marine microalgae: systems biology from 'omics'. pp. 207–221. *In*: Systems Biology of Marine Ecosystems. Springer International Publishing. https://doi.org/10.1007/978-3-319-62094-7_10.

Asnicar, D., Novoa-Abelleira, A., Minichino, R., Badocco, D., Pastore, P., Finos, L. et al. (2021). When site matters: Metabolic and behavioural responses of adult sea urchins from different environments during long-term exposure to seawater acidification. Marine Environmental Research 169(105372): 105372. https://doi.org/10.1016/j.marenvres.2021.105372.

Assan, D., Kuebutornye, F. K. A., Mustapha, U. F., Chen, H. and Li, G. (2020). Effects of climate change on marine organisms. American Journal of Climate Change 09(03): 204–216. https://doi.org/10.4236/ajcc.2020.93013.

Axworthy, J. B., Timmins-Schiffman, E., Brown, T., Rodrigues, L. J., Nunn, B. L. and Padilla-Gamiño, J. L. (2022). Shotgun proteomics identifies active metabolic pathways in bleached coral tissue and intraskeletal compartments. Frontiers in Marine Science 9. https://doi.org/10.3389/fmars.2022.797517.

Ayata, S.-D., Lazure, P. and Thiébaut, É. (2010). How does the connectivity between populations mediate range limits of marine invertebrates? A case study of larval dispersal between the Bay of Biscay and the English Channel (North-East Atlantic). Progress in Oceanography 87(1–4): 18–36. https://doi.org/10.1016/j.pocean.2010.09.022.

Azra, M. N., Chen, J.-C., Ikhwanuddin, M. and Abol-Munafi, A. B. (2018). Thermal tolerance and locomotor activity of blue swimmer crab Portunus pelagicus instar reared at different temperatures. Journal of Thermal Biology 74: 234–240. https://doi.org/10.1016/j.jtherbio.2018.04.002.

Baker, M. R. (2021). Contrast of warm and cold phases in the Bering Sea to understand spatial distributions of Arctic and sub-Arctic gadids. Polar Biology 44(6): 1083–1105. https://doi.org/10.1007/s00300-021-02856-x.

Barshis, D. J., Ladner, J. T., Oliver, T. A., Seneca, F. O., Traylor-Knowles, N. and Palumbi, S. R. (2013). Genomic basis for coral resilience to climate change. Proceedings of the National Academy of Sciences of the United States of America 110(4): 1387–1392. https://doi.org/10.1073/pnas.1210224110.

Basu, N., Todgham, A. E., Ackerman, P. A., Bibeau, M. R., Nakano, K., Schulte, P. M. et al. (2002). Heat shock protein genes and their functional significance in fish. Gene 295(2): 173–183. https://doi.org/10.1016/s0378-1119(02)00687-x.

Bautista, N. M. and Crespel, A. (2021). Within- and trans-generational environmental adaptation to climate change: Perspectives and new challenges. Frontiers in Marine Science 8. https://doi.org/10.3389/fmars.2021.729194.

Beaugrand, G., Brander, K. M., Alistair Lindley, J., Souissi, S. and Reid, P. C. (2003). Plankton effect on cod recruitment in the North Sea. Nature 426(6967): 661–664. https://doi.org/10.1038/nature02164.

Beaugrand, G. and Kirby, R. R. (2018). How do marine pelagic species respond to climate change? Theories and observations. Annual Review of Marine Science 10(1): 169–197. https://doi.org/10.1146/annurev-marine-121916-063304.

Bednaršek, N. and Ohman, M. D. (2015). Changes in pteropod distributions and shell dissolution across a frontal system in the California Current System. Marine Ecology Progress Series 523: 93–103. https://doi.org/10.3354/meps11199.

Beemelmanns, A., Zanuzzo, F. S., Sandrelli, R. M., Rise, M. L. and Gamperl, A. K. (2021a). The Atlantic salmon's stress- and immune-related transcriptional responses to moderate hypoxia, an incremental temperature increase, and these challenges combined. G3 (Bethesda, Md.) 11(7). https://doi.org/10.1093/g3journal/jkab102.

Beemelmanns, A., Zanuzzo, F. S., Xue, X., Sandrelli, R. M., Rise, M. L. and Gamperl, A. K. (2021b). The transcriptomic responses of Atlantic salmon (*Salmo salar*) to high temperature stress alone, and in combination with moderate hypoxia. BMC Genomics 22(1): 261. https://doi.org/10.1186/s12864-021-07464-x.

Bernal, M. A., Schunter, C., Lehmann, R., Lightfoot, D. J., Allan, B. J. M., Veilleux, H. D. et al. (2020). Species-specific molecular responses of wild coral reef fishes during a marine heatwave. Science Advances 6(12): eaay3423. https://doi.org/10.1126/sciadv.aay3423.

Bertolini, C. and Pastres, R. (2021). Tolerance landscapes can be used to predict species-specific responses to climate change beyond the marine heatwave concept: Using tolerance landscape models for an ecologically meaningful classification of extreme climate events. Estuarine, Coastal and Shelf Science 252(107284): 107284. https://doi.org/10.1016/j.ecss.2021.107284.

Beusen, A., Breitburg, D., Comeau, S., Dupont, S., Isensee, K., Kudela, R. M. et al. (2022). Multiple ocean stressors: A scientific summary for policy makers. UNESCO-IOC. https://doi.org/10.25607/OBP-1724.

Bewick, S., Cantrell, R. S., Cosner, C. and Fagan, W. F. (2016). How resource phenology affects consumer population dynamics. The American Naturalist 187(2): 151–166. https://doi.org/10.1086/684432.

Bhuiyan, M. K. A., Rodríguez, B. M., Billah, M. M., Pires, A., Freitas, R. and Conradi, M. (2022). Effects of ocean acidification on the biochemistry, physiology and parental transfer of Ampelisca brevicornis (Costa, 1853). Environmental Pollution (Barking, Essex: 1987) 293(118549): 118549. https://doi.org/10.1016/j.envpol.2021.118549.

Bjornsson, B. (2001). Optimal temperature for growth and feed conversion of immature cod (Gadus morhua L.). ICES Journal of Marine Science: Journal Du Conseil 58(1): 29–38. https://doi.org/10.1006/jmsc.2000.0986.

Blöcker, A. M., Gutte, H. M., Bender, R. L., Otto, S. A., Sguotti, C. and Möllmann, C. (2023). Regime shift dynamics, tipping points and the success of fisheries management. Scientific Reports 13(1). https://doi.org/10.1038/s41598-022-27104-y.

Bonaviri, C., Graham, M., Gianguzza, P. and Shears, N. T. (2017). Warmer temperatures reduce the influence of an important keystone predator. The Journal of Animal Ecology 86(3): 490–500. https://doi.org/10.1111/1365-2656.12634.

Borges, F. O., Sampaio, E., Figueiredo, C., Rosa, R. and Grilo, T. F. (2018). Hypercapnia-induced disruption of long-distance mate-detection and reduction of energy expenditure in a coastal keystone crustacean. Physiology & Behavior 195: 69–75. https://doi.org/10.1016/j.physbeh.2018.07.023.

Borowiec, B. G. and Scott, G. R. (2021). Rapid and reversible modulation of blood haemoglobin content during diel cycles of hypoxia in killifish (Fundulus heteroclitus). Comparative Biochemistry and Physiology. Part A, Molecular & Integrative Physiology 261(111054): 111054. https://doi.org/10.1016/j.cbpa.2021.111054.

Botsford, Hastings and Gaines. (2001). Dependence of sustainability on the configuration of marine reserves and larval dispersal distance. Ecology Letters 4(2): 144–150. https://doi.org/10.1046/j.1461-0248.2001.00208.x.

Boulais, M., Chenevert, K. J., Demey, A. T., Darrow, E. S., Robison, M. R., Roberts, J. P. et al. (2017). Oyster reproduction is compromised by acidification experienced seasonally in coastal regions. Scientific Reports 7(1). https://doi.org/10.1038/s41598-017-13480-3.

Boutou, E., Louka, M., Pappa, V., Stürzbecher, H.-W., Knippschild, U., Vlachodimitropoulos, D. et al. (2015). Ubiquitination—an evolving role in DNA repair. In Advances in DNA Repair. InTech. https://doi.org/10.5772/61582.

Boyd, P., Collins, S., Dupont, S., Fabricius, K., Gattuso, J.-P., Havenhand, J. et al. (2019). SCOR WG149 handbook to support the SCOR best practice guide for 'multiple drivers' marine research. University of Tasmania, on behalf of Scientific Committee on Oceanic Research (SCOR). https://doi.org/10.25959/5C92FDF0D3C7A.

Boyd, P. W., Collins, S., Dupont, S., Fabricius, K., Gattuso, J.-P., Havenhand, J. et al. (2018). Experimental strategies to assess the biological ramifications of multiple drivers of global ocean change—A review. Global Change Biology 24(6): 2239–2261. https://doi.org/10.1111/gcb.14102.

Bradbury, I. R. and Snelgrove, P. V. R. (2001). Contrasting larval transport in demersal fish and benthic invertebrates: the roles of behaviour and advective processes in determining spatial pattern. Canadian Journal of Fisheries and Aquatic Sciences 58(4): 811–823. https://doi.org/10.1139/f01-031.

Bresolin de Souza, K., Jutfelt, F., Kling, P., Förlin, L. and Sturve, J. (2014). Effects of increased CO_2 on fish gill and plasma proteome. PloS One 9(7): e102901. https://doi.org/10.1371/journal.pone.0102901.

Brito-Morales, I., Schoeman, D. S., Molinos, J. G., Burrows, M. T., Klein, C. J., Arafeh-Dalmau, N. et al. (2020). Climate velocity reveals increasing exposure of deep-ocean biodiversity to future warming. Nature Climate Change 10(6): 576–581. https://doi.org/10.1038/s41558-020-0773-5.

Brothers, C. J., Van Der Pol, W. J., Morrow, C. D., Hakim, J. A., Koo, H. and McClintock, J. B. (2018). Ocean warming alters predicted microbiome functionality in a common sea urchin. Proceedings. Biological Sciences 285(1881): 20180340. https://doi.org/10.1098/rspb.2018.0340.

Brown, J. H., Whitham, T. G., Morgan Ernest, S. K. and Gehring, C. A. (2001). Complex species interactions and the dynamics of ecological systems: long-term experiments. Science (New York, N.Y.) 293(5530): 643–650. https://doi.org/10.1126/science.293.5530.643.

Brown, James H., Gillooly, J. F., Allen, A. P., Savage, V. M. and West, G. B. (2004). Toward a metabolic theory of ecology. Ecology 85(7): 1771–1789. https://doi.org/10.1890/03-9000.

Bryan-Brown, D. N., Brown, C. J., Hughes, J. M. and Connolly, R. M. (2017). Patterns and trends in marine population connectivity research. Marine Ecology Progress Series 585: 243–256. https://doi.org/10.3354/meps12418.

Bulman, C., Althaus, F., He, X., Bax, N. J. and Williams, A. (2001). Diets and trophic guilds of demersal fishes of the south-eastern Australian shelf. Marine & Freshwater Research 52(4): 537. https://doi.org/10.1071/mf99152.

Burge, C. A., Mark Eakin, C., Friedman, C. S., Froelich, B., Hershberger, P. K., Hofmann, E. E. et al. (2014). Climate change influences on marine infectious diseases: Implications for management and society. Annual Review of Marine Science 6(1): 249–277. https://doi.org/10.1146/annurev-marine-010213-135029.

Burrows, M. T., Schoeman, D. S., Buckley, L. B., Moore, P., Poloczanska, E. S., Brander, K. M. et al. (2011). The pace of shifting climate in marine and terrestrial ecosystems. Science (New York, N.Y.) 334(6056): 652–655. https://doi.org/10.1126/science.1210288.

Byrne, M. and Przeslawski, R. (2013). Multistressor impacts of warming and acidification of the ocean on marine invertebrates' life histories. Integrative and Comparative Biology 53(4): 582–596. https://doi.org/10.1093/icb/ict049.

Cai, W., Wang, G., Santoso, A., McPhaden, M. J., Wu, L., Jin, F.-F. et al. (2015). Increased frequency of extreme La Niña events under greenhouse warming. Nature Climate Change 5(2): 132–137. https://doi.org/10.1038/nclimate2492.

Caldwell, I. R. and Gergel, S. E. (2013). Thresholds in seascape connectivity: influence of mobility, habitat distribution, and current strength on fish movement. Landscape Ecology 28(10): 1937–1948. https://doi.org/10.1007/s10980-013-9930-9.

Calosi, P., Turner, L. M., Hawkins, M., Bertolini, C., Nightingale, G., Truebano, M. et al. (2013). Multiple physiological responses to multiple environmental challenges: an individual approach. Integrative and Comparative Biology 53(4): 660–670. https://doi.org/10.1093/icb/ict041.

Calosi, Piero, De Wit, P., Thor, P. and Dupont, S. (2016). Will life find a way? Evolution of marine species under global change. Evolutionary Applications 9(9): 1035–1042. https://doi.org/10.1111/eva.12418.

Canadell, J. G., Monteiro, P. M. S., Costa, M. H., Cotrim da Cunha, L., Cox, P. M., Eliseev, A. V. et al. (2021). Global carbon and other biogeochemical cycles and feedbacks. pp. 673–816. *In*: Masson-Delmotte, V., Zhai, P., Pirani, A., Connors, S.L., Péan, C., Berger, S. et al. (eds.). Climate Change 2021: The Physical Science Basis. Contribution of Working Group I to the Sixth Assessment Report of the Intergovernmental Panel on Climate Change. Cambridge University Press.

Cao, R., Liu, Y., Wang, Q., Yang, D., Liu, H., Ran, W. et al. (2018). Seawater acidification reduced the resistance of Crassostrea gigas to Vibrio splendidus challenge: an energy metabolism perspective. Frontiers in Physiology 9: 880. https://doi.org/10.3389/fphys.2018.00880.

Cao, R., Wang, D., Wei, Q., Wang, Q., Yang, D., Liu, H. et al. (2018). Integrative biomarker assessment of the influence of saxitoxin on marine bivalves: A comparative study of the two bivalve species oysters, Crassostrea gigas, and scallops, Chlamys farreri. Frontiers in Physiology 9: 1173. https://doi.org/10.3389/fphys.2018.01173.

Carneiro, A. P., Soares, C. H. L., Manso, P. R. J. and Pagliosa, P. R. (2020). Impact of marine heat waves and cold spell events on the bivalve Anomalocardia flexuosa: A seasonal comparison. Marine Environmental Research 156(104898): 104898. https://doi.org/10.1016/j.marenvres.2020.104898.

Cattano, C., Fine, M., Quattrocchi, F., Holzman, R. and Milazzo, M. (2019). Behavioural responses of fish groups exposed to a predatory threat under elevated CO_2. Marine Environmental Research 147: 179–184. https://doi.org/10.1016/j.marenvres.2019.04.011.

Cattano, Carlo, Claudet, J., Domenici, P. and Milazzo, M. (2018). Living in a high CO_2 world: a global meta-analysis shows multiple trait-mediated fish responses to ocean acidification. Ecological Monographs 88(3): 320–335. https://doi.org/10.1002/ecm.1297.

Chainy, G. B. N., Paital, B. and Dandapat, J. (2016). An overview of seasonal changes in oxidative stress and antioxidant defence parameters in some invertebrate and vertebrate species. Scientifica 2016: 6126570. https://doi.org/10.1155/2016/6126570.

Chen, B., Feder, M. E. and Kang, L. (2018). Evolution of heat-shock protein expression underlying adaptive responses to environmental stress. Molecular Ecology 27(15): 3040–3054. https://doi.org/10.1111/mec.14769.

Cheng, C.-H., Yang, F.-F., Liao, S.-A., Miao, Y.-T., Ye, C.-X., Wang, A.-L. et al. (2015). High temperature induces apoptosis and oxidative stress in pufferfish (Takifugu obscurus) blood cells. Journal of Thermal Biology 53: 172–179. https://doi.org/10.1016/j.jtherbio.2015.08.002.

Cheung, W. W. L., Close, C., Lam, V., Watson, R. and Pauly, D. (2008). Application of macroecological theory to predict effects of climate change on global fisheries potential. Marine Ecology Progress Series 365: 187–197. https://doi.org/10.3354/meps07414.

Cheung, William W. L., Lam, V. W. Y., Sarmiento, J. L., Kearney, K., Watson, R. and Pauly, D. (2009). Projecting global marine biodiversity impacts under climate change scenarios. Fish and Fisheries (Oxford, England) 10(3): 235–251. https://doi.org/10.1111/j.1467-2979.2008.00315.x.

Ciechanover, A. (1998). The ubiquitin-proteasome pathway: on protein death and cell life. The EMBO Journal 17(24): 7151–7160. https://doi.org/10.1093/emboj/17.24.7151.

Clark, M. S., Husmann, G., Thorne, M. A. S., Burns, G., Truebano, M., Peck, L. S. et al. (2013). Hypoxia impacts large adults first: consequences in a warming world. Global Change Biology 19(7): 2251–2263. https://doi.org/10.1111/gcb.12197.

Clark, M. S., Sommer, U., Sihra, J. K., Thorne, M. A. S., Morley, S. A., King, M. et al. (2017). Biodiversity in marine invertebrate responses to acute warming revealed by a comparative multi-omics approach. Global Change Biology 23(1): 318–330. https://doi.org/10.1111/gcb.13357.

Clarke, A. (2004). Is there a Universal Temperature Dependence of metabolism? Functional Ecology 18(2): 252–256. https://doi.org/10.1111/j.0269-8463.2004.00842.x.

Clarke, A. and Fraser, K. P. P. (2004). Why does metabolism scale with temperature? Functional Ecology 18(2): 243–251. https://doi.org/10.1111/j.0269-8463.2004.00841.x.

Coleman, M. A., Cetina-Heredia, P., Roughan, M., Feng, M., van Sebille, E. and Kelaher, B. P. (2017). Anticipating changes to future connectivity within a network of marine protected areas. Global Change Biology 23(9): 3533–3542. https://doi.org/10.1111/gcb.13634.

Coleman, M. A., Roughan, M., Macdonald, H. S., Connell, S. D., Gillanders, B. M., Kelaher, B. P. et al. (2011). Variation in the strength of continental boundary currents determines continent-wide connectivity in kelp. The Journal of Ecology 99(4): 1026–1032. https://doi.org/10.1111/j.1365-2745.2011.01822.x.

Collins, M., Peck, L. S. and Clark, M. S. (2021). Large within, and between, species differences in marine cellular responses: Unpredictability in a changing environment. The Science of the Total Environment 794(148594): 148594. https://doi.org/10.1016/j.scitotenv.2021.148594.

Cominassi, L., Moyano, M., Claireaux, G., Howald, S., Mark, F. C., Zambonino-Infante, J.-L. et al. (2020). Food availability modulates the combined effects of ocean acidification and warming on fish growth. Scientific Reports 10(1). https://doi.org/10.1038/s41598-020-58846-2.

Conaway, R. C., Brower, C. S. and Conaway, J. W. (2002). Emerging roles of ubiquitin in transcription regulation. Science (New York, N.Y.) 296(5571): 1254–1258. https://doi.org/10.1126/science.1067466.

Connell, S. D., Doubleday, Z. A., Hamlyn, S. B., Foster, N. R., Harley, C. D. G., Helmuth, B. et al. (2017). How ocean acidification can benefit calcifiers. Current Biology: CB 27(3): R95–R96. https://doi.org/10.1016/j.cub.2016.12.004.

Connell, S. D. and Russell, B. D. (2010). The direct effects of increasing CO_2 and temperature on non-calcifying organisms: increasing the potential for phase shifts in kelp forests. Proceedings. Biological Sciences 277(1686): 1409–1415. https://doi.org/10.1098/rspb.2009.2069.

Cooley, S., Schoeman, D., Bopp, L., Boyd, P., Donner, S., Ghebrehiwet, D. Y. et al. (2022). Oceans and coastal ecosystems and their services. pp. 379–550. *In*: Pörtner, D.C., Roberts, M. Tignor, E.S. Poloczanska, K. Mintenbeck, A. Alegría, M. Craig et al. (eds.). Climate Change 2022 – Impacts, Adaptation and Vulnerability: Working Group II Contribution to the Sixth Assessment Report of the Intergovernmental Panel on Climate Change. Cambridge University Press. https://doi.org/10.1017/9781009325844.005.

Costantini, D. (2019). Understanding diversity in oxidative status and oxidative stress: the opportunities and challenges ahead. The Journal of Experimental Biology 222(Pt 13): jeb194688. https://doi.org/10.1242/jeb.194688.

Cowen, R. K. and Sponaugle, S. (2009). Larval dispersal and marine population connectivity. Annual Review of Marine Science 1(1): 443–466. https://doi.org/10.1146/annurev.marine.010908.163757.

Cui, D., Liu, L., Zhao, T., Zhan, Y., Song, J., Zhang, W. et al. (2022). Responses of sea urchins (Strongylocentrotus intermedius) with different sexes to CO_2-induced seawater acidification: Histology, physiology, and metabolomics. Marine Pollution Bulletin 178(113606): 113606. https://doi.org/10.1016/j.marpolbul.2022.113606.

Cury, P. (2000). Small pelagics in upwelling systems: patterns of interaction and structural changes in "wasp-waist" ecosystems. ICES Journal of Marine Science: Journal Du Conseil 57(3): 603–618. https://doi.org/10.1006/jmsc.2000.0712.

Cushing, D. H. (1969). The regularity of the spawning season of some fishes. ICES Journal of Marine Science: Journal Du Conseil 33(1): 81–92. https://doi.org/10.1093/icesjms/33.1.81.

Cushing, D. H. (1990). Plankton production and year-class strength in fish populations: An update of the match/mismatch hypothesis. pp. 249–293. In Advances in Marine Biology. Elsevier. https://doi.org/10.1016/s0065-2881(08)60202-3.

Dahlke, F. T., Wohlrab, S., Butzin, M. and Pörtner, H.-O. (2020). Thermal bottlenecks in the life cycle define climate vulnerability of fish. Science (New York, N.Y.) 369(6499): 65–70. https://doi.org/10.1126/science.aaz3658.

Del Giudice, M., Buck, C. L., Chaby, L. E., Gormally, B. M., Taff, C. C., Thawley, C. J. et al. (2018). What is stress? A systems perspective. Integrative and Comparative Biology 58(6): 1019–1032. https://doi.org/10.1093/icb/icy114.

Delhey, K., Dale, J., Valcu, M. and Kempenaers, B. (2020). Why climate change should generally lead to lighter coloured animals. Current Biology: CB 30(23): R1406–R1407. https://doi.org/10.1016/j.cub.2020.10.070.

DeSalvo, M. K., Sunagawa, S., Voolstra, C. R. and Medina, M. (2010). Transcriptomic responses to heat stress and bleaching in the elkhorn coral Acropora palmata. Marine Ecology Progress Series 402: 97–113. https://doi.org/10.3354/meps08372.

Deutsch, C., Ferrel, A., Seibel, B., Pörtner, H.-O. and Huey, R. B. (2015). Climate change tightens a metabolic constraint on marine habitats. Science (New York, N.Y.) 348(6239): 1132–1135. https://doi.org/10.1126/science.aaa1605.

deYoung, B., Barange, M., Beaugrand, G., Harris, R., Perry, R. I., Scheffer, M. et al. (2008). Regime shifts in marine ecosystems: detection, prediction and management. Trends in Ecology & Evolution 23(7): 402–409. https://doi.org/10.1016/j.tree.2008.03.008.

Dias, M., Ferreira, A., Gouveia, R., Madeira, C., Jogee, N., Cabral, H. et al. (2019). Long-term exposure to increasing temperatures on scleractinian coral fragments reveals oxidative stress. Marine Environmental Research 150(104758): 104758. https://doi.org/10.1016/j.marenvres.2019.104758.

Dias, M., Madeira, C., Jogee, N., Ferreira, A., Gouveia, R., Cabral, H. et al. (2020). Integrative indices for health assessment in reef corals under thermal stress. Ecological Indicators 113(106230): 106230. https://doi.org/10.1016/j.ecolind.2020.106230.

Dickinson, G. H., Ivanina, A. V., Matoo, O. B., Pörtner, H. O., Lannig, G., Bock, C. et al. (2012). Interactive effects of salinity and elevated CO_2 levels on juvenile eastern oysters, *Crassostrea virginica*. The Journal of Experimental Biology 215(1): 29–43. https://doi.org/10.1242/jeb.061481.

Dillon, M. E., Wang, G. and Huey, R. B. (2010). Global metabolic impacts of recent climate warming. Nature 467(7316): 704–706. https://doi.org/10.1038/nature09407.

Dilly, G. F., Young, C. R., Lane, W. S., Pangilinan, J. and Girguis, P. R. (2012). Exploring the limit of metazoan thermal tolerance via comparative proteomics: thermally induced changes in protein abundance by two hydrothermal vent polychaetes. Proceedings. Biological Sciences 279(1741): 3347–3356. https://doi.org/10.1098/rspb.2012.0098.

Dimos, B. A., Mahmud, S. A., Fuess, L. E., Mydlarz, L. D. and Pellegrino, M. W. (2019). Uncovering a mitochondrial unfolded protein response in corals and its role in adapting to a changing world. Proceedings. Biological Sciences 286(1905): 20190470. https://doi.org/10.1098/rspb.2019.0470.

Ding, J., Liu, C., Luo, S., Zhang, Y., Gao, X., Wu, X. et al. (2020). Transcriptome and physiology analysis identify key metabolic changes in the liver of the large yellow croaker (Larimichthys crocea) in response to acute hypoxia. Ecotoxicology and Environmental Safety 189(109957): 109957. https://doi.org/10.1016/j.ecoenv.2019.109957.

Dobson, K. L., Newton, J. A., Widdicombe, S., Schoo, K. L., Acquafredda, M. P., Kitch, G. et al. (2022). Ocean acidification research for sustainability: co-designing global action on local scales. ICES Journal of Marine Science: Journal Du Conseil. https://doi.org/10.1093/icesjms/fsac158.

Domenici, P. (2010). Context-dependent variability in the components of fish escape response: integrating locomotor performance and behavior. Journal of Experimental Zoology. Part A, Ecological Genetics and Physiology 313A(2): 59–79. https://doi.org/10.1002/jez.580.

Domenici, P., Allan, B. J. M., Lefrançois, C. and McCormick, M. I. (2019). The effect of climate change on the escape kinematics and performance of fishes: implications for future predator–prey interactions. Conservation Physiology 7(1). https://doi.org/10.1093/conphys/coz078.

Donelson, J. M., Salinas, S., Munday, P. L. and Shama, L. N. S. (2018). Transgenerational plasticity and climate change experiments: Where do we go from here? Global Change Biology 24(1): 13–34. https://doi.org/10.1111/gcb.13903.

Donelson, J. M., Sunday, J. M., Figueira, W. F., Gaitán-Espitia, J. D., Hobday, A. J., Johnson, C. R. et al. (2019). Understanding interactions between plasticity, adaptation and range shifts in response to marine environmental change. Philosophical Transactions of the Royal Society of London. Series B, Biological Sciences 374(1768): 20180186. https://doi.org/10.1098/rstb.2018.0186.

Doney, S. C., Fabry, V. J., Feely, R. A. and Kleypas, J. A. (2009). Ocean acidification: The other CO_2 problem. Annual Review of Marine Science 1(1): 169–192. https://doi.org/10.1146/annurev.marine.010908.163834.

Doney, S. C., Ruckelshaus, M., Emmett Duffy, J., Barry, J. P., Chan, F., English, C. A. et al. (2012). Climate change impacts on marine ecosystems. Annual Review of Marine Science 4(1): 11–37. https://doi.org/10.1146/annurev-marine-041911-111611.

Dong, Y., Miller, L. P., Sanders, J. G. and Somero, G. N. (2008). Heat-shock protein 70 (Hsp70) expression in four limpets of the genus Lottia: interspecific variation in constitutive and inducible synthesis correlates with *in situ* exposure to heat stress. The Biological Bulletin 215(2): 173–181. https://doi.org/10.2307/25470698.

Doubleday, Z. A., Nagelkerken, I., Coutts, M. D., Goldenberg, S. U. and Connell, S. D. (2019). A triple trophic boost: How carbon emissions indirectly change a marine food chain. Global Change Biology 25(3): 978–984. https://doi.org/10.1111/gcb.14536.

Doxa, C. K., Sfakianakis, D., Sterioti, A. and Kentouri, M. (2021). Effect of temperature on the development of deformities during the embryonic stages of Charonia seguenzae (Aradas & Benoit, 1870). Journal of Thermal Biology 100(103046): 103046. https://doi.org/10.1016/j.jtherbio.2021.103046.

Driver, L. J. and Hoeinghaus, D. J. (2016). Fish metacommunity responses to experimental drought are determined by habitat heterogeneity and connectivity. Freshwater Biology 61(4): 533–548. https://doi.org/10.1111/fwb.12726.

du Pontavice, H., Gascuel, D., Reygondeau, G., Maureaud, A. and Cheung, W. W. L. (2020). Climate change undermines the global functioning of marine food webs. Global Change Biology 26(3): 1306–1318. https://doi.org/10.1111/gcb.14944.

du Pontavice, H., Gascuel, D., Reygondeau, G., Stock, C. and Cheung, W. W. L. (2021). Climate-induced decrease in biomass flow in marine food webs may severely affect predators and ecosystem production. Global Change Biology 27(11): 2608–2622. https://doi.org/10.1111/gcb.15576.

Dulvy, N. K., Rogers, S. I., Jennings, S., Stelzenmüller, V., Dye, S. R. and Skjoldal, H. R. (2008). Climate change and deepening of the North Sea fish assemblage: a biotic indicator of warming seas. The Journal of Applied Ecology 45(4): 1029–1039. https://doi.org/10.1111/j.1365-2664.2008.01488.x.

Dupont, S., Gattuso, J.-P., Pörtner, H.-O. and Widdicombe, S. (2021). Ocean acidification science stands strong [Review of Ocean acidification science stands strong]. Science (New York, N.Y.) 372(6547): 1160–1161. American Association for the Advancement of Science (AAAS). https://doi.org/10.1126/science.abj4129.

Dur, G., Won, E.-J., Han, J., Lee, J.-S. and Souissi, S. (2021). An individual-based model for evaluating post-exposure effects of UV-B radiation on zooplankton reproduction. Ecological Modelling 441(109379): 109379. https://doi.org/10.1016/j.ecolmodel.2020.109379.

Durant, J. M., Hjermann, D. Ø., Ottersen, G. and Stenseth, N. C. (2007). Climate and the match or mismatch between predator requirements and resource availability. Climate Research 33: 271–283. https://doi.org/10.3354/cr033271.

Durant, Joë M., Anker-Nilssen, T. and Stenseth, N. C. (2003). Trophic interactions under climate fluctuations: the Atlantic puffin as an example. Proceedings. Biological Sciences 270(1523): 1461–1466. https://doi.org/10.1098/rspb.2003.2397.

Durant, Joël M., Hjermann, D. Ø., Anker-Nilssen, T., Beaugrand, G., Mysterud, A., Pettorelli, N. et al. (2005). Timing and abundance as key mechanisms affecting trophic interactions in variable environments. Ecology Letters 8(9): 952–958. https://doi.org/10.1111/j.1461-0248.2005.00798.x.

Durant, Joël M., Molinero, J.-C., Ottersen, G., Reygondeau, G., Stige, L. C. and Langangen, Ø. (2019). Contrasting effects of rising temperatures on trophic interactions in marine ecosystems. Scientific Reports 9(1). https://doi.org/10.1038/s41598-019-51607-w.

Edwards, M. and Richardson, A. J. (2004). Impact of climate change on marine pelagic phenology and trophic mismatch. Nature 430(7002): 881–884. https://doi.org/10.1038/nature02808.

Eisner, L. B., Zuenko, Y. I., Basyuk, E. O., Britt, L. L., Duffy-Anderson, J. T., Kotwicki, S. et al. (2020). Environmental impacts on walleye pollock (Gadus chalcogrammus) distribution across the Bering Sea shelf. Deep-Sea Research. Part II, Topical Studies in Oceanography 181–182(104881): 104881. https://doi.org/10.1016/j.dsr2.2020.104881.

Englund, G., Öhlund, G., Hein, C. L. and Diehl, S. (2011). Temperature dependence of the functional response. Ecology Letters 14(9): 914–921. https://doi.org/10.1111/j.1461-0248.2011.01661.x.

Evans, T. G. and Kültz, D. (2020). The cellular stress response in fish exposed to salinity fluctuations. Journal of Experimental Zoology. Part A, Ecological and Integrative Physiology 333(6): 421–435. https://doi.org/10.1002/jez.2350.

Fabricius, K. E., Langdon, C., Uthicke, S., Humphrey, C., Noonan, S., De'ath, G. et al. (2011). Losers and winners in coral reefs acclimatized to elevated carbon dioxide concentrations. Nature Climate Change 1(3): 165–169. https://doi.org/10.1038/nclimate1122.

Fabry, V. J., Seibel, B. A., Feely, R. A. and Orr, J. C. (2008). Impacts of ocean acidification on marine fauna and ecosystem processes. ICES Journal of Marine Science: Journal Du Conseil 65(3): 414–432. https://doi.org/10.1093/icesjms/fsn048.

Facey, S. L., Ellsworth, D. S., Staley, J. T., Wright, D. J. and Johnson, S. N. (2014). Upsetting the order: how climate and atmospheric change affects herbivore–enemy interactions. Current Opinion in Insect Science 5: 66–74. https://doi.org/10.1016/j.cois.2014.09.015.

Falfushynska, H., Piontkivska, H. and Sokolova, I. M. (2020). Effects of intermittent hypoxia on cell survival and inflammatory responses in the intertidal marine bivalves Mytilus edulis and Crassostrea gigas. The Journal of Experimental Biology 223(Pt 4): jeb217026. https://doi.org/10.1242/jeb.217026.

Feary, D. A., Pratchett, M. S., J Emslie, M., Fowler, A. M., Figueira, W. F., Luiz, O. J. et al. (2014). Latitudinal shifts in coral reef fishes: why some species do and others do not shift. Fish and Fisheries (Oxford, England) 15(4): 593–615. https://doi.org/10.1111/faf.12036.

Figueira, W. F. and Booth, D. J. (2010). Increasing ocean temperatures allow tropical fishes to survive overwinter in temperate waters. Global Change Biology 16(2): 506–516. https://doi.org/10.1111/j.1365-2486.2009.01934.x.

Figueiredo, J., Thomas, C. J., Deleersnijder, E., Lambrechts, J., Baird, A. H., Connolly, S. R. et al. (2022). Global warming decreases connectivity among coral populations. Nature Climate Change 12(1): 83–87. https://doi.org/10.1038/s41558-021-01248-7.

Finley, D. and Chau, V. (1991). Ubiquitination. Annual Review of Cell Biology 7(1): 25–69. https://doi.org/10.1146/annurev.cb.07.110191.000325.

Foo, S. A., Byrne, M., Ricevuto, E. and Gambi, M. C. (2018). The carbon dioxide vents of ischia, Italy, A natural system to assess impacts of ocean acidification on marine ecosystems: An overview of research and comparisons with other vent systems. pp. 237–310. *In*: Oceanography and Marine Biology. CRC Press. https://doi.org/10.1201/9780429454455-4.

Fossheim, M., Primicerio, R., Johannesen, E., Ingvaldsen, R. B., Aschan, M. M. and Dolgov, A. V. (2015). Recent warming leads to a rapid borealization of fish communities in the Arctic. Nature Climate Change 5(7): 673–677. https://doi.org/10.1038/nclimate2647.

Fox-Kemper, B., Hewitt, H. T., Xiao, C., Aðalgeirsdóttir, G., Drijfhout, S. S., Edwards, T. L. et al. (2021). Ocean, cryosphere and sea level change. pp. 1211–1362. *In*: Masson-Delmotte, V., Zhai, P., Pirani, A., Connors, S. L., Péan, C., Berger, S. et al. (eds.). Climate Change 2021 – The Physical Science Basis. Contribution of Working Group I to the Sixth Assessment Report of the Intergovernmental Panel on Climate Change. Cambridge University Press. https://doi.org/10.1017/9781009157896.011.

Franzova, V. A., MacLeod, C. D., Wang, T. and Harley, C. D. G. (2019). Complex and interactive effects of ocean acidification and warming on the life span of a marine trematode parasite. International Journal for Parasitology 49(13–14): 1015–1021. https://doi.org/10.1016/j.ijpara.2019.07.005.

Friesen, S. K., Martone, R., Rubidge, E., Baggio, J. A. and Ban, N. C. (2019). An approach to incorporating inferred connectivity of adult movement into marine protected area design with limited data. Ecological Applications: A Publication of the Ecological Society of America 29(4). https://doi.org/10.1002/eap.1890.

Friesen, S. K., Rubidge, E., Martone, R., Hunter, K. L., Peña, M. A. and Ban, N. C. (2021). Effects of changing ocean temperatures on ecological connectivity among marine protected areas in northern British Columbia. Ocean & Coastal Management 211(105776): 105776. https://doi.org/10.1016/j.ocecoaman.2021.105776.

Frisk, M. G., Martell, S. J. D., Miller, T. J. and Sosebee, K. (2010). Exploring the population dynamics of winter skate (Leucoraja ocellata) in the Georges Bank region using a statistical catch-at-age model incorporating length, migration, and recruitment process errors. Canadian Journal of Fisheries and Aquatic Sciences 67(5): 774–792. https://doi.org/10.1139/f10-008.

Frisk, M. G., Miller, T. J., Martell, S. J. D. and Sosebee, K. (2008). New hypothesis helps explain elasmobranch "outburst" on georges bank in the 1980s. Ecological Applications: A Publication of the Ecological Society of America 18(1): 234–245. https://doi.org/10.1890/06-1392.1.

Frisk, Michael G., Jordaan, A. and Miller, T. J. (2014). Moving beyond the current paradigm in marine population connectivity: are adults the missing link? Fish and Fisheries (Oxford, England) 15(2): 242–254. https://doi.org/10.1111/faf.12014.

Frölicher, T. L., Fischer, E. M. and Gruber, N. (2018). Marine heatwaves under global warming. Nature 560(7718): 360–364. https://doi.org/10.1038/s41586-018-0383-9.

Fullerton, A. H., Burnett, K. M., Steel, E. A., Flitcroft, R. L., Pess, G. R., Feist, B. E. et al. (2010). Hydrological connectivity for riverine fish: measurement challenges and research opportunities. Freshwater Biology 55(11): 2215–2237. https://doi.org/10.1111/j.1365-2427.2010.02448.x.

Galasso, N. M., Bonaviri, C., Trapani, F. D., Picciotto, M., Gianguzza, P., Agnetta, D. et al. (2015). Fish-seastar facilitation leads to algal forest restoration on protected rocky reefs. Scientific Reports 5(1). https://doi.org/10.1038/srep12409.

Garcia, R. A., Cabeza, M., Rahbek, C. and Araújo, M. B. (2014). Multiple dimensions of climate change and their implications for biodiversity. Science (New York, N.Y.) 344(6183). https://doi.org/10.1126/science.1247579.

Garland, M. A., Stillman, J. H. and Tomanek, L. (2015). The proteomic response of cheliped myofibril tissue in the eurythermal porcelain crab Petrolisthes cinctipes to heat shock following acclimation to daily temperature fluctuations. The Journal of Experimental Biology 218(Pt 3): 388–403. https://doi.org/10.1242/jeb.112250.

Gattuso, J.-P., Magnan, A., Billé, R., Cheung, W. W. L., Howes, E. L., Joos, F. et al. (2015). Oceanography. Contrasting futures for ocean and society from different anthropogenic CO_2 emissions scenarios. Science (New York, N.Y.) 349(6243): aac4722. https://doi.org/10.1126/science.aac4722.

Gavel, I., Byrne, M. and Thomson, M. (2022). Effects of raised temperature on viviparous reproduction in the marine isopod Cirolana harfordi. Journal of Experimental Marine Biology and Ecology 546(151648): 151648. https://doi.org/10.1016/j.jembe.2021.151648.

Gaylord, B., Kroeker, K. J., Sunday, J. M., Anderson, K. M., Barry, J. P., Brown, N. E. et al. (2015). Ocean acidification through the lens of ecological theory. Ecology 96(1): 3–15. https://doi.org/10.1890/14-0802.1.

Geist, J. and Hawkins, S. J. (2016). Habitat recovery and restoration in aquatic ecosystems: current progress and future challenges. Aquatic Conservation: Marine and Freshwater Ecosystems 26(5): 942–962. https://doi.org/10.1002/aqc.2702.

Gerber, L. R., Mancha-Cisneros, M. D. M., O'Connor, M. I. and Selig, E. R. (2014). Climate change impacts on connectivity in the ocean: Implications for conservation. Ecosphere (Washington, D.C) 5(3): art33. https://doi.org/10.1890/es13-00336.1.

Gilman, S. E., Urban, M. C., Tewksbury, J., Gilchrist, G. W. and Holt, R. D. (2010). A framework for community interactions under climate change. Trends in Ecology and Evolution 25(6): 325–331. https://doi.org/10.1016/j.tree.2010.03.002.

Gissi, E., Manea, E., Mazaris, A. D., Fraschetti, S., Almpanidou, V., Bevilacqua, S. et al. (2021). A review of the combined effects of climate change and other local human stressors on the marine environment. The Science of the Total Environment 755(142564): 142564. https://doi.org/10.1016/j.scitotenv.2020.142564.

Giuliani, M. E., Nardi, A., Di Carlo, M., Benedetti, M. and Regoli, F. (2021). Transcriptional and catalytic responsiveness of the Antarctic fish Trematomus bernacchii antioxidant system toward multiple stressors. Antioxidants (Basel, Switzerland) 10(3): 410. https://doi.org/10.3390/antiox10030410.

Gleason, D. F. and Hofmann, D. K. (2011). Coral larvae: From gametes to recruits. Journal of Experimental Marine Biology and Ecology 408(1–2): 42–57. https://doi.org/10.1016/j.jembe.2011.07.025.

Gobler, C. J., Merlo, L. R., Morrell, B. K. and Griffith, A. W. (2018). Temperature, acidification, and food supply interact to negatively affect the growth and survival of the forage fish, Menidia beryllina (inland Silverside), and Cyprinodon variegatus (Sheepshead minnow). Frontiers in Marine Science 5. https://doi.org/10.3389/fmars.2018.00086.

Goh, B. P. L. and Lai, C. H. (2014). Establishing the thermal threshold of the tropical mussel Perna viridis in the face of global warming. Marine Pollution Bulletin 85(2): 325–331. https://doi.org/10.1016/j.marpolbul.2013.10.041.

González, P. M., Malanga, G. and Puntarulo, S. (2015). Cellular oxidant/antioxidant network: Update on the environmental effects over marine organisms. The Open Marine Biology Journal 9(1): 1–13. https://doi.org/10.2174/1874450801509010001.

Gracey, A. Y., Troll, J. V. and Somero, G. N. (2001). Hypoxia-induced gene expression profiling in the euryoxic fish Gillichthys mirabilis. Proceedings of the National Academy of Sciences of the United States of America 98(4): 1993–1998. https://doi.org/10.1073/pnas.98.4.1993.

Gravinese, P. M., Enochs, I. C., Manzello, D. P. and van Woesik, R. (2018). Warming and pCO_2 effects on Florida stone crab larvae. Estuarine, Coastal and Shelf Science 204: 193–201. https://doi.org/10.1016/j.ecss.2018.02.021.

Griffith, A. W. and Gobler, C. J. (2017). Transgenerational exposure of North Atlantic bivalves to ocean acidification renders offspring more vulnerable to low pH and additional stressors. Scientific Reports 7(1): 11394. https://doi.org/10.1038/s41598-017-11442-3.

Griffith, G. P., Hop, H., Vihtakari, M., Wold, A., Kalhagen, K. and Gabrielsen, G. W. (2019). Ecological resilience of Arctic marine food webs to climate change. Nature Climate Change 9(11): 868–872. https://doi.org/10.1038/s41558-019-0601-y.

Grigaltchik, V. S., Ward, A. J. W. and Seebacher, F. (2012). Thermal acclimation of interactions: differential responses to temperature change alter predator–prey relationship. Proceedings. Biological Sciences 279(1744): 4058–4064. https://doi.org/10.1098/rspb.2012.1277.

Grottoli, A. G., Tchernov, D. and Winters, G. (2017). Physiological and biogeochemical responses of super-corals to thermal stress from the northern gulf of Aqaba, red sea. Frontiers in Marine Science 4. https://doi.org/10.3389/fmars.2017.00215.

Guinotte, J. M. and Fabry, V. J. (2008). Ocean acidification and its potential effects on marine ecosystems. Annals of the New York Academy of Sciences 1134(1): 320–342. https://doi.org/10.1196/annals.1439.013.

Gulland, F. M. D., Baker, J. D., Howe, M., LaBrecque, E., Leach, L., Moore, S. E. et al. (2022). A review of climate change effects on marine mammals in United States waters: Past predictions, observed impacts, current research and conservation imperatives. Climate Change Ecology 3(100054): 100054. https://doi.org/10.1016/j.ecochg.2022.100054.

Habary, A., Johansen, J. L., Nay, T. J., Steffensen, J. F. and Rummer, J. L. (2017). Adapt, move or die – how will tropical coral reef fishes cope with ocean warming? Global Change Biology 23(2): 566–577. https://doi.org/10.1111/gcb.13488.

Hallegraeff, G. M. (2010). Ocean climate change, phytoplankton community responses, and harmful algal blooms: A formidable predictive challenge. Journal of Phycology 46(2): 220–235. https://doi.org/10.1111/j.1529-8817.2010.00815.x.

Hall-Spencer, J. M., Rodolfo-Metalpa, R., Martin, S., Ransome, E., Fine, M., Turner, S. M. et al. (2008). Volcanic carbon dioxide vents show ecosystem effects of ocean acidification. Nature 454(7200): 96–99. https://doi.org/10.1038/nature07051.

Hamilton, S. L., Regetz, J. and Warner, R. R. (2008). Postsettlement survival linked to larval life in a marine fish. Proceedings of the National Academy of Sciences of the United States of America 105(5): 1561–1566. https://doi.org/10.1073/pnas.0707676105.

Hastings, A. and Harrison, S. (1994). Metapopulation dynamics and genetics. Annual Review of Ecology and Systematics 25(1): 167–188. https://doi.org/10.1146/annurev.es.25.110194.001123.

Hattab, T., Leprieur, F., Ben Rais Lasram, F., Gravel, D., Loc'h, F. L. and Albouy, C. (2016). Forecasting fine-scale changes in the food-web structure of coastal marine communities under climate change. Ecography 39(12): 1227–1237. https://doi.org/10.1111/ecog.01937.

He, M., Zhou, Z., Shah, A. A., Zou, H., Tao, J., Chen, Q. et al. (2016). The emerging role of deubiquitinating enzymes in genomic integrity, diseases, and therapeutics. Cell & Bioscience 6(1): 62. https://doi.org/10.1186/s13578-016-0127-1.

He, X., Pan, Z., Zhang, L. and Han, D. (2021). Physiological and behavioral responses of the copepod Temora turbinata to hypoxia. Marine Pollution Bulletin 171(112692): 112692. https://doi.org/10.1016/j.marpolbul.2021.112692.

Heinze, C., Blenckner, T., Martins, H., Rusiecka, D., Döscher, R., Gehlen, M. et al. (2021). The quiet crossing of ocean tipping points. Proceedings of the National Academy of Sciences of the United States of America 118(9). https://doi.org/10.1073/pnas.2008478118.

Hemraj, D. A., Posnett, N. C., Minuti, J. J., Firth, L. B. and Russell, B. D. (2020). Survived but not safe: Marine heatwave hinders metabolism in two gastropod survivors. Marine Environmental Research 162(105117): 105117. https://doi.org/10.1016/j.marenvres.2020.105117.

Henson, S. A., Beaulieu, C., Ilyina, T., John, J. G., Long, M., Séférian, R. et al. (2017). Rapid emergence of climate change in environmental drivers of marine ecosystems. Nature Communications 8(1): 14682. https://doi.org/10.1038/ncomms14682.

Hermes-Lima, M., Moreira, D. C., Rivera-Ingraham, G. A., Giraud-Billoud, M., Genaro-Mattos, T. C. and Campos, É. G. (2015). Preparation for oxidative stress under hypoxia and metabolic depression: Revisiting the proposal two decades later. Free Radical Biology & Medicine 89: 1122–1143. https://doi.org/10.1016/j.freeradbiomed.2015.07.156.

Hernández-Elizárraga, V. H., Olguín-López, N., Hernández-Matehuala, R., Ocharán-Mercado, A., Cruz-Hernández, A., Guevara-González, R. G. et al. (2019). Comparative analysis of the soluble proteome and the cytolytic activity of unbleached and bleached Millepora complanata ('fire coral') from the Mexican Caribbean. Marine Drugs 17(7): 393. https://doi.org/10.3390/md17070393.

Heupel, M. R., Knip, D. M., Simpfendorfer, C. A. and Dulvy, N. K. (2014). Sizing up the ecological role of sharks as predators. Marine Ecology Progress Series 495: 291–298. https://doi.org/10.3354/meps10597.

Hines, C. W., Fang, Y., Chan, V. K. S., Stiller, K. T., Brauner, C. J. and Richards, J. G. (2019). The effect of salinity and photoperiod on thermal tolerance of Atlantic and coho salmon reared from smolt to adult in recirculating aquaculture systems. Comparative Biochemistry and Physiology. Part A, Molecular & Integrative Physiology 230: 1–6. https://doi.org/10.1016/j.cbpa.2018.12.008.

Hirst, A. and López-Urrutia, A. (2006). Effects of evolution on egg development time. Marine Ecology Progress Series 326: 29–35. https://doi.org/10.3354/meps326029.

Hobday, A. J., Alexander, L. V., Perkins, S. E., Smale, D. A., Straub, S. C., Oliver, E. C. J. et al. (2016). A hierarchical approach to defining marine heatwaves. Progress in Oceanography 141: 227–238. https://doi.org/10.1016/j.pocean.2015.12.014.

Hoegh-Guldberg, O. and Bruno, J. F. (2010). The impact of climate change on the world's marine ecosystems. Science (New York, N.Y.) 328(5985): 1523–1528. https://doi.org/10.1126/science.1189930.

Hoffmann, A. A. and Sgrò, C. M. (2011). Climate change and evolutionary adaptation. Nature 470(7335): 479–485. https://doi.org/10.1038/nature09670.

Hofmann, G. E., Buckley, B. A., Airaksinen, S., Keen, J. E. and Somero, G. N. (2000). Heat-shock protein expression is absent in the antarctic fish Trematomus bernacchii (family Nototheniidae). The Journal of Experimental Biology 203(Pt 15): 2331–2339. https://doi.org/10.1242/jeb.203.15.2331.

Hofmann, Gretchen E. (2005). Patterns of Hsp gene expression in ectothermic marine organisms on small to large biogeographic scales. Integrative and Comparative Biology 45(2): 247–255. https://doi.org/10.1093/icb/45.2.247.

Horn, H. G., Boersma, M., Garzke, J., Sommer, U. and Aberle, N. (2020). High CO_2 and warming affect microzooplankton food web dynamics in a Baltic Sea summer plankton community. Marine Biology 167(5). https://doi.org/10.1007/s00227-020-03683-0.

Howard, E. M., Penn, J. L., Frenzel, H., Seibel, B. A., Bianchi, D., Renault, L. et al. (2020). Climate-driven aerobic habitat loss in the California Current System. Science Advances 6(20): eaay3188. https://doi.org/10.1126/sciadv.aay3188.

Huang, M., Ding, L., Wang, J., Ding, C. and Tao, J. (2021). The impacts of climate change on fish growth: A summary of conducted studies and current knowledge. Ecological Indicators 121(106976): 106976. https://doi.org/10.1016/j.ecolind.2020.106976.

Hue, T., Chateau, O., Lecellier, G., Marin, C., Coulombier, N., Le Dean, L. et al. (2022). Impact of near-future ocean warming and acidification on the larval development of coral-eating starfish Acanthaster cf. solaris after parental exposure. Journal of Experimental Marine Biology and Ecology 548(151685): 151685. https://doi.org/10.1016/j.jembe.2021.151685.

Hughes, J. M., Stewart, J., Lyle, J. M., McAllister, J., Stocks, J. R. and Suthers, I. M. (2013). Latitudinal, ontogenetic, and historical shifts in the diet of a carnivorous teleost, Arripis trutta, in a coastal pelagic ecosystem altered by climate change. Canadian Journal of Fisheries and Aquatic Sciences 70(8): 1209–1230. https://doi.org/10.1139/cjfas-2013-0083.

Hughes, T. P., Kerry, J. T., Connolly, S. R., Baird, A. H., Eakin, C. M., Heron, S. F. et al. (2019). Ecological memory modifies the cumulative impact of recurrent climate extremes. Nature Climate Change 9(1): 40–43. https://doi.org/10.1038/s41558-018-0351-2.

Hunt, G. L., Jr, Blanchard, A. L., Boveng, P., Dalpadado, P., Drinkwater, K. F., Eisner, L. et al. (2013). The Barents and Chukchi Seas: Comparison of two Arctic shelf ecosystems. Journal of Marine Systems: Journal of the European Association of Marine Sciences and Techniques 109–110: 43–68. https://doi.org/10.1016/j.jmarsys.2012.08.003.

Huret, M., Petitgas, P. and Woillez, M. (2010). Dispersal kernels and their drivers captured with a hydrodynamic model and spatial indices: A case study on anchovy (Engraulis encrasicolus) early life stages in the Bay of Biscay. Progress in Oceanography 87(1–4): 6–17. https://doi.org/10.1016/j.pocean.2010.09.023.

Huth, T. J. and Place, S. P. (2016). Transcriptome wide analyses reveal a sustained cellular stress response in the gill tissue of Trematomus bernacchii after acclimation to multiple stressors. BMC Genomics 17(1): 127. https://doi.org/10.1186/s12864-016-2454-3.

Hyndes, G. A., Nagelkerken, I., McLeod, R. J., Connolly, R. M., Lavery, P. S. and Vanderklift, M. A. (2014). Mechanisms and ecological role of carbon transfer within coastal seascapes. Biological Reviews of the Cambridge Philosophical Society 89(1): 232–254. https://doi.org/10.1111/brv.12055.

Intergovernmental Panel on Climate Change. (2019). IPCC Special Report on the Ocean and Cryosphere in a Changing Climate (Pörtner, H.-O., Roberts, D. C., Masson-Delmotte, V., Zhai, P., Tignor, M., Poloczanska, E. et al. eds.). Cambridge University Press.

Intergovernmental Panel on Climate Change (IPCC). (2023a). Summary for Policymakers. pp. 3–34. In Climate Change 2022 – Impacts, Adaptation and Vulnerability: Working Group II Contribution to the Sixth Assessment Report of the Intergovernmental Panel on Climate Change. Cambridge University Press. https://doi.org/10.1017/9781009325844.001.

Intergovernmental Panel on Climate Change (IPCC). (2023b). Climate Change 2022 – Impacts, Adaptation and Vulnerability. Cambridge University Press. https://doi.org/10.1017/9781009325844

Irlandi, E. A. and Crawford, M. K. (1997). Habitat linkages: the effect of intertidal saltmarshes and adjacent subtidal habitats on abundance, movement, and growth of an estuarine fish. Oecologia 110(2): 222–230. https://doi.org/10.1007/s004420050154.

Ivanina, A. V., Hawkins, C. and Sokolova, I. M. (2016). Interactive effects of copper exposure and environmental hypercapnia on immune functions of marine bivalves Crassostrea virginica and Mercenaria mercenaria. Fish & Shellfish Immunology 49: 54–65. https://doi.org/10.1016/j.fsi.2015.12.011.

Jagers, S. C., Harring, N., Löfgren, Å., Sjöstedt, M., Alpizar, F., Brülde, B. et al. (2020). On the preconditions for large-scale collective action. Ambio 49(7): 1282–1296. https://doi.org/10.1007/s13280-019-01284-w.

Jayasundara, N., Tomanek, L., Dowd, W. W. and Somero, G. N. (2015). Proteomic analysis of cardiac response to thermal acclimation in the eurythermal goby fish Gillichthys mirabilis. The Journal of Experimental Biology 218(Pt 9): 1359–1372. https://doi.org/10.1242/jeb.118760.

Ji, R., Edwards, M., Mackas, D. L., Runge, J. A. and Thomas, A. C. (2010). Marine plankton phenology and life history in a changing climate: current research and future directions. Journal of Plankton Research 32(10): 1355–1368. https://doi.org/10.1093/plankt/fbq062.

Johnson, K. M. and Hofmann, G. E. (2017). Transcriptomic response of the Antarctic pteropod Limacina helicina antarctica to ocean acidification. BMC Genomics 18(1): 812. https://doi.org/10.1186/s12864-017-4161-0.

Johnson, M. D., Scott, J. J., Leray, M., Lucey, N., Bravo, L. M. R., Wied, W. L. et al. (2021). Rapid ecosystem-scale consequences of acute deoxygenation on a Caribbean coral reef. Nature Communications 12(1): 4522. https://doi.org/10.1038/s41467-021-24777-3.

Johnston, M. A., Nuttall, M. F., Eckert, R. J., Blakeway, R. D., Sterne, T. K., Hickerson, E. L. et al. (2018). Localized coral reef mortality event at East Flower Garden Bank, Gulf of Mexico. Bulletin of Marine Science 95(2): 239–250. https://doi.org/10.5343/bms.2018.0057.

Johnstone, J., Nash, S., Hernandez, E. and Rahman, M. S. (2019). Effects of elevated temperature on gonadal functions, cellular apoptosis, and oxidative stress in Atlantic sea urchin Arbacia punculata. Marine Environmental Research 149: 40–49. https://doi.org/10.1016/j.marenvres.2019.05.017.

Kaniewska, P., Campbell, P. R., Kline, D. I., Rodriguez-Lanetty, M., Miller, D. J., Dove, S. et al. (2012). Major cellular and physiological impacts of ocean acidification on a reef building coral. PloS One 7(4): e34659. https://doi.org/10.1371/journal.pone.0034659.

Kaymak, N., Winemiller, K. O., Akin, S., Altuner, Z., Polat, F. and Dal, T. (2018). Spatial and temporal variation in food web structure of an impounded river in Anatolia. Marine & Freshwater Research 69(9): 1453. https://doi.org/10.1071/mf17270.

Keeling, R. E., Körtzinger, A. and Gruber, N. (2010). Ocean deoxygenation in a warming world. Annual Review of Marine Science 2(1): 199–229. https://doi.org/10.1146/annurev.marine.010908.163855.

Ketelut-Carneiro, N. and Fitzgerald, K. A. (2022). Apoptosis, pyroptosis, and necroptosis-oh my! The many ways a cell can die. Journal of Molecular Biology 434(4): 167378. https://doi.org/10.1016/j.jmb.2021.167378.

Khan, F. U., Chen, H., Gu, H., Wang, T., Dupont, S., Kong, H. et al. (2021). Antioxidant responses of the mussel Mytilus coruscus co-exposed to ocean acidification, hypoxia and warming. Marine Pollution Bulletin 162(111869): 111869. https://doi.org/10.1016/j.marpolbul.2020.111869.

Kieffer, J., Currie, S. and Tufts, B. (1994). Effects of environmental temperature on the metabolic and acid-base responses of rainbow trout to exhaustive exercise. The Journal of Experimental Biology 194(1): 299–317. https://doi.org/10.1242/jeb.194.1.299.

Kindlmann, P. and Burel, F. (2008). Connectivity measures: a review. Landscape Ecology. https://doi.org/10.1007/s10980-008-9245-4.

Kinlan, B. P. and Gaines, S. D. (2003). Propagule dispersal in marine and terrestrial environments: A community perspective. Ecology 84(8): 2007–2020. https://doi.org/10.1890/01-0622.

Kleypas, J. and Yates, K. (2009). Coral Reefs and Ocean Acidification. Oceanography (Washington, D.C.) 22(4): 108–117. https://doi.org/10.5670/oceanog.2009.101.

Kroeker, K. J., Kordas, R. L., Crim, R., Hendriks, I. E., Ramajo, L., Singh, G. S. et al. (2013). Impacts of ocean acidification on marine organisms: quantifying sensitivities and interaction with warming. Global Change Biology 19(6): 1884–1896. https://doi.org/10.1111/gcb.12179.

Krumhansl, K. A., Okamoto, D. K., Rassweiler, A., Novak, M., Bolton, J. J., Cavanaugh, K. C. et al. (2016). Global patterns of kelp forest change over the past half-century. Proceedings of the National Academy of Sciences of the United States of America 113(48): 13785–13790. https://doi.org/10.1073/pnas.1606102113.

Krumschnabel, G. and Podrabsky, J. E. (2009). Fish as model systems for the study of vertebrate apoptosis. Apoptosis: An International Journal on Programmed Cell Death 14(1): 1–21. https://doi.org/10.1007/s10495-008-0281-y.

Kuhn, A. A. and Darnell, M. Z. (2019). Elevated temperature induces a decrease in intermolt period and growth per molt in the lesser blue crab Callinectes similis Williams, 1966 (Decapoda: Brachyura: Portunidae). Journal of Crustacean Biology: A Quarterly of the Crustacean Society for the Publication of Research on Any Aspect of the Biology of Crustacea 39(1): 22–27. https://doi.org/10.1093/jcbiol/ruy089.

Kühnhold, H., Kamyab, E., Novais, S., Indriana, L., Kunzmann, A., Slater, M. et al. (2017). Thermal stress effects on energy resource allocation and oxygen consumption rate in the juvenile sea cucumber, Holothuria scabra (Jaeger, 1833). Aquaculture (Amsterdam, Netherlands) 467: 109–117. https://doi.org/10.1016/j.aquaculture.2016.03.018.

Kültz, D. (2003). Evolution of the cellular stress proteome: from monophyletic origin to ubiquitous function. The Journal of Experimental Biology 206(Pt 18): 3119–3124. https://doi.org/10.1242/jeb.00549.

Kültz, D. (2020). Evolution of cellular stress response mechanisms. Journal of Experimental Zoology. Part A, Ecological and Integrative Physiology 333(6): 359–378. https://doi.org/10.1002/jez.2347.

Kumagai, N. H., García Molinos, J., Yamano, H., Takao, S., Fujii, M. and Yamanaka, Y. (2018). Ocean currents and herbivory drive macroalgae-to-coral community shift under climate warming. Proceedings of the National Academy of Sciences of the United States of America 115(36): 8990–8995. https://doi.org/10.1073/pnas.1716826115.

Kvitt, H., Rosenfeld, H. and Tchernov, D. (2016). The regulation of thermal stress induced apoptosis in corals reveals high similarities in gene expression and function to higher animals. Scientific Reports 6(1). https://doi.org/10.1038/srep30359.

Kwiatkowski, L., Torres, O., Bopp, L., Aumont, O., Chamberlain, M., Christian, J. R. et al. (2020). Twenty-first century ocean warming, acidification, deoxygenation, and upper-ocean nutrient and primary production decline from CMIP6 model projections. Biogeosciences 17(13): 3439–3470. https://doi.org/10.5194/bg-17-3439-2020.

Lauder, G. V. and Di Santo, V. (2015). Swimming mechanics and energetics of elasmobranch fishes. pp. 219–253. In Fish Physiology. Elsevier. https://doi.org/10.1016/b978-0-12-801289-5.00006-7.

Laufkötter, C., Zscheischler, J. and Frölicher, T. L. (2020). High-impact marine heatwaves attributable to human-induced global warming. Science (New York, N.Y.) 369(6511): 1621–1625. https://doi.org/10.1126/science.aba0690.

Lee, J. S. F., Cook, M. A., Luckenbach, J. A., Berejikian, B. A., Simchick, C. A., Oden, S. M. et al. (2017). Investigation of long-term effects of larval rearing temperature on growth, deformities, flesh quality, and phenotypic sex of cultured sablefish (Anoplopoma fimbria). Aquaculture 479: 91–99. https://doi.org/10.1016/j.aquaculture.2017.05.010.

Lee, Y. H., Jeong, C.-B., Wang, M., Hagiwara, A. and Lee, J.-S. (2020). Transgenerational acclimation to changes in ocean acidification in marine invertebrates. Marine Pollution Bulletin 153(111006): 111006. https://doi.org/10.1016/j.marpolbul.2020.111006.

Lee, Y. H., Kang, H.-M., Kim, M.-S., Lee, J.-S., Wang, M., Hagiwara, A. et al. (2020). Multigenerational mitigating effects of ocean acidification on *in vivo* endpoints, antioxidant defense, DNA damage response, and epigenetic modification in an asexual monogonont rotifer. Environmental Science & Technology 54(13): 7858–7869. https://doi.org/10.1021/acs.est.0c01438.

Leggat, W., Seneca, F., Wasmund, K., Ukani, L., Yellowlees, D. and Ainsworth, T. D. (2011). Differential responses of the coral host and their algal symbiont to thermal stress. PloS One 6(10): e26687. https://doi.org/10.1371/journal.pone.0026687.

Lehtonen, T. K., Wong, B. B. M. and Kvarnemo, C. (2016). Effects of salinity on nest-building behaviour in a marine fish. BMC Ecology 16(1). https://doi.org/10.1186/s12898-016-0067-y.

Leiva, F. P., Calosi, P. and Verberk, W. C. E. P. (2019). Scaling of thermal tolerance with body mass and genome size in ectotherms: a comparison between water- and air-breathers. Philosophical Transactions of the Royal Society of London. Series B, Biological Sciences 374(1778): 20190035. https://doi.org/10.1098/rstb.2019.0035.

Lesser, M. P. (2006). Oxidative stress in marine environments: biochemistry and physiological ecology. Annual Review of Physiology 68(1): 253–278. https://doi.org/10.1146/annurev.physiol.68.040104.110001.

Lesser, M. P. (2016). Climate change stressors cause metabolic depression in the blue mussel,Mytilus edulis, from the Gulf of Maine. Limnology and Oceanography 61(5): 1705–1717. https://doi.org/10.1002/lno.10326.

Lett, C., Ayata, S.-D., Huret, M. and Irisson, J.-O. (2010). Biophysical modelling to investigate the effects of climate change on marine population dispersal and connectivity. Progress in Oceanography 87(1–4): 106–113. https://doi.org/10.1016/j.pocean.2010.09.005.

Leung, J. Y. S., Connell, S. D. and Russell, B. D. (2017). Heatwaves diminish the survival of a subtidal gastropod through reduction in energy budget and depletion of energy reserves. Scientific Reports 7(1). https://doi.org/10.1038/s41598-017-16341-1.

Leung, J. Y. S., Nagelkerken, I., Russell, B. D., Ferreira, C. M. and Connell, S. D. (2018). Boosted nutritional quality of food by CO_2 enrichment fails to offset energy demand of herbivores under ocean warming, causing energy depletion and mortality. The Science of the Total Environment 639: 360–366. https://doi.org/10.1016/j.scitotenv.2018.05.161.

Leung, J. Y. S., Russell, B. D. and Connell, S. D. (2020). Linking energy budget to physiological adaptation: How a calcifying gastropod adjusts or succumbs to ocean acidification and warming. The Science of the Total Environment 715(136939): 136939. https://doi.org/10.1016/j.scitotenv.2020.136939.

Leung, J. Y. S., Russell, B. D., Coleman, M. A., Kelaher, B. P. and Connell, S. D. (2021). Long-term thermal acclimation drives adaptive physiological adjustments of a marine gastropod to reduce sensitivity to climate change. The Science of the Total Environment 771(145208): 145208. https://doi.org/10.1016/j.scitotenv.2021.145208.

Levin, L. A. (2006). Recent progress in understanding larval dispersal: new directions and digressions. Integrative and Comparative Biology 46(3): 282–297. https://doi.org/10.1093/icb/icj024.

Li, L., Hollowed, A. B., Cokelet, E. D., Barbeaux, S. J., Bond, N. A., Keller, A. A. et al. (2019). Subregional differences in groundfish distributional responses to anomalous ocean bottom temperatures in the northeast Pacific. Global Change Biology 25(8): 2560–2575. https://doi.org/10.1111/gcb.14676.

Li, Q., Sun, S., Zhang, F., Wang, M. and Li, M. (2019). Effects of hypoxia on survival, behavior, metabolism and cellular damage of Manila clam (Ruditapes philippinarum). PloS One 14(4): e0215158. https://doi.org/10.1371/journal.pone.0215158.

Li, R., Jia, Z. and Trush, M. A. (2016). Defining ROS in biology and medicine. Reactive Oxygen Species (Apex, N.C.) 1(1): 9–21. https://doi.org/10.20455/ros.2016.803.

Liang, X., Martyniuk, C. J. and Simmons, D. B. D. (2021). Current topics in omics, ecotoxicology, and environmental science. Comparative Biochemistry and Physiology. Part D, Genomics and Proteomics 38(100782): 100782. https://doi.org/10.1016/j.cbd.2020.100782.

Liao, H., Yang, Z., Dou, Z., Sun, F., Kou, S., Zhang, Z. et al. (2018). Impact of ocean acidification on the energy metabolism and antioxidant responses of the Yesso scallop (Patinopecten yessoensis). Frontiers in Physiology 9: 1967. https://doi.org/10.3389/fphys.2018.01967.

Ling, S. D. (2008). Range expansion of a habitat-modifying species leads to loss of taxonomic diversity: a new and impoverished reef state. Oecologia 156(4): 883–894. https://doi.org/10.1007/s00442-008-1043-9.

Ling, S. D., Scheibling, R. E., Rassweiler, A., Johnson, C. R., Shears, N., Connell, S. D. et al. (2015). Global regime shift dynamics of catastrophic sea urchin overgrazing. Philosophical Transactions of the Royal Society of London. Series B, Biological Sciences 370(1659): 20130269. https://doi.org/10.1098/rstb.2013.0269.

Lopes, A. R., Borges, F. O., Figueiredo, C., Sampaio, E., Diniz, M., Rosa, R. et al. (2019). Transgenerational exposure to ocean acidification induces biochemical distress in a keystone amphipod species (Gammarus locusta). Environmental Research 170: 168–177. https://doi.org/10.1016/j.envres.2018.12.040.

López-Pedrouso, M., Lorenzo, J. M., Varela, Z., Fernández, J. Á. and Franco, D. (2022). Finding biomarkers in antioxidant molecular mechanisms for ensuring food safety of bivalves threatened by marine pollution. Antioxidants (Basel, Switzerland) 11(2): 369. https://doi.org/10.3390/antiox11020369.

Lotze, H. K., Tittensor, D. P., Bryndum-Buchholz, A., Eddy, T. D., Cheung, W. W. L., Galbraith, E. D. et al. (2019). Global ensemble projections reveal trophic amplification of ocean biomass declines with climate change. Proceedings of the National Academy of Sciences of the United States of America 116(26): 12907–12912. https://doi.org/10.1073/pnas.1900194116.

Lowe, W. H., Martin, T. E., Skelly, D. K. and Woods, H. A. (2021). Metamorphosis in an era of increasing climate variability. Trends in Ecology and Evolution 36(4): 360–375. https://doi.org/10.1016/j.tree.2020.11.012.

Lu, Y., Zhang, J., Cao, J., Liu, P., Li, J. and Meng, X. (2022). Long-term ammonia toxicity in the hepatopancreas of swimming crab Portunus trituberculatus: Cellular stress response and tissue damage. Frontiers in Marine Science 8. https://doi.org/10.3389/fmars.2021.757602.

Lushchak, V. I. (2011). Environmentally induced oxidative stress in aquatic animals. Aquatic Toxicology (Amsterdam, Netherlands) 101(1): 13–30. https://doi.org/10.1016/j.aquatox.2010.10.006.

Lynam, C. P., Llope, M., Möllmann, C., Helaouët, P., Bayliss-Brown, G. A. and Stenseth, N. C. (2017). Interaction between top-down and bottom-up control in marine food webs. Proceedings of the National Academy of Sciences of the United States of America 114(8): 1952–1957. https://doi.org/10.1073/pnas.1621037114.

Lynch, A. J., McDowell, J. E. and Graves, J. (2010). A molecular genetic investigation of the population structure of Atlantic menhaden (Brevoortia Tyrannus). Fishery Bulletin 108(1): 87. https://scholarworks.wm.edu/vimsarticles/554.

Madeira, C., Mendonça, V., Leal, M. C., Diniz, M. S., Cabral, H. N., Flores, A. A. V. et al. (2019). Present and future invasion perspectives of an alien shrimp in South Atlantic coastal waters: an experimental assessment of functional biomarkers and thermal tolerance. Biological Invasions 21(5): 1567–1584. https://doi.org/10.1007/s10530-019-01921-1.

Madeira, C., Mendonça, V., Leal, M. C., Flores, A. A. V., Cabral, H. N., Diniz, M. S. et al. (2018). Environmental health assessment of warming coastal ecosystems in the tropics – Application of integrative physiological indices. The Science of the Total Environment 643: 28–39. https://doi.org/10.1016/j.scitotenv.2018.06.152.

Madeira, D., Narciso, L., Cabral, H. N., Vinagre, C. and Diniz, M. S. (2012). HSP70 production patterns in coastal and estuarine organisms facing increasing temperatures. Journal of Sea Research 73: 137–147. https://doi.org/10.1016/j.seares.2012.07.003.

Madeira, D., Vinagre, C., Costa, P. M. and Diniz, M. S. (2014). Histopathological alterations, physiological limits, and molecular changes of juvenile Sparus aurata in response to thermal stress. Marine Ecology Progress Series 505: 253–266. https://doi.org/10.3354/meps10794.

Madeira, D., Araújo, J. E., Vitorino, R., Costa, P. M., Capelo, J. L., Vinagre, C. et al. (2017). Molecular plasticity under ocean warming: Proteomics and fitness data provides clues for a better understanding of the thermal tolerance in fish. Frontiers in Physiology 8. https://doi.org/10.3389/fphys.2017.00825.

Madeira, D., Costa, P. M., Vinagre, C. and Diniz, M. S. (2016). When warming hits harder: survival, cellular stress and thermal limits of Sparus aurata larvae under global change. Marine Biology 163(4). https://doi.org/10.1007/s00227-016-2856-4.

Madeira, D., Vinagre, C. and Diniz, M. S. (2016). Are fish in hot water? Effects of warming on oxidative stress metabolism in the commercial species Sparus aurata. Ecological Indicators 63: 324–331. https://doi.org/10.1016/j.ecolind.2015.12.008.

Madeira, D., Araújo, J. E., Madeira, C., Mendonça, V., Vitorino, R., Vinagre, C. et al. (2020a). Seasonal proteome variation in intertidal shrimps under a natural setting: Connecting molecular networks with environmental fluctuations. The Science of the Total Environment 703(134957): 134957. https://doi.org/10.1016/j.scitotenv.2019.134957.

Madeira, D., Madeira, C., Costa, P. M., Vinagre, C., Pörtner, H.-O. and Diniz, M. S. (2020b). Different sensitivity to heatwaves across the life cycle of fish reflects phenotypic adaptation to environmental niche. Marine Environmental Research 162(105192): 105192. https://doi.org/10.1016/j.marenvres.2020.105192.

Madeira, D., Fernandes, J. F., Jerónimo, D., Ricardo, F., Santos, A., Domingues, M. R. et al. (2021). Calcium homeostasis and stable fatty acid composition underpin heatwave tolerance of the keystone polychaete Hediste diversicolor. Environmental Research 195(110885): 110885. https://doi.org/10.1016/j.envres.2021.110885.

Magris, R. A., Treml, E. A., Pressey, R. L. and Weeks, R. (2016). Integrating multiple species connectivity and habitat quality into conservation planning for coral reefs. Ecography 39(7): 649–664. https://doi.org/10.1111/ecog.01507.

Maia, S., Marques, S. C., Dupont, S., Neves, M., Pinto, H. J., Reis, J. et al. (2022). Effects of ocean acidification and warming on the development and biochemical responses of juvenile shrimp Palaemon elegans (Rathke, 1837). Marine Environmental Research 176(105580): 105580. https://doi.org/10.1016/j.marenvres.2022.105580.

Makrinos, D. L. and Bowden, T. J. (2016). Natural environmental impacts on teleost immune function. Fish and Shellfish Immunology 53: 50–57. https://doi.org/10.1016/j.fsi.2016.03.008.

Malhi, Y., Franklin, J., Seddon, N., Solan, M., Turner, M. G., Field, C. B. et al. (2020). Climate change and ecosystems: threats, opportunities and solutions. Philosophical Transactions of the Royal Society of London. Series B, Biological Sciences 375(1794): 20190104. https://doi.org/10.1098/rstb.2019.0104.

Marchini, C., Gizzi, F., Pondrelli, T., Moreddu, L., Marisaldi, L., Montori, F. et al. (2021). Decreasing pH impairs sexual reproduction in a Mediterranean coral transplanted at a CO_2 vent. Limnology and Oceanography 66(11): 3990–4000. https://doi.org/10.1002/lno.11937.

Marcogliese, D. J. (2008). The impact of climate change on the parasites and infectious diseases of aquatic animals. Revue Scientifique et Technique (International Office of Epizootics) 27(2): 467–484. https://www.ncbi.nlm.nih.gov/pubmed/18819673.

Marsh, J. M. and Mueter, F. J. (2020). Influences of temperature, predators, and competitors on polar cod (Boreogadus saida) at the southern margin of their distribution. Polar Biology 43(8): 995–1014. https://doi.org/10.1007/s00300-019-02575-4.

Marshall, D. J. and McQuaid, C. D. (2020). Metabolic regulation, oxygen limitation and heat tolerance in a subtidal marine gastropod reveal the complexity of predicting climate change vulnerability. Frontiers in Physiology 11. https://doi.org/10.3389/fphys.2020.01106.

Martínez-Reyes, I. and Chandel, N. S. (2020). Mitochondrial TCA cycle metabolites control physiology and disease. Nature Communications 11(1): 102. https://doi.org/10.1038/s41467-019-13668-3.

Martins, J. C., Domínguez-Pérez, D., Azevedo, C., Braga, A. C., Costa, P. R., Osório, H. et al. (2020). Molecular responses of mussel Mytilus galloprovincialis associated to accumulation and depuration of marine biotoxins okadaic acid and dinophysistoxin-1 revealed by shotgun proteomics. Frontiers in Marine Science 7. https://doi.org/10.3389/fmars.2020.589822.

Marzloff, M. P., Melbourne-Thomas, J., Hamon, K. G., Hoshino, E., Jennings, S., van Putten, I. E. et al. (2016). Modelling marine community responses to climate-driven species redistribution to guide monitoring and adaptive ecosystem-based management. Global Change Biology 22(7): 2462–2474. https://doi.org/10.1111/gcb.13285.

Mathis, J. T., Cooley, S. R., Lucey, N., Colt, S., Ekstrom, J., Hurst, T. et al. (2015). Ocean acidification risk assessment for Alaska's fishery sector. Progress in Oceanography 136: 71–91. https://doi.org/10.1016/j.pocean.2014.07.001.

Maulvault, A. L., Santos, L. H. M. L. M., Paula, J. R., Camacho, C., Pissarra, V., Fogaça, F. et al. (2018a). Differential behavioural responses to venlafaxine exposure route, warming and acidification in juvenile fish (Argyrosomus regius). The Science of the Total Environment 634: 1136–1147. https://doi.org/10.1016/j.scitotenv.2018.04.015.

Maulvault, A. L., Barbosa, V., Alves, R., Anacleto, P., Camacho, C., Cunha, S. et al. (2018b). Integrated multi-biomarker responses of juvenile seabass to diclofenac, warming and acidification co-exposure. Aquatic Toxicology (Amsterdam, Netherlands) 202: 65–79. https://doi.org/10.1016/j.aquatox.2018.06.016.

Maulvault, A. L., Camacho, C., Barbosa, V., Alves, R., Anacleto, P., Pousão-Ferreira, P. et al. (2019). Living in a multi-stressors environment: An integrated biomarker approach to assess the ecotoxicological response of meagre (Argyrosomus regius) to venlafaxine, warming and acidification. Environmental Research 169: 7–25. https://doi.org/10.1016/j.envres.2018.10.021.

Maureaud, A., Gascuel, D., Colléter, M., Palomares, M. L. D., Du Pontavice, H., Pauly, D. et al. (2017). Global change in the trophic functioning of marine food webs. PloS One 12(8): e0182826. https://doi.org/10.1371/journal.pone.0182826.

Maxwell, B. A., Gwon, Y., Mishra, A., Peng, J., Zhang, K., Kim, H. J. et al. (2021). Ubiquitination is essential for recovery of cellular activities following heat shock. In bioRxiv. https://doi.org/10.1101/2021.04.22.440934.

Mayor, D. J., Everett, N. R. and Cook, K. B. (2012). End of century ocean warming and acidification effects on reproductive success in a temperate marine copepod. Journal of Plankton Research 34(3): 258–262. https://doi.org/10.1093/plankt/fbr107.

McLean, M., Mouillot, D., Maureaud, A. A., Hattab, T., MacNeil, M. A., Goberville, E. et al. (2021). Disentangling tropicalization and deborealization in marine ecosystems under climate change. Current Biology: CB 31(21): 4817–4823.e5. https://doi.org/10.1016/j.cub.2021.08.034.

McLeod, D. J., Hobday, A. J., Lyle, J. M. and Welsford, D. C. (2012). A prey-related shift in the abundance of small pelagic fish in eastern Tasmania? ICES Journal of Marine Science: Journal Du Conseil 69(6): 953–960. https://doi.org/10.1093/icesjms/fss069.

Menge, B. A. and Sutherland, J. P. (1987). Community regulation: Variation in disturbance, competition, and predation in relation to environmental stress and recruitment. The American Naturalist 130(5): 730–757. https://doi.org/10.1086/284741.

Metcalfe, J. D., Le Quesne, W. J. F., Cheung, W. W. L. and Righton, D. A. (2012). Conservation physiology for applied management of marine fish: an overview with perspectives on the role and value of telemetry. Philosophical Transactions of the Royal Society of London. Series B, Biological Sciences 367(1596): 1746–1756. https://doi.org/10.1098/rstb.2012.0017.

Milazzo, M., Mirto, S., Domenici, P. and Gristina, M. (2013). Climate change exacerbates interspecific interactions in sympatric coastal fishes. The Journal of Animal Ecology 82(2): 468–477. https://doi.org/10.1111/j.1365-2656.2012.02034.x.

Miller, L. P., Matassa, C. M. and Trussell, G. C. (2014). Climate change enhances the negative effects of predation risk on an intermediate consumer. Global Change Biology 20(12): 3834–3844. https://doi.org/10.1111/gcb.12639.

Minuti, J. J., Byrne, M., Hemraj, D. A. and Russell, B. D. (2021). Capacity of an ecologically key urchin to recover from extreme events: Physiological impacts of heatwaves and the road to recovery. The Science of the Total Environment 785(147281): 147281. https://doi.org/10.1016/j.scitotenv.2021.147281.

Missionário, M., Fernandes, J. F., Travesso, M., Freitas, E., Calado, R. and Madeira, D. (2022). Sex-specific thermal tolerance limits in the ditch shrimp Palaemon varians: Eco-evolutionary implications under a warming ocean. Journal of Thermal Biology 103(103151): 103151. https://doi.org/10.1016/j.jtherbio.2021.103151.

Mittermayer, F. H., Stiasny, M. H., Clemmesen, C., Bayer, T., Puvanendran, V., Chierici, M. et al. (2019). Transcriptome profiling reveals exposure to predicted end-of-century ocean acidification as a stealth stressor for Atlantic cod larvae. Scientific Reports 9(1): 16908. https://doi.org/10.1038/s41598-019-52628-1.

Moffitt, E. A., Botsford, L. W., Kaplan, D. M. and O'Farrell, M. R. (2009). Marine reserve networks for species that move within a home range. Ecological Applications: A Publication of the Ecological Society of America 19(7): 1835–1847. https://doi.org/10.1890/08-1101.1.

Montánchez, I. and Kaberdin, V. R. (2020). Vibrio harveyi: A brief survey of general characteristics and recent epidemiological traits associated with climate change. Marine Environmental Research 154(104850): 104850. https://doi.org/10.1016/j.marenvres.2019.104850.

Morikawa, M. K. and Palumbi, S. R. (2019). Using naturally occurring climate resilient corals to construct bleaching-resistant nurseries. Proceedings of the National Academy of Sciences of the United States of America 116(21): 10586–10591. https://doi.org/10.1073/pnas.1721415116.

Muir, W. D. and Williams, J. G. (2012). Improving connectivity between freshwater and marine environments for salmon migrating through the lower Snake and Columbia River hydropower system. Ecological Engineering 48: 19–24. https://doi.org/10.1016/j.ecoleng.2011.06.034.

Multiple stressors - forces that combine to worsen deoxygenation and its effects. in Ocean deoxygenation: Everyone's problem Causes, impacts, consequences and solutions. (n.d.).

Munday, P. L., Leis, J. M., Lough, J. M., Paris, C. B., Kingsford, M. J., Berumen, M. L. et al. (2009). Climate change and coral reef connectivity. Coral Reefs 28(2): 379–395. https://doi.org/10.1007/s00338-008-0461-9.

Munsch, S. H., Greene, C. M., Johnson, R. C., Satterthwaite, W. H., Imaki, H. and Brandes, P. L. (2019). Warm, dry winters truncate timing and size distribution of seaward-migrating salmon across a large, regulated watershed. Ecological Applications: A Publication of the Ecological Society of America 29(4): e01880. https://doi.org/10.1002/eap.1880.

Nagelkerken, I. and Munday, P. L. (2016). Animal behaviour shapes the ecological effects of ocean acidification and warming: moving from individual to community-level responses. Global Change Biology 22(3): 974–989. https://doi.org/10.1111/gcb.13167.

Nikinmaa, M. and Anttila, K. (2020). 3 Responses of marine animals to ocean acidification. pp. 107–132. *In*: Climate Change and Marine and Freshwater Toxins. De Gruyter. https://doi.org/10.1515/9783110625738-003.

Nilsson, G. E., Dixson, D. L., Domenici, P., McCormick, M. I., Sørensen, C., Watson, S.-A. et al. (2012). Near-future carbon dioxide levels alter fish behaviour by interfering with neurotransmitter function. Nature Climate Change 2(3): 201–204. https://doi.org/10.1038/nclimate1352.

Nissling, A. and Larsson, R. (2018). Population specific sperm production in European flounder *Platichthys flesus*: Adaptation to salinity at spawning. Journal of Fish Biology 93(1): 47–52. https://doi.org/10.1111/jfb.13667.

Niu, P., Liu, L., Gong, Z., Tan, H., Wang, F., Yuan, J. et al. (2006). Overexpressed heat shock protein 70 protects cells against DNA damage caused by ultraviolet C in a dose-dependent manner. Cell Stress and Chaperones 11(2): 162–169. https://doi.org/10.1379/csc-175r.1.

Noisette, F., Calosi, P., Madeira, D., Chemel, M., Menu-Courey, K., Piedalue, S. et al. (2021). Tolerant larvae and sensitive juveniles: Integrating metabolomics and whole-organism responses to define life-stage specific sensitivity to ocean acidification in the American lobster. Metabolites 11(9): 584. https://doi.org/10.3390/metabo11090584.

Oakley, C. A., Durand, E., Wilkinson, S. P., Peng, L., Weis, V. M., Grossman, A. R. et al. (2017). Thermal shock induces host proteostasis disruption and endoplasmic reticulum stress in the model symbiotic cnidarian Aiptasia. Journal of Proteome Research 16(6): 2121–2134. https://doi.org/10.1021/acs.jproteome.6b00797.

O'Connor, C. and Booth, D. J. (2021). Response of estuarine fishes to elevated temperatures within temperate Australia: Implications of climate change on fish growth and foraging performance. Journal of Experimental Marine Biology and Ecology 544(151626): 151626. https://doi.org/10.1016/j.jembe.2021.151626.

O'Connor, M. I., Bruno, J. F., Gaines, S. D., Halpern, B. S., Lester, S. E., Kinlan, B. P. et al. (2007). Temperature control of larval dispersal and the implications for marine ecology, evolution, and conservation. Proceedings of the National Academy of Sciences of the United States of America 104(4): 1266–1271. https://doi.org/10.1073/pnas.0603422104.

O'Connor, M. I., Piehler, M. F., Leech, D. M., Anton, A. and Bruno, J. F. (2009). Warming and resource availability shift food web structure and metabolism. PLoS Biology 7(8): e1000178. https://doi.org/10.1371/journal.pbio.1000178.

Olds, A. D., Vargas-Fonseca, E., Connolly, R. M., Gilby, B. L., Huijbers, C. M., Hyndes, G. A. et al. (2018). The ecology of fish in the surf zones of ocean beaches: A global review. Fish and Fisheries (Oxford, England) 19(1): 78–89. https://doi.org/10.1111/faf.12237.

Palumbi, S., Stanford University, Evans, T., Pespeni, M. and Somero, G. (2019). Present and future adaptation of marine species assemblages: DNA-based insights into climate change from studies of physiology, genomics, and evolution. Oceanography (Washington, D.C.) 32(3): 82–93. https://doi.org/10.5670/oceanog.2019.314.

Pandori, L. L. M. and Sorte, C. J. B. (2019). The weakest link: sensitivity to climate extremes across life stages of marine invertebrates. Oikos (Copenhagen, Denmark) 128(5): 621–629. https://doi.org/10.1111/oik.05886.

Pankhurst, N. W. and Porter, M. J. R. (2003). Cold and dark or warm and light: variations on the theme of environmental control of reproduction. Fish Physiology and Biochemistry 28(1–4): 385–389. https://doi.org/10.1023/b:fish.0000030602.51939.50.

Pankhurst, Ned W. and Munday, P. L. (2011). Effects of climate change on fish reproduction and early life history stages. Marine and Freshwater Research 62(9): 1015. https://doi.org/10.1071/mf10269.

Park, S., Ahn, I.-Y., Sin, E., Shim, J. and Kim, T. (2020). Ocean freshening and acidification differentially influence mortality and behavior of the Antarctic amphipod Gondogeneia antarctica. Marine Environmental Research 154(104847): 104847. https://doi.org/10.1016/j.marenvres.2019.104847.

Parker, L. M., Scanes, E., O'Connor, W. A., Coleman, R. A., Byrne, M., Pörtner, H.-O. et al. (2017). Ocean acidification narrows the acute thermal and salinity tolerance of the Sydney rock oyster Saccostrea glomerata. Marine Pollution Bulletin 122(1–2): 263–271. https://doi.org/10.1016/j.marpolbul.2017.06.052.

Parmesan, C. (2006). Ecological and evolutionary responses to recent climate change. Annual Review of Ecology, Evolution, and Systematics 37(1): 637–669. https://doi.org/10.1146/annurev.ecolsys.37.091305.110100.

Paula, J. R., Baptista, M., Carvalho, F., Repolho, T., Bshary, R. and Rosa, R. (2019). The past, present and future of cleaner fish cognitive performance as a function of CO_2 levels. Biology Letters 15(12): 20190618. https://doi.org/10.1098/rsbl.2019.0618.

Pauly, D. and Liang, C. (2022). Temperature and the maturation of fish: a simple sine-wave model for predicting accelerated spring spawning. Environmental Biology of Fishes 105(10): 1481–1487. https://doi.org/10.1007/s10641-022-01212-0.

Pechenik, J. A. (2006). Larval experience and latent effects - Metamorphosis is not a new beginning. Integrative and Comparative Biology 46(3): 323–333. https://doi.org/10.1093/icb/icj028.

Perez, A. U., Schmitter-Soto, J. J., Adams, A. J. and Heyman, W. D. (2019). Connectivity mediated by seasonal bonefish (Albula vulpes) migration between the Caribbean Sea and a tropical estuary of Belize and Mexico. Environmental Biology of Fishes 102(2): 197–207. https://doi.org/10.1007/s10641-018-0834-z.

Pernice, M., Dunn, S. R., Miard, T., Dufour, S., Dove, S. and Hoegh-Guldberg, O. (2011). Regulation of apoptotic mediators reveals dynamic responses to thermal stress in the reef building coral Acropora millepora. PloS One 6(1): e16095. https://doi.org/10.1371/journal.pone.0016095.

Perry, A. L., Low, P. J., Ellis, J. R. and Reynolds, J. D. (2005). Climate change and distribution shifts in marine fishes. Science (New York, N.Y.) 308(5730): 1912–1915. https://doi.org/10.1126/science.1111322.

Perry, D., Staveley, T. A. B. and Gullström, M. (2018). Habitat connectivity of fish in temperate Shallow-Water seascapes. Frontiers in Marine Science 4. https://doi.org/10.3389/fmars.2017.00440.

Peruzza, L., Thatje, S. and Hauton, C. (2021). Acclimation to cyclic hypoxia improves thermal tolerance and copper survival in the caridean shrimp Palaemon varians. Comparative Biochemistry and Physiology. Part A, Molecular and Integrative Physiology 259(111010): 111010. https://doi.org/10.1016/j.cbpa.2021.111010.

Petchey, O. L., McPhearson, P. T., Casey, T. M. and Morin, P. J. (1999). Environmental warming alters food-web structure and ecosystem function. Nature 402(6757): 69–72. https://doi.org/10.1038/47023.

Petes, L. E., Mouchka, M. E., Milston-Clements, R. H., Momoda, T. S. and Menge, B. A. (2008). Effects of environmental stress on intertidal mussels and their sea star predators. Oecologia 156(3): 671–680. https://doi.org/10.1007/s00442-008-1018-x.

Petrik, C. M., Stock, C. A., Andersen, K. H., van Denderen, P. D. and Watson, J. R. (2020). Large pelagic fish are most sensitive to climate change despite pelagification of ocean food webs. Frontiers in Marine Science 7. https://doi.org/10.3389/fmars.2020.588482.

Petrou, K., Nunn, B. L., Padula, M. P., Miller, D. J. and Nielsen, D. A. (2021). Broad scale proteomic analysis of heat-destabilised symbiosis in the hard coral Acropora millepora. Scientific Reports 11(1): 19061. https://doi.org/10.1038/s41598-021-98548-x.

Pimentel, M. S., Faleiro, F., Diniz, M., Machado, J., Pousão-Ferreira, P., Peck, M. A. et al. (2015). Oxidative stress and digestive enzyme activity of flatfish larvae in a changing ocean. PloS One 10(7): e0134082. https://doi.org/10.1371/journal.pone.0134082.

Pineda, M. C., McQuaid, C. D., Turon, X., López-Legentil, S., Ordóñez, V. and Rius, M. (2012). Tough adults, frail babies: an analysis of stress sensitivity across early life-history stages of widely introduced marine invertebrates. PloS One 7(10): e46672. https://doi.org/10.1371/journal.pone.0046672.

Pinsky, M. L., Selden, R. L. and Kitchel, Z. J. (2020). Climate-driven shifts in marine species ranges: Scaling from organisms to communities. Annual Review of Marine Science 12(1): 153–179. https://doi.org/10.1146/annurev-marine-010419-010916.

Pinsky, M. L., Worm, B., Fogarty, M. J., Sarmiento, J. L. and Levin, S. A. (2013). Marine taxa track local climate velocities. Science (New York, N.Y.) 341(6151): 1239–1242. https://doi.org/10.1126/science.1239352.

Pittman, S. J., Monaco, M. E., Friedlander, A. M., Legare, B., Nemeth, R. S., Kendall, M. S. et al. (2014). Fish with chips: Tracking reef fish movements to evaluate size and connectivity of Caribbean marine protected areas. PloS One 9(5): e96028. https://doi.org/10.1371/journal.pone.0096028.

Place, S. P. and Hofmann, G. E. (2005). Constitutive expression of a stress-inducible heat shock protein gene, hsp70, in phylogenetically distant Antarctic fish. Polar Biology 28(4): 261–267. https://doi.org/10.1007/s00300-004-0697-y.

Polis, G. A. and Strong, D. R. (1996). Food web complexity and community dynamics. The American Naturalist 147(5): 813–846. https://doi.org/10.1086/285880.

Poloczanska, E., Hoegh-Guldberg, O., Cheung, W., Pörtner, H. O. and Burrows, M. T. (2014). Crosss-chapter box on observed global responses of marine biogeography, abundance, and phenology to climate change. pp. 123–127. *In*: Field, C. B., Barros, V. R., Dokken, D. J., Mach, K. J., Mastrandrea, M. D., Bilir, T. E. et al. (eds.). Climate Change 2014: Impacts, Adaptation, and Vulnerability. Part A: Global and Sectoral Aspects. Contribution of Working Group II to the Fifth Assessment Report of the Intergovernmental Panel on Climate Change. Cambridge University Press. https://epic.awi.de/id/eprint/37500/.

Poloczanska, E. S., Brown, C. J., Sydeman, W. J., Kiessling, W., Schoeman, D. S., Moore, P. J. et al. (2013). Global imprint of climate change on marine life. Nature Climate Change 3(10): 919–925. https://doi.org/10.1038/nclimate1958.

Poloczanska, E. S., Burrows, M. T., Brown, C. J., García Molinos, J., Halpern, B. S., Hoegh-Guldberg, O. et al. (2016). Responses of marine organisms to climate change across oceans. Frontiers in Marine Science 3. https://doi.org/10.3389/fmars.2016.00062.

Pörtner, H. O. (2012). Integrating climate-related stressor effects on marine organisms: unifying principles linking molecule to ecosystem-level changes. Marine Ecology Progress Series 470: 273–290. https://doi.org/10.3354/meps10123.

Pörtner, Hans O. and Knust, R. (2007). Climate change affects marine fishes through the oxygen limitation of thermal tolerance. Science (New York, N.Y.) 315(5808): 95–97. https://doi.org/10.1126/science.1135471.

Prince, E. D. and Goodyear, C. P. (2006). Hypoxia-based habitat compression of tropical pelagic fishes. Fisheries Oceanography 15(6): 451–464. https://doi.org/10.1111/j.1365-2419.2005.00393.x.

Prince, E. D. and Goodyear, C. P. (2007). Consequences of ocean scale hypoxia constrained habitat for tropical pelagic fishes. Gulf and Caribbean Research 19. https://doi.org/10.18785/gcr.1902.04.

Rahman, M. S. and Rahman, M. S. (2021). Effects of elevated temperature on prooxidant-antioxidant homeostasis and redox status in the American oyster: Signaling pathways of cellular apoptosis during heat stress. Environmental Research 196(110428): 110428. https://doi.org/10.1016/j.envres.2020.110428.

Rathburn, C. K., Sharp, N. J., Ryan, J. C., Neely, M. G., Cook, M., Chapman, R. W. et al. (2013). Transcriptomic responses of juvenile Pacific whiteleg shrimp, Litopenaeus vannamei, to hypoxia and hypercapnic hypoxia. Physiological Genomics 45(17): 794–807. https://doi.org/10.1152/physiolgenomics.00043.2013.

Richier, S., Sabourault, C., Courtiade, J., Zucchini, N., Allemand, D. and Furla, P. (2006). Oxidative stress and apoptotic events during thermal stress in the symbiotic sea anemone, Anemonia viridis. The FEBS Journal 273(18): 4186–4198. https://doi.org/10.1111/j.1742-4658.2006.05414.x.

Ridgway, K. R. (2007). Long-term trend and decadal variability of the southward penetration of the East Australian Current. Geophysical Research Letters 34(13). https://doi.org/10.1029/2007gl030393.

Riebesell, U. and Gattuso, J.-P. (2015). Lessons learned from ocean acidification research. Nature Climate Change 5(1): 12–14. https://doi.org/10.1038/nclimate2456.

Righton, D. A., Andersen, K. H., Neat, F., Thorsteinsson, V., Steingrund, P., Svedäng, H. et al. (2010). Thermal niche of Atlantic cod Gadus morhua: limits, tolerance and optima. Marine Ecology Progress Series 420: 1–13. https://doi.org/10.3354/meps08889.

Rijnsdorp, A. D., Peck, M. A., Engelhard, G. H., Möllmann, C. and Pinnegar, J. K. (2009). Resolving the effect of climate change on fish populations. ICES Journal of Marine Science: Journal Du Conseil 66(7): 1570–1583. https://doi.org/10.1093/icesjms/fsp056.

Rivera-Ingraham, G. A., Nommick, A., Blondeau-Bidet, E., Ladurner, P. and Lignot, J.-H. (2016). Salinity stress from the perspective of the energy-redox axis: Lessons from a marine intertidal flatworm. Redox Biology 10: 53–64. https://doi.org/10.1016/j.redox.2016.09.012.

Rodrigues, L. J. and Grottoli, A. G. (2007). Energy reserves and metabolism as indicators of coral recovery from bleaching. Limnology and Oceanography 52(5): 1874–1882. https://doi.org/10.4319/lo.2007.52.5.1874.

Rodriguez-Dominguez, A., Connell, S. D., Leung, J. Y. S. and Nagelkerken, I. (2019). Adaptive responses of fishes to climate change: Feedback between physiology and behaviour. The Science of the Total Environment 692: 1242–1249. https://doi.org/10.1016/j.scitotenv.2019.07.226.

Rodríguez-Martínez, M., Nielsen, J., Dupont, S., Vamathevan, J., Glover, B. J., Crosswell, L. C. et al. (2022). Molecular biology for green recovery-A call for action. PLoS Biology 20(4): e3001623. https://doi.org/10.1371/journal.pbio.3001623.

Roessig, J. M., Woodley, C. M., Cech, J. J., Jr and Hansen, L. J. (2004). Effects of global climate change on marine and estuarine fishes and fisheries. Reviews in Fish Biology and Fisheries 14(2): 251–275. https://doi.org/10.1007/s11160-004-6749-0.

Rosa, R., Lopes, A. R., Pimentel, M., Faleiro, F., Baptista, M., Trübenbach, K. et al. (2014). Ocean cleaning stations under a changing climate: biological responses of tropical and temperate fish-cleaner shrimp to global warming. Global Change Biology 20(10): 3068–3079. https://doi.org/10.1111/gcb.12621.

Ross, P. M., Parker, L. and Byrne, M. (2016). Transgenerational responses of molluscs and echinoderms to changing ocean conditions. ICES Journal of Marine Science: Journal Du Conseil 73(3): 537–549. https://doi.org/10.1093/icesjms/fsv254.

Ruiz-Jarabo, I., Gregório, S. F., Alves, A., Mancera, J. M. and Fuentes, J. (2021). Ocean acidification compromises energy management in Sparus aurata (Pisces: Teleostei). Comparative Biochemistry and Physiology. Part A, Molecular and Integrative Physiology 256(110911): 110911. https://doi.org/10.1016/j.cbpa.2021.110911.

Ruiz-Jones, L. J. and Palumbi, S. R. (2017). Tidal heat pulses on a reef trigger a fine-tuned transcriptional response in corals to maintain homeostasis. Science Advances 3(3): e1601298. https://doi.org/10.1126/sciadv.1601298.

Rundle, S. D. and Spicer, J. I. (2016). Heterokairy: a significant form of developmental plasticity? Biology Letters 12(9): 20160509. https://doi.org/10.1098/rsbl.2016.0509.

Rungrassamee, W., Leelatanawit, R., Jiravanichpaisal, P., Klinbunga, S. and Karoonuthaisiri, N. (2010). Expression and distribution of three heat shock protein genes under heat shock stress and under exposure to Vibrio harveyi in Penaeus monodon. Developmental and Comparative Immunology 34(10): 1082–1089. https://doi.org/10.1016/j.dci.2010.05.012.

Rutherford, S. L. (2003). Between genotype and phenotype: protein chaperones and evolvability. Nature Reviews. Genetics 4(4): 263–274. https://doi.org/10.1038/nrg1041.

Saha, N., Koner, D. and Sharma, R. (2022). Environmental hypoxia: A threat to the gonadal development and reproduction in bony fishes. Aquaculture and Fisheries 7(5): 572–582. https://doi.org/10.1016/j.aaf.2022.02.002.

Sampaio, E., Lopes, A. R., Francisco, S., Paula, J. R., Pimentel, M., Maulvault, A. L. et al. (2018). Ocean acidification dampens physiological stress response to warming and contamination in a commercially-important fish (Argyrosomus regius). The Science of the Total Environment 618: 388–398. https://doi.org/10.1016/j.scitotenv.2017.11.059.

Sano, R. and Reed, J. C. (2013). ER stress-induced cell death mechanisms. Biochimica et Biophysica Acta. Molecular Cell Research 1833(12): 3460–3470. https://doi.org/10.1016/j.bbamcr.2013.06.028.

Saravia, J., Paschke, K., Oyarzún-Salazar, R., Cheng, C.-H. C., Navarro, J. M. and Vargas-Chacoff, L. (2021). Effects of warming rates on physiological and molecular components of response to CTMax heat stress in the Antarctic fish Harpagifer antarcticus. *Journal of Thermal Biology*, *99*(103021), 103021. https://doi.org/10.1016/j.jtherbio.2021.103021

Scanes, E., Parker, L. M., Seymour, J. R., Siboni, N., King, W. L., Danckert, N. P. et al. (2021). Climate change alters the haemolymph microbiome of oysters. Marine Pollution Bulletin 164(111991): 111991. https://doi.org/10.1016/j.marpolbul.2021.111991.

Scapini, F., Innocenti Degli, E. and Defeo, O. (2019). Behavioral adaptations of sandy beach macrofauna in face of climate change impacts: A conceptual framework. Estuarine, Coastal and Shelf Science 225(106236): 106236. https://doi.org/10.1016/j.ecss.2019.05.018.

Schoepf, V., Grottoli, A. G., Warner, M. E., Cai, W.-J., Melman, T. F., Hoadley, K. D. et al. (2013). Coral energy reserves and calcification in a high-CO_2 world at two temperatures. PloS One 8(10): e75049. https://doi.org/10.1371/journal.pone.0075049.

Schultz, J. A., Cloutier, R. N. and Côté, I. M. (2016). Evidence for a trophic cascade on rocky reefs following sea star mass mortality in British Columbia. PeerJ 4(e1980): e1980. https://doi.org/10.7717/peerj.1980.

Seebacher, F., White, C. R. and Franklin, C. E. (2015). Physiological plasticity increases resilience of ectothermic animals to climate change. Nature Climate Change 5(1): 61–66. https://doi.org/10.1038/nclimate2457.

Servili, A., Canario, A. V. M., Mouchel, O. and Muñoz-Cueto, J. A. (2020). Climate change impacts on fish reproduction are mediated at multiple levels of the brain-pituitary-gonad axis. General and Comparative Endocrinology 291(113439): 113439. https://doi.org/10.1016/j.ygcen.2020.113439.

Sguotti, C., Gokhale, S., Lai, T.-Y., Schuch, E., Möllmann, C. and Richter, A. (2023). Telecouplings in Atlantic cod—The role of global trade and climate change. Marine Policy 157(105818): 105818. https://doi.org/10.1016/j.marpol.2023.105818.

Shang, Y., Wang, X., Deng, Y., Wang, S., Gu, H., Wang, T. et al. (2020). Diel-cycling seawater acidification and hypoxia impair the physiological and growth performance of marine mussels. The Science of the Total Environment 722(138001): 138001. https://doi.org/10.1016/j.scitotenv.2020.138001.

Sharma, D. and Ravindran, C. (2020). Diseases and pathogens of marine invertebrate corals in Indian reefs. Journal of Invertebrate Pathology 173(107373): 107373. https://doi.org/10.1016/j.jip.2020.107373.

Sheaves, M. (2009). Consequences of ecological connectivity: the coastal ecosystem mosaic. Marine Ecology Progress Series 391: 107–115. https://doi.org/10.3354/meps08121.

Shi, D., Zhao, C., Chen, Y., Ding, J., Zhang, L. and Chang, Y. (2020). Transcriptomes shed light on transgenerational and developmental effects of ocean warming on embryos of the sea urchin Strongylocentrotus intermedius. Scientific Reports 10(1): 7931. https://doi.org/10.1038/s41598-020-64872-x.

Shlesinger, T. and Loya, Y. (2019). Breakdown in spawning synchrony: A silent threat to coral persistence. Science (New York, N.Y.) 365(6457): 1002–1007. https://doi.org/10.1126/science.aax0110.

Sibly, R. M., Grimm, V., Martin, B. T., Johnston, A. S. A., Kułakowska, K., Topping, C. J. et al. (2013). Representing the acquisition and use of energy by individuals in agent-based models of animal populations. Methods in Ecology and Evolution 4(2): 151–161. https://doi.org/10.1111/2041-210x.12002.

Siegel, D. A., Kinlan, B. P., Gaylord, B. and Gaines, S. D. (2003). Lagrangian descriptions of marine larval dispersion. Marine Ecology Progress Series 260: 83–96. https://doi.org/10.3354/meps260083.

Siegel, D. A., Mitarai, S., Costello, C. J., Gaines, S. D., Kendall, B. E., Warner, R. R. et al. (2008). The stochastic nature of larval connectivity among nearshore marine populations. Proceedings of the National Academy of Sciences of the United States of America 105(26): 8974–8979. https://doi.org/10.1073/pnas.0802544105.

Sleadd, I. M., Lee, M., Hassumani, D. O., Stecyk, T. M. A., Zeitz, O. K. and Buckley, B. A. (2014). Sub-lethal heat stress causes apoptosis in an Antarctic fish that lacks an inducible heat shock response. Journal of Thermal Biology 44: 119–125. https://doi.org/10.1016/j.jtherbio.2014.06.007.

Sleight, V. A., Peck, L. S., Dyrynda, E. A., Smith, V. J. and Clark, M. S. (2018). Cellular stress responses to chronic heat shock and shell damage in temperate Mya truncata. Cell Stress and Chaperones 23(5): 1003–1017. https://doi.org/10.1007/s12192-018-0910-5.

Smale, D. A. and Wernberg, T. (2013). Extreme climatic event drives range contraction of a habitat-forming species. Proceedings. Biological Sciences 280(1754): 20122829. https://doi.org/10.1098/rspb.2012.2829.

Smale, D. A., Wernberg, T., Oliver, E. C. J., Thomsen, M., Harvey, B. P., Straub, S. C. et al. (2019). Marine heatwaves threaten global biodiversity and the provision of ecosystem services. Nature Climate Change 9(4): 306–312. https://doi.org/10.1038/s41558-019-0412-1.

Smith, K. E., Burrows, M. T., Hobday, A. J., Sen Gupta, A., Moore, P. J., Thomsen, M. et al. (2021). Socioeconomic impacts of marine heatwaves: Global issues and opportunities. Science (New York, N.Y.) 374(6566): eabj3593. https://doi.org/10.1126/science.abj3593.

Smolinský, R. and Gvoždík, L. (2014). Effect of temperature extremes on the spatial dynamics of predator–prey interactions: A case study with dragonfly nymphs and newt larvae. Journal of Thermal Biology 39: 12–16. https://doi.org/10.1016/j.jtherbio.2013.11.004.

Snoeijs-Leijonmalm, P., Flores, H., Sakinan, S., Hildebrandt, N., Svenson, A., Castellani, G. et al. (2022). Unexpected fish and squid in the central Arctic deep scattering layer. Science Advances 8(7). https://doi.org/10.1126/sciadv.abj7536.

Sokolova, I. M. (2013). Energy-limited tolerance to stress as a conceptual framework to integrate the effects of multiple stressors. Integrative and Comparative Biology 53(4): 597–608. https://doi.org/10.1093/icb/ict028.

Sokolova, I. M., Frederich, M., Bagwe, R., Lannig, G. and Sukhotin, A. A. (2012). Energy homeostasis as an integrative tool for assessing limits of environmental stress tolerance in aquatic invertebrates. Marine Environmental Research 79: 1–15. https://doi.org/10.1016/j.marenvres.2012.04.003.

Somero, G. N. (2020). The cellular stress response and temperature: Function, regulation, and evolution. Journal of Experimental Zoology. Part A, Ecological and Integrative Physiology 333(6): 379–397. https://doi.org/10.1002/jez.2344.

Song, J. A. and Choi, C. Y. (2021). Temporal changes in physiological responses of bay scallop: Performance of antioxidant mechanism in Argopecten irradians in response to sudden changes in habitat salinity. Antioxidants (Basel, Switzerland) 10(11): 1673. https://doi.org/10.3390/antiox10111673.

Song, M., Zhao, J., Wen, H.-S., Li, Y., Li, J.-F., Li, L.-M. et al. (2019). The impact of acute thermal stress on the metabolome of the black rockfish (Sebastes schlegelii). PloS One 14(5): e0217133. https://doi.org/10.1371/journal.pone.0217133.

Sørensen, J. G., Kristensen, T. N. and Loeschcke, V. (2003). The evolutionary and ecological role of heat shock proteins. Ecology Letters 6(11): 1025–1037. https://doi.org/10.1046/j.1461-0248.2003.00528.x.

Sottile, M. L. and Nadin, S. B. (2018). Heat shock proteins and DNA repair mechanisms: an updated overview. Cell Stress and Chaperones 23(3): 303–315. https://doi.org/10.1007/s12192-017-0843-4.

Spencer, L. H., Horkan, E., Crim, R. and Roberts, S. B. (2021). Latent effects of winter warming on Olympia oyster reproduction and larval viability. Journal of Experimental Marine Biology and Ecology 542–543(151604): 151604. https://doi.org/10.1016/j.jembe.2021.151604.

Spies, I., Gruenthal, K. M., Drinan, D. P., Hollowed, A. B., Stevenson, D. E., Tarpey, C. M. et al. (2020). Genetic evidence of a northward range expansion in the eastern Bering Sea stock of Pacific cod. Evolutionary Applications 13(2): 362–375. https://doi.org/10.1111/eva.12874.

Staveley, T. A. B., Jacoby, D. M. P., Perry, D., van der Meijs, F., Lagenfelt, I., Cremle, M. et al. (2019). Sea surface temperature dictates movement and habitat connectivity of Atlantic cod in a coastal fjord system. Ecology and Evolution 9(16): 9076–9086. https://doi.org/10.1002/ece3.5453.

Stevenson, D. E. and Lauth, R. R. (2019). Bottom trawl surveys in the northern Bering Sea indicate recent shifts in the distribution of marine species. Polar Biology 42(2): 407–421. https://doi.org/10.1007/s00300-018-2431-1.

Stevenson-Holt, C. D., Watts, K., Bellamy, C. C., Nevin, O. T. and Ramsey, A. D. (2014). Defining landscape resistance values in least-cost connectivity models for the invasive grey squirrel: A comparison of approaches using expert-opinion and habitat suitability modelling. PloS One 9(11): e112119. https://doi.org/10.1371/journal.pone.0112119.

Strano, F., Micaroni, V., Davy, S. K., Woods, L. and Bell, J. J. (2022). Near-future extreme temperatures affect physiology, morphology and recruitment of the temperate sponge Crella incrustans. The Science of the Total Environment 823(153466): 153466. https://doi.org/10.1016/j.scitotenv.2022.153466.

Sunday, J. M., Pecl, G. T., Frusher, S., Hobday, A. J., Hill, N., Holbrook, N. J. et al. (2015). Species traits and climate velocity explain geographic range shifts in an ocean-warming hotspot. Ecology Letters 18(9): 944–953. https://doi.org/10.1111/ele.12474.

Sydeman, W. J., Poloczanska, E., Reed, T. E. and Thompson, S. A. (2015). Climate change and marine vertebrates. Science (New York, N.Y.) 350(6262): 772–777. https://doi.org/10.1126/science.aac9874.

Szegezdi, E., Logue, S. E., Gorman, A. M. and Samali, A. (2006). Mediators of endoplasmic reticulum stress-induced apoptosis. EMBO Reports 7(9): 880–885. https://doi.org/10.1038/sj.embor.7400779.

Takle, H. and Andersen, Ø. (2007). Caspases and apoptosis in fish. Journal of Fish Biology 71(sc): 326–349. https://doi.org/10.1111/j.1095-8649.2007.01665.x.

Talukder, B., Ganguli, N., Matthew, R., vanLoon, G. W., Hipel, K. W. and Orbinski, J. (2022). Climate change-accelerated ocean biodiversity loss and associated planetary health impacts. The Journal of Climate Change and Health 6: 100114. https://doi.org/10.1016/j.joclim.2022.100114.

Thangal, S. H., Nivetha, M., Muttharasi, C., Anandhan, K. and Muralisankar, T. (2021). Effects of acidified seawater on biological and physiological responses of Artemia franciscana. Marine Pollution Bulletin 169(112476): 112476. https://doi.org/10.1016/j.marpolbul.2021.112476.

Thibault, C., Massamba-N'Siala, G., Noisette, F., Vermandele, F., Babin, M. and Calosi, P. (2020). Within- and trans-generational responses to combined global changes are highly divergent in two congeneric species of marine annelids. Marine Biology 167(4). https://doi.org/10.1007/s00227-019-3644-8.

Thomas, A., Ramkumar, A. and Shanmugam, A. (2022). CO_2 acidification and its differential responses on aquatic biota—a review. Environmental Advances 8(100219): 100219. https://doi.org/10.1016/j.envadv.2022.100219.

Thomas, Y., Flye-Sainte-Marie, J., Chabot, D., Aguirre-Velarde, A., Marques, G. M. and Pecquerie, L. (2019). Effects of hypoxia on metabolic functions in marine organisms: Observed patterns and modelling assumptions within the context of Dynamic Energy Budget (DEB) theory. Journal of Sea Research 143: 231–242. https://doi.org/10.1016/j.seares.2018.05.001.

Thompson, R. M., Hemberg, M., Starzomski, B. M. and Shurin, J. B. (2007). Trophic levels and trophic tangles: The prevalence of omnivory in real food webs. Ecology 88(3): 612–617. https://doi.org/10.1890/05-1454.

Tian, L. and Benton, M. J. (2020). Predicting biotic responses to future climate warming with classic ecogeographic rules. Current Biology: CB 30(13): R744–R749. https://doi.org/10.1016/j.cub.2020.06.003.

Tilbrook, B., Jewett, E. B., DeGrandpre, M. D., Hernandez-Ayon, J. M., Feely, R. A., Gledhill, D. K. et al. (2019). An enhanced ocean acidification observing network: From people to technology to data synthesis and information exchange. Frontiers in Marine Science 6. https://doi.org/10.3389/fmars.2019.00337.

Tittensor, D. P., Beger, M., Boerder, K., Boyce, D. G., Cavanagh, R. D., Cosandey-Godin, A. et al. (2019). Integrating climate adaptation and biodiversity conservation in the global ocean. Science Advances 5(11). https://doi.org/10.1126/sciadv.aay9969.

Tomanek, L. (2011). Environmental proteomics: changes in the proteome of marine organisms in response to environmental stress, pollutants, infection, symbiosis, and development. Annual Review of Marine Science 3(1): 373–399. https://doi.org/10.1146/annurev-marine-120709-142729.

Tomanek, L. (2014). Proteomics to study adaptations in marine organisms to environmental stress. Journal of Proteomics 105: 92–106. https://doi.org/10.1016/j.jprot.2014.04.009.

Tomanek, L., Zuzow, M. J., Ivanina, A. V., Beniash, E. and Sokolova, I. M. (2011). Proteomic response to elevated PCO2 level in eastern oysters, Crassostrea virginica: evidence for oxidative stress. The Journal of Experimental Biology 214(Pt 11): 1836–1844. https://doi.org/10.1242/jeb.055475.

Tracey, S. R., Hartmann, K. and Hobday, A. J. (2012). The effect of dispersal and temperature on the early life history of a temperate marine fish. Fisheries Oceanography 21(5): 336–347. https://doi.org/10.1111/j.1365-2419.2012.00628.x.

Tracy, A. M., Pielmeier, M. L., Yoshioka, R. M., Heron, S. F. and Harvell, C. D. (2019). Increases and decreases in marine disease reports in an era of global change. Proceedings. Biological Sciences 286(1912): 20191718. https://doi.org/10.1098/rspb.2019.1718.

Treml, E. A., Halpin, P. N., Urban, D. L. and Pratson, L. F. (2008). Modeling population connectivity by ocean currents, a graph-theoretic approach for marine conservation. Landscape Ecology 23(S1): 19–36. https://doi.org/10.1007/s10980-007-9138-y.

Turgeon, K., Robillard, A., Grégoire, J., Duclos, V. and Kramer, D. L. (2010). Functional connectivity from a reef fish perspective: behavioral tactics for moving in a fragmented landscape. Ecology 91(11): 3332–3342. https://doi.org/10.1890/09-2015.1.

Ullah, H., Fordham, D. A. and Nagelkerken, I. (2021). Climate change negates positive CO_2 effects on marine species biomass and productivity by altering the strength and direction of trophic interactions. The Science of the Total Environment 801(149624): 149624. https://doi.org/10.1016/j.scitotenv.2021.149624.

Ullah, H., Nagelkerken, I., Goldenberg, S. U. and Fordham, D. A. (2018). Climate change could drive marine food web collapse through altered trophic flows and cyanobacterial proliferation. PLoS Biology 16(1): e2003446. https://doi.org/10.1371/journal.pbio.2003446.

United Nations: Department of Economic and Social Affairs. (2022). The sustainable development goals report 2022 The sustainable development goals report 2022. United Nations.

Valdivieso, A., Wilson, C. A., Amores, A., da Silva Rodrigues, M., Nóbrega, R. H., Ribas, L. et al. (2022). Environmentally-induced sex reversal in fish with chromosomal vs. polygenic sex determination. Environmental Research 213(113549): 113549. https://doi.org/10.1016/j.envres.2022.113549.

Valiñas, M. S. and Walter Helbling, E. (2016). Metabolic and behavioral responses of the reef fish Patagonotothen cornucola to ultraviolet radiation: Influence of the diet. Journal of Experimental Marine Biology and Ecology 474: 180–184. https://doi.org/10.1016/j.jembe.2015.10.011.

van der Meer, J. (2019). Metabolic theories in ecology: The dynamic energy budget theory and the metabolic theory of ecology. pp. 463–471. *In*: Encyclopedia of Ecology. Elsevier. https://doi.org/10.1016/b978-0-12-409548-9.10568-8.

Van der Putten, W. H., Macel, M. and Visser, M. E. (2010). Predicting species distribution and abundance responses to climate change: why it is essential to include biotic interactions across trophic levels. Philosophical Transactions of the Royal Society of London. Series B, Biological Sciences 365(1549): 2025–2034. https://doi.org/10.1098/rstb.2010.0037.

van Gennip, S. J., Popova, E. E., Yool, A., Pecl, G. T., Hobday, A. J. and Sorte, C. J. B. (2017). Going with the flow: the role of ocean circulation in global marine ecosystems under a changing climate. Global Change Biology 23(7): 2602–2617. https://doi.org/10.1111/gcb.13586.

van Raaij, M. T., Pit, D. S., Balm, P. H., Steffens, A. B. and van den Thillart, G. E. (1996). Behavioral strategy and the physiological stress response in rainbow trout exposed to severe hypoxia. Hormones and Behavior 30(1): 85–92. https://doi.org/10.1006/hbeh.1996.0012.

Vargas, C. A., Cuevas, L. A., Broitman, B. R., San Martin, V. A., Lagos, N. A., Gaitán-Espitia, J. D. et al. (2022). Upper environmental pCO_2 drives sensitivity to ocean acidification in marine invertebrates. Nature Climate Change 12(2): 200–207. https://doi.org/10.1038/s41558-021-01269-2.

Vargas, C. A., Lagos, N. A., Lardies, M. A., Duarte, C., Manríquez, P. H., Aguilera, V. M. et al. (2017). Species-specific responses to ocean acidification should account for local adaptation and adaptive plasticity. Nature Ecology and Evolution 1(4): 84. https://doi.org/10.1038/s41559-017-0084.

Vasquez, M. C., Houston, C. T., Alcantar, C. Y., Milshteyn, L., Brazil, C. A. and Zepeda, O. G. (2022). Interactive effects of multiple stressors on the physiological performance of the invasive mussel Mytilus galloprovincialis. Marine Environmental Research 178(105665): 105665. https://doi.org/10.1016/j.marenvres.2022.105665.

Vasseur, D. A. and McCann, K. S. (2005). A mechanistic approach for modeling temperature-dependent consumer-resource dynamics. The American Naturalist 166(2): 184–198. https://doi.org/10.1086/431285.

Vedor, M., Queiroz, N., Mucientes, G., Couto, A., Costa, I. da, Santos, A. D. et al. (2021). Climate-driven deoxygenation elevates fishing vulnerability for the ocean's widest ranging shark. ELife 10. https://doi.org/10.7554/eLife.62508.

Veilleux, H. D., Ryu, T., Donelson, J. M., Ravasi, T. and Munday, P. L. (2018). Molecular response to extreme summer temperatures differs between two genetically differentiated populations of a coral reef fish. Frontiers in Marine Science 5. https://doi.org/10.3389/fmars.2018.00349.

Veilleux, H. D., Ryu, T., Donelson, J. M., van Herwerden, L., Seridi, L., Ghosheh, Y. et al. (2015). Molecular processes of transgenerational acclimation to a warming ocean. Nature Climate Change 5(12): 1074–1078. https://doi.org/10.1038/nclimate2724.

Venn, A. A., Tambutté, E., Holcomb, M., Laurent, J., Allemand, D. and Tambutté, S. (2013). Impact of seawater acidification on pH at the tissue-skeleton interface and calcification in reef corals. Proceedings of the National Academy of Sciences of the United States of America 110(5): 1634–1639. https://doi.org/10.1073/pnas.1216153110.

Vergés, A., Steinberg, P. D., Hay, M. E., Poore, A. G. B., Campbell, A. H., Ballesteros, E. et al. (2014). The tropicalization of temperate marine ecosystems: climate-mediated changes in herbivory and community phase shifts. Proceedings. Biological Sciences 281(1789): 20140846. https://doi.org/10.1098/rspb.2014.0846.

Verlecar, X. N., Jena, K. B. and Chainy, G. B. N. (2007). Biochemical markers of oxidative stress in Perna viridis exposed to mercury and temperature. Chemico-Biological Interactions 167(3): 219–226. https://doi.org/10.1016/j.cbi.2007.01.018.

Vinagre, C., Madeira, D., Mendonça, V., Madeira, C. and Diniz, M. S. (2021). Warming in shallow waters: Seasonal response of stress biomarkers in a tide pool fish. Estuarine, Coastal and Shelf Science 251(107187): 107187. https://doi.org/10.1016/j.ecss.2021.107187.

Vizzini, S., Martínez-Crego, B., Andolina, C., Massa-Gallucci, A., Connell, S. D. and Gambi, M. C. (2017). Ocean acidification as a driver of community simplification via the collapse of higher-order and rise of lower-order consumers. Scientific Reports 7(1). https://doi.org/10.1038/s41598-017-03802-w.

Voigt, W., Perner, J., Davis, A. J., Eggers, T., Schumacher, J., Bährmann, R. et al. (2003). Trophic levels are differentially sensitive to climate. Ecology 84(9): 2444–2453. https://doi.org/10.1890/02-0266.

Wang, C., Yue, X., Lu, X. and Liu, B. (2013). The role of catalase in the immune response to oxidative stress and pathogen challenge in the clam Meretrix meretrix. Fish and Shellfish Immunology 34(1): 91–99. https://doi.org/10.1016/j.fsi.2012.10.013.

Wang, P., Liu, H., Zhao, S., Yu, S., Xie, S., Hua, S. et al. (2022). Hypoxia stress affects the physiological responses, apoptosis and innate immunity of Kuruma shrimp, Marsupenaeus japonicus. Fish and Shellfish Immunology 122: 206–214. https://doi.org/10.1016/j.fsi.2022.02.016.

Wang, Q., Cao, R., Ning, X., You, L., Mu, C., Wang, C. et al. (2016). Effects of ocean acidification on immune responses of the Pacific oyster Crassostrea gigas. Fish and Shellfish Immunology 49: 24–33. https://doi.org/10.1016/j.fsi.2015.12.025.

Wang, X., Li, C., Jia, Z., Xu, T., Wang, Y., Sun, M. et al. (2021). Regulation of apoptosis by Pacific oyster Crassostrea gigas reveals acclimation strategy to CO_2 driven acidification. Ecotoxicology and Environmental Safety 217(112235): 112235. https://doi.org/10.1016/j.ecoenv.2021.112235.

Wassmann, P., Duarte, C. M., Agustí, S. and Sejr, M. K. (2011). Footprints of climate change in the Arctic marine ecosystem. Global Change Biology 17(2): 1235–1249. https://doi.org/10.1111/j.1365-2486.2010.02311.x.

Watson, S.-A., Fields, J. B. and Munday, P. L. (2017). Ocean acidification alters predator behaviour and reduces predation rate. Biology Letters 13(2): 20160797. https://doi.org/10.1098/rsbl.2016.0797.

Watson, S.-A., Lefevre, S., McCormick, M. I., Domenici, P., Nilsson, G. E. and Munday, P. L. (2014). Marine mollusc predator-escape behaviour altered by near-future carbon dioxide levels. Proceedings. Biological Sciences 281(1774): 20132377. https://doi.org/10.1098/rspb.2013.2377.

Wernberg, T., Bennett, S., Babcock, R. C., de Bettignies, T., Cure, K., Depczynski, M. et al. (2016). Climate-driven regime shift of a temperate marine ecosystem. Science (New York, N.Y.) 353(6295): 169–172. https://doi.org/10.1126/science.aad8745.

Wernberg, T., Smale, D. A., Tuya, F., Thomsen, M. S., Langlois, T. J., de Bettignies, T. et al. (2013). An extreme climatic event alters marine ecosystem structure in a global biodiversity hotspot. Nature Climate Change 3(1): 78–82. https://doi.org/10.1038/nclimate1627.

White, J. W., Oregon State University, Carr, M., Caselle, J., Washburn, L., Woodson, C. B., Palumbi, S. et al. (2019). Connectivity, dispersal, and recruitment: Connecting benthic communities and the coastal ocean. Oceanography (Washington, D.C.) 32(3): 50–59. https://doi.org/10.5670/oceanog.2019.310.

Whitley, D., Goldberg, S. P. and Jordan, W. D. (1999). Heat shock proteins: a review of the molecular chaperones. Journal of Vascular Surgery 29(4): 748–751. https://doi.org/10.1016/s0741-5214(99)70329-0.

Widdicombe, S., Isensee, K., Artioli, Y., Gaitán-Espitia, J. D., Hauri, C., Newton, J. A. et al. (2022). Unifying biological field observations to detect and compare ocean acidification impacts across marine species and ecosystems: What to monitor and why. https://doi.org/10.5194/egusphere-2022-907.

Williams, J. J., Papastamatiou, Y. P., Caselle, J. E., Bradley, D. and Jacoby, D. M. P. (2018). Mobile marine predators: an understudied source of nutrients to coral reefs in an unfished atoll. Proceedings. Biological Sciences 285(1875): 20172456. https://doi.org/10.1098/rspb.2017.2456.

Wilson, R. A. and Barker, M. J. (2021). Biological Individuals. *In*: Zalta, E. N. (ed.). The Stanford Encyclopedia of Philosophy (Winter 2021). Metaphysics Research Lab, Stanford University. https://plato.stanford.edu/archives/win2021/entries/biology-individual/.

Worm, B., Barbier, E. B., Beaumont, N., Duffy, J. E., Folke, C., Halpern, B. S. et al. (2006). Impacts of biodiversity loss on ocean ecosystem services. Science (New York, N.Y.) 314(5800): 787–790. https://doi.org/10.1126/science.1132294.

Xin, Y., Yang, Z., Zhu, Y., Li, Y., Yu, J., Zhong, W. et al. (2022). Hypoxia induces oxidative injury and apoptosis via mediating the Nrf-2/hippo pathway in blood cells of largemouth bass (Micropterus salmoides). Frontiers in Ecology and Evolution 10. https://doi.org/10.3389/fevo.2022.841318.

Xu, G., Kong, H., Chang, X., Dupont, S., Chen, H., Deng, Y. et al. (2021). Gonadal antioxidant responses to seawater acidification and hypoxia in the marine mussel Mytilus coruscus. Environmental Science and Pollution Research International 28(38): 53847–53856. https://doi.org/10.1007/s11356-021-14584-0.

Yan, W., Qiao, Y., He, J., Qu, J., Liu, Y., Zhang, Q. et al. (2022). Molecular mechanism based on histopathology, antioxidant system and transcriptomic profiles in heat stress response in the gills of japanese flounder. International Journal of Molecular Sciences 23(6). https://doi.org/10.3390/ijms23063286.

Yap, K. N., Yamada, K., Zikeli, S., Kiaris, H. and Hood, W. R. (2021). Evaluating endoplasmic reticulum stress and unfolded protein response through the lens of ecology and evolution. Biological Reviews of the Cambridge Philosophical Society 96(2): 541–556. https://doi.org/10.1111/brv.12667.

Young, E. F., Tysklind, N., Meredith, M. P., de Bruyn, M., Belchier, M., Murphy, E. J. et al. (2018). Stepping stones to isolation: Impacts of a changing climate on the connectivity of fragmented fish populations. Evolutionary Applications 11(6): 978–994. https://doi.org/10.1111/eva.12613.

Zanuzzo, F. S., Beemelmanns, A., Hall, J. R., Rise, M. L. and Gamperl, A. K. (2020). The innate immune response of Atlantic salmon (Salmo salar) is not negatively affected by high temperature and moderate hypoxia. Frontiers in Immunology 11. https://doi.org/10.3389/fimmu.2020.01009.

Zhang, H., Gong, W., Wu, S. and Perrett, S. (2022). Hsp70 in redox homeostasis. Cells (Basel, Switzerland) 11(5): 829. https://doi.org/10.3390/cells11050829.

Zhao, X., Li, L., Li, C., Liu, E., Zhu, H. and Ling, Q. (2022). Heat stress-induced endoplasmic reticulum stress promotes liver apoptosis in largemouth bass (Micropterus salmoides). Aquaculture (Amsterdam, Netherlands) 546(737401): 737401. https://doi.org/10.1016/j.aquaculture.2021.737401.

Zheng, C., Zhao, Q., Li, E., Zhao, D. and Sun, S. (2022). Role of hypoxia in the behaviour, physiology, immunity and response mechanisms of crustaceans: A review. Reviews in Aquaculture 14(2): 676–687. https://doi.org/10.1111/raq.12618.

Zou, D., Ning, J., Lu, X., Wang, X., Chen, M., Liu, B. et al. (2021). Physiological and transcriptional responses to acute and chronic thermal stress in the ark shell Scapharca subcrenata. Frontiers in Marine Science 8. https://doi.org/10.3389/fmars.2021.739662.

CHAPTER 2

Impact of Fisheries on Marine Ecosystems

Raúl A. González,[1,2,*] *Maite A. Narvarte,*[1,2] *Magdalena Arias,*[1,2] *M. Soledad Avaca*[1,2] and *Enrique A. Crespo*[1,3]

Introduction: Fishing around the world

On a global level, the agrifood systems produce some 11 billion tons of food each year to form the backbone of many economies (FAO 2021). Within these systems, fisheries contribute an estimated 179 million tons, with a total first-sale value estimated at $US 401 billion (FAO 2020). Marine fishing constitutes 88% of global fish catches, and fish provides approximately 17% of the world population's intake of animal protein.

In the last six decades, the global consumption of seafood increased at an average annual rate of 3.1%, one higher than that of all other animal-protein foods and almost twice that of the annual world-population growth (1.6%), whereas *per-capita* food-fish consumption grew from 9.0 kg to 20.5 kg (about 1.5% per year) for the same period (FAO 2020).

Around 60 million people are directly engaged in the primary sector of capture fishing (40 million people) plus aquaculture (20 million people), while women accounted for 14% of the total. When the data from postharvest operations are included, half of the workers in the sector have been estimated to be women. Most of the people involved in primary production are in developing countries, and most are devoted to small-scale, artisanal fishing and aquaculture (FAO 2020). A recent report from the United Nations (United Nations 2022) pointed out that about 35–38% of total fishing and aquaculture production is exported, valued at $173 billion annually, with seafood trade having increased 6.6% in nominal terms and 4.1% in real terms annually between 1976 and the present.

These basic data illustrate not only the significance of fishing on a global level, but also the strategic role that that occupation has had in developing countries in providing essential income and food to those societies. In many nations and regions of the world, particularly in the developing countries, capture fishing constitutes the basic source of economic livelihood and food security for the people, while the industrialization of fish production contributes to the generation of intensive employment and exports.

Without a doubt, the great impact that fishing has on the well-being of humanity at a global level explains to a large extent the reason why fishing is a sector of growing importance on the world economy.

[1] Centro de Investigación Aplicada y Transferencia Tecnológica en Recursos Marinos, Almirante Storni - Escuela Superior de Ciencias Marinas, Universidad Nacional del Comahue, San Martín 247, (R8520BNE) San Antonio Oeste, Río Negro, Argentina.

[2] Consejo Nacional de Investigaciones Científicas y Técnicas (CONICET).

[3] Centro para el Estudio de Sistemas Marinos, Boulevard Alte. Brown 2915, (U9120ACD) Puerto Madryn - Chubut - Argentina.

* Corresponding author: racg05@gmail.com

Global fishing, however, currently faces a variety of threats that puts its continuity and sustainability at risk. Human stressors, including overfishing, habitat destruction, biodiversity loss, excessive nutrient loads, marine pollution and also irresponsible tourism, are affecting the entire ocean and fishery productivity. In addition, climate change increases the intensity and frequency of natural hazards—i.e., ocean warming, acidity, and deoxygenation—and causes shifts in the distribution and abundance of fish populations, thus adversely affecting fishing-dependent developing countries (IPCC 2019).

The effects of fishing on marine ecosystems have been the subject of controversy in recent decades, with the debate mainly focused on two main themes: the so called "depredatory fishing"—i.e., Chinese jiggers fishing fleets targeting squid in Southwest Atlantic Ocean, in the Patagonian Continental Shelf, and near the Galapagos Islands—and the effects of bottom trawling on the structure of the seabed and benthic biodiversity. Although the first of these issues has been dominating the environmental media debate, the problem of bottom trawling has received more exhaustive attention from the scientific community (Kaiser et al. 1998, National Research Council 2002, Løkkeberg 2005, Hiddink et al. 2017, Pitcher et al. 2022 among others).

In the history of bottom trawling, two stages have been recognized: the first from the beginning of the 18th Century to 1920, dominated by sailboats, and the second from 1930 onwards, with the adoption of the internal-combustion engine. Bottom trawling began to be used in Europe around 1820 (Thurstan et al. 2013) and in the early 20th Century in the USA (National Research Council 2002). After the Second World War, trawl fishing (bottom and midwater) underwent a massive development around the world, fostered by new technologies: the manufacture of nets with synthetic fibers and the development of echosounders and the GPS for the respective improvement of fish-finding and navigation. This expansion has meant that bottom trawling is currently the fishing method *par excellence*, contributing about 24% worldwide (*ca.* 19 million tons) of global landings of caught wild fish and shellfish (Kaiser 2019).

The impact of fishing on marine ecosystems, and particularly the effects of bottom trawling, has generated so much concern in the fisheries administration of certain countries that the practice has been the subject of analyses by academia for the last four decades (National Research Council 2002). During this time, the issue has been the subject of a continuous debate at the global level, which is still ongoing, including the establishment of specific regulations[1] in developed countries (i.e., the USA and the EU) to limit operations with bottom-trawling gear.

Fishing affects marine habitats and ecosystems in many ways, depending on the spatial extent, the level of the effort, and the type of gear used (Auster and Langton 1999, National Research Council 2002). Beyond the type of fishing gear used, the effects of fishing on the marine ecosystems may be direct—for example, through the removal of part of the biomass or of certain age classes of target species or through the bycatch (unintentionally caught associated species), or when mobile fishing gear (e.g., bottom-trawling nets and dredges) disturbs the structure of the seafloor so as to affect the physical environment (e.g., relief, sedimentology, the habitat complexity and availability) and the composition and biodiversity of benthic communities (Fig. 1).

Indirect effects result when fishing causes a disruption of the food web, an alteration in the rate of decomposition of organic matter, and/or a recycling of nutrients through resuspension of bottom sediments (National Research Council 2002). The debate on this topic seems endless, with a recent study (Oberle et al. 2016) extending the controversy of trawling and dredging beyond the fishing itself and any ecological interest, to the point of making this issue a problem from a geological point of view.

In this chapter, we present a review of the present knowledge dealing with the impact of fishing on marine ecosystems, focusing on the main aspects that have been at the center of the debate in recent decades. For that purpose, we first explore the available information referring to the direct effects—namely, the magnitude of the catches in relation to the biologic productivity of the stocks, the problem of bycatch and discards, and the mechanical effect and disturbance of bottom trawling on the seafloor structure and dynamics. Thereafter, we refer to the indirect impacts of fishing, such as the alteration of trophic webs and the effects on top predators. Certain instances and examples are commented on

---

[1] https://oceans-and-fisheries.ec.europa.eu/news/fisheries-end-bottom-fishing-protected-deep-sea-ecosystems-eu-waters-2022-09-15_en.

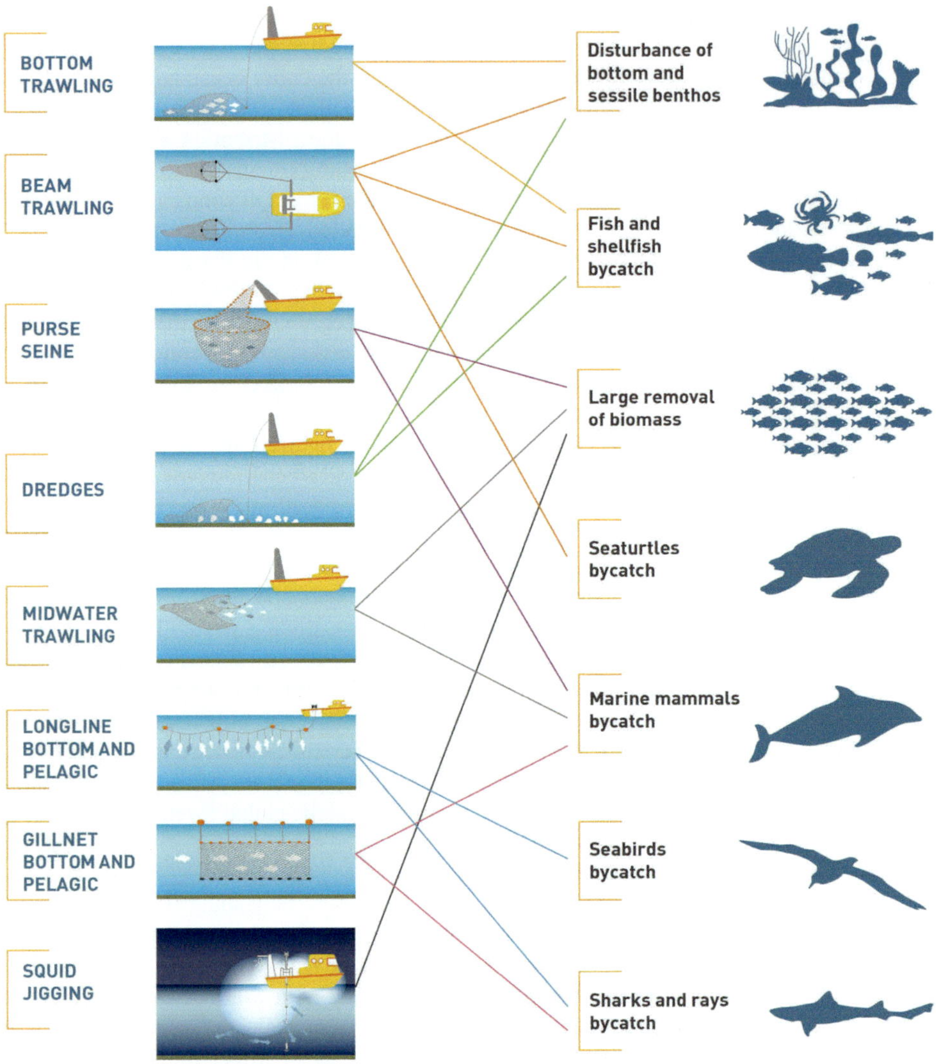

Figure 1. Main interactions and direct impacts of different types of fishing gear with marine ecosystem components. The fishing modalities on the left can cause any one or more of the adverse effects on the fishery environment exemplified on the right.

and analyzed in order to expose the scope of the problem within the context of the relevant regional fisheries. Finally, the socioeconomic implications are discussed along with the efforts that science and management are making at the global level to prevent or mitigate the impacts of fishing in general on marine ecosystems.

Directs effects of fishing on the marine ecosystems

Biomass removal

The most direct and immediate effect of fishing on marine ecosystems is the removal of biomass of one or more species by the fishing gear. In certain instances, fishing can be so selective and specific that catches are made up almost exclusively of a single species. This is the case of the squid fishing using jiggers, or of the small pelagic fishing using purse seines, in which incidental catches of other species are either very low or practically null. In other instances, fishing is so nonselective that catches of the target species are usually accompanied by a significant number of other species, and especially in so-called multispecies fishing, where the catch targets themselves are multiple species. For example, bottom and midwater trawling respectively target demersal or demersal-pelagic species around the world, in which the fishing gears used (mainly nets) have a very low selectivity[2] for species or size. This scenario also applies to the fishing targeting shrimp and prawn with beam trawls and the dredging fisheries targeting bivalve species (e.g., scallops, mussels, clams).

The global biomass removal by different types of fishing reached 84.4 million tons in 2018, with the total marine catches having been relatively stable since the mid-2000s and ranging from 78 million tons to 81 million tons per year, with a decline in the peak catches occurring in the late 1990s (FAO 2020). Beyond these official data provided by the FAO, recent reconstructions of world-fishing statistics (Zeller et al. 2016) have indicated that global catches have been above 100 million tons per year since the 1990s. Thus, the total global fish catches for the period 1950–2010 have been underreported by around 50% by what is documented in the official data, though the statistics have decreased considerably since the peak in the mid-1990s (Zeller et al. 2016).

On a global scale, the greatest removal of biomass from a stock by fishing has occurred for several decades with the anchovy *Engraulis ringens* in the Humboldt Current System off the coasts of Perú and Chile. In 2018, catches of anchovy reached just over 7.0 million tons, amounting to 10% of global finfish catches, where the latter reached 71.9 million tons per year, constituting 85% of the annual catches of all species (FAO 2020). Other notable biomass removal on a global level is the Alaska pollock *Gadus chalcogrammus* at 3.4 million tons, the Skipjack tuna *Katsuwonus pelamis* at 3.1 million tons, and the Atlantic herring *Clupea harengus* at 1.8 million tons (5%, 4%, and 3% of global finfish catches, respectively). The magnitude of these catches on a worldwide scale is significant in relation to the rest of the marine species, which all together add up to 12.4 million tons, in addition to 7.1 and 7.0 million tons of crustaceans and mollusks, respectively (FAO 2020).

These amounts of biomass captured in these fisheries do not attest to anything by themselves about the impact of fishing on marine ecosystems if they are not stated with reference to the biomass of their stocks—that is, whether or not the extraction of biomass is duly maintained within the limits of biologic productivity. Therefore, the information provided periodically by the FAO on the state of fishery resources (SOFIA reports) includes references to the number of stocks that are exploited within the limits of their biologic capabilities, as well as those that are overexploited. According to the last of these reports (FAO 2020), 78.7% of current landings come from biologically sustainable stocks. Nevertheless, the fish stocks exploited within biologically sustainable levels have decreased from 90% in 1974 to 65.8% in 2017. Likewise, the number of stocks fished at biologically unsustainable levels have increased from 10% (1974) to 34.2% (2017).

Other recent studies (Palomares et al. 2020) analyzed the long-term (1950–2014) patterns of fishery biomasses for 1,320 fish and invertebrate populations, including 483 species exploited by fishing in the 232 coastal Marine Ecoregions around the world. For that purpose, as an indicator of fish-population health, they used the population-biomass status (B) relative to the optimal biomass level for Maximum Sustainable Yield (BMSY), the B/BMSY. A most notable result was that they found a consistent decline in the biomass of fishing-exploited populations that encompassed virtually all the climatic zones and ocean basins in the world. Consequently, they found that for the last five-years of the data series analyzed (2010–2014), the B/BMSY indicated that only 18% of all the assessed populations might be deemed "healthy"—i.e., biomasses of exploited populations above the level deemed optimal for the MSY—whereas the 82% of the populations remaining are in various states of depletion relative to the biomass levels regarding MSY. More than 8% of these populations are currently in the "very bad" category with biomass levels less than 20% of the level that might maximize sustainable fishing catches.

Beyond the information on the biomass removed by fishing at a global level, knowing how much biomass is removed annually by each type of fishing gear worldwide is essential. Information on the amount of biomass captured by the different types of fishing gear also enables an estimate of the extent of

[2] A definition of selectivity is provided later (see section "Efforts to Prevent the Negative Impacts of Fishing").

other direct effects on the fisheries and their environment associated with the typical operations of certain types of fishing gear (e.g., bottom trawling). In this respect, Watson et al. (2006) reported that during the second half of the 20th century, a dramatic rise occurred in the catch taken with both seines and midwater trawling gear, with both types producing 50–60% of catches worldwide during that period. Mobile gear contacting the seabed (e.g., bottom trawling nets and dredging gear) has risen steadily, accounting 20–25% of the total catches, whereas hook-and-line gear have increased to levels similar to those of gillnets, with each contributing around 10% of the global catches (Watson et al. 2006).

Bycatch and discards

Up to this point, we have focused on the problem of biomass removal by fishing a given target species exclusively. With a few exceptions, however, fishing gear also catches specimens of other species unintentionally (the bycatch), some of which are unwanted because they have no commercial value. In addition to this side effect, certain fishing gear (e.g., trawling nets) have a very low selectivity by size and capture juvenile specimens of the commercially valuable species. In general, these fish neither have commercial value nor is their capture prohibited by technical regulations (i.e., the imposition of a legal minimum size). Unwanted bycatch species, juveniles without commercial value, damaged catches, or prohibited species make up the fishing discards (Fig. 2). This means that even in well-managed fisheries, bycatch discarding is a multicausal practice, as fishermen cannot allocate it to individual catch quotas. Discarding practices negatively affect the sustainable exploitation of marine-biologic resources, marine ecosystems, and the financial viability of fishing.[3]

Perez Roda et al. (2019) estimated that the annual discards from global marine-capture fishing between 2010 and 2014 amounted to 9.1 million tons. Other estimates independent of the official FAO data indicated that global discards peaked at 19 million tons in 1989, only declining to levels below 10 million tons in recent decades, while the largest part (93%) of the global discard was caused by industrial fleets (Zeller et al. 2018). Discard rates vary widely among gear types (Table 1) and around 45.5% (4.2 million tons) of total annual discards have come from bottom-trawling gear—including otter trawls, shrimp trawls, pair-bottom trawls, twin otter trawls, and beam trawls. In contrast, relatively low discard rates (4.7%) were associated with purse seines, but because of the large volume of catches from these types of gear, the total discards were estimated at up to 1.02 million tons.

Likewise, Gray and Kennelly (2018) reported at least 20 million individuals of endangered, threatened, and/or protected (ETP) species interacted with capture fishing annually. ETP species captured

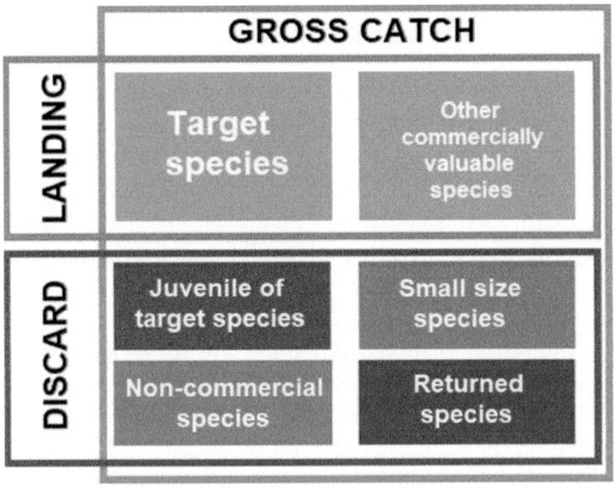

Figure 2. Composition of gross catch, landing, and discard in fishing.

[3] https://oceans-and-fisheries.ec.europa.eu/fisheries/rules/discarding-fisheries_en.

Table 1. Discard rates (ordered from smallest to largest) and annually discarded volumes for the most important gears categories. Modified from Perez Roda et al. (2019).

Fishing gear	Discard rate[1]	Discard (millions of tons)
Purse seine	4.7	1.02
Otter trawl, midwater	12.1	0.881
Longline (mainly bottom)	12.3	0.36
Dredges	13.6	0.2
Gillnets (surface and bottom)	17.4	0.1
Trawl, pair, midwater	19.2	0.06
Boat seines	23.5	0.48
Otter trawl, bottom	30.9	2.39
Trawl, shrimp	54.9	0.836

[1] Tons of discards*100/tons of total catch

and discarded each year in fishing worldwide comprises one million seabirds, 8.5 million sea turtles, 225,000 sea snakes, 650,000 marine mammals, and 10 million sharks (Perez Roda et al. 2019).

A review by Pott and Wiedenfeld (2017) reported for the period 1974–2015 that global seabird bycatch affected 378 marine and coastal bird species in 18 different types of gear. Four families (Gaviidae, Podicipedidae, Diomedeidae, and Sulidae) were the most accidentally caught groups, and gillnets (both set and drifting) and hook gear (longline and handlines) had the greatest number of documented species interactions (the four added together constituted the bycatch of 193 seabird species). Certain types of fishing gear, however—such as industrial seine nets—and the artisanal fishing sector are deemed poorly studied by these authors in relation to the seabird-bycatch patterns worldwide.

Marine reptilians are usually caught by different types of fishing gear throughout the world. Wallace et al. (2010) estimated the global annual bycatch of sea turtles to be 8.5 million individuals. Seven species of these turtles, from which three are considered endangered or critically endangered, are accidentally caught in trawling, longline, and gillnet fishing. Likewise, 12–14 sea-snake species are regularly caught by bottom trawls targeting penaeid shrimps in tropical regions throughout the Indo-Pacific Ocean (Gray and Kennelly 2018). The number of specimens captured annually was estimated between 80,000 and 120,000 between 1990 and 2000 (Wassenberg et al. 1994, 2001, Courtney et al. 2010).

Marine mammals interact with many different types of human activities of which fishing is perhaps the most important. The interaction occurs either by accident or because those mammals pursue the same resources as man. Primitive societies used to catch marine mammals with harpoons, spears, clubs, traps, and a variety of designs were used in most coastal cultures (Crespo 2022). Although whaling started very early, it was not until the industrial period when most of the whale populations everywhere decreased, and more recently in the mid-20th century, the introduction of gillnets started the risk of entanglement and incidental mortality of coastal cetaceans and pinnipeds. The entanglement in gillnets and other passive gear around the world in artisanal fishing, in addition to the incidents recorded in large-scale pelagic fishing as with purse-seines and driftnets, became the major cause of mortality via suffocation for stocks of dolphins, whales, and seals in the 1960–1980 period and in certain locations remains so today.

Marine-mammal bycatch occurs throughout the world and throughout various fisheries, with hotspots identified in the eastern Pacific, Mediterranean, and Atlantic oceans (Lewison et al. 2014). The most notorious example of incidental catches of marine mammals, known as the "tuna-dolphin problem in the eastern Pacific Ocean", occurred in the mid-60s, when half million of individuals of several dolphin species were taken in the purse-seine fishery targeting yellow-fin tuna *Thunnus albacares* (Francis et al. 1992, Hall et al. 2000).

Seals, sea lions and dolphins, interact with fishing gears to forage for prey, behavior which becomes dangerous increasing the risk of entanglement in gillnets and trapped in trawling, where they finally die by drowning (Read 2008, Reeves et al. 2013). Gillnets appear widely as the most dangerous type of fishing gear for marine mammals around the world, including sirenians (Reeves et al. 2013). Read et al. (2006) estimated that, in the early 90s, around 650,000 marine mammals (mainly seals and dolphins) were caught per year as bycatch in gillnet fisheries, while Reeves et al. (2013) reported that several hundred thousand marine mammals were accidentally captured in gillnets annually. Despite the implementation of marine-mammal–bycatch-monitoring programs in several important fisheries, the information worldwide is still both incomplete and fragmented. Furthermore, except for specific examples, the mortality of these species in small-scale fishing is still unknown (Gray and Kennelly 2018).

Chondrichthyan (sharks, skates, rays, sawfish, and chimaeras) are usually caught by bottom trawling along with pelagic longline and gillnet fisheries around the world. The bycatch of sharks and rays occurs globally both in large- and small-scale fisheries, and according to Gray and Kennelly (2018), the total annually discarded number of ETP sharks would approximate 176,000 tons (9.76 million individuals). We need to note that in the example of groups of species such as sharks, rays, and chimaeras, reliable statistics are still lacking worldwide since a significant proportion of the catches are not reported at the species level (FAO 2021). Thus, the number of these ETP sharks that is being discarded globally is still significant in relation to the total catches of sharks which, according to Clarke et al. (2006), would have reached 1.7 million tons each year in the 1996–2000 period.

Impact of fishing on bottom structure and benthic communities

The benthos comprises the set of the flora and fauna occurring on, in, or close to the bottom substrata of the ocean floor. As a primary selective criterion, the nature of the substratum can determine the biotic composition of the benthos (Encyclopedia of Environmetrics 2012). Benthic organisms may be sessile, attached to a firm surface such as rocks, or mobile, moving freely on or in the bottom sediment. Among the latter, the benthic fish may move freely in the water immediately above the substratum, exploiting that surface for food in a foraging strategy or as a refuge against predators. Benthos may be autotrophic such as seagrasses and algae, or heterotrophic preying upon other organisms, or maybe filter feeders (ingesting organic matter or phytoplankton), or decomposers such as bacteria. Benthos is present in all habitats, from the intertidal to the abyssal plain. Among the benthic organisms, those habitat-forming species like shellfish, macroalgae, plants, corals, or sponges add structure to the seafloor and increase habitat complexity (Airoldi et al. 2008), which improves the survival of many fish species, also serving as a nursery for juvenile fish and providing refuge and food for adults (Collie et al. 2000a, b, Kaiser et al. 2017, Sciberras et al. 2018). Thus, the ecologic roles of the benthos include the formation habitat structures, the oxygenation of the seafloor, and the reworking of sediments.

Because bottom trawling causes adverse effects on seabed habitats, notable concerns have arisen over the resulting impact of this form of fishing on the structure and functioning of benthic ecosystems (National Research Council 2002, Sciberras et al. 2018, and references therein). These effects include, among others—and through the progression of state changes (National Research Council 2002, Smith et al. 2016)—a decrease in the biomass and biodiversity of the benthic ecosystem, a reduction in the complexity of seabed habitats (Collie et al. 2000a, Kaiser et al. 2006, Buhl-Mortensen et al. 2016), and alterations in the functioning and productivity of the benthic ecosystem (Jennings et al. 2001a, Hiddink et al. 2011, van Denderen et al. 2015, Pusceddu et al. 2014). Recent studies have reviewed our current understanding of the extent of trawling and its impacts on seabed fauna around the world (Kaiser et al. 2018, Sciberras et al. 2018, Hiddink et al. 2020, Mazor et al. 2021).

Benthic species are commonly used as indicators for measuring ecosystem health or disturbance (Solan et al. 2004, Hiddink et al. 2020, Mazor et al. 2021). The Ecosystem Approach to Fisheries management requires managers to consider the environmental impacts of fishing in management planning (Pikitch et al. 2004, Rice et al. 2012). To that end, information on the distribution and impact of bottom trawling and the status of biota and habitats are needed in order to develop indicators of seafloor integrity. In that regard, different indicators have been evaluated (Hiddink et al. 2020) with the conclusion that the effects of trawling are stronger on whole-community numbers and biomass than for individual taxa, and that certain commonly used community attributes such as species richness and evenness are not good (or even bad) as indicators for monitoring the impact of bottom trawling on the seafloor.

Benthic bottom gears are highly variable in their dimensions between regions and types of fishing, both reflecting the diversity of fishing fleets using them (Amoroso et al. 2018) and making the possibility to draw common conclusions difficult. Nonetheless, most organisms living in shallow, tide-swept and wave-impacted areas are well adapted to high rates of mortality and disturbance, making the effects of trawling relatively minor (Kaiser et al. 2006). In addition to the type of bottom gear, the effects of bottom trawling depend on the nature of the seabed, the footprint of the trawling, and the frequency of trawling and sensitivity of the seabed and benthic ecosystem (Jennings et al. 2005, Lucchetti and Sala 2012, Rijnsdorp et al. 2018). Seabed habitats can be categorized according to the characteristics of the substratum as hard (bedrock to cobble), soft (gravel to mud), and biogenic (habitats formed by living organisms). Scallop dredging produces the most severe impact on biogenic habitats, followed by beam trawling in sandy bottoms and otter trawling in muddy substrates (Collie et al. 2000b, Kaiser et al. 2006). For example, in sandy sediments, deposit-feeding macrofauna were reduced by 20% by beam trawling and otter trawling and 40% by scallop dredging, whereas suspension feeders were reduced by 70% by beam trawling, 45% by scallop dredging, and 5% by otter trawling (Collie et al. 2000b). The recovery rate of the species, in turn, depends on the life-history characteristics, mainly the rates of reproduction and dispersal (Bolam et al. 2014), and may be affected by environmental conditions such as temperature and hydrodynamics (Lambert et al. 2014). Even though bedrock and biogenic reefs are the habitats most familiar to the public, sediments (muds and sands) dominate the present-day continental shelf seabed (Hanebuth et al. 2015).

Recent efforts recapitulated the current understanding of the impact of trawling on the ecologic footprint (Eigaard et al. 2017, Amoroso et al. 2018), pointing out that even in heavily trawled regions, the effort is not evenly distributed—i.e., some areas may be trawled several times per year, while others much less frequently, if at all, thus leaving undisturbed trawling areas in almost every region. Last, but not least, many interrelated factors other than bottom trawling also influence the community structure of benthos at trawled sites—e g., seasonal variation in species diversity and abundance due to particular life cycles of different species and physical conditions such as storms and currents, among others; thus, it is difficult to separate the effects of non-anthropic factors from those of fishing.

Direct trawling effects—i.e., those resulting from the direct contact of the fishing gear with the seabed—are the most common impacts studied in fisheries on a global scale, and include mortality, increased food availability, and loss of habitats. The disturbance paradigm (Holling 1973, National Research Council 2002, Bolam et al. 2014, Sciberras et al. 2018) predicts that short-lived, highly mobile, or dispersing taxa with high reproduction rates will recover from disturbance faster than will long-lived, sessile, low-dispersing taxa. Focusing on the direct disturbance to the seabed and associated biota, Sciberras et al. (2018) reported that sessile and low-mobility biota with longer life-spans—such as sponges, soft corals, and bivalves—took much longer to recover after fishing (> 3 years) than did mobile biota with shorter lifespans such as polychaetes and malacostracans (< 1 year).

The effects on seabed communities occur on different temporal scales. Over the short term (two to three days), the carrion generated because of direct mortality of organisms in the seabed, and by discarded bycatch, produces food subsidies for scavenging species (Ramsay et al. 1997, Kaiser et al. 2018). This surplus of food can lead to an influx of scavengers in recently fished areas (Collie et al. 2017). Over the longer term, the disturbance by chronic bottom fishing can lead to a reduction in community production; changes in trophic structure and function due to decreases in faunal biomass, numbers, and diversity; alterations in the body-size and age structure of benthic populations; and a shift toward communities dominated by fauna with faster life histories (Duplisea et al. 2001, Hiddink et al. 2006, van Denderen et al. 2015). We need to stress that most of the studies included in the meta-analysis by Sciberras et al. (2018) were carried out at depths < 40 m, where levels of natural disturbance from waves and tidal currents are expected to be high, and many of the environments in the studies reviewed will favor smaller species with faster life histories that are more resilient to fishing. Both experimental and comparative studies reported smaller effects of fishing in high-energy environments and dynamic habitats (Kaiser et al. 2017, 2018, Hall-Spencer and Moore 2000, Narvarte et al. 2012, van Denderen et al. 2015).

Different meta-analyses also indicate that at the community level, recovery rates are faster in areas with higher frequencies of trawling (Hiddink et al. 2017, Sciberras et al. 2018). This effect seems to be a consequence of shifts in community composition towards species with faster life histories in fished areas (Jennings et al. 2001a, 2005). For example, despite the large impacts of fishing on their abundance, polychaeta recovered within a few months of the disturbance, resilience is not surprising since most studies (75%) involved free-living polychaeta species with high intrinsic growth rates, thus facilitating a rapid colonization and a prompt recovery from the disturbance (Jennings et al. 2001b, Asch and Collie 2008). Malacostracans, especially mobile-crab genera, also recovered quickly from fishing disturbances, presumably because those crabs could recolonize impacted areas more quickly than other sessile or low-mobility groups. Conversely, bivalves and sessile taxa such as ascidians and bryozoans evidenced little or no short-term recovery within the 3-year time frame of the studies examined. For those species, recovery largely depended positively on recruitment, settlement, and growth in impacted areas and negatively on the extent of substrate modification following the fishing disturbance. For example, many sessile invertebrates such as anemones, tunicates, and soft corals in sandy disturbed areas depended on empty shells for attachment, where trawling can bury those shells and hamper recovery of those taxa (McConnaughey and Syrjala 2014).

The indirect effects of bottom trawling—certainly no less important than the direct ones—include a disturbance of the food web, an alteration in the rate of decomposition of the organic matter, and a recycling of nutrients through the resuspension of bottom sediments. Indeed, reductions in the total biomass of target fish, along with the direct effects noted above, could be expected to simultaneously affect the predators, prey, and competitors of a target species, as well as the overall seafloor-community structure. Indirect effects also encompass potential changes in the flow of materials and energy through ecosystems, plus shifts in the balance among the processes of primary production, primary consumption, and secondary production. Trawling can be considered a disturbance to environments, with the effects often being compared to natural disturbances that occur in the same ecosystems.

Indirect effects of fishing on marine ecosystems

The organisms caught by marine fishing, both the target species and the bycatch, are part of trophic webs—that is, the interactions between species determined by predator-prey relationships. In marine ecosystems, herbivores (zooplankton) are usually smaller than first-order carnivores (small fish), which in turn are consumed by other larger piscivorous organisms (fish, birds, mammals), i.e., the higher up in a marine food web a predator is, the larger its body size becomes (Pauly et al. 2002).

With few exceptions, fishing usually removes the largest organisms from marine food webs. One way to visualize this is through the trophic level (TL) of a certain species or group of species, which is defined as "1 + the mean TL of their prey" (Pauly et al. 1998, 2002). Marine trophic webs are made up of phytoplankton (TL = 1), herbivorous zooplankton that feed on zooplankton (TL = 2), large zooplankton or small fish that feed on herbivorous zooplankton (TL = 3) and large carnivorous fish whose food tends to be a mix of low and high TL organisms (TL = 3.5–4.5) (Pauly et al. 2002).

The mean TL of the fishing landings, also known as Marine Trophic Index (i.e., the mean trophic level of the catch), is deemed as a robust indicator of biodiversity loss (CBD 2004, Pauly and Watson 2005). Fishing tends to eliminate large, slower-growing fish at first and thus reduce the mean trophic level of organisms remaining in an ecosystem. This eventually leads to decreasing trends in the average trophic levels of catches taken from that ecosystem, a process now known as "fishing down the marine food webs" (Pauly et al. 1998).

In a first global-scale exploration of the problem, Pauly et al. (1998) demonstrated that global landings had changed over the last 45 years from large piscivorous fish to small invertebrates and planktivorous fish, especially in the northern hemisphere. This could imply major changes in the structure of marine food webs. In recent years, the occurrence of this fishing down of the marine food webs has been identified globally as one of the most significant effects of fishing on the structure of marine ecosystems, while the MTI assessments became a practical and revelatory proxy for the state of health of the fishery ecosystem (Pauly and Watson 2005, Pauly and Palomares 2005, Jaureguizar and Milessi 2008, Freire and Pauly 2010, among others).

Notwithstanding, the manifestation of these effects of fishing on marine trophic webs may be masked in certain ecosystems because of the expansion of fishing towards new fisheries. Although fishing down marine food webs implies the sequential replacement of high-value upper-trophic-level species with less valuable lower-trophic-level species, a sequential addition of lower-trophic-level fisheries could possibly also occur within an ecosystem—that is, while new fisheries are added to those already exploited, the fishing for high-trophic-level species are maintained despite a decline in the overall mean trophic level of the landings. This process, in turn, has been called "fishing *through* the food web" (Essington et al. 2006), and its occurrence has been demonstrated in several ecosystems in the world (Essington et al. 2006, Bhathal and Pauly 2008, Freire and Pauly 2010).

An economic approach to the problem

So far, we have reviewed the main impacts generated by fishing on marine ecosystems worldwide. On the one hand, we emphasized on the economic and social benefits of fishing, which include employment and subsistence opportunities, the development of the industrial economy, and the enhancement of food security. In the opposite way, we underscored the negative effects of fishing on the components of the ecosystems such as the overexploitation of fish stocks, the destruction of critical habitats, the disturbance of food webs, and the loss of biodiversity.

Regarding the negative effects of fishing, the focus is usually on the depletion of the natural resources; however, we find an exploration of how the unsustainable management of fisheries—because of inappropriate fishing practices and deficient management systems—has also been leading to significant economic losses at a global level. In the realm of fisheries economics, the World Bank (2017) has reported that for 2012 the currently unsustainable fishing-management practices have led globally overexploited fish stocks to produce $83 billion US dollars less in annual net benefits than they could have produced if the fishing had been managed sustainably (that loss in profit is called the "sunken billions"). This result emphasizes that the adoption of more sustainable fishing practices would pay for itself many times over and greatly improve the livelihoods of millions of people (World Bank 2017).

When this economic loss is analyzed by region, the sustainability of current fishing practices and therefore the economic benefits could be seen to vary considerably from one region to another. For example, Asia and Africa have the most highly overexploited fisheries, while in Oceania the total catch is likely to be below the maximum sustainable yield. Thus, the regional analysis suggests that the adoption of more sustainable fishing practices would benefit most the developing countries that have a significant portion of their populations living in poverty (World Bank 2017).

Efforts to prevent the negative impacts of fishing

The direct interactions between trawling gear and the elements of the marine ecosystem occur basically in two ways: the effects on the species vulnerable to the fishing gear and the effects on the seabed structure. As to the first, most fishing gear can catch certain species and not others or can retain fish of certain sizes while excluding others. This property of the fishing gear is called selectivity. Thompson and Ben-Yami (1984) considered selectivity as "... the capacity of any method of gear type to capture certain fractions or sections of the fish population whether grouped by species, age, size, or behavior and to exclude others". Regarding the effects on the bottom elements of the ecosystem, mobile trawling gear comes in different designs, sizes, components, and materials, which interact in distinct ways on the surface of the seabed. Therefore, the resulting effects will depend on both the characteristics of the fishing gear and the nature of the seabed.

Accordingly, the selectivity by species or size, and seabed disturbance by fishing gear (mainly mobile gear) have been the two issues dominating the debate on the effects of fishing on marine ecosystems in recent decades. Both problems have historically been approached from a mechanistic perspective, emphasizing improving the selectivity of the nets and minimizing the contact of the fishing gear with the seabed. Nevertheless, given the multiplicity of variables involved in the selection of mobile gears—i.e., dynamic changes in shape and size of the meshes and varying behavior of the species, among others, as well as in the mechanical interaction between the trawling parts and the seabed—the achievement of generalized solutions has not always been possible. Because of this shortcoming, in recent years other perspectives for mitigating the problem have begun to be considered based on alternative management approaches.

The mechanistic perspective

Selectivity by species and size. Improving gear selectivity has been a goal of the scientists and technologists during recent decades, with the effort being accordingly focused both on the innovation in designs and materials of the net panels and on the development of selective devices. The modifications in the net panels are aimed at improving size selectivity, while the selective devices are designed to exclude unwanted species from the catch.

Fishing nets have always been built with diamond-shaped mesh, which in trawl nets undergo dynamic deformations during trawling as the load on the codend of the net increases, where the weight usually results in a reduction in the opening of the mesh and therefore in the net's selective properties. To solve this problem, one of the most widely studied options has been the use of square mesh panels placed in different parts of the nets, usually in the tunnel or extension, as well as in the codend (Fig. 3A).

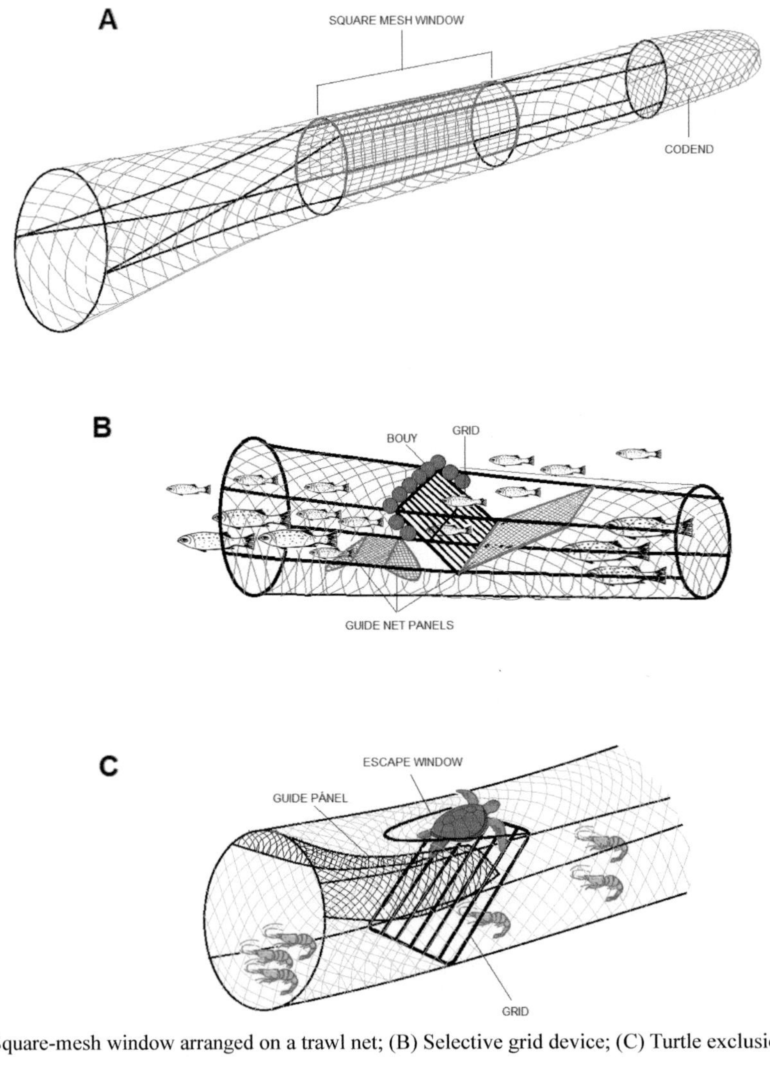

Figure 3. (A) Square-mesh window arranged on a trawl net; (B) Selective grid device; (C) Turtle exclusion device (TED).

Square-mesh windows in their different versions and sizes have generally produced improvements in selectivity, though the efficacy of that approach varies depending on different variables such as the type and size of meshes, the trawling speed and, mainly, the behavior of the species involved.

Selective devices are based on the use of metal grids or other materials (rigid or flexible), generally placed in some part of the tunnel or extension of the trawl net. These grids are used both to exclude juvenile fish of small sizes (Fig. 3B), retaining only adults of commercial value and to exclude individuals of certain species that are otherwise part of the bycatch, as is the case of sea turtles and fish species associated with catches in prawn and shrimp fishing (Fig. 3C).

Other modifications in fishing gear aimed at avoiding or reducing the mortality of unwanted species have been developed in certain types of fishing, in which other fishing gear such as longlines, and purse seines are used. In pelagic or bottom longline operations, incidental catches of seabirds (mainly albatrosses, petrels and shearwater) occur during the setting of the gear (birds get caught when they take the baited hooks that fall from the boat). To avoid this, the use of bird scaring lines and hook shielding devices has been proposed, which keep birds away from the stern of the ship while the fishing gear is set.

The bycatch of dolphins in purse-seine fishing for tuna in the eastern Pacific Ocean is a well-known and highly studied problem (Hall et al. 2000). The high number of dolphins captured in the 1970s and 1980s caused several stocks of those mammals to be declared depleted in the early 1990s. The implementation of mortality-assessment programs and certain technical measures in fishing operations (changes in the gear and in the procedures) enabled a reduction in the bycatch and mortality of dolphins by the end of the 1990s (Hall et al. 2000). For some gear types and species groups, significant progress has been made to improve the handling and release practices of unintended animals in order to maximize the post-catch survival. These methods have been most effectively developed for seabirds and sea turtles captured in longline fishing, as well as the dolphins caught in purse seines (Hall 1996, Perez Roda et al. 2019, among others).

Interactions of gears with seabed. Contact with the seabed is an essential and unavoidable condition for the operation of certain fishing gear. During bottom trawling (Fig. 4A), the leading parts of the trawl (footrope, trawl doors, bridles, and cables) maintain contact with the bottom, provide spread, and herd target species. Consequently, these parts are essential for proper equipment performance and capture efficiency (Valdemarsen et al. 2007). In the example of benthic dredges (Fig. 4B), teeth of different sizes

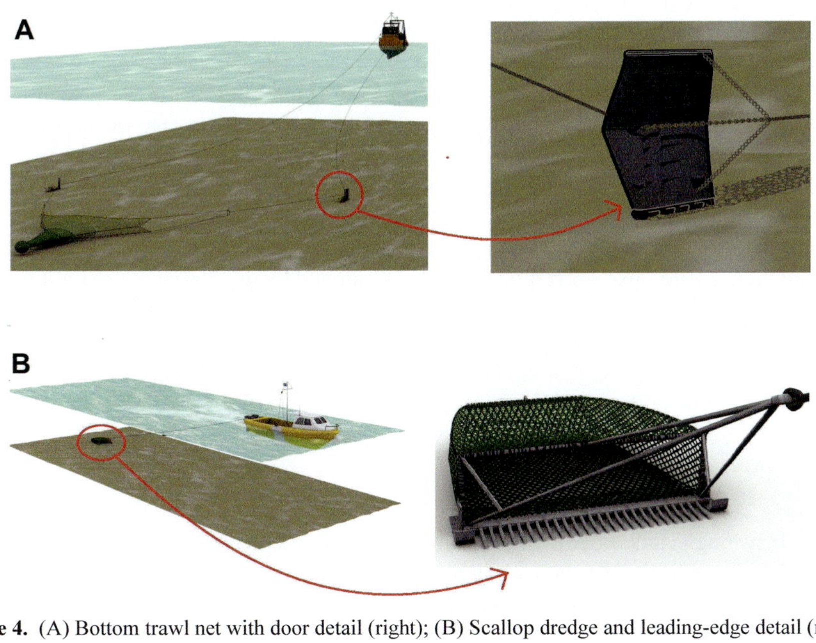

Figure 4. (A) Bottom trawl net with door detail (right); (B) Scallop dredge and leading-edge detail (right).

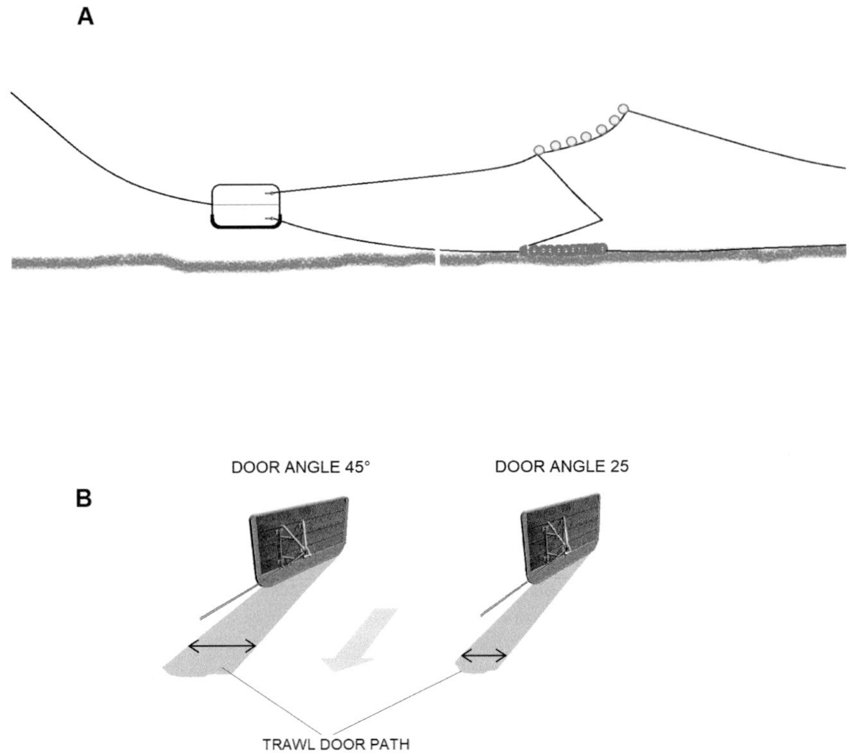

Figure 5. (A) Off-bottom rigging of the trawling doors (right); (B) Reduced door-attack angles (45° left, 25° right). Adapted from Valdemarsen et al. 2007.

along the lower leading edge are also used, which vary according to the target species and the seafloor type. In both examples, the basic mitigation measures are aimed at reducing pressure on the bottom of various trawling components and minimizing the impacted area while trawling.

Since trawl doors are one of the components of bottom trawling that exert the greatest contact with the seabed, modifications to the doors have been prioritized (Broadhurst et al. 2015). In that regard, the options include the use of lighter doors as well as the shortening of the warp-length-to-depth ratio (Fig. 5A), to the extent that the doors work separated from the seabed (He et al. 2006, Valdemarsen et al. 2007). Likewise, the sea-bottom area directly affected by trawl doors can be reduced (Fig. 5B) using high-aspect-ratio doors and by reducing the shoe angle relative to the towing direction (Sterling and Eayrs 2006, Valdemarsen et al. 2007). Other parts of bottom trawling that can be modified to reduce contact are the sweeps and/or the bridles and the ground gear. On the former, discs or bobbins can be arranged along the cable to raise everything off the bottom and by off-bottom rigging of the lower bridle. Likewise, pressure of the ground gear can be reduced through the use of lighter gear components and/or gear components with lifting capabilities (Valdemarsen et al. 2007). In the example of benthic dredges targeting scallops and other shellfish species, innovative designs have been proposed with an aim at reducing contact with the bottom and the consequent disturbance of the benthos (Shephard et al. 2009).

The management approach

Fishing closures. As emphasized above, certain sessile benthic communities can be irreversibly altered by the effect of fishing. Thus, one of the extreme actions implemented from the perspective of fishery management is the prohibition of the use of bottom-trawling gear in sectors of the marine ecosystems considered critical habitats for those communities. This regulation therefore excludes the use of mobile trawling gear, and even in some extreme instances the use of other nonmobile gear that contacts the bottom (i.e., bottom longlines and gillnets). Although fishing closures have been implemented in various fisheries around the world for some time, the use of that measure has become increasingly frequent in recent years.

An example that has recently generated public debate is the closure of 87 sensitive zones to all bottom gears in the seas of the northeast Atlantic by the European Union,[4] based on the Deep-sea Access Regulation (EU 2016). This ban applies to species that occur in deep-sea water and that are characterized by a combination of the following biologic characteristics: maturation at relatively old ages, slow growth, long life expectancies, low natural-mortality rates, intermittent recruitment of successful year classes, and spawning that may not occur every year (EU 2016). The total area of the closures encompasses 16,419 km^2 for the protection of vulnerable marine ecosystems below 400 meters, representing the 1.16% of the EU waters of the northeast Atlantic. The closures include all bottom gear—meaning bottom trawls, dredges, bottom-set gill nets, bottom-set longlines, pots, and traps.

Balanced harvesting, a disruptive approach. This novel (but still debated) approach to fishery management is based on the recognition that fishing extraction is essentially a species-size–selective practice that unavoidably causes changes in the species composition and in the demographic structure of the population in a community. That selective removal of specific species and age groups in preference to others may contribute to changes in the species and size composition of fish communities, alter trophic interactions, and lead to phenotypic or genotypic changes in harvested stocks (Breen et al. 2016). Likewise, these changes could affect the ecosystem structure and functioning—namely, energy flow, element recycling, species interactions, productivity, and resilience—thus leading marine ecosystems to unknown trajectories and scenarios.

The Balanced Harvesting (BH) has been proposed as a potential alternative strategy to current management practices (García et al. 2011). BH considers a moderate fishing mortality distributed across the widest possible range of species, stocks, and sizes in an ecosystem, in proportion to their natural productivity, with the intention of maintaining the relative size and species composition throughout the ecosystem components (Breen et al. 2016). The BH concept (García et al. 2011, Kolding et al. 2015, among others) implies that this strategy may contribute to improving fishing yields and jointly achieving ecosystem objectives (e.g., biodiversity maintenance). Although the conceptual validity of management based on BH has been demonstrated through modelling tools, its implementation in practice is still incipient worldwide (Breen et al. 2016, García et al. 2016). Nevertheless, beyond the difficulties and complexities that may arise in the implementation of BH, the debate on this issue has been strongly installed, especially because the concept questions the traditional management models based on monospecific population dynamics and selective fishing.

Concluding remarks

From the above review, we can deduce that fishing practices producing direct or indirect negative effects on marine ecosystems—such as overfishing, excessive bycatch and discards, fishing down the trophic food webs, and habitat destruction, among others—constitute a true dead end to ecologic sustainability with respect to the fundamental interests of the societies whose livelihood depend on fishing. The complex causality of the negative impacts of fishing on marine ecosystems, however, strongly argues against the possibility of a resolution or mitigation on the basis of a single approach.

Depending on the circumstances and types of effects of fishing gear on the components of the marine ecosystem, science and technology have already provided certain specific principles of solution. For example, from a mechanistic perspective, some of the proposed modifications in mobile fishing gear to reduce bottom contact can be widely applied to a broad spectrum of demersal and benthic fisheries. Likewise, square-mesh panels and selectivity or exclusion devices are currently technologically and economically viable options in many forms of fishing. Nonetheless, to suppose that the implementation

[4] https://oceans-and-fisheries.ec.europa.eu/news/fisheries-end-bottom-fishing-protected-deep-sea-ecosystems-eu-waters-2022-09-15_en.

of these options in fishing will be easy and exempt from difficulties and resistance to change would be naïve. Customs and ways of working are so deeply ingrained in many fishing practices that the people who do the fishing will very likely be reluctant to consider other possibilities in their customary way of fishing.

Therefore, the application of technologies to reduce the negative effects of fishing on marine ecosystems must be promoted on the basis of a new approach to management. In addition, the mitigation of the indirect effects of fishing on marine ecosystems—such as the structural changes induced by fishing down the marine food webs—does not constitute a plausible approach from a mechanistic perspective but could be considered only after a change in management perspective. Thus, the current management of fishing on a global scale, and particularly in some regions, should be reconsidered with a view toward a sustainable management model from an ecosystem point of view. Therefore, the holistic principles of the Ecosystem Approach to Fisheries (EAF) and the corresponding operational framework (García et al. 2003) should be progressively adopted to improve the fishing management worldwide.

The EAF is a broad concept that considers the influence of the environment on the fishery, the impact of fishing on the environment, and the socioeconomic benefits that can be gained from fishing and postharvest activities (Staples and Funge-Smith 2009). The most significant reform that is needed to achieve the potential of the EAF is a shift in the paradigm of the policy from one that is production-oriented to one that is benefits-oriented with respect to both economics and the society (Staples and Funge-Smith 2009)—indeed, an approach that focuses on human and environmental well-being and sustainability rather than to maximize the gross domestic product. The potential of the EAF as a conceptual and operational framework to achieve the sustainable management of fisheries involves two key issues: the active role of the stakeholders in the decision-making and the inclusion of economic and social considerations (Narvarte et al. 2022).

In summary, although various alternatives have been proposed to mitigate the direct and indirect negative effects of fishing on marine ecosystems, the only strategy that seems to have any chance of success is to manage fisheries under the principles and criteria of management based on the ecosystem. This is essential to understand because the EAF not only considers the ecologic aspects of fishery management but also places a great emphasis on economic and social considerations. This multiple focus is particularly relevant to the search for solutions to the negative effects of fishing on ecosystems since many proven technological options for mitigation—e.g., selectivity devices, gear modification, changes in fishing operation—imply changes in economic benefits and in the culture of the workers who do the fishing. The experience produced in the implementation of this form of management worldwide in recent years has proven positive so that, along with continuing the search for technological solutions, the greatest efforts should be focused on making this management approach equally effective throughout all scales of fisheries and all forms of fishing.

Acknowledgements

This study was funded by two grants (PICT-2017-4299 and PICT-2018-3107) from the Agencia Nacional de Promoción de la Investigación, el Desarrollo Tecnológico y la Innovación (to RG and MN, respectively). Figures 1 and 4 were provided by Guillermo Soria and Agustina Neglia respectively. Donald Haggerty reviewed the writing and grammar of the manuscript. The authors are also grateful for the suggestions of an anonymous reviewer that contributed to improve the quality of the final version.

References

Airoldi, L., Balata, D. and Beck, M. W. (2008). The Gray Zone: Relationships between habitat loss and marine diversity and their applications in conservation. J. Exp. Mar. Biol. Ecol. 366: 8–15.

Amoroso, R. O., Pitcher, C. R., Rijnsdorp, A. D., McConnaughey, R. A., Parma, A. M., Suuronen, P. et al. (2018). Bottom trawl fishing footprints on the world's continental shelves. Proceedings of the National Academy of Sciences 115(43): E10275–E10282.

Asch, R. G. and Collie, J. S. (2008). Changes in a benthic megafaunal community due to disturbance from bottom fishing and the establishment of a fishery closure. Fish. Bull. 106: 438–456.

Auster, P. J. and Langton, R. W. (1999). The effects of fishing on fish habitat. *In*: Benaka, L. (ed.). Fish Habitat: Essential Fish Habitat and Rehabilitation. American Fisheries Society, Bethesda, MD.
Bhathal, B. and Pauly, D. (2008). 'Fishing down marine food webs' and spatial expansion of coastal fisheries in India, 1950–2000. Fish. Res. 91: 26–34.
Bolam, S. G., Coggan, R. C., Eggleton, J. E., Diesing, M. and Stephens, D. (2014). Sensitivity of macrobenthic secondary production to trawling in the English sector of the Greater North Sea: a biological traits approach. J. Sea Res. 85: 162–177.
Breen, M., Graham, N., Pol, M., He, P., Reid, D. and Suuronen, P. (2016). Selective fishing and balanced harvesting. Fisheries Research 184: 2–8. https://doi.org/10.1016/j.fishres.2016.03.014.
Broadhurst, M. K., Sterling, D. J. and Millar, R. B. (2015). Modifying otter boards to reduce bottom contact: effects on catches and efficiencies of triple-rigged penaeid trawls. Fisheries Management and Ecology 22: 407–418.
Buhl-Mortensen, L., Ellingsen, K. E., Buhl-Mortensen, P., Skaar, K. L. and Gonzalez-Mirelis, G. (2016). Trawling disturbance on mega-benthos and sediment in the Barents Sea: chronic effects on density, diversity, and composition. ICES J. Mar. Sci. 73(Suppl. 1): i98–i114.
CBD. (2004). Annex I, decision VII/30. The 2020 biodiversity target: a framework for implementation. *In*: Decisions from the Seventh Meeting of the Conference of the Parties of the Convention on Biological Diversity, Montreal: Secretariat of the CBD, Kuala Lumpur, 9–10 and 27 February, p. 351.
Clarke, S. C., McAllister, M. K., Milner-Gulland, E. J., Kirkwood, G. P., Michielsens, C. G. J., Agnew, D. J. et al. (2006). Global estimates of shark catches using trade records from commercial markets. Ecology Letters 9(10): 1115–1126. doi:10.1111/j.1461-0248.2006.00968.x.
Collie, J. S., Hall, S. J., Kaiser, M. J. and Poiner, I. R. (2000a). A quantitative analysis of fishing impacts on shelf-sea benthos. J. Anim. Ecol. 69: 785–798.
Collie, J. S., Escanero, G. A. and Valentine, P. C. (2000b). Photographic evaluation of the impacts of bottom fishing on benthic epifauna. ICES J. Mar. Sci. 57: 987–1001.
Collie, J., Hiddink, J. G., van Kooten, T., Rijnsdorp, A. D., Kaiser, M. J., Jennings, S. et al. (2017). Indirect effects of bottom fishing on the productivity of marine fish. Fish and Fisheries 18: 619–637.
Courtney, A. J., Schemel, B. C., Wallace, R., Campbell, M. J., Mayer, D. G. and Young, B. (2010). Reducing the impact of Queensland's trawl fisheries on protected sea snakes. Fisheries Research and Development Coorporation (FRDC) Project No. 2005/053 Final Report, p. 132.
Crespo, E. A. (2022). Long-term population trends of patagonian marine mammals and their ecosystem interactions in the context of climate change. *In*: Helbling, W., Narvarte, M., González, R. and Villafañe, V. (eds.). Global Change in Atlantic Coastal Patagonian Ecosystems: A Journey Through Time. Springer, Cham.
Duplisea, D. E., Jennings, S., Malcolm, S. J., Parker, R. and Sivyer, D. B. (2001). Modelling potential impacts of bottom trawl fisheries on soft sediment biogeochemistry in the North Sea. Geochem. Trans. 14: 1–6.
Eigaard, O. R., Bastardie, F., Hintzen, N. T., Buhl-Mortensen, L., Buhl-Mortensen, P. and Catarino, R. (2017). The footprint of bottom trawling in European waters: distribution, intensity, and seabed integrity. ICES J. Mar. Sci. 74: 847–865.
Encyclopedia of Environmetrics. (2012). John Wiley & Sons, Ltd. Second Edition. DOI: 10.1002/9780470057339.vab010.pub2.
Essington, T. E., Beaudreau, A. H. and Wiedenmann, J. (2006). Fishing through marine food webs. PNAS 103(9): 3171–3175.
EU. (2016). Regulation (EU) 2016/2336 of the European Parliament and of the Council. Official Journal of the European Union.
FAO. (2020). The State of World Fisheries and Aquaculture 2020. Sustainability in action. Rome. https://doi.org/10.4060/ca9229en.
FAO. (2021). The State of Food and Agriculture 2021. Making agrifood systems more resilient to shocks and stresses. Rome. https://doi.org/10.4060/cb4476en.
Francis, R. C., Awbrey, F. T., Goudey, C. L., Hall, M. A., King, D. M., Medina, H. et al. (1992). Dolphins and the Tuna Industry. Washington, DC, National Research Council, National Academy Press.
Freire, K. and Pauly, D. (2010). Fishing down Brazilian marine food webs, with emphasis on the east Brazil large marine ecosystem. Fisheries Research 105(1): 57–62. https://doi.org/10.1016/j.fishres.2010.02.008.
García, S., Zerbi, A., Aliaume, C., Do Chi, T. and Lasserre, G. (2003). The ecosystem approach to fisheries. Issues, terminology, principles, institutional foundations, implementation and outlook, FAO Fisheries Technical Paper. no. 443. FAO, Rome.
García, S. M., Kolding, J., Rice, J., Rochet, M., Zhou, S., Arimoto, T. et al. (2011). Selective Fishing and Balanced Harvest in Relation to Fisheries and Ecosystem Sustainability. Report of a scientific workshop organized by

the IUCN-CEM Fisheries Expert Group (FEG) and the European Bureau for Conservation and Development (EBCD) in Nagoya (Japan). Gland, Switzerland and Brussels, Belgium: IUCN and EBCD.

García, S. M., Rice, J. and Charles, A. (2016). Balanced harvesting in fisheries: a preliminary analysis of management implications. ICES Journal of Marine Science 73(6): 1668–1678. doi:10.1093/icesjms/fsv156.

Gray, C. A. and Kennelly, S. J. (2018). Bycatches of endangered, threatened and protected species in marine fisheries. Reviews in Fish Biology and Fisheries. https://doi.org/10.1007/s11160-018-9520-7.

Hall, M. A. (1996). On bycatches. Review of Fish Biology and Fisheries 6: 319–352.

Hall, M., Alverson, D. and Metuzals, K. (2000). By-Catch: Problems and Solutions. Marine Pollution Bulletin 41(1-6): 204–219.

Hall-Spencer, J. M. and Moore, P. G. (2000). Scallop dredging has profound, long-term impacts on maerl habitats. ICES. J. Mar. Sci. 57: 1407–1415.

Hanebuth, T. J. J., Lantzsch, H. and Nizou, J. (2015). Mud depocenters on continental shelves-appearance, initiation times, and growth dynamics. Geo-Mar. Lett. 35: 487–503.

He, P., Hamilton, R., Littlefield, G. and Syphers, R. H. (2006). Design and test of a semi-pelagic shrimp trawl to reduce seabed impact. Final report submitted to the Northeast Consortium. University of New Hampshire, Durham, NH.

Hiddink, J., Jennings, S., Sciberrasa, M., Szosteka, C., Hughes, K., Ellis, N. et al. (2017). Global analysis of depletion and recovery of seabed biota after bottom trawling disturbance. PNAS 114 (31): 8301–8306.

Hiddink, J. G., Jennings, S., Kaiser, M. J., Queiros, A. M., Duplisea, D. E. and Piet, G. J. (2006). Cumulative impacts of seabed trawl disturbance on benthic biomass, production, and species richness in different habitats. Can. J. Fish. Aquat. Sci. 63: 721–736.

Hiddink, J. G., Johnson, A. F., Kingham, R. and Hinz, H. (2011). Could our fisheries be more productive? Indirect negative effects of bottom trawl fisheries on fish condition. J. Appl. Ecol. 48: 1441–1449.

Hiddink, J. G., Kaiser, M. J., Sciberras, M., McConnaughey, R. A., Mazor, T., Hilborn, R. et al. (2020). Selection of indicators for assessing and managing the impacts of bottom trawling on seabed habitats. J. Appl. Ecol. 57: 1199–1209.

Holling, C. S. (1973). Resilience and stability of Ecological Systems. Annual Review of Ecol Syst. 4: 1–23.

IPCC. (2019). The Second World Ocean Assessment, vol. I (United Nations publication, 2021); and Intergovernmental Panel on Climate Change, The Ocean and Cryosphere in a Changing Climate: A Special Report of the Intergovernmental Panel on Climate Change.

Jaureguizar, A. and Milessi, A. (2008). Assessing the sources of the fishing down marine food web process in the Argentinean-Uruguayan Common Fishing Zone. Scientia Marina 72(1): 25–36. https://doi.org/10.3989/scimar.2008.72n125.

Jennings, S., Dinmore, T. A., Duplisea, D. E., Warr, K. J. and Lancaster, J. E. (2001a). Trawling disturbance can modify benthic production processes. J. Anim. Ecol. 70: 459–475.

Jennings, S., Pinnegar, J. K., Polunin, N. V. C. and Warr, K. J. (2001b). Impacts of trawling disturbance on the trophic structure of benthic invertebrate communities. Mar. Ecol. Prog. Ser. 213: 127–142.

Jennings, S., Freeman, S., Parker, R., Duplisea, D. E. and Dinmore, T. A. (2005). Ecosystem consequences of bottom fishing disturbance. Am. Fish. Soc. Symp. 41: 73–90.

Kaiser, M. J., Edwards, D. B., Armstrong, P. J., Radford, K., Lough, N. E. L., Flat, R. P. et al. (1998). Changes in megafauna benthic communities in different habitats after trawling disturbance. ICES J. Mar. Sci. 55: 353–361.

Kaiser, M. J., Clarke, K. R., Hinz, H., Austen, M. C. V., Somerfield, P. J. and Karakassis, I. (2006). Global analysis of response and recovery of benthic biota to fishing. Mar. Ecol. Prog. Ser. 311: 1–14.

Kaiser, M. J. (2017). Global analysis of depletion and recovery of seabed biota following bottom trawling disturbance. PNAS 114: 8301–8306.

Kaiser, M. J., Hormbrey, S., Booth, J. R., Hinz, H. and Hiddink, J. G. (2018). Recovery linked to life history of sessile epifauna following exclusion of towed mobile fishing gear. J. Appl. Ecol. 55(3): 1060–1070.

Kaiser, M. J. (2019). Recent advances in understanding the environmental footprint of trawling on the seabed. Annual Meeting of the Canadian Society of Zoologists at Memorial University of Newfoundland, St. John's.

Kolding, J., Jacobsen, N. S., Andersen, K. H. and van Zwieten, P. A. M. (2015). Maximizing fisheries yields while maintaining community structure. Can. J. Fish. Aquat. Sci. http://dx.doi.org/10.1139/cjfas-2015-0098.

Lambert, G. I., Jennings, S., Kaiser, M. J., Davies, T. W. and Hiddink, J. G. (2014). Quantifying recovery rates and resilience of seabed habitats impacted by bottom fishing. Journal of Applied Ecology 51: 1326–1336.

Lewison, R. L., Crowder, L. B., Wallace, B. P., Moore, J. E., Cox, T., Zydelis, R. et al. (2014). Global patterns of marine mammal, seabird, and sea turtle bycatch reveal taxa-specific and cumulative megafauna hotspots. Proc. Natl. Acad. Sci. 111: 5271–5276.

Løkkeberg, S. (2005). Impacts of trawling and scallop dredging on benthic communities. FAO Fish. Tech. Paper No. 472, Rome, Italy. http://www.fao.org/docrep/008/y7135e/y7135e00.htm.

Lucchetti, A. and Sala, A. (2012). Impact and performance of Mediterranean fishing gear by side-scan sonar technology. Can. J. Fish. Aquat. Sci. 69: 1806–1816.

Mazor, T., Pitcher, C. R., Rochester, W., Kaiser, M. J., Hiddink, J. G., Jennings, S. et al. (2021) Trawl fishing impacts on the status of seabed fauna in diverse regions of the globe. Fish Fish. 22(1): 72–86.

McConnaughey, R. A. and Syrjala, S. E. (2014). Short-term effects of bottom trawling and a storm event onsoft-bottom benthos in the eastern Bering Sea. ICES J. Mar. Sci. 71(9): 2469–2483.

Narvarte, M., González, R., Medina, A., Avaca, M. S., Guinsberg, S. and Aliotta, S. (2012). Short term impact of artisanal dredges in a Patagonian mussel fishery: comparisons with commercial diving and control sites. Mar. Environ. Res. 73: 53–61.

Narvarte, M., Avaca, M. S., de la Barra, P., Góngora, M. E., Jaureguízar, A. J., Ocampo Reinaldo, M. et al. (2022). The Patagonian fisheries over time: Facts and lessons to be learned to face global change. *In*: Helbling, W., Narvarte, M., González, R. and Villafañe, V. (eds.). Global Change in Atlantic Coastal Patagonian Ecosystems: A Journey Through Time. Springer, Cham.

National Research Council. (2002). Effects of Trawling and Dredging on Seafloor Habitat. Washington, DC: The National Academies Press. https://doi.org/10.17226/10323.

Oberle, F., Storlazzi, C. and Hanebuth, T. (2016). What a drag: Quantifying the global impact of chronic bottom trawling on continental shelf sediment. Journal of Marine Systems 159: 109–119. https://doi.org/10.1016/j.jmarsys.2015.12.007.

Palomares, M. L. D., Froese, R., Derrick, B., Meeuwig, J. J., Nöel, S.-L., Tsui, G. et al. (2020). Fishery biomass trends of exploited fish populations in marine ecoregions, climatic zones and ocean basins. Estuarine, Coastal and Shelf Science. https://doi.org/10.1016/j.ecss.2020.106896.

Pauly, D., Christensen, V., Dalsgaard, J., Froese, R. and Torres, F.Jr. (1998). Fishing down marine food webs. Science 279: 860–863.

Pauly, D., Christensen, V., Guénette, S., Pitcher, T. J., Sumaila, R., Walters, C. J. et al. (2002). Towards sustainability in world fisheries. Nature 418(8): 689–695.

Pauly, D. and Palomares, M. L. (2005). Fishing Down Marine Food Web: It is Far More Pervasive Than We Thought. Bulletin of Marine Science 76(2): 197–212.

Pauly, D. and Watson, R. (2005). Background and interpretation of the 'Marine Trophic Index' as a measure of biodiversity. Phil. Trans. R. Soc. B. 360: 415–423.

Perez Roda, M. A., Gilman, E., Huntington, T., Kennelly, S. J., Suuronen, P., Chaloupka, M. et al. (2019). A third assessment of global marine fisheries discards. FAO Fisheries and Aquaculture Technical Paper No. 633. Rome, FAO.

Pikitch, E. K., Santora, C., Babcock, E. A., Bakun, A., Bonfil, R., Conover, D. O. et al. (2004). Ecosystem-based fishery management. Science 305: 346–347.

Pitcher, C. R., Hiddink, J., Jennings, S., Collie, J., Parma, A., Amoroso, R. et al. (2022). Trawl impacts on the relative status of biotic communities of seabed sedimentary habitats in 24 regions worldwide. PNAS 119: 2. http://doi.org/10.1073/pnas.2109449119.

Pott, C. and Wiedenfeld, D. A. (2017). Information gaps limit our understanding of seabird bycatch in global Fisheries. Biological Conservation 210: 192–204.

Pusceddu, A., Bianchelli, S., Martin, J., Puig, P., Palanques, A., Masque, P. et al. (2014). Chronic and intensive bottom trawling impairs deep-sea biodiversity and ecosystem functioning. PNAS 111: 8861–8866.

Ramsay, K., Kaiser, M. J., Moore, P. G. and Hughes, R. N. (1997). Consumption of fisheries discards by benthic scavengers: Utilization of energy subsidies in different marine habitats. J. Anim. Ecol. 66: 884–896.

Read, A. J. (2008). The looming crisis: interactions between marine mammals and fisheries. J. Mamm. 89: 541–548.

Reeves, R. R., McClellan, K. and Werner, T. B. (2013). Marine mammal bycatch in gillnet and other entangling net fisheries, 1990 to 2011. Endanger Species Res. 20: 71–97.

Rice, J., Arvanitidis, C., Borja, A., Frid, C., Hiddink, J. G., Krause, J. et al. (2012). Indicators for sea-floor integrity under the European Marine Strategy Framework Directive. Ecol. Ind. 12: 174–184.

Rijnsdorp, A. D., Bolam, S. G., Garcia, C., Hiddink, J. G., Hintzen, N. and Kooten, T. V. (2018). Estimating the sensitivity seafloor habitats to disturbance by bottom trawling impacts based on the longevity of benthic fauna. Ecol. Appl. 28: 1302–1312.

Sciberras, M., Hiddink, J. G., Jennings, S., Szostek, C. L., Hughes, K. M., Kneafsey, B. et al. (2018). Response of benthic fauna to experimental bottom fishing: A global meta-analysis. Fish Fish. 19: 698–715.

Shephard, S., Goudey, C., Read, A. and Kaiser, M. (2009). Hydrodredge: Reducing the negative impacts of scallop dredging. Fisheries Research 95: 206–209.

Smith, C. J., Papadopoulou, K. N., Barnard, S., Mazik, K., Elliott, M., Patrıcio, J. et al. (2016). Managing the marine environment, conceptual models and assessment considerations for the european marine strategy framework directive. Front. Mar. Sci. 3: 19.

Solan, M., Cardinale, B. J., Downing, A. L., Engelhardt, K. A. M., Ruesink, J. L. and Srivastava, D. S. (2004). Extinction and ecosystem function in the marine benthos. Science 641(306): 1177–1180.

Staples, D. and Funge-Smith, S. (2009). Ecosystem approach to fisheries and aquaculture: Implementing the FAO code of conduct for responsible fisheries. FAO Regional Office for Asia and the Pacific, Bangkok, Thailand. RAP Publication 2009/11.

Sterling, D. and Eayrs, S. (2006). Designs and assessment of two gear modifications to reduce the benthic impact and fuel intensity of prawn trawling in Australia. ICES Symposium. Fishing Technology in the 21st Century: Integrating Fishing and Ecosystem Conservation, Boston.

Thompson, D. B. and Ben Yami, M. (1984). FAO Fisheries Report No. 289 Supplement 2.

Thurstan, R. H., Hawkins, J. P. and Roberts, C. M. (2013). Origins of the bottom trawling controversy in the British Isles: 19th century witness testimonies reveal evidence of early fishery declines. Fish and Fisheries 15(3): 506–522.

United Nations. (2022). Interactive dialogue 4: Making fisheries sustainable and providing access for small-scale artisanal fishers to marine resources and markets. United Nations Conference to Support the Implementation of Sustainable Development Goal 14: Conserve and sustainably use the oceans, seas and marine resources for sustainable development. Concept paper.

Valdemarsen, J. W., Jørgensen, T. and Engås, A. (2007). Options to mitigate bottom habitat impact of dragged gear. FAO Fisheries Technical Paper N°506. FAO, Rome.

Van Denderen, P. D., Bolam, S. G., Hiddink, J. G., Jennings, S., Kenny, A., Rijnsdorp, A. D. et al. (2015). Similar effects of bottom trawling and natural disturbance on composition and function of benthic communities across habitats. Mar. Ecol. Prog. Ser. 541: 31–43.

Wallace, B. P., Lewison, R. L., McDonald, S., McDonald, R. T., Bjorkland, R. K., Kelez, S. et al. (2010). Global patterns of marine turtle bycatch. Conserv Lett. 3: 131–142.

Wassenberg, T. J., Salini, J. P., Heatwole, H. and Kerr, J. D. (1994). Incidental capture of sea-snakes (Hydrophiidae) by prawn trawlers in the Gulf of Carpentaria, Australia. Mar Freshw Res. 45: 429–443.

Wassenberg, T. J., Milton, D. A. and Burridge, C. Y. (2001). Survival rates of sea snakes caught by demersal trawlers in northern and eastern Australia. Biol. Conserv. 100: 271–280.

Watson, R., Revenga, C. and Kura, Y. (2006). Fishing gear associated with global marine catches I. Database development. Fisheries Research 79: 97–102.

World Bank. (2017). The Sunken Billions Revisited: Progress and Challenges in Global Marine Fisheries. Washington, DC: World Bank. Environment and Sustainable Development series. doi:10.1596/978-1-4648-0919-4.

Zeller, D., Palomares, M. L. D., Tavakolie, A., Ang, M., Belhabib, D., Cheung, W. W. L. et al. (2016). Still catching attention: sea Around Us reconstructed global catch data, their spatial expression and public accessibility. Mar. Pol. 70: 145–152.

Zeller, D., Cashion, T., Palomares, M. L. D. and Pauly, D. (2018). Global marine fisheries discards: a synthesis of reconstructed data. Fish Fish. 19: 30–39.

CHAPTER 3

Options for Low-technology and/or Low-cost IMTA Systems

Paula Senff[1,2] and *Andreas Kunzmann*[2,*]

Introduction

General

The aquatic realm, covering 71% percent of our planet, is full of life and productivity. Humans have long since benefitted from the resources it provides directly through gleaning or fishing (Richter et al. 2008). Throughout the second half of the 20th century, the "Blue Revolution" has seen the widespread intensification of the sector, as aquaculture was seen left to fill the gap between an increasing global demand for fish protein and static capture fishery production (MacKay 1983, Thia-Eng 1997, Costa-Pierce 2002, FAO 2018). In 2012, United Nations Secretary General, Kofi Annan, stated that Aquaculture is crucial for supplying the world's food needs for the next 50 years (Annan 2012). In the process of intensification, aquaculture operations are increasingly moving into closed, land-based systems that provide greater control over environmental parameters, optimize the use of space and water and minimize discharge (Martins et al. 2010, Badiola et al. 2012, Verdegem 2013). However, Belton et al. (2020) are very critical of this development, and claim that farming fish in the sea will not nourish the world, because most authors tend to overstate marine finfish aquaculture's potential to deliver food security and environmental sustainability. In addition, they claim that maritime space allocation usually excludes fishers and therefore recommend that in order to seek to increase the availability and accessibility of affordable and sustainable farmed aquatic foods, future aquaculture should focus on freshwater aquaculture. These statements again are heavily opposed by other authors such as Costa-Pierce et al. (2022) and this conflict is taken up again in the discussion of this chapter.

Half of the feed supplied in open cage aquaculture remains uneaten and is directly lost as particulate and dissolved waste (Ballester-Moltó et al. 2017). Around 75% of N and P contained in fish feed is lost to surrounding waters (Holby and Hall 1991, Handy and Poxton 1993, Hargreaves 1998). Feed, feces and other organic material released from aquaculture elevate water turbidity, decrease photic depth and are finally deposited, modifying the biochemistry of the sediment (Holmer and Kristensen 1992, Price et al. 2014, Ballester-Moltó et al. 2017). Under high nutrient loads, microbial communities become overwhelmed and sediments turn increasingly anoxic (Holmer et al. 2003).

As discharge limits are put in place by governments to minimize the environmental impact of aquaculture operations, effluents can no longer just be discarded and need to be treated (van Rijn 2013).

[1] French Agricultural Research and International Cooperation Organization (CIRAD), Montpellier, France.
[2] Leibniz-Centre for Tropical Marine Research (ZMT), Bremen, Germany.
* Corresponding author: andreas.kunzmann@leibniz-zmt.de

For the treatment of inorganic nitrogen, closed systems commonly include bacterial biofilters (Van Rijn 1996, Troell et al. 2003, Summerfelt and Sharrer 2004, Chen et al. 2006, Lyssenko and Wheaton 2006, Malone and Pfeiffer 2006). Concentrations of inorganic nitrogen can also be controlled through biofloc technology. In this approach, microbial growth in the water column of fish tanks or ponds is stimulated through the addition of carbohydrates in the form of sugar, starch or cellulose (Avnimelech 2012).

The successful application of water treatment technology, however, requires knowledge of the system, close monitoring to maintain appropriate C:N ratios and microbial densities, and a correct upscaling from laboratory to economic applications (Kuhn et al. 2010, Hende et al. 2014). In an "ecosystem approach to aquaculture", expectations for the growth of the industry are for it to use resources sustainably and promote comprehensive nutrient recycling within the culture system to reduce effluent volume and minimize feed input (FAO 2010). The disposal of uneaten feed, fecal matter and dissolved nutrients pose a challenge for the growth of the aquaculture sector, but in the last decade it was propagated to regard these leftovers as resources high in nutrients and essential fatty acids that should be recycled, not discarded (Bischoff et al. 2009). Recycling the nutrients in that fashion is increasingly becoming popular in the so-called "circular economy".

The principle of reusing waste of one livestock for the production of another crop has a long history of application (Lu and Li 2006, Birhanu and Natarajan 2019). The development of multi-trophic aquaculture systems that are suitable for the sustainable intensification of this sector is therefore of great research interest. In the last decade, the concept of integrated multi trophic aquaculture (IMTA) was coined, striving to apply a simplified food web structure by combining species at different trophic levels for a balanced-ecosystem approach (Neori et al. 2004, Chopin et al. 2007). The IMTA concept is a direct application of the ecosystem approach at the farm level (Brugère et al. 2019). Nutrient load is reduced with the additional aim to produce higher yields than mono-species systems, while satisfying rising consumer demands for environmental standards and improving the image of aquaculture (Neori et al. 2004, Barrington et al. 2010). In a fully optimized system, IMTA combines the feeding modes of deposit-feeders, filter-feeders and autotrophs that take up dissolved nutrients in addition to the fed organism (Buck et al. 2018). This co-culture approach finds application in a range of operations, from open-water mariculture to closed Recirculation Aquaculture Systems (RAS) with aquaponics (Gunning et al. 2016, Buck et al. 2018, Stenton-Dozey et al. 2020).

In open-water IMTA, hydrodynamic and seasonal conditions influence efficient and direct nutrient uptake (Kerrigan and Suckling 2018). In more closed pond- or tank-based systems, concentrated dissolved effluents can be directly valorized by growing primary producers (Wei et al. 2017, Li et al. 2019). IMTA provides an additional opportunity also in RAS, where effluents that are to be recycled are in controlled and accessible environments (Guerrero and Cremades 2012). Particulate organic waste collected by the filtration systems in RAS constitutes a readily available feed for deposit feeding invertebrates, for example polychaetes or crustaceans (Bischoff 2007, Brown et al. 2011, Palmer et al. 2014).

Sea cucumbers and deposit feeding, bioremediation

Deposit-feeding sea cucumbers are excellent candidates for integrated aquaculture with high market value and nutritional quality (Macdonald et al. 2013, Cubillo et al. 2016, Zamora et al. 2018). Their body wall is dried into *bêche-de-mer*, a Chinese delicacy and luxury food in the ranks of shark fin, fish maw and abalone (Robinson 2013, Purcell et al. 2014).

Globally, the cultivation of *Apostichopus japonicus* dominates sea cucumber aquaculture production, but *Holothuria scabra* is a tropical species (Fig. 1) with fully developed aquaculture technology and increasing production (Purcell et al. 2012, Robinson 2013). Culture production of *H. scabra* is practiced in tropical countries throughout Asia, Africa, Australia and the Central Pacific (Ivy et al. 2006, Robinson and Pascal 2009, Hair 2012, Mills et al. 2012). *H. scabra* has been successfully cultivated in extensive mariculture in marine ponds, sea pens, and sea ranches (Fig. 1A) (Juinio-Meñez et al. 2017, Pitt 2001, Purcell et al. 2012). Also, recently, two local species in the Caribbean have been explored for its aquaculture suitability (Acosta et al. 2020, 2021) and at least one species has been reproduced in captivity (Acosta et al. 2020).

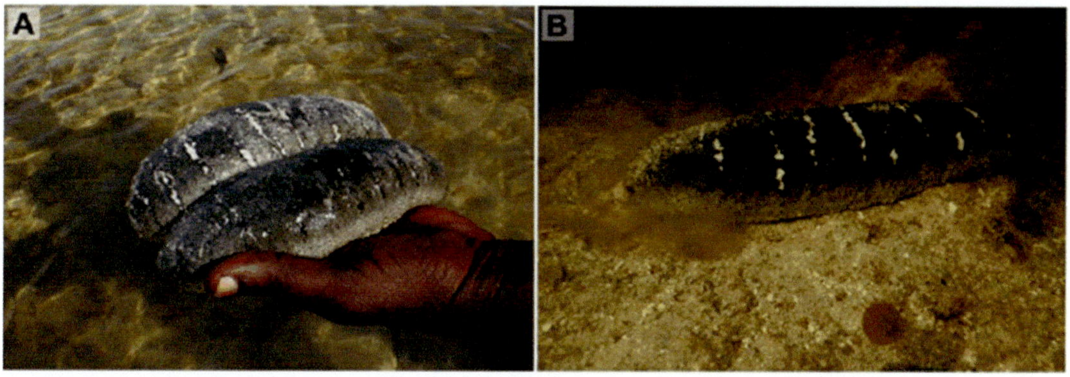

Figure 1. (A) Two adult *Holothuria scabra* sea cucumbers at a sea ranching site on Zanzibar, Tanzania and (B) *H. scabra* feeding in a seagrass meadow on Zanzibar.

As detritivorous animals, sea cucumbers can feed on debris from fish or shellfish cultivation (Ahlgren 1998, Slater et al. 2011, Hannah et al. 2013). Co-cultivation of sea cucumbers with fish culture has been investigated with temperate and tropical species (review by Zamora et al. 2018). Experiments have shown that sea cucumbers *Parastichopus californicus* and *A. japonicus* can utilize organic waste from finfish aquaculture farms and even achieve higher growth rates than when reared in natural habitats (Ahlgren 1998, Hannah et al. 2013, Yokoyama 2013, Yu et al. 2013, 2014). In tank experiments, *H. scabra* successfully fed on waste from shrimp farming (Pitt et al. 2004, Purcell et al. 2006, Watanabe et al. 2012). In direct co-cultivation, shrimps can however predate on the sea cucumbers (Pitt et al. 2004). A feasible possibility seems to be a rotational system in ponds, where one production cycle of sea cucumbers alternates with one production cycle of shrimps (Mills et al. 2012, Hochard et al. 2016).

The interplay between deposit-feeding sea cucumbers and sedimentary microbial communities has great potential for the recycling of nutrients in aquaculture waste treatment (Robinson et al. 2016, Robinson 2017). Adding nutrients that promote the growth of preferred bacteria is already an established method in controlling concentrations of Dissolved Inorganic Nitrogen (DIN) in aquaculture (Crab et al. 2007, Avnimelech 2014). Modification of C:N ratios also improves performance of sediment-based bioremediation systems (Robinson et al. 2018, 2019).

Algae and halophytes for remediation and IMTA

Macroalgae are the extractive organism most commonly used in open-water IMTA to take up dissolved nutrients (Kerrigan and Suckling 2018). Algae have fast growth rates, are easy to cultivate and while producing valuable biomass, they increase concentrations of dissolved oxygen in the water (Gao and McKinley 1994, Chopin et al. 2001, Neori et al. 2003). Many plants grow well in hydroponics, with their roots submerged in soilless medium, and aquaponics have been explored with a wide array of fruits, vegetables and herbs as extractive species (Diver 2006, Danner et al. 2019, Ogah et al. 2020). Salt-tolerant halophytes hold great potential as they can be irrigated with salt water, produce yields similar to conventional crops and can be integrated into saline aquaculture systems (Glenn et al. 1999, Shpigel et al. 2013). Nutrient removal from aquaculture wastewater in constructed wetlands planted with halophytes is highly variable, very much depending on location, size and plant species (review by Custódio et al. 2017).

The implementation of IMTA is seen as an important approach to meet current and future challenges, but the majority of IMTA research has been undertaken in temperate regions, with research in tropical countries lagging behind (Granada et al. 2015, Ahmed and Glaser 2016, Felaco et al. 2020, Putro et al. 2020). There is a clear lack of research on IMTA systems in the tropics, particularly about simple and low-cost systems and also particularly for small-scale family- or village-type systems. As a consequence, there is an increasing need for effluent treatment in coastal aquaculture systems in the tropics. Reasons for the low proliferation of IMTA systems in the tropics are assumed to be issues of system complexity and high prices for components. Therefore, simple and low-cost systems are badly needed.

A general overview about potential impacts of aquaculture to the marine environment is given in Belton et al. (2020) and Mizuta et al. (2022). This chapter aims to meet the need for more information on the performance of tropical extractive species in saline IMTA systems. As a case study and first proof-of-concept, a simple and low-cost deposit feeding sea cucumber species and an edible halophyte were combined for the first time as extractive species in a closed RAS of fed finfish in a small-scale system in Tanzania. In addition, different methods of stable isotope analysis were applied to examine the faith of bioremediation by two extractive species in more detail, particularly about the role of carbon supplementation in order to achieve a balanced C:N ratio.

Case example in Zanzibar

With the emphasis on tropical species, low-cost, simple technology and family- or village-systems, a number of experiments on bioremediation (combination of RAS with IMTA) were performed in tropical aquaria in Tanzania and individual aspects were co-checked in aquaria systems in ZMT in Bremen.

Two experiments were conducted over a period of 70 and 104 days, respectively, in 2018/9 at the Zanzibar Mariculture hatchery, Tanzania ((6°06'54"S 39°12'45"E) with milkfish (*Chanos chanos*), sea cucumbers (*Holothuria scabra*) and halophytes (*Sesuvium portulacastrum*). A schematic setup is given in Fig. 2. More detailed Information is given in Senff et al. (2020).

Three different treatments were applied. The "Sludge" treatment received aquaculture waste only, while for the "Cellulose" and "Bagasse" treatments, the sludge was mixed with microcrystalline cellulose or finely ground bagasse, respectively (Table 1). For further details, see Senff et al. (2022).

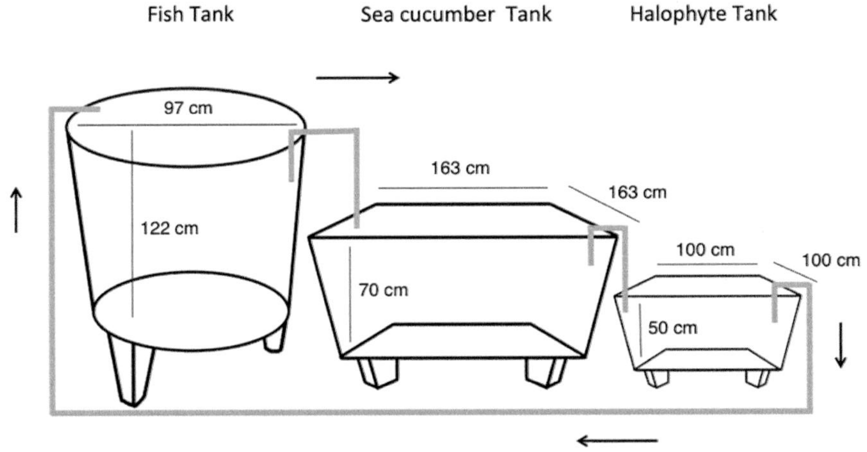

Figure 2. Schematic set-up of the experimental RAS. Grey lines represent water pipes, black arrows indicate direction of water flow.

Table 1. Experimental treatments receiving waste from milkfish aquaculture with or without the supplementation with different types of carbon sources and with or without the presence of sea cucumber *Holothuria scabra*.

Treatment	Feed type	C:N	n	mmol C m^{-2} d^{-1}	mmol N m^{-2} d^{-1}	Sea cucumbers (n)
A	Sludge	8.5	4	43.6	4.4	3
B	Sludge + bagasse	30	4	160.4	4.6	3
C	Sludge + cellulose	30	4	153.4	4.4	3
AC	Sludge	8.5	4	43.6	4.4	0
BC	Sludge + bagasse	30	4	160.4	4.6	0
CC	Sludge + cellulose	30	4	153.4	4.4	0

Table 2. Growth (mean ± SD) of the organisms in the RAS (n = 3), - = not applicable.

Tank	Fish	Sea cucumbers	Halophytes
Absolute growth (g day^{-1})	1.9 ± 0.2	1.2 ±1.1	0.3 ± 0.02
SGR (% day^{-1})	1.1 ± 0.1	1.1 ± 0.7	1.3 ± 0.1
FCR	2.3 ± 0.2	-	-
Final density (g m^{-2})	-	67 ± 25	5425 ± 216

The organisms in all three systems increased in biomass at an SGR above 1% per day for all species (Table 2) and survival was 100%. The nitrogen content in the surface sediments of the sea cucumber tanks was overall very low, but increased over the course of the experiment.

Over the course of the experiment, each RAS received an input of 132 g N via the fish feed. Figure 3 shows a conceptual representation of the three experimental low-cost, low-tech RAS systems.

Although RAS in combination with IMTA are considered best management practices, high material costs and difficult maintenance still hinder their implementation, especially in developing countries and the tropics. Few case studies of such systems with tropical species exist. Here we tested for the first time an extremely low budget system by combining the halophyte sea purslane and a detritivore - sandfish, with finfish milkfish over eight weeks on Zanzibar, Tanzania. Milkfish and sea purslane showed good growth, while sea cucumber growth was low. The system operated without filter units and did not discharge any solid or dissolved waste and provides a proof-of-concept of a simple pilot-scale RAS, integrating tropical species at three trophic levels.

In order to control N turnover, additional carbon was added to the experiment (cellulose, bagasse, sludge). At the start of the experiment, sediment nitrogen content in all tanks was below the detection limit, then increased over the course of the experiment in all treatments. The bagasse control treatment had the highest increase in nitrogen content, and significantly higher than the sludge and sludge control treatments. Content of organic carbon in the sediment also increased in all treatments and reached

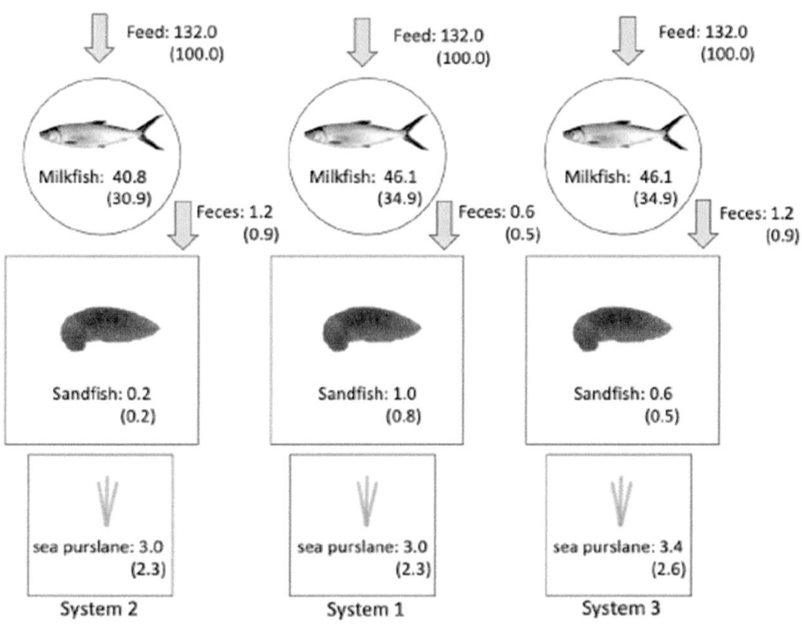

Figure 3. Conceptual representation of the three experimental RAS, each including a milkfish tank (round), sea cucumber tank (large square) and hydroponic sea purslane tank (small square). Numbers indicate the nitrogen (N) content (g) of the different compartments, with the corresponding % N relative to the input (feed) in parentheses.

significantly higher values in the two treatments receiving bagasse. The sea cucumbers grew in all treatments over the course of the experiment. However, the sea cucumbers in the cellulose treatment subsequently showed negative growth rates for the remainder of the experiment, while growth in the sludge and bagasse treatments was relatively stable.

It is **concluded** that carbon supplementation to increase the carbon to nitrogen ratio (C:N) of the nitrogen-rich aquaculture waste supports higher deposit feeder biomass. Carbon supplementation also prevented the accumulation of dissolved inorganic nitrogen. Sea cucumbers were able to assimilate carbon directly from the cellulose. Bagasse supported higher uptake of carbon from aquaculture waste, thus identifying it as a viable resource to improve bioremediation by deposit feeding sea cucumbers in integrated aquaculture. While illustrating the importance of nutrient balances for bioremediation, questions remain about the roles played by deposit feeders vs. microbial communities. A well-established bacterial community in the sediment seems to be a key factor for successful bioremediation.

It is also **concluded** that a simple and low-cost RAS system, integrating valuable tropical species of three different trophic levels, is viable. For the first time, sea cucumbers and halophytes were combined with these units replacing a separate biofilter. The system operated without producing solid waste or discharging polluted water and optimized the use of water and space. This kind of sustainable aquaculture model has the potential to provide food security and income, and the simple system used here also allows for application in developing coastal communities.

In order to trace nutrient removal and nitrogen recovery from aquaculture effluents by the edible halophyte *Sesuvium portulacastrum,* another experiment was set up in hydroponics and in sand substrate in the laboratories of ZMT in Bremen (Fig. 4). Removal from milkfish wastewater by systems planted with *S. portulacastrum* in hydroponics, with the roots in water, versus planted in sand, was investigated to determine the mode of cultivation, which most efficiently assimilates nutrients from wastewater into new valuable biomass.

The assimilation of fixed N by *S. portulacastrum* was traced by adding a defined amount of $^{15}NH_4$ or $^{15}NO_3$ in experiments with planted or unplanted (control) systems. Overall, the recovery of wastewater N into new plant biomass was several-fold higher in hydroponics than in sand substrate. These results provide useful details for the successful integration of *S. portulacastrum* into aquaculture systems with the hydroponic cultivation identified as a far more efficient way to turn aquaculture effluents into a valuable resource.

It was noted that the substrate showed clear differences in its microbial and/or algal community. While the sand surface in the boxes planted with halophytes was overgrown with green and filamentous

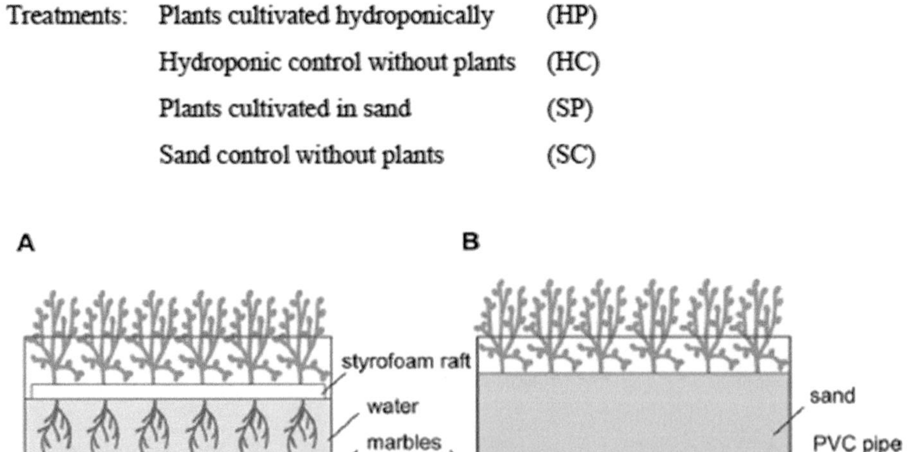

Figure 4. Schematic drawing of the two different boxes planted with *Sesuvium portulacastrum* grown (A) in hydroponics or (B) in sand substrate in the experiments.

Figure 5. Example of the sand in the boxes that had received RAS water (A) planted with *S. portulacastrum* and (B) unplanted control treatment after black cover was removed upon termination of the experiment.

algae, the control treatments only showed small patches of brown or red microalgae and/or cyanobacteria. The control treatments also showed a clear anoxic layer at an approximate depth of 3 cm, while the boxes with plants did not (Fig. 5).

From this experiment, it can be **concluded** that integrating extractive species in aquaculture to treat wastewater and increase nutrient recovery into usable biomass is an important step towards sustainability. Cultivating the edible halophyte *Sesuvium portulacastrum* in hydroponics or sand substrate effectively treated milkfish effluents and removed dissolved inorganic nutrients. Determining uptake rates for different N forms could aid in further optimizing the hydroponic growth method applied in this study and move towards more complete resource recovery in closed recirculation systems and integrated aquaculture.

These three experiments provide examples of integrated multi-trophic aquaculture (IMTA) with edible tropical species at three different trophic levels. A popular fish (milkfish) can be cultivated with minimal water exchange in a low technology RAS alongside a high-value deposit-feeder and an edible halophytic plant to allow the efficient reuse of water and dissolved nutrients under controlled conditions. Extractive species can be adjusted to improve the recovery of nutrients from aquaculture effluents. Nitrogen uptake from aquaculture effluents was traced by *Sesuvium portulacastrum*, showing superior nitrogen recovery in hydroponic cultivation compared to sand substrate.

For future IMTA setups in the tropics, it was important to show that a high value deposit-feeding sea cucumber species and an edible halophyte can be integrated as extractive species with fed finfish in a small RAS. The closed RAS operated without water exchange and without investing into maintenance or space for a nitrifying/denitrifying biofilter. Concentrations of ammonia, nitrite and nitrate were kept consistently low and below harmful levels and good biomass production was achieved. The halophyte *S. portulacastrum* removed nitrogen from the system and assimilated it into harvestable biomass. Juvenile sea cucumbers *H. scabra* were cultivated on sediment and their tanks double-functioned as a settlement basin for solid waste. The sand also likely provided substrate for bacterial nitrification-denitrification activity.

Carbon amendments have previously been shown to improve the growth of sea cucumbers in microbial-deposit feeder remediation systems (Robinson et al. 2019). The effect of adding lab-grade microcrystalline cellulose was compared to bagasse, an agricultural waste product. Bagasse significantly improved sea cucumber final biomass, which cellulose failed to do. Stable isotope analysis of feed components and sea cucumber tissues indicated that this improved growth was not due to direct assimilation of nutrients from bagasse, but a higher assimilation of sludge-derived nutrients. The carbon amendments also decreased concentrations of dissolved nitrate in the systems, indicating that they had the desired effect of stimulating microbial growth leading to immobilization of DIN into microbial biomass.

In integrated aquaculture or aquaponics, a high biomass ratio of extractive vs. fed species is required to ensure complete nutrient uptake (Rakocy 2012). If space for plant cultivation is limited, the approach may need to find a balance between nitrogen removal and recycling into usable biomass.

Role of microbes and probiotics

Target species in aquaculture, even more so than terrestrial livestock, are permanently in close contact with their surrounding medium and its microbes, directly shaping their internal microbiome, which is an important driver of animal health (Huang et al. 2020). Microbial communities in aquaculture systems thus influence a range of factors from water quality over biomass production to disease outbreaks (Zhou et al. 2009, Huang et al. 2020). In this study, the role of microbial processes was likely evidenced in the form of (a) nitrification activity in sea cucumber tanks, (b) lower nitrogen concentrations in treatments receiving carbon amendments to aquaculture sludge and (c) considerable nitrogen removal, even in unplanted systems receiving milkfish effluents. In future research, the role of microbes needs to be examined for a more complete understanding of system dynamics and to determine how probiotics could further improve nutrient recovery by extractive species.

Probiotics are whole or parts of microorganisms such as bacteria, microalgae and yeast which, when administered in adequate amounts, confer health benefit to the host, including improved disease resistance, growth, feed utilization and adequate stress response (Marrifield et al. 2010, Morelli and Capurso 2012). Often administered via feed to the gastrointestinal tract, they can improve the intestinal microbial balance (Fuller 1989, Gatesoupe 1999). Especially in aquatic animals, whose internal microbiota constantly interact with the surrounding medium, their application can also aim at improving the quality of the ambient environment to support the health of the target species (Verschuere et al. 2000, Hai 2015).

In the environment, beneficial microorganisms fulfill important functions of preventing disease outbreaks, improving water quality and producing bioactive compounds that improve growth and feed use efficiency in aquaculture (Zhou et al. 2009). In its most basic form, this is practiced widely in traditional "green water aquaculture", where fertilization of pond water promotes the growth of microalgae, bacteria, fungi, protozoa and zooplankton that provide feed for filter-feeding fish and shrimp, while at the same time inhibiting the growth of pathogens (Neori 2011, Yarnold et al. 2019). Optimized in the so-called "bio-floc technology", the concept is applied to promote the assimilation of inorganic nitrogen by heterotrophic bacteria, which remove nitrogen from the water and convert it into microbial protein (Schneider et al. 2005, De Schryver and Verstraete 2009).

Researchers and aquaculture practitioners are becoming increasingly aware of how water parameters drive microbial community compositions and how sometimes modifying even a single factor such as pH or salinity can control their community composition or prevent a disease outbreak (Hou et al. 2017a, Alfiansah et al. 2020). C:N ratios can for example be modified by dissolving glucose as a carbon source and alter the bacterial community composition to boost their positive effect on water quality (Zheng et al. 2018). Deng et al. (2020) found that feeding formulated feed instead of trash fish to crabs in an IMTA system with bivalves and shrimp had a similar effect on water quality, improved animal growth and achieved higher cost efficiency. The higher carbohydrate content of the formulated feed influenced the microbial community in the pond water toward pathways of carbon and lipid metabolism.

An ecosystem approach to microbes in IMTA

In future systems, plants and animals should not only be seen as individual organisms, but also as the holobiont, the system that they constitute with their microbial communities (McFall-Ngai et al. 2013). It is then only a logical next step to consider aquaculture not only as cultivating one or more aquatic organisms, but also as "cultivating ecosystems" (Bartelme 2018). To learn how they can be manipulated brings us closer to managing disease resistance and productivity based on ecological principles (De Schryver and Vadstein 2014).

Besides abiotic factors, the biotic components of IMTA systems, namely the cultivation of extractive species, could provide a strategy to harness not only the direct function of those species, but also the benefits of their associated microbiome or their influence on the microbial community in water or sediment. IMTA systems can inhibit the growth of potential pathogens and increase beneficial bacteria (Lin et al. 2016). Compared to monoculture systems, IMTA systems with shellfish and seaweed were found to prevent the development of drug resistance and maintain the integrity of the microbial community in the seawater (Ying et al. 2018). The role played by extractive species in controlling the circulating gene flow between aquaculture systems, the environment and human consumers still needs to be investigated in more depth (Ying et al. 2018).

Sediment feeding sea cucumbers are in constant exchange with the sediment, and the overlying water column. They depend on detritus and microbial biomass and in turn influence the microphytobenthos and bacterial abundance (Yingst 1976, Moriarty 1982, Uthicke 2001, Yamazaki et al. 2019). Via their fecal pellets, sea cucumbers enrich sediments with their heterotrophic gut microbiota and thus support decomposition of organic matter not only with their own digestion, but also by supporting fermentation and mineralization in the sediment (MacTavish et al. 2012, Plotieau et al. 2013, Yamazaki et al. 2019).

In the cultivation of plants, growth-promoting rhizobacteria have positive effects on plants both in soil and hydroponic culture through nitrogen fixation, regulating plant hormones, reducing stress and outcompeting pathogens (Lee and Lee 2015). The roots of halophytes grown in aquaponics were recently found to harbor a rich community of nutrient cycling bacteria involved in nitrogen fixation, denitrification and mineralization. The diversity of the bacterial community differs depending on plant species, thus influencing the potential to treat aquaculture effluents (Oliveira et al. 2020).

The small size of an aquaculture farm can be the reason why an IMTA design is not profitable: it is especially these small-scale ventures that account for much of the valuable contribution of aquaculture to nutrition in the developing tropics (Greenfeld et al. 2018, Knowler et al. 2020, Brugere et al. 2021). This in turn provides an opportunity for IMTA to benefit smallholders directly with waste reduction and/or increased food security (Knowler et al. 2020). The diversification of crops in these systems and their multiple harvest cycles lead to increased and/or more steady income and improve household nutrition and resilience (Haque et al. 2015, Tran et al. 2020).

Policy, social and economic considerations

Political will and administrative agendas alone are not enough to change the behavior of aquaculture practitioners and shape the sustainable development of the sector (Brugère et al. 2019, Milewski and Smith 2019). Distributing information about IMTA is an important step to educate practitioners and other stakeholders, address the public perception of re-using fish feces for food production and raise awareness about its role in sustainable aquaculture (Fig. 2) (Alexander et al. 2016, Gunning et al. 2016, Correia et al. 2020, Knowler et al. 2020).

The economic viability of IMTA RAS could be improved through value-added products. Edible halophytes are suitable extractive plants for integration into multitrophic RAS (Pinheiro et al. 2017, Senff et al. 2020). Their comparatively high price as a gourmet food can make their production in RAS profitable (Barreira et al. 2017, Castilho-Barros et al. 2018). It has been shown recently that the fatty acid and lipid profile of edible coastal halophytes from integrated aquaculture are superior to wild harvests (Maciel et al. 2020). Such arguments further underline the value of biomass production in IMTA, especially in terms of recycling fatty acids and amino acids that are specific to and especially valuable in aquafeeds.

Final considerations

Coming back to the conflicting views of various authors on the questions "Who/What is going to feed the world?", it becomes clear that Belton et al. (2020) only highlighted the high-tech, high-energy, high-investment parts of aquaculture and therefore came to their negative conclusion. They were focusing on finfish only and here exclusively on carnivorous fish. In addition, they concentrated on offshore production at large scale. Small scale, community-based (CB) production in family or village systems is not mentioned at all.

Their recommendation to concentrate on freshwater aquaculture neglects the fact that we already have a severe conflict with regard to available land and available freshwater in the efforts to increase the output of agricultural production worldwide. Introduction of more freshwater aquaculture, either in lakes, rivers or land-based aquaculture would only increase that competition. Along those lines, also Costa-Pierce et al. (2022) state: "They (Belton et al. 2020) discredit marine aquaculture rather than explain how all aquaculture sectors are significantly more efficient and sustainable for the future of food than nearly all land-based animal protein alternatives."

In this chapter, we already mentioned the so-called CB production systems, which are common in many tropical countries, such as Indonesia, Vietnam, Tanzania, and Madagascar (Kunzmann et al. 2023). By multiplying the amount of such CB systems along the long coastlines of tropical countries in Southeast Asia, Africa and Latin-America, a significant amount of seafood can be produced, in a more environmentally friendly way, and also more socially inclusive. From our point of view, there are also options in developed countries for co-using space, as for example the North-Sea wind power plant projects show (Buck et al. 2018). This does not exclude fishing, but fishermen need to be restricted to areas where their fishing methods cannot cause major damage. We strongly suggest to do more research on CB production systems, considering low-cost and low-tech IMTA approaches.

References

Acosta, E. J., Rodriguez-Forero, A., Werding, B. and Kunzmann, A. (2020). Effect of density, temperature and diet on the growth, survival and development of larvae and juveniles of *Isostichopus* sp.. Aquaculture Research 52(2): 611–624. doi: 10.1111/are.14918.

Acosta, E. J., Rodríguez-Forero, A., Werding, B. and Kunzmann, A. (2021). Ecological and reproductive characteristics of holothuroids *Isostichopus badionotus* and *Isostichopus* sp. in Colombia. PLoS ONE 16(2): e0247158. doi: 10.1371/journal.pone.0247158.

Ahlgren, M. O. (1998). Consumption and assimilation of salmon net pen fouling debris by the red sea cucumber parastichopus californicus: implications for polyculture. J. World Aquac. Soc. 29: 133–139.

Ahmed, N. and Glaser, M. (2016). Can "Integrated Multi-Trophic Aquaculture (IMTA)" adapt to climate change in coastal Bangladesh? Ocean Coast Manag. 132: 120–131.

Alexander, K. A., Angel, D., Freeman, S., Israel, D. and Johansen, J. (2016). Improving sustainability of aquaculture in Europe: Stakeholder dialogues on Integrated Multi-trophic Aquaculture (IMTA). Environmental Science and Policy 55: 96–106.

Alfiansah, Y. R., Peters, S., Harder, J., Hassenrück, C. and Gärdes, A. (2020). Structure and co-occurrence patterns of bacterial communities associated with white faeces disease outbreaks in Pacific white-leg shrimp *Penaeus vannamei* aquaculture. Scientific Reports 10: 11980.

Annan, K. 2012. AquaVision, Stavanger, Norway, June 2012.

Avnimelech, Y. (2012). Biofloc Technology—A Practical Guide Book, 2nd ed. The World Aquaculture Society, Baton Rouge, Louisiana, United States.

Avnimelech, Y. (2014). Biofloc Technology—A Practical Guidebook, 3rd ed. World Aquaculture Society.

Badiola, M., Mendiola, D. and Bostock, J. (2012). Aquacultural Engineering Recirculating Aquaculture Systems (RAS) analysis: Main issues on management and future challenges. Aquac Eng. 51: 26–35.

Ballester-Moltó, M., Sanchez-Jerez, P., Cerezo-Valverde, J. and Aguado-Giménez, F. (2017). Particulate waste outflow from fish-farming cages. How much is uneaten feed? Mar. Pollut. Bull 119: 23–30.

Barreira, L., Resek, E., João, M., Isabel, M., Pereira, H., Bandarra, N. et al. (2017). Halophytes: Gourmet food with nutritional health benefits? J. Food Compos Anal. 59: 35–42.

Barrington, K., Ridler, N., Chopin, T., Robinson, S. and Robinson, B. (2010). Social aspects of the sustainability of integrated multi-trophic aquaculture. Aquac Int.

Bartelme, R. P., Smith, M. C., Sepulveda-Villet, O. J. and Newton, R. J. (2019). Component microenvironments and system biogeography structure microorganism distributions in recirculating aquaculture and aquaponic systems. mSphere 4: 1–15.

Belton, B., Little, D. C., Zhang, W., Edwards, P., Skladany, M. and Thilsted, S. H. (2020). Farming fish in the sea will not nourish the world. Nature Communications 11: 5804. doi: 10.1038/s41467-020-19679-9.

Birhanu, S. M. and Natarajan, P. (2019). Status of integrated aquaculture-Agriculture systems in Africa. Int. J. Fish Aquat. Stud. 7: 263–269.

Bischoff, A. A. (2007). Solid waste reduction of closed recirculated aquaculture systems by secondary culture of detritivorous organisms xxxx.

Bischoff, A. A., Fink, P. and Waller, U. (2009). The fatty acid composition of *Nereis diversicolor* cultured in an integrated recirculated system: Possible implications for aquaculture. Aquaculture 296: 271–276.

Brown, N., Eddy, S. and Plaud, S. (2011). Utilization of waste from a marine recirculating fish culture system as a feed source for the polychaete worm, *Nereis virens*. Aquaculture 322–323: 177–183.

Brugère, C., Aguilar-Manjarrez, J., Beveridge, M. C. M. and Soto, D. (2019). The ecosystem approach to aquaculture 10 years on—a critical review and consideration of its future role in blue growth. Rev. Aquac. 11: 493–514.019.

Brugere, C., Troell, M. and Eriksson, H. (2021). More than fish: Policy coherence and benefit sharing as necessary conditions for equitable aquaculture development. Marine Policy 123: 104271.

Buck, B. H., Troell, M. F., Krause, G., Angel, D. L., Grote, B. and Chopin, T. (2018). State of the art and challenges for offshore integrated multi-trophic aquaculture (IMTA). Front. Mar. Sci. 5: 165. doi: 10.3389/fmars.2018.00165.

Castilho-Barros, L., Almeida, F. H., Henriques, M. B. and Seiffert, W. Q. (2018). Economic evaluation of the commercial production between Brazilian samphire and whiteleg shrimp in an aquaponics system. Aquacult Int. 26: 1187–1206.

Chen, S., Ling, J. and Blancheton, J. P. (2006). Nitrification kinetics of biofilm as affected by water quality factors. Aquac Eng. 34: 179–197.

Chopin, T., Buschmann, A. H., Halling, C., Troell, M., Kautsky, N., Neori, A. et al. (2001). Integrating seaweeds into marine aquaculture systems: A key toward sustainability. J. Phycol. 37: 975–986.

Chopin, T., Yarish, C. and Sharp, G. (2007). Beyond the monospecific approach to animal aquaculture—the light of integrated multi-trophic aquaculture. Ecol. Genet Implic Aquac Act Methods Technol. Fish Biol. Fish 6: 447–458.

Chopin, T. (2017). Challenges of moving integrated multi-trophic aquaculture along the RandD and commercialization continuum in the Western World. J. Ocean Technol. 12: 34–47.

Correia, M., Azevedo, I. C., Peres, H., Magalhães, R., Oliva-Teles, A., Almeida, C. M. R. et al. (2020). Integrated multi-trophic aquaculture: a laboratory and hands-on experimental activity to promote environmental sustainability awareness and value of aquaculture products. Front. Mar. Sci. 7: 156. doi: 10.3389/fmars.2020.00156.

Costa-Pierce, B. A. (2002). Ecological Aquaculture: The Evolution of the Blue Revolution. John Wiley and Sons Ltd.

Costa-Pierce, B. A., Bockus, A. B., Buck, B. H., van den Burg, S. W. K., Chopin, T., Ferreira, J. G. et al. (2021). A fishy story promoting a false dichotomy to policy makers: it is not freshwater vs. marine aquaculture. Reviews in Fisheries Science and Aquaculture 30(4): 429–446. doi: 10.1080/23308249.2021.2014175.

Crab, R., Avnimelech, Y., Defoirdt, T., Bossier, P. and Verstraete, W. (2007). Nitrogen removal techniques in aquaculture for a sustainable production. Aquaculture 270: 1–14.

Cubillo, A. M., Ferreira, J. G., Robinson, S. M. C., Pearce, C. M., Corner, R. A. and Johansen, J. (2016). Role of deposit feeders in integrated multi-trophic aquaculture—A model analysis. Aquaculture 453: 54–66.

Custódio, M., Villasante, S., Cremades, J., Calado, R. and Lillebø, A. I. (2017). Unravelling the potential of halophytes for marine integrated multi-trophic aquaculture (IMTA)—A perspective on performance, opportunities and challenges. Aquac Environ Interact. 9: 445–460.

Danner, R. I., Mankasingh, U., Anamthawat-Jonsson, K. and Thorarinsdottir, R. I. (2019). Designing aquaponic production systems towards integration into greenhouse farming. Water (Switzerland) 11.

De Schryver, P. and Verstraete, W. (2009). Nitrogen removal from aquaculture pond water by heterotrophic nitrogen assimilation in lab-scale sequencing batch reactors. Bioresource Technology 100: 1162–1167.

De Schryver, P. and Vadstein, O. (2014). Ecological theory as a foundation to control pathogenic invasion in aquaculture. The ISME Journal 8: 2360–2368.

Deng, Y., Zhou, F., Ruan, Y., Ma, B., Ding, X., Yue, X. et al. (2020). Feed types driven differentiation of microbial community and functionality in marine integrated multitrophic aquaculture system. Water 12: 95.

Diver, S. (2006). Aquaponics—Integration of Hydroponics with Aquaculture. ATTRA: 1–28.

FAO. (2010). Aquaculture development. 4. Ecosystem approach to aquaculture. In: FAO Technical Guidelines for Responsible Fisheries, No. 5, Supp. Rome, Italy, p 53.

FAO. (2018). The State of World Fisheries and Aquaculture 2018 - Meeting the sustainable development goals. Rome. Licence: CC BY-NC-SA 3.0 IGO.

Felaco, L., Olvera-Novoa, M. A. and Robledo, D. (2020). Multitrophic integration of the tropical red seaweed Solieria filiformis with sea cucumbers and fish. Aquaculture 527.

Fuller, R. (1989). A Review: Probiotics in man and animals. J. Appl. Bact. 66: 365–378.

Gao, K. and McKinley, K. R. (1994). Use of macroalgae for marine biomass production and CO_2 remediation: a review. J. Appl. Phycol. 6: 45–60.

Gatesoupe, F. J. (1999). Review: The use of probiotics in aquaculture. Aquaculture 180: 147–165.

Glenn, E. P., Brown, J. J. and Blumwald, E. (1999). Salt Tolerance and Crop Potential of Halophytes. CRC Crit. Rev. Plant Sci. 18: 227–255.

Granada, L., Sousa, N., Lopes, S. and Lemos, M. F. L. (2015). Is integrated multitrophic aquaculture the solution to the sectors' major challenges?—a review. Rev. Aquac. 6: 6–18.

Greenfeld, A., Becker, N., McIlwain, J., Fotedar, R. and Bornman, J.F. (2018). Economically viable aquaponics? Identifying the gap between potential and current uncertainties. Reviews in Aquaculture: 1–15. doi: 10.1111/raq.12269.

Guerrero, S. and Cremades, J. (2012). Integrated Multi-trophic Aquaculture (IMTA): A sustainable, pioneering alternative for marine cultures in Galicia. 111: 978–84.

Gunning, D., Maguire, J. and Burnell, G. (2016). The development of sustainable saltwater-based food production systems: a review of established and novel concepts. Water 8: 598.

Hai, N. V. (2015). The use of medicinal plants as immunostimulants in aquaculture: A review. Aquaculture 446: 88–96.

Hair, C. (2012). Sandfish (*Holothuria scabra*) production and sea-ranching trial in Fiji. Asia–Pacific Trop sea cucumber Aquac:129.

Handy, R. D. and Poxton, M. G. (1993). Nitrogen pollution in mariculture: toxicity and excretion of nitrogenous compounds by marine fish. Rev. Fish Biol. Fish 3: 205–241.

Hannah, L., Pearce, C. M. and Cross, S. F. (2013). Growth and survival of California sea cucumbers (*Parastichopus californicus*) cultivated with sable fish (*Anoplopoma fimbria*) at an integrated multi-trophic aquaculture site. Aquaculture 406–407: 34–42.

Haquea, M. M., Alamb, M. R., Alamc, M. M., Basakd, B., Sumib, K. R., Beltond, B. et al. (2015). Integrated floating cage aquageoponics system (IFCAS): An innovation in fish and vegetable production for shaded ponds in Bangladesh. Aquaculture Reports 2: 1–5.

Hargreaves, J. A. (1998). Nitrogen biogeochemistry of aquaculture ponds. Aquaculture 166: 181–212.

Hochard, S., Ducroq, M., Lemonnier, H., Royer, F., Hubert, M., Michaut, H. et al. (2016). Projet Hobical. Holothuries: Lien entre performances de production et de bioremédiation dans le contexte de la crevetticulture en Nouvelle-Calédonie. pp. 163.

Holby, O. and Hall, P. O. J. (1991). Chemical fluxes and mass balances in a marine fish cage farm. II Phosphorus. Mar. Ecol. Prog. Ser. 70: 263–272.

Holmer, M. and Kristensen, E. (1992). Impact of marine fish cage farming on metabolism and sulfate reduction of underlying sediments. 80: 191–201.

Holmer, M., Duarte, C. M., Heilskov, A., Olesen, B. and Terrados, J. (2003). Biogeochemical conditions in sediments enriched by organic matter from net-pen fish farms in the Bolinao area, Philippines. Mar Pollut Bull 46: 1470–1479.

Hou, Y., Sun, Y., Gao, Q., Dong, S., Wen, B. and Yu, H. (2017). Optimal mud to sand ratios in sediment for sea cucumber aquaculture as revealed by carbon stable isotopes. Aquac Environ Interact. 9: 281–291.

Huang, Y., Ciais, P., Goll, D. S., Cresto-aleina, F. and Zhang, H. (2020). The shift of phosphorous transfers in global fisheries and aquaculture. Nat. Commun: 1–10.

Ivy, G., Azari, D. and Giraspy, B. (2006). Development of large-scale hatchery production techniques for the commercially important sea cucumber *Holothuria scabra* var. versicolor (Conand, 1986) in Queensland, Australia. 28–34.

Juinio-Meñez, Marie Antonette, Elsie D. Tech, Isidora P. Ticao, Jay RC Gorospe, Christine Mae A. Edullantes, and Rose Angeli V. Rioja. (2017). Adaptive and integrated culture production systems for the Tropical Sea Cucumber Holothuria Scabra. Fisheries Research 186: 502–13. https://doi.org/10.1016/j.fishres.2016.07.017.

Kerrigan, D. and Suckling, C. C. (2018). A meta-analysis of integrated multitrophic aquaculture: extractive species growth is most successful within close proximity to open-water fish farms. Rev. Aquac. 10: 560–572.

Knowler, D., Chopin, T., Martınez-Espineira, R., Neori, A., Nobre, A., Noce, A. et al. (2020). The economics of Integrated Multi-Trophic Aquaculture: where are we now and where do we need to go? Reviews in Aquaculture 12: 1579–1594.

Kunzmann, A., Todinanahary, G., Msuya, F. E. and Alfiansah, A. (2023). Review: Comparative Environmental Impacts and Development Benefits of Coastal Aquaculture in Three Tropical Countries: Madagascar, Tanzania and Indonesia (early view). Tropical Life Sciences Research. https://ejournal.usm.my/tlsr/article/view/4536.

Lee, S. and Lee, J. (2015). Beneficial bacteria and fungi in hydroponic systems: Types and characteristics of hydroponic food production methods. Scientia Horticulturae 195: 206–215.

Lin, J., Ju, B., Yao, Y., Lin, X., Xing, R., Teng, L. et al. (2016). Microbial community in a multi-trophic aquaculture system of *Apostichopus japonicus, Styela clava* and microalgae. Aquacult. Int. 24: 1119–1140.

Lu, J. and Li, X. (2006). Review of rice-fish-farming systems in China - One of the Globally Important Ingenious Agricultural Heritage Systems (GIAHS). Aquaculture 260: 106–113.

Lyssenko, C. and Wheaton, F. (2006). Impact of positive ramp short-term operating disturbances on ammonia removal by trickling and submerged-upflow biofilters for intensive recirculating aquaculture. Aquac Eng. 35: 26–37.

Macdonald, C. L. E., Stead, S. M. and Slater, M. (2013). Consumption and remediation of European Seabass (Dicentrarchus labrax) waste by the sea cucumber *Holothuria forskali*. Aquac. Int. 21: 1279–1290.

Maciel, E., Domingues, P., Domingues, M. R. M., Calado, R. and Lillebø, A. I. (2020). Halophyte plants from sustainable marine aquaponics are a valuable source of omega-3 polar lipids. Food Chem. 0: 0–32.

MacKay, K. T. (1983). Ecological aquaculture, new approaches to aquaculture in North America. J. World Maric Soc. 14: 704–713.

MacTavish, T., Stenton-Dozey, J., Vopel, K. and Savage, C. (2012). Deposit-feeding sea cucumbers enhance mineralization and nutrient cycling in organically-enriched coastal sediments. PLoS One 7: e50031.

Malone, R. F. and Pfeiffer, T. J. (2006). Rating fixed film nitrifying biofilters used in recirculating aquaculture systems. Aquac Eng. 34: 389–402.

Marrifield, D. L., Dimitroglou, A., Foey, A., Davies, S. J., Baker, R. T. M., Bogwald, J. et al. (2010). Review: The current status and future focus of probiotic and prebiotic applications for salmonids. Aquaculture 302: 1–18.

Martins, C. I. M., Eding, E. H., Verdegem, M. C. J., Heinsbroek, L. T. N., Schneider, O., Blancheton, J. P. et al. (2010). New developments in recirculating aquaculture systems in Europe: A perspective on environmental sustainability. Aquac Eng. 43: 83–93.

McFall-Ngaia, M., Hadfieldb, M. G., Bosch, T. C. G., Careyd, H. V., Domazet-Lo, T., Douglas, A. E. et al. (2013). Animals in a bacterial world, a new imperative for the life sciences. PNAS 110(9): 3229–3236.

Milewski, I. and Smith, R. E. (2019). Sustainable aquaculture in Canada: Lost in translation. Marine Policy 107: 103571.

Mills, D., Duy, N. D. Q., Juinio-Menez, M. a., Raison, C. M. and Zarate, J. M. (2012). Overview of sea cucumber aquaculture and sea-ranching research in the South-East Asian region. Proc an Int Symp held Noumea, New Caledonia, 15–17 Febr 2011 ACIAR Proc No. 136 Aust Cent Int Agric Res Canberra 209: 22–31.

Mizuta, D. D., Froehlich, H. E. and Wilson, J. R. (2022). The changing role and definitions of aquaculture for environmental purposes. Reviews in Aquaculture: 1–12. doi: 10.1111/raq.12706.

Morelli, L. and Capurso, L. (2012). FAO/WHO Guidelines on Probiotics, 10 years later. J. Clin. Gastroenterol. 46(Suppl 1): 2.

Moriarty, D. J. W. (1982). Feeding of Holothuria atra and Stichopus chloronotus on bacteria, organic carbon and organic nitrogen in sediments of the Great Barrier Reef. Mar. Freshw. Res. 33: 255–263.

Neori, A. (2011). "Green water" microalgae: the leading sector in world aquaculture. J. Appl. Phycol. 23: 143–149.

Neori, A., Msuya, F. E., Shauli, L., Schuenhoff, A., Kopel, F. and Shpigel, M. (2003). A novel three-stage seaweed (*Ulva lactuca*) biofilter design for integrated mariculture. J. Appl. Phycol. 15: 543–553.

Neori, A., Chopin, T., Troell, M., Buschmann, A. H., Kraemer, G. P., Halling, C. et al. (2004). Integrated aquaculture: Rationale, evolution and state of the art emphasizing seaweed biofiltration in modern mariculture. Aquaculture 231: 361–391.

Ogah, S. I., Kamarudin, M. S., Amin, S. M. N. and Edaroyati, M. W. P. (2020). Biological filtration properties of selected herbs in an aquaponic system. Aquac Res: 1–9.

Oliveira, V., Martins, P., Marques, B., Cleary, D. F. R., Lillebø, A. I. and Calado, R. (2020). Aquaponics using a fish farm effluent shifts bacterial communities profile in halophytes rhizosphere and endosphere. Sci. Rep. 10: 1–11.

Palmer, P.J., Wang, S., Houlihan, A. and Brock, I. (2014). Nutritional status of a nereidid polychaete cultured in sand filters of mariculture wastewater. Aquac. Nutr. 20: 675–691.

Pinheiro, I., Arantes, R., do Espírito Santo, C.M., do Nascimento Vieira, F., Lapa, K.R., Gonzaga, L.V. et al. (2017). Production of the halophyte *Sarcocornia ambigua* and Pacific white shrimp in an aquaponic system with biofloc technology. Ecol. Eng. 100: 261–267.

Pitt, R. (2001). Review of sandfish breeding and rearing methods SPC Beche-de-mer Information Bulletin 14: 14–21.

Pitt, R., Duy, N., Duy, T. and Long, H. (2004). Sandfish (*Holothuria scabra*) with shrimp (*Penaeus monodon*) co-culture tank trials. SPC beche-de-mer Inf Bull:12–22.

Plotieau, T., Lavitra, T., Gillan, D.C. and Eeckhaut, I. (2013). Bacterial diversity of the sediments transiting through the gut of *Holothuria scabra* (Holothuroidea; Echinodermata). Mar. Biol. 160: 3087–3101.

Price, C., Black, K. D., Hargrave, B. T. and Morris, J. A. (2014). Marine cage culture and the environment: Effects on water quality and primary production. Aquac Environ. Interact 6: 151–174.

Purcell, S. W., Patrois, J. and Fraisse, N. (2006). Experimental evaluation of co-culture of juvenile sea cucumbers, *Holothuria scabra* (Jaeger), with juvenile blue shrimp, *Litopenaeus stylirostris* (Stimpson). Aquac. Res. 37: 515–522.

Purcell, S. W., Hair, C. A. and Mills, D. J. (2012). Sea cucumber culture, farming and sea ranching in the tropics: Progress, problems and opportunities. Aquaculture 368–369: 68–81.

Purcell, S. W., Polidoro, B. a., Hamel, J.-F., Gamboa, R. U. and Mercier, A. (2014). The cost of being valuable: predictors of extinction risk in marine invertebrates exploited as luxury seafood. Proc R Soc B 281: 20133296.

Putro, S. P., Sharani, J., Widowati, Adhy, S. and Suryono. (2020). Biomonitoring of the application of monoculture and integrated multi-trophic aquaculture (IMTA) using macrobenthic structures at tembelas island, kepulauan riau province, Indonesia. J. Mar. Sci. Eng. 8: 1–13.

Rakocy, J. E. (2012). Aquaponics-Integrating fish and plant culture. pp. 343–386. *In*: Tidwell, J. H. (ed.). Aquaculture Productions Systems. WAS Willey-Blackwell.

Richter, C., Roa-Quiaoit, H., Jantzen, C., Al-Zibdah, M. and Kochzius, M. (2008). Collapse of a new living species of giant clam in the Red Sea. Curr. Biol. 18: 1349–1354.

Robinson, G. (2013). A Bright Future for Sandfish Aquaculture. World Aquac March: 18–24.

Robinson, G. (2017). Microbial – Deposit Feeder Aquaculture Bioremediation Systems. Newcastle University.

Robinson, G. and Pascal, B. (2009). From hatchery to community – Madagascar's first village-based holothurian mariculture programme. SPC Beche-de-mer Inf. Bull 29: 38–42.

Robinson, G., Caldwell, G. S., Wade, M. J., Free, A., Jones, C. L. W. and Stead, S. M. (2016). Profiling bacterial communities associated with sediment-based aquaculture bioremediation systems under contrasting redox regimes. Sci. Rep.

Robinson, G., MacTavish, T., Savage, C., Caldwell, G. S., Jones, C. L. W., Probyn, T. et al. (2018). Carbon amendment stimulates benthic nitrogen cycling during the bioremediation of particulate aquaculture waste. Biogeosciences 15: 1863–1878.

Robinson, G., Caldwell, G. S., Jones, C. L. W. and Stead, S. M. (2019). The effect of resource quality on the growth of *Holothuria scabra* during aquaculture waste bioremediation. Aquaculture 499: 101–108.

Schneider, O., Sereti, V., Eding, E. H. and Verreth, J. A. J. (2005). Analysis of nutrient flows in integrated intensive aquaculture systems. Aquac Eng. 32: 379–401.

Senf, P., Blanc, P., Slater, M. and Kunzmann, A. (2020). Low-technology recirculating aquaculture system integrating milkfish *Chanos chanos*, sea cucumber *Holothuria scabra* and sea purslane *Sesuvium portulacastrum*. Aquaculture Environment Interactions 12: 471–484. doi:10.3354/aei00377.

Senff, P. (2021). Improving bioremediation with extractive species in integrated aquaculture: Towards Application in Tropical Recirculating Systems. PhD Thesis, University of Bremen, 152 pp.

Senff, P., Elba, B., Kunzmann, A., Gillis, L. G. and Robinson, G. (2022). Carbon supplementation promotes assimilation of aquaculture waste by the sea cucumber *Holothuria scabra*: Evidence from stable isotope analysis. Aquaculture 547: 7373295. doi: 10.1016/j.aquaculture.2021.737295.

Shpigel, M., Ben-Ezra, D., Shauli, L., Sagi, M., Ventura, Y., Samocha, T. et al. (2013). Constructed wetland with Salicornia as a biofilter for mariculture effluents. Aquaculture 413: 52–63.

Slater, M., Jeffs, A. G. and Sewell, M. A. (2011). Organically selective movement and depositfeeding in juvenile sea cucumber, *Australostichopus mollis* determined in situ and in the laboratory. J. Exp. Mar. Bio. Ecol. 409: 315–323.

Stenton-Dozey, J. M. E., Heath, P., Ren, J. S. and Zamora, L. N. (2020). New Zealand aquaculture industry: research, opportunities and constraints for integrative multitrophic farming. New Zeal J. Mar. Freshw. Res: 1–21.

Summerfelt, S. T. and Sharrer, M. J. (2004). Design implication of carbon dioxide production within biofilters contained in recirculating salmonid culture systems. Aquac. Eng. 32: 171–182.

Thia-Eng, C. (1997). Sustainable aquaculture and integrated coastal management. pp. 177–200. *In*: Bardach, J. E. (ed.). Sustainable Aquaculture. John Wiley and Sons Ltd, New York.

Tran, N., Rodriguez, U.-P., Chan, C. Y., Phillips, M. J., Mohan, C. V., Henriksson, P. J. G. et al. (2017). Indonesian aquaculture futures: An analysis of fish supply and demand in Indonesia to 2030 and role of aquaculture using the Asia Fish model. Mar Policy 79: 25–32.

Troell, M., Halling, C., Neori, A., Chopin, T., Buschmann, A. H., Kautsky, N. et al. (2003). Integrated mariculture: Asking the right questions. Aquaculture 226: 69–90.

Uthicke, S. (2001). Interactions between sediment-feeders and microalgae on coral reefs: Grazing losses versus production enhancement. Mar. Ecol. Prog. Ser. 210: 125–138.

Van Rijn, J. (1996). The potential for integrated biological treatment systems in recirculating fish culture—A review. Aquaculture 139: 181–201.

Van Rijn, J. (2013). Waste treatment in recirculating aquaculture systems. Aquac Eng. 53: 49–56.

Verdegem, M. C. J. (2013). Nutrient discharge from aquaculture operations in function of system design and production environment. Rev. Aquac. 5: 158–171.

Verschuere, L., Rombaut, G., Sorgeloos, P. and Verstraete, W. (2000). Probiotic bacteria as biological control agents in aquaculture. Microbiology and Molecular Biology Reviews 64(4): 655–671.

Watanabe, S., Zarate, J. M., Lebata-Ramos, M. J. H., Nievales, M. F. J. and Kodama, M. (2012). Utilization of Organic Waste from Black Tiger Shrimp, *Penaeus monodon*, by Sandfish, *Holothuria scabra*. pp. 81–86. *In*: Tanaka, K., Morioka, S. and Watanabe, S. (eds.). Sustainable Stock Management and Development of Aquaculture Technology Suitable for Southeast Asia (JIRCAS Working Report No. 75). Japan International Research Center for Agricultural Sciences, Tsukuba, Ibaraki, Japan.

Yamazaki, Y., Sakai, Y., Mino, S. and Sawabe, T. (2020). An annual faecal 16S amplicon sequencing of individual sea cucumber (*Apostichopus japonicus*) demonstrates the feeding behaviours against eukaryotes in natural environments. Aquac Res: 1–7.

Yarnold, J., Karan, H., Oey, M. and Hankamer, B. (2019). Review: Microalgal aquafeeds as part of a circular bioeconomy. Trends in Plant Science 24(10): 959–970.

Ying, C., Chang, M. J., Hua, C. H., Chang, Y. T., Chao, W. L., Yeh, S. L. et al. (2018). The effects of marine farm-scale sequentially integrated multi-trophic aquaculture systems on microbial community composition, prevalence of sulfonamide-resistant bacteria and sulfonamide resistance gene sul1. STOTEN 643: 681–691.

Yingst, J. Y. (1976). The utilization of organic matter in shallow marine sediments by an epibenthic deposit-feeding holothurian. J. Exp. Mar. Bio. Ecol. 23: 55–69.

Zamora, L. N., Yuan, X., Carton, A. G. and Slater, M. (2018). Role of deposit-feeding sea cucumbers in integrated multitrophic aquaculture: progress, problems, potential and future challenges. Rev. Aquac. 10: 57–74.

Zheng, X., Zhang, D., Qin, J. and Wang, Y. (2018). The effect of C/N ratio on bacterial community and water quality in a mussel-fish integrated system. Aquac Res. 49: 1699–1708.

Zhou, Q., Li, K., Jun, X. and Bo, L. (2009). Role and functions of beneficial microorganisms in sustainable aquaculture. Bioresource Technology 100: 3780–3786.

CHAPTER 4

Organic Pollutants and Plastics Threatening the Ocean's Health

An Urgent Issue

Andrés H. Arias,[1,2,*] *Lautaro Girones,*[1] *Ana L. Oliva,*[1] *Jorge E. Marcovecchio,*[1] *Adriana Larrea Valdivia*[3] and *Guilherme Malafaia*[4]

Introduction

The oceans cover 70.8% of the Earth's surface, i.e., approximately 361.3 million km^2 for a total surface area of 510 million km^2. By volume, the oceans and sea ice account for 97.95% of the planet's water. It is estimated that 60% of the world's population (3.6 billion people) live within 60 km of the coast and that this proportion will increase to 75% (6.4 billion) in the next 30 years. 80% of ocean resources are concentrated on the continental shelf; the ocean holds 80% of the world's biological biodiversity, with an uncertain percentage still undiscovered. In this regard, it is estimated that the seabed, which remains largely unexplored, is home to an estimated 10 million species about which we still know nothing.

Over three-quarters of the materials that degrade the oceans come from the land. They infiltrate the soil and reappear in the rivers and basins that flow into the sea through direct runoff or through the atmosphere. The main causes are the lack of urban sanitation, intensive agriculture, and industry. For example, 70% of the waste discharged into the Pacific is not pre-treated (UNEP 1997), while 1.7 billion cubic meters of municipal sewage is discharged directly into the Mediterranean Sea each year, three-quarters of which is untreated (UNEP), and comes along with some 66 billion cubic meters of industrial waste. Developing countries accumulate these disadvantages, as strong population growth, lack of financial means, and often lack of management expertise prevent them from coping with this reality.

The 20th century saw countless changes in the pace of growth and production of human civilization. The development of intensive agriculture, the large-scale concentration of urban areas, and increasing industrialization have contributed to the global pollution of marine ecosystems. The main substances human activity introduces include organochlorine compounds, petroleum products, and heavy metals (Fowler 1990, Tatsukawa et al. 1990). Most of these compounds have a conservative behavior, and

[1] IADO, Instituto Argentino de Oceanografía, CCT-CONICET, Bahía Blanca, Argentina.
[2] Departamento de Química, Universidad Nacional del Sur (UNS)-CONICET, Bahía Blanca, Argentina.
[3] Universidad Nacional de San Agustin de Arequipa – UNSA, Arequipa, Perú.
[4] Biological Sciences Department, Laboratory of Toxicology Applied to the Environment, Goiano Federal Institute, Urutaí, GO, Brazil.
* Corresponding author: aharias@iado-conicet.gov.ar

many are essentially permanently deposited despite their recognized toxicity (Richardson et al. 1995). Therefore, they are commonly included in persistent organic compounds (POPs).

Although oil spills are spectacular in terms of public impact, the marine environment is the victim of silent outrages with much more severe consequences, such as the disposal of gas from ships or the disposal of used lubricants. This negligence is estimated to account for 33% of marine oil pollution. Evidence shows that the concentration of offshore oil has been reduced by improved techniques for transporting and handling oil products (12% spills). However, approximately 40% of the pollution of oil products comes from internal systems. This is mainly due to spills by urban pipelines and lubricants used by engines. The remaining 9% came mainly from offshore oil and gas pipelines (Haynes and Johnson 2000).

Polycyclic aromatic hydrocarbons

Polycyclic aromatic hydrocarbons (PAHs) are a large class of organic compounds comprised of two or more fused benzene rings or, in some cases, a pentagonal ring. Although PAHs describe more than hundreds of individual chemical compounds, most regulations, guidelines, and published data focus on only a few PAHs, typically between 15 and 30 individual PAH compounds. High melting and boiling points, low vapor pressures, and low solubility in water characterize them. Also, PAHs present high octanol/water coefficients (K_{OW}). In general, their hydrophobicity, adsorption to particulate matter, toxicity, and environmental persistence tend to increase with increasing molecular weight. These characteristics are severely important because they are essential in the distribution of PAHs in the environment, facilitating bioaccumulation processes, persistence, and mobility in different environmental matrices (Oliva et al. 2020a).

Although PAHs can originate from natural and anthropogenic sources, anthropogenic activities are generally considered the major source of PAH release into the environment. Thus, pressure from continental regions could be one of the most significant sources of PAHs in the marine environment (Oliva et al. 2020b and references therein). Examples of natural PAH sources include forest fires and volcanic eruptions. Anthropogenic sources of PAHs include, among others, burning fossil fuels, carbon and wood, industrial emissions, and oil and petroleum spills.

PAHs are ubiquitous in marine environments. Due to their persistence, toxicity, carcinogenicity, and mutagenicity, as well as their effects on reproductive impairments and immune function, growth, and development in lower and higher trophic levels, they are regulated by several international environmental agencies. For example, the United States Environmental Protection Agency (USEPA) has designated 16 PAHs as priority control pollutants (USEPA 2002). According to both the USEPA and the International Agency for Research on Cancer, over half of these PAHs are potentially carcinogenic to humans (IARC 2010).

PAHs can enter marine environments through various pathways, for instance, spillage of fossil fuels, ship traffic, urban runoff, municipal and industrial wastewater discharge, and wet/dry atmospheric depositions. For example, oil spill accidents are one of the most visible events of releasing PAHs into the marine environment (Honda and Suzuki 2020). Several oil spills have occurred in the last year, and tons of crude oil have been released into the aquatic/marine environment (Beyer et al. 2016, Silva et al. 2022). For instance, one of the most recent oil spills was the Deepwater Horizon Oil Spill, in which approximately 4 million barrels of oil were discharged in the northern Gulf of Mexico between April 20 and July 15, 2010 (Botello et al. 2015, Honda and Suzuki 2020). The toxicity of PAHs introduced in this spill was well demonstrated in a wide range of organisms such as fish, invertebrates, plankton, birds, and mammals, causing broad adverse effects including effects in growth and reproduction, impaired physiological health, and mortality (Beyer et al. 2016, Daly et al. 2016, Honda and Suzuki 2020). Moreover, several PAHs exceeding the toxicity threshold occurred in surface waters and sediments (Beyer et al. 2016). Due to their persistence, PAHs associated with sediments can be re-suspended, turning bioavailable for aquatic organisms and entering the food chain, posing risks to marine life and human health. So, efforts to mitigate the impacts of such spills often involve monitoring and managing the presence of PAHs in affected ecosystems.

PAHs can be transferred among environmental compartments. They are widely distributed in the atmosphere and are one of the first atmospheric pollutants designated as suspected carcinogens, with atmospheric transport being the most critical pathway for their environmental distribution (Kim et al. 2013). For instance, domestic combustion represents approximately 84% of the total atmospheric load (ref). PAHs are usually sorbed onto airborne particulate matter and can be transported over long distances before deposition through precipitation and dry deposition (Ravindra et al. 2008). Once in the aquatic media, due to their hydrophobic properties, PAHs are transported from surface waters by sorption to settling particles and ultimately deposited in sediments, which represent the most important reservoir for PAHs in the marine environment (Culotta et al. 2006, Maletić et al. 2019). Then, due to changes in aquatic conditions or hydrodynamic factors, PAHs can be remobilized into the water column, becoming bioavailable to organisms (Zhang et al. 2019). Consequently, for decades, PAHs in marine sediments have garnered significant attention from the scientific community, as mentioned earlier, due to their potential adverse effects on both marine biota and humans. Most studies have been conducted in anthropogenically impacted coastal ecosystems, such as estuaries, ports, and bays (Hong et al. 2016, Rajpara et al. 2017, Lin et al. 2018, Oliva et al. 2020a, Eça et al. 2021, Dai et al. 2022, Shi et al. 2022). These studies suggest that PAHs heavily pollute coastal sediments in several worldwide locations, and high ecological risks associated with these sediments have been observed, particularly in areas near anthropogenic activity (Li et al. 2016, Liu et al. 2017, Merhaby et al. 2019, Oliva et al. 2020a, Dai et al. 2022, Bianco et al. 2023).

Persistent organic pollutants

Persistent Organic Pollutants (POPs) constitute a diverse group of chemical compounds known for their remarkable resistance to natural degradation processes (Stockholm Convention 2001). This category encompasses a wide range of substances, including organochlorine pesticides like DDT, lindane, dieldrin, aldrin, and chlordane, industrial chemicals such as polychlorinated biphenyls (PCBs) and polybrominated diphenyl ethers (PBDEs), as well as unintentionally produced compounds like polychlorinated dibenzo-p-dioxins (PCDDs) and polychlorinated dibenzofurans (PCDFs). Their enduring presence in the environment and tendency to accumulate within living organisms pose a dual and significant concern for environmental and human health (DeBruyn et al. 2009). This characteristic leads to a phenomenon known as biomagnification, where their concentration increases as they move through trophic levels in the food chain (Fig. 1; Kajiwara et al. 2004, Kleivane et al. 2004, Tanabe et al. 1994). This process presents a substantial threat to marine ecosystems. At the top of the food web, organisms such as marine mammals, sharks, and large fish species can accumulate dangerous levels of POPs (Noël et al. 2009, Mos et al. 2010). This bioaccumulation and biomagnification directly affect wildlife and human populations, particularly coastal communities whose livelihoods are closely tied to marine resources for sustenance and economic prosperity.

The global community has responded to this environmental and social challenge by instituting critical regulatory frameworks. The Stockholm Convention, a pivotal international agreement adopted in 2001, stands as a testament to the effectiveness of international cooperation in addressing the challenge of POPs. Its primary objective is to eliminate or strictly control the production and use of key POPs, thereby safeguarding human health and the environment. Over the years, it has incorporated new substances with characteristic POP features: high environmental persistence, bioaccumulative potential, toxicity, long-range transport capabilities, and large-scale production (Stockholm Convention 2001). Simultaneously, the Basel Convention, which came into force in 1992, seeks to regulate and reduce cross-border movements of hazardous wastes, including specific POPs, and their proper disposal (Basel Convention 1992). By establishing strict guidelines for hazardous waste management, this convention reinforces efforts to mitigate the release of POPs into the environment and safeguard ecosystems, including our oceans.

Although there is no dedicated international convention exclusively addressing POPs in the oceans, the provisions of the Stockholm and Basel Conventions hold profound relevance for marine environments. By regulating the movement and disposal of hazardous wastes, these conventions indirectly contribute

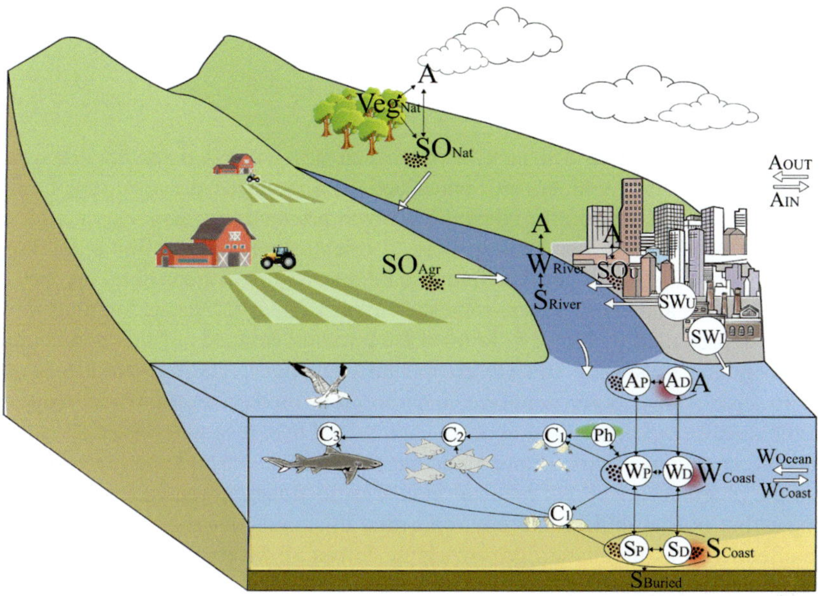

Figure 1. Schematic representation of the main entry pathways into the ocean for Persistent Organic Pollutants (POPs) and associated environmental fate processes. A: air; AP: air-borne particle-associated fraction; AD: air-dissolved fraction; W: water; WP: water-borne particle-associated fraction; WD: water-dissolved fraction; S: sediment or soil; SP: sediment-borne particle-associated fraction; SD: sediment-dissolved fraction; C: consumer (C1, C2, C3: trophic levels); Phyto: primary producers; Agri.S: agricultural soil; Nat.S: natural soil; Urb.S: urban soil; WW and IW: urban and industrial wastewater discharges; Veg: vegetation.

to protecting oceanic ecosystems. In doing so, they play a fundamental role in mitigating the impact of POPs on our oceans.

The sources of POPs entering the marine environment are numerous and diverse. These can include direct discharges from industrial processes, runoff from pesticide-laden agricultural areas, and atmospheric depositions, where POPs can be transported over great distances before finally settling into the ocean (Fig. 1; Girones et al. 2020, Takeoka et al. 1991). Furthermore, legacy POPs, such as certain persistent pesticides like DDT, continue to be released from historically contaminated sites, increasing the burden of these contaminants in marine ecosystems (Fig. 1; Mishra et al. 2022).

Once within the marine environment, these contaminants can be transported over long distances through ocean currents, affecting even the most remote and seemingly pristine regions (Ilyina et al. 2006, Mackay et al. 2006, Morales et al. 2022, Riget et al. 2019). This interconnectedness of marine environments means that the impacts of POPs can be far-reaching, with consequences felt even in areas far removed from their original sources.

Currently, numerous globally implemented measures aimed at controlling well-established POPs have demonstrated positive results. Concentrations of certain legacy POPs have decreased in specific regions, showcasing the effectiveness of regulatory efforts (Jang et al. 2022, Rigét et al. 2010, 2019, Tsygankov et al. 2022). However, challenges persist, particularly in the case of emerging POPs and those with broad exemptions in their use and production (Jang et al. 2022, Muir and Miaz 2021, Rigét et al. 2019). Additionally, the ongoing identification of new POPs underscores the complex nature of this issue. Many chemical substances remain unidentified, underscoring the need for sustained research and monitoring efforts.

In conclusion, the presence of POPs constitutes a significant threat to the vitality of our oceans, with far-reaching implications for both the environment and human societies. While strides have been made in mitigating the impacts of established POPs, new challenges persist in the form of emerging contaminants and those with broad exemptions. It is imperative that we remain vigilant, committed to regulation, and dedicated to ongoing research efforts in order to safeguard our oceans and the diverse array of species dependent upon them.

Plastics in the marine system

The unprecedented increase in world population, which has witnessed remarkable growth over the past century, has ushered in a new era of human activity that spans the spectrum from urbanization to industrialization, from agriculture to innovation-driven productivity. This exponential rise in global population has propelled an ever-expanding wave of urbanization, with millions flocking to cities in search of better opportunities and improved living standards. As urban areas burgeon, they demand more infrastructure, resources, and energy, increasing the environmental footprint. Simultaneously, industrial activity has surged, propelled by the need to meet the growing demands of this burgeoning population, creating both economic prosperity and environmental challenges. The agricultural sector has had to revolutionize to sustain this larger populace, incorporating modern farming techniques, which have boosted food production but also raised concerns about land use, resource depletion, and the impact of agrochemicals on ecosystems. Furthermore, pursuing enhanced healthcare and increased life expectancy has prompted the pharmaceutical and medical industries to flourish, generating life-saving treatments and technologies but posing ethical and environmental dilemmas. As the global population continues to surge, the effects of this population growth ripple across all facets of human activity, necessitating a delicate balance between progress and sustainability. This challenge underscores the imperative of adopting innovative and sustainable practices in urban development, industry, agriculture, and healthcare to ensure that the benefits of an increased population do not come at the cost of ecological and social well-being.

Due to poor management, many human activities release thousands of pollutants and xenobiotic compounds, exerting direct or indirect pressures on aquatic and terrestrial ecosystems worldwide. In particular, the production and disposal of plastics and related materials have recently surfaced as a major global pollution concern, as highlighted by Beaumont et al. in 2019. Numerous scientific investigations and publications from governmental and non-governmental sources have consistently demonstrated the widespread presence of plastics within the global aquatic ecosystem. This pervasive distribution has raised a spectrum of toxicological concerns for nearly all aquatic life forms (Anbumani et al. 2018). The persistent release and enduring presence of these contaminants in marine environments have emerged as a pressing worldwide challenge, demanding immediate implementation of management strategies to mitigate potential threats to organism survival, protect human health, and preserve the aesthetic integrity of ecosystems.

The marine environment is a saltwater water body that covers most of the Earth's surface and supports the biodiversity of almost half of the world's primary products. Additionally, it provides important support to foster economic growth in many countries through ecotourism and natural resource management (fisheries, energy, etc.). More than 23% of the world's population (1.2 billion people) lives along 100 kilometers of coastline and depends on it for its functions, highlighting the importance of the marine environment for human survival (Ramsh et al. 2015). Despite this explicit recognition of critical importance, human activities have meant that the destination of most xenobiotics produced is seawater, making plastic debris a large part of marine litter in our seas and oceans (Galgani et al. 2019). Plastics are prevalent in nearly all marine ecosystems worldwide, and their structure and chemical composition allow them to penetrate many species and cause numerous toxic effects (Guzzeti et al. 2018). In addition to the chemical components of plastics (complex monomer mixtures), other molecules reconfigure their functionality (stabilizers, flame retardants, pigments, etc.), and many other pollutants in the oceans (e.g., persistent organic compounds, trace amounts of metals) also be adsorbed and circulate with plastics. This significantly increases the risks of plastics to marine life and human health (Hermabessiere et al. 2017). Over time, plastic breaks down into different sizes and shapes through mechanical and thermal processes.

In the formation of secondary microplastics, plastic fragments, and fibers are continuously produced and released into the marine environment by photothermal degradation, oxidation, and/or mechanical degradation (Li et al. 2018). Microplastics are ubiquitous in most marine environments around the world, containing more than 95 percent of waste, including surface water, water column, coastal lines, sea sediments, marine layers, and biological layers, and their abundance and distribution are variable in space (Free et al. 2017).

During manufacturing, various chemicals, such as flame retardants, plastics, antioxidants, UV stabilizers, and pigments, are added to change the final characteristics of materials. For example, plastic packaging contains 906 chemical compounds (approximately 40% of the world's annual plastic production). In addition to these additives and chemical compounds, plastic particles can absorb pollutants from the water column and form a potential source of transport materials and toxic chemicals. The adsorption and decomposition processes are complex and depend on many factors such as pollutants' water absorption, polymer type, period of residence in the medium, medium type (water, soil, sediments, etc.), temperature, pressure, and salt. In these processes, sediments and organic matter compete in the adsorption of pollutants, and plastics are not the only agents exposed to these pollutants, leading to complex systems. In particular, microplastics can absorb many persistent organic pollutants (POPs), such as polycyclic aromatic hydrocarbons (PAHs), polycyclic biphenyl (PCBs) and dichlorodiphenyltrichloroethane (DDTs). Similarly, they can absorb trace metals (such as copper and lead) (Pozo et al. 2020). As the size range of plastics decreases, nanoplastics have high surface versus volume ratios, and extremely high absorbing affinities, change exposure and associated risk, and increase the importance of contributing to chemical exposure.

Mass balance and transportation routes

It is estimated that the current plastic content of the ocean is between 75 million and 199 million tons, 80% of which comes from the earth due to poor management of municipal solid waste (MSWs). Plastic waste contributes to the oceans of each continent; for example, Asia contributes between 42 and 58 million tons (MTon), Africa contributes between 10 and 20 tons, Latin America and the Caribbean contribute between 6.7 and 8.3 tons, Europe contributes between 1.3 and 9.1 tons, North America contributes between 0.03 and 0.3 tons, and Oceania contributes between 0.05 and 0.32 tons (Jambeck et al. 2015). These values have been updated recently, indicating that about 47% of plastic waste produced annually by municipalities around the world (between 60 and 90 tons) is not managed correctly and is likely to end up in the oceans. Latin America accounts for 4% of world plastics production and 8% of global plastic consumption (UNEP 2018). In Latin America, the main structures of microplastics are fibers (62%), fragments (17%), and pellets (9%), and the main polymers are PE (80% polyethylene), PP (polypropylene), PET (polyethylene terephthalate) and PS (polystyrene).

Plastic channels to the ocean are extensive and varied, including land wash, river, river discharge, wastewater discharge, air transport, and direct discharge. In addition, storms and natural disasters are places where large quantities of plastic enter the oceans (GESAMP 2015). The largest amount of plastics entering the ocean comes from land sources. A primary microplastic model indicates that about two-thirds (63.1%) of the contribution comes from washing (33.8% from synthetic textiles) and tire abrasion (28.3%). The third major contribution (24.2%) came from urban dust. Personal care products account for only 2% of the world's oceans (Bouscher and Friot 2017). Waste and plastics move on different transport routes or in different parts of the marine environment, depending on their chemical and physical properties, speeds, surface properties, and particle sizes. It is estimated that there are significant delays between land deposits and their arrival in the ocean, ranging from several years to decades. Furthermore, they showed that the current generation of secondary microplastics in the ocean results from the degradation of objects produced in the 1990s and early 1990s.

Changes must be applied

The business-as-usual model shows that the current population and plastic market growth will lead to a 70% increase in plastic usage and waste comparing 2040 to 2020 (OECD 2023). Presently, the end-of-life ends of plastic waste are split between landfilling (50%), incineration (17%), recycling (14%),

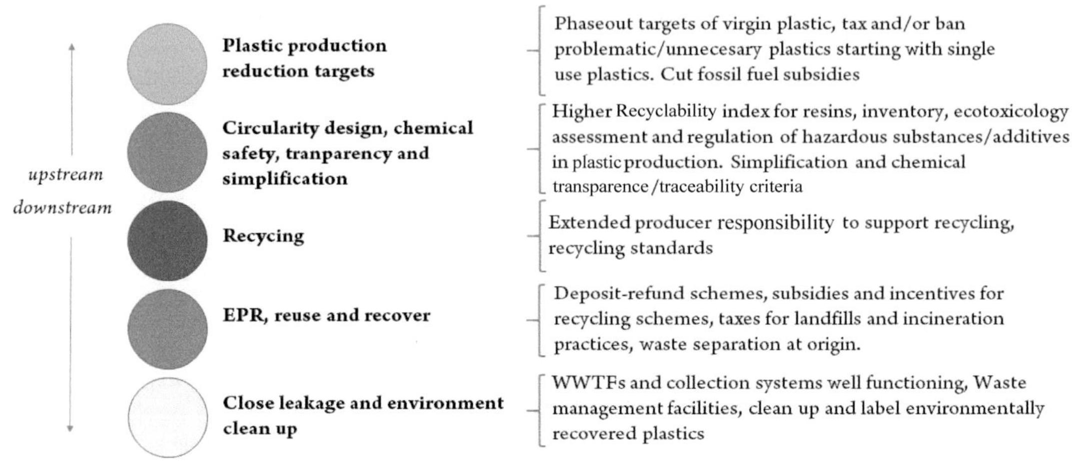

Figure 2. World measures and policy interventions are needed to transition towards a plastic-free environment/oceans. Note - WWTFs: wastewater treatment plants; EPR: Extended Producer Responsibility.

and mismanagement (19%); if changes are not applied, a major role for landfilling and incineration is expected to persist at the end of the life cycle of plastics (OECD 2023). Plastic waste currently makes up 41% of marine debris, with an estimated 12 million tons of plastic entering the ocean each year and 142 million tons of trash currently stored in the ocean (Horton 2022). As projected in numerous models, the plastic leakage to water systems and oceans will be accelerated in the coming years (OECD 2023); indeed, without a significant improvement of waste management facilities, a tenfold increase in ocean dumps is expected for 2025 (Pincelli et al. 2021).

The plastic life cycle needs to be addressed as a whole. The release of chemicals associated with plastics through leaching into humans and the environment should be addressed since there are many chemicals of concern or even well-known endocrine-disrupting compounds presently used in plastics during their production. The last report of the Basel, Rotterdam, and Stockholm Convention shows more than 13000 chemicals used during their manufacturing, with 46% of them without hazard data in the regulatory databases. Micro and nanoplastics can enter the human body through inhalation and absorption via the skin and accumulate in organs, including the placenta. Human uptake of microplastics via seafood will likely pose severe threats to coastal and indigenous populations, where marine species are the main food source. Therefore, we urge the application of safety, sustainability, essentiality, and transparency criteria for all plastic types: bio- and fossil-based feedstock, chemicals, polymers, alternatives, substitutes, etc. This must be accompanied by an independent, trusted science-policy interface to review and update reduction targets, chemical hazard assessment criteria, guidelines, protocols, and procedures for monitoring and reporting (Fig. 2).

References

Anbumani, S. and Kakkar, P. (2018). Ecotoxicological effects of microplastics on biota: a review. Environmental Science and Pollution Research 25: 14373–14396.

Basel Convention on the Control of Transboundary Movements of Hazardous Wastes and their Disposal, March 22, 1989, 1673 U.N.T.S. 126, entered into force on May 5, 1992.

Beaumont, N. J., Aanesen, M., Austen, M. C., Borger, T., Clark, J. R., Cole, M. et al. (2019). Global ecological, social and economic impacts of marine plastic. Marine Pollution Bulletin 142: 189–195.

Beyer, J., Trannum, H. C., Bakke, T., Hodson, P. V. and Collier, T. K. (2016). Environmental effects of the Deepwater Horizon oil spill: a review. Marine Pollution Bulletin 110(1): 28–51.

Bianco, F., Race, M., Papirio, S. and Esposito, G. (2023). A critical review of the remediation of PAH-polluted marine sediments: current knowledge and future perspectives. Resources, Environment and Sustainability 11: 100101.

Botello, A. V., Soto, L. A., Ponce-Vélez, G. and Villanueva, S. (2015). Baseline for PAHs and metals in NW Gulf of Mexico related to the Deepwater Horizon oil spill. Estuarine, Coastal and Shelf Science 156: 124–133.

Boucher, J. and Friot D. (2017). Primary Microplastics in the Oceans: A Global Evaluation of Sources. Gland, Switzerland: IUCN. 43: 2017–002.

Culotta, L., De Stefano, C., Gianguzza, A., Mannino, M. R. and Orecchio, S. (2006). The PAH composition of surface sediments from Stagnone coastal lagoon, Marsala (Italy). Marine Chemistry 99(1-4): 117–127.

Dai, C., Han, Y., Duan, Y., Lai, X., Fu, R., Liu, S. et al. (2022). Review on the contamination and remediation of polycyclic aromatic hydrocarbons (PAHs) in coastal soil and sediments. Environmental Research 205: 112423.

Daly, K. L., Passow, U., Chanton, J. and Hollander, D. (2016). Assessing the impacts of oil-associated marine snow formation and sedimentation during and after the Deepwater Horizon oil spill. Anthropocene 13: 18–33.

Debruyn, A. M., Meloche, L. M. and Lowe, C. J. (2009). Patterns of bioaccumulation of polybrominated diphenyl ether and polychlorinated biphenyl congeners in marine mussels. Environmental Science and Technology 43(10): 3700–3704.

Eça, G. F., Albergaria-Barbosa, A. C., de Souza, M. M., Costa, P. G., Leite, A. S., Fillmann, G. et al. (2021). Polycyclic aromatic hydrocarbons in sediments and shellfish from Todos os Santos bay, Brazil. Marine Pollution Bulletin 173: 112944.

Fowler, S. W. (1990). Critical review of selected heavy metal and chlorinated hydrocarbon concentrations in the marine environment. Marine Environmental Research 29(1): 1–64.

Fowler, S. W., Readman, J. W., Oregioni, B., Villeneuve, J. P. and McKay, K. (1993). Petroleum hydrocarbons and trace metals in near shore Gulf sediments and biota before and after the 1991 war: An assessment of temporal and spatial trends. Mar. Pollut. Bull. 27: 171–182.

Frère, Paul-Pont, I., Rinnert, E., Petton, S., Jaffré, J., Bihannic, I. et al. (2017). Influence of environmental and anthropogenic factors on the composition, concentration and spatial distribution of microplastics: a case study of the Bay of Brest (Brittany, France). Environmental Pollution 225: 211–222.

Galgani, L., Beiras, R., Galgani, F., Panti, C. and Borja, A. (2019). Impacts of marine litter. Frontiers in Marine Science 6: 208.

GESAMP. (2015). Sources, Fate and Effects of Microplastics in the Marine Environment: A Global Assessment, pp 98.

Guzetti, Sureda, A., Tejada, S. and Faggio, C. (2018). Microplastic in marine organism: Environmental and toxicological effects. Environmental Toxicology and Pharmacology 64: 164–171.

Hahladakis, J. N., Velis, C. A., Weber, R., Iacovidou, E. and Purnell, P. (2018). Closing the loop on plastic packaging materials: What is quality and how does it affect their circularity? Science of the Total Environment 630: 1394–1400.

Hermabessiere, L., Dehaut, A., Paul-Pont, I., Lacroix, C., Jezequel, R., Soudant, P. et al. (2017). Occurrence and effects of plastic additives on marine environments and organisms: a review. Chemosphere 182: 781–793.

Honda, M. and Suzuki, N. (2020). Toxicities of polycyclic aromatic hydrocarbons for aquatic animals. International Journal of Environmental Research and Public Health 17(4): 1363.

Hong, W. J., Jia, H., Li, Y. F., Sun, Y., Liu, X. and Wang, L. (2016). Polycyclic aromatic hydrocarbons (PAHs) and alkylated PAHs in the coastal seawater, surface sediment and oyster from Dalian, Northeast China. Ecotoxicology and Environmental Safety 128: 11–20.

Horton, A. A. (ed.). (2022). Plastic Pollution in the Global Ocean (Vol. 1). World Scientific.

IARC. (2010). International Agency for Research on Cancer. Some Non-heterocyclic Polycyclic Aromatic Hydrocarbons and Some Related Exposures, Vol 92. http://monographs.iarc.fr/ENG/Monographs/vol92/mono92.pdf.

Ilyina, T., Pohlmann, T., Lammel, G. and Sündermann, J. (2006). A fate and transport ocean model for persistent organic pollutants and its application to the North Sea. Journal of Marine Systems 63(1-2): 1–19.

Jambeck, J. R., Geyer, R., Wilcox, C., Siegler, T. R., Perryman, M., Andrady, A. et al. (2015). Plastic waste inputs from land into the ocean. Science 347(6223): 768–771.

Jang, M., Shim, W. J., Han, G. M., Ha, S. Y., Cho, Y., Kim, M. et al. (2022). Spatial distribution and temporal trends of classical and emerging persistent organic pollutants (POPs) in black-tailed gull (Larus crassirostris) eggs from Korea. Science of The Total Environment 845: 157244.

Kajiwara, N., Matsuoka, S., Iwata, H., Tanabe, S., Rosas, F. C. W., Fillmann, G. et al. (2004). Contamination by persistent organochlorines in cetaceans incidentally caught along Brazilian coastal waters. Archives of Environmental Contamination and Toxicology 46: 124–134.

Kim, K. H., Jahan, S. A., Kabir, E. and Brown, R. J. (2013). A review of airborne polycyclic aromatic hydrocarbons (PAHs) and their human health effects. Environment International 60: 71–80.

Kleivane, L., Severinsen, T., Lydersen, C., Berg, V. and Skaare, J.U. (2004). Total blubber burden of organochlorine pollutants in phocid seals; methods and suggested standardization. Science of the Total Environment 320: 109–119.

Komyakova, V., Vince, J. and Haward, M. (2020). Rep. Natl. Env. Sci. Progr - Univ. Tasmania, 40 pp.
Li, W. C. (2018). The occurrence, fate, and effects of microplastics in the marine environment. pp. 133–173. en: Zeng, E. Y. (ed.). Microplastic Contamination in Aquatic Environments. Elsevier, Amsterdam (Holanda).
Li, X., Hou, L., Li, Y., Liu, M., Lin, X. and Cheng, L. (2016). Polycyclic aromatic hydrocarbons and black carbon in intertidal sediments of China coastal zones: concentration, ecological risk, source and their relationship. Science of the Total Environment 566: 1387–1397.
Lin, F., Han, B., Ding, Y., Li, Q., Gao, W. and Zheng, L. (2018). Distribution characteristics, sources, and ecological risk assessment of polycyclic aromatic hydrocarbons in sediments from the Qinhuangdao coastal wetland, China. Marine Pollution Bulletin 127: 788–793.
Liu, N., Li, X., Zhang, D., Liu, Q., Xiang, L., Liu, K. et al. (2017). Distribution, sources, and ecological risk assessment of polycyclic aromatic hydrocarbons in surface sediments from the Nantong Coast, China. Marine Pollution Bulletin 114(1): 571–576.
Mackay, D., Shiu, W. Y., Ma, K. C. and Lee, S. C. (2006). Handbook of Physical-Chemical Properties and Environmental Fate for Organic Chemicals. CRC Press.
Maletić, S. P., Beljin, J. M., Rončević, S. D., Grgić, M. G. and Dalmacija, B. D. (2019). State of the art and future challenges for polycyclic aromatic hydrocarbons is sediments: sources, fate, bioavailability and remediation techniques. Journal of Hazardous Materials 365: 467–482.
Merhaby, D., Rabodonirina, S., Net, S., Ouddane, B. and Halwani, J. (2019). Overview of sediments pollution by PAHs and PCBs in mediterranean basin: Transport, fate, occurrence, and distribution. Marine Pollution Bulletin 149: 110646.
Mishra, A., Kumari, M., Kumar, R., Iqbal, K. and Thakur, I. S. (2022). Persistent organic pollutants in the environment: Risk assessment, hazards, and mitigation strategies. Bioresource Technology Reports 101143.
Morales, P., Roscales, J. L., Muñoz-Arnanz, J., Barbosa, A. and Jiménez, B. 2022. Evaluation of PCDD/Fs, PCBs and PBDEs in two penguin species from Antarctica. Chemosphere 286: 131871.
Mos, L., Cameron, M., Jeffries, S. J., Koop, B. F. and Ross, P. S. (2010). Risk-based analysis of polychlorinated biphenyl toxicity in harbor seals. Integrated Environmental Assessment and Management 6(4): 631–640.
Muir, D. and Miaz, L. T. (2021). Spatial and temporal trends of perfluoroalkyl substances in global ocean and coastal waters. Environmental Science and Technology 55(14): 9527–9537.
Noël, M., Barrett-Lennard, L., Guinet, C., Dangerfield, N. and Ross, P. S. (2009). Persistent organic pollutants (POPs) in killer whales (Orcinus orca) from the Crozet Archipelago, southern Indian Ocean. Marine Environmental Research 68(4): 196–202.
Oliva, A. L., Ronda, A. C., Girones, L., Orazi, M., Recabarren-Villalón, T., Marcovecchio, J. E. et al. (2020a). Polycyclic aromatic hydrocarbons. Coastal and Deep Ocean Pollution 288–316.
Oliva, A. L., Quintas, P. Y., Ronda, A. C., Marcovecchio, J. E. and Arias, A. H. (2020b). First evidence of polycyclic aromatic hydrocarbons in sediments from a marine protected area within Argentinean Continental Shelf. Marine Pollution Bulletin 158: 111385.
Pincelli, I. P., de Castilhos Júnior, A. B., Matias, M. S. and Rutkowski, E. W. (2021). Post-consumer plastic packaging waste flow analysis for Brazil: The challenges moving towards a circular economy. Waste Management 126: 781–790.
Pozo, K., Urbina, W., Gómez, V., Torres, M., Nuñez, D., Přibylová, P. et al. (2020). Persistent organic pollutants sorbed in plastic resin pellet—"Nurdles" from coastal areas of Central Chile. Marine Pollution Bulletin 151: 110786.
Rajpara, R. K., Dudhagara, D. R., Bhatt, J. K., Gosai, H. B. and Dave, B. P. (2017). Polycyclic aromatic hydrocarbons (PAHs) at the Gulf of Kutch, Gujarat, India: Occurrence, source apportionment, and toxicity of PAHs as an emerging issue. Marine Pollution Bulletin 119(2): 231–238.
Ramesh, R., Chen, Z., Cummins, V., Day, J., D'Elia, C., Dennison, B. et al. (2015). Land–ocean interactions in the coastal zone: Past, present & future. Anthropocene 12: 85–98.
Ravindra, K., Wauters, E. and Van Grieken, R. (2008). Variation in particulate PAHs levels and their relation with the transboundary movement of the air masses. Science of the Total Environment 396(2-3): 100–110.
Rigét, F., Bignert, A., Braune, B., Dam, M., Dietz, R., Evans, M. et al. (2019). Temporal trends of persistent organic pollutants in Arctic marine and freshwater biota. Science of the Total Environment 649: 99–110.
Rigét, F., Bignert, A., Braune, B., Stow, J. and Wilson, S. (2010). Temporal trends of legacy POPs in Arctic biota, an update. Science of the Total Environment 408(15): 2874–2884.
Shi, W., Xu, M., Liu, Q. and Xie, S. (2022). Polycyclic aromatic hydrocarbons in seawater, surface sediment, and marine organisms of Haizhou Bay in Yellow Sea, China: Distribution, source apportionment, and health risk assessment. Marine Pollution Bulletin 174: 113280.
Stockholm Convention on Persistent Organic Pollutants, May 22, 2001, 40 I.L.M. 532: 2001, revised in 2019.

Takeoka, H., Ramesh, A., Iwata, H., Tanabe, S., Subramanian, A. N., Mohan, D. et al. (1991). Fate of the insecticide HCH in the tropical coastal area of South India. Marine Pollution Bulletin 22: 290–297.

Tanabe, S., Iwata, H. and Tatsukawa, R. (1994). Global contamination by persistent organochlorines and their ecotoxicological impact on marine mammals. Science of the Total Environment 154: 163–177.

Tatsukawa, R., Hashimoto, S. and Wakimoto, T. (1990). PCDDs in the sediments accumulated about 8120 years ago from Japanese coastal areas. Chemosphere 21(7): 825–835.

Towards Eliminating Plastic Pollution by 2040: A Policy Scenario Analysis. OECD, November 2023.

Tsygankov, V. Y., Donets, M. M., Gumovskiy, A. N. and Khristoforova, N. K. (2022). Temporal trends of persistent organic pollutants biotransport by Pacific salmon in the Northwest Pacific (2008–2018). Marine Pollution Bulletin 185: 114256.

UNEP. (2018). Waste Management Outlook for Latin America and the Caribbean.

USEPA. (2002). Polycyclic Organic Matter. US Environmental Protection Agency. Avalible from: https://www.epa.gov/sites/production/files/2016-09/documents/polycyclic-organic-matter.pdf.

Zhang, M., He, P., Qiao, G., Wang, J., Huang, J., Yuan, X. et al. (2019). Distribution, sources, and risk assessment of polycyclic aromatic hydrocarbons (PAHs) in surface sediments of the Subei Shoal, China. Marine Pollution Bulletin 149: 110640.

CHAPTER 5

The Role of Driftwood in the Trapping of Plastics in Beach Environments

Nelson Rangel-Buitrago,[1,2,*] *Adriana Gracia C.,*[2] *Juan Rangel,*[2] *Mohamed Ben-Haddad*[3] and *William J. Neal*[4]

Introduction

Beaches are unique environments of planet Earth. These sediment accumulations range in particle size from sand to boulders and are formed by coastal waves, tides, and currents along a coastline (Pilkey et al. 2011). These areas are among the most dynamic and changing earth features, serving as home to environments of extraordinary biodiversity; beaches are also areas of significant importance due to their unique ecological functions (Jackson and Short 2020).

Beaches can support a significant biodiversity, including organisms that do not inhabit other environments. The beach intertidal zone is a habitat for many organisms, such as bacteria, protozoans, microalgae, and macrobenthic invertebrates (Gracia et al. 2018). These environments are also strategic areas that provide multiple environmental services such as the provision of raw materials, geochemical regulation, ecologic support, and cultural services, particularly recreation (Rangel-Buitrago 2019).

Beaches account for about 35% of the world's coasts (Luijendik et al. 2018). For marine biologists, a beach is an area that serves as a habitat for different species that interact as complex networks or interconnected systems (Gracia et al. 2021). For an economist, a beach is a resource to be used with the primary goal of attracting customers (tourists) interested in the beach's different amenities (Lynch 2022). Conversely, for geologists, a beach is the accumulation of sediments found along marine, lacustrine and estuarine shorelines and deposited by waves, currents, and tides (Davis and Fitzgerald 2004). These sediments now include plastics and DW materials.

[1] Programa de Física, Facultad de Ciencias Básicas, Universidad del Atlántico, Barranquilla, Atlántico, Colombia.
[2] Programa de Biología, Facultad de Ciencias Básicas, Universidad del Atlántico, Barranquilla, Atlántico, Colombia.
[3] Laboratory of Aquatic Systems, Marine and Continental Environments (AQUAMAR), Faculty of Sciences, Ibn Zohr University, Agadir, Morocco.
[4] Department of Geology, Grand Valley State University, The Seymour K. & Esther R. Padnos Hall of Science 213A, Allendale, Michigan, U.S.A. 4940.
* Corresponding author: nelsonrangel@mail.uniatlantico.edu.co

Along many world beaches, especially those located in tropical areas such as the central Caribbean coast of Colombia, the swash zone is often occupied by diverse materials deposited by winds, waves, currents, and tidal forces. These materials, named Beach-Cast (Mateo et al. 2003, Cardona and Garcia 2008), are a combination of diverse types of items that come from specific sources:

- **Marine Wrack (MW):** Organic marine-origin items washed up, transported, and deposited onto the beach shore. These are composed of marine autotrophs (micro and macro algae and aquatic higher plants), animals, and/or their shells (Macreadie et al. 2017).
- **Terrestrial debris (TD):** Organic terrestrial-origin debris (including plants and animals) that occur naturally in the beach environment, or are introduced into river basins, lodging in the river channel for subsequent transport to and final accumulation along the beach (Grilliot et al. 2019).
- **Anthropogenic Litter (AL):** All items that have been made and used by humans, accidentally lost and/or deliberately discarded into the environment, or brought there indirectly by rivers, sewage outfalls, storms, winds, waves, currents, or tides (Williams and Rangel-Buitrago 2019 and 2022).

Included among these items, and explicitly being part of the terrestrial debris, is Driftwood. This material corresponds with any dead, woody plant material, including logs, branches, standing dead trees, and root wads (root balls) that are introduced into a basin, and are transported until their final accumulation in the coastal-marine system (Stevens 1997, Scheungrab et al. 2000, Terich 2019, Murphy et al. 2021, Rangel-Buitrago et al. 2021).

Driftwood is a functional and structural component of beach ecosystems. Such material is critical to support healthy beaches by providing habitat, structure, nutrients, and carbon storage (Maser et al. 1988), supplying links between aquatic and terrestrial ecosystems (Grilliot 2019), such as substrate and nursery sites for seeds and vegetation (Hood 2007), nesting and perches for animals (Bull 2002), shade and habitat for a variety of flora and fauna (Storry et al. 2006), nutrients (de Jonge et al. 2019), carbon and water storage (Kramer 2016), solid substrates for fish spawn (Gonor et al. 1988), a refuge from predation (Everett and Ruiz 1993), and resting areas for marine mammals (Kwak et al. 2015). Driftwood also can trap sediments to stabilize beach and dune environments (Heathfield and Walker 2011, Grilliot 2019, Davidson et al. 2020).

Driftwood has ecological benefits within the beach system. However, in some cases, its presence can be a problem. Driftwood accumulations generate complex three-dimensional frameworks that serve as traps for substantial amounts of litter, especially plastics. Existing information about litter magnitudes in the central Caribbean coast of Colombia (Rangel-Buitrago et al. 2017, 2018, 2019, 2020, 2021, 2022) suggests a close relationship between high litter densities and the driftwood presence. Under the conditions of this coast, the transport and accumulation of driftwood are continuously converting this shoreline into a perfect setting for plastic accumulation. In this context, the primary purpose of this work is to evaluate the role of driftwood as a trap for plastic accumulation.

The 'wind shadows' behind big pieces of driftwood cause sand accumulation (and plastic litter trapping) with potential embryonic dune formation and burial of plastic litter. This brings up another point to consider—litter burial by beach or dune sand accumulation, followed by erosional re-exposure. So plastic litter accumulation is also time dependent. A second mode of entrapment occurs when storm waves break over driftwood on the beach, allowing transported plastic litter to be trapped in the 3D structure generated by the driftwood.

The Central Caribbean coast of Colombia

The study area corresponds with the Central Caribbean coast of Colombia. This area is located between the Magdalena River delta in the Atlantico Department and Galerazamba in the Bolivar Department (Fig. 1). This coastal reach includes eight beaches (sampling areas) distributed along a 75 km long coastal reach (Fig. 1 and Table 1).

The study area is located inside a strategic region of the country, which is divided into six municipalities, with 1,314,159 inhabitants (Barranquilla, Puerto Colombia, Tubará, Juan de Acosta, Piojo and Santa Catalina). This population (3% of the Colombian total) is mainly established at Barranquilla city, the largest urban agglomeration on the Colombian Caribbean coast, with the remainder in small villages or rural locales (Rangel-Buitrago et al. 2022).

The study area can be classified as a semi-arid tropical environment strongly dominated by the Intertropical Convergence Zone (ITCZ), where the mean annual temperature is 28°C with annual extremes of 20° and 39°C (Sánchez-Páez et al. 1997). The rainfall regime of this area shows a bimodal pattern with two dry periods (Dec.-Mar. and Jul.-Sept.) and two rainy periods (Apr.-May and Oct.-Nov.) that record total precipitation values up to 2,550 mm/yr (Rangel-Buitrago et al. 2013). Wind velocity has average values lower than 15 m/s. High wind velocity values are sporadic and are related to winds blowing from N-NE during the dry period. Lower wind velocity values are observed during the rainy period associated with winds blowing from E (Anfuso et al. 2015). Winds generate significant sand transport as well as generating waves and currents that transport and deposit driftwood, from large to small pieces. This DW creates effective wind-shadow sand traps that also trap finer wind-borne litter. Down-drift sand trapping can form embryonic dunes that bury litter, that may be re-exposed by later erosion to create a time-dependent aspect to plastic accumulation. At the back of some beaches, a foredune can form because of an abundance of driftwood – again burying litter/plastic. Similarly, storm waves may generate overwash when breaking over large pieces of driftwood to trap litter behind such obstacles.

Along the central Caribbean coast of Colombia, average significant wave height is 1.5 m, and the peak period average is 7 s. From November to July, the wave system is dominated by NE swells, and for the remainder of the time, waves from NW, WSW, and even SW occur. This shift of wave direction

Figure 1. The central Caribbean coast of Colombia with the locations of surveyed beaches.

Table 1. Location and general characteristics of beaches located on the central Caribbean coast of Colombia.

Beach Code	Name	Type	Exposure	Condition
A7	Puerto Velero - Exposed	Remote	Exposed	Non-Cleaned
A17	Playa linda	Remote	Sheltered	Non-Cleaned
A20	Salinas del Rey	Village	Sheltered	Non-Cleaned
A21	Loma de Piedra	Remote	Sheltered	Non-Cleaned
A23	Bocatocinos	Rural	Exposed	Non-Cleaned
A24	Punta Astilleros	Rural	Sheltered	Non-Cleaned
A25	Salinas de Galerazamba	Remote	Exposed	Non-Cleaned
A26	Galerazamba	Village	Sheltered	Non-Cleaned

corresponds with a decrease in significant wave heights, with the lowest values occurring between August and October (1.5 m), whereas the highest energy conditions occur from November to July when wave heights can exceed 2.5 m (Rangel-Buitrago et al. 2015). Extreme wave events, Cold Fronts, and Hurricanes occur from January to March, and July to November, respectively, affecting the study area with heavy rains and high winds. Waves generate a longshore current with a prevailing southwestward component. Some minor reversals (to the northeast) can occur during rainy periods when southerly winds predominate in some areas, and set up short, high-frequency waves. Tides are mixed semi-diurnal, with maximum amplitudes of 65 cm (Andrade 2008).

Methodological approach

The eight beaches surveyed were categorized in one of the five typologies described in the Bathing Area Registration and Evaluation System (BARE - Williams and Micallef 2009). This categorization allows beach classification into five classes (remote, village, rural, urban, and resort), considering their accessibility, facilities, environment, accommodation grade, and safety. Also, beaches were categorized according to their degree of exposure to waves (i.e., sheltered, or exposed - Table 1).

Within each of the eight beaches surveyed, a sampling area of 50 m^2 was defined. This sampling area was located between the low-tide mark and the backshore (covering the central part of the beach). Within each of these sampling units, all driftwood and litter available was collected. All plastic items (> 200 mm) found were separated according to one of the 54 plastic/polystyrene typologies existing in the OSPAR (2010) classification. Results were presented as "number of items" and associated "densities." Densities were calculated as the number of items collected along the sampling unit divided by its area (50 m^2) and were presented as items/m^2.

Plastic typologies were grouped and used to determine the possible source (economic sector or human activity from which it originated). Plastic typologies were grouped according to the Ocean Conservancy Classification (Ocean Conservancy 2010). This classification relates each plastic item with the most probable activity or economic sector responsible for generation (origin/source).

Three indexes were used to determine driftwood and plastic influence along the study area. The first index used was the **Driftwood Abundance Index (DWAI)** proposed by Rangel-Buitrago et al. (2021). This Index determines the DW abundance (presence) in a beach by means of the calculation of the existing relation between the amount of DW and the total amount of refuse items collected. The DWAI for each of the 25 beaches surveyed was calculated using the formula:

$$DWAI = \frac{\dfrac{\sum DW\ items}{log\,log\,10 \sum Total\ Refuse\ Items}}{Area} * 20$$

Where DWAI is the number of DW items per square meter, considering the existing relationship between DW and the log10 of the total amount of all refuse items collected along the sampling area. This index allows the categorization of the beach in terms of the WD presence according to five classes that range from "very low abundance" to "very high abundance" (Table 2).

The second index used was the **Driftwood Coverage Index (DWCI).** This index, proposed by Rangel-Buitrago et al. (2021), determines the DW coverage in relation with the beach area. In essence, this index measures the beach state of being free of DW, and if the beach in question can reach and maintain this state. The DWCI was calculated using the formula:

$$DWCI = \frac{\sum DW\ Items}{Length(m) * Width(m)} * 20$$

Where DWCI is the number of DW items collected per square meter, considering the existing relation between all DW collected in the sampling area. The DWCI determines the DW coverage according to five categories that range from "clean" to "fully covered by WD" (Table 3).

The third index used was the Plastic Abundance Index (PAI - Rangel-Buitrago et al. 2021). This Index determines the plastic abundance (presence) on a beach by calculating the existing relationship between the amount of plastic and the total amount of anthropogenic litter items collected. The PAI for each of the 8 beaches surveyed was calculated using the formula:

$$PAI = \frac{\dfrac{\sum Plastic\ items}{log\, log\, 10 \sum Total\ MAL\ Items}}{Area} * K20$$

Where PAI is the number of plastic items per square meter, considering the existing relationship between plastics and the log10 of all litter items collected along the sampling area. The PAI allows the categorization of the beach in terms of plastic presence according to five classes that range from "very low abundance" to "very high abundance" (Table 4).

Obtained data were subjected to Hierarchical Cluster Analysis (HCA) intended to build an iterative classification map of beaches from a proximity matrix between them, based on DW and AL magnitudes, and its beach characteristics. Hierarchical Cluster Analysis (HCA) was performed using Ward's method with Squared Euclidean distance, without standardized variables (categories). With non-standardized variables, categories are weighted in the analysis by their individual numerical values. Hence, those with high abundance at sites have a strong influence on derived clusters relative to those with low occurrences. A dendrogram graph was used to illustrate hierarchical relations between both site and category clusters. Analyses were performed using PRIMER and XLSTAT.

Table 2. Driftwood Abundance Index (DWAI). This index determines the DW abundance (presence) on a beach by means of calculating the existing relation between the amount of DW and the total amount of refuse collected.

WDAI	Type	Description
0 to 2	Very low abundance/Absence	No DW is seen
0.1 to 1	Low abundance	Some DW in the sample
1.1 to 4	Moderate abundance	A considerable amount of DW is seen
4.1 to 8	High abundance	A lot of DW in the sample
More than 8	Very high abundance	Most of the sample is composed of DW

Table 3. Driftwood Coverage Index (DWCI). This index determines the DW coverage in relation to the beach area. In essence, this index measures the beach state of being free of DW, and if this beach can reach and maintain this state.

WDCI	Type	Description
0 to 2	Clean	No DW is seen over the beach
2 to 5	Low	Some DW pieces are seen over a large area
5 to 10	Moderate	A considerable amount of DW pieces cover the beach
10 to 20	High	A significant amount of DW pieces cover the beach
More than 20	Fully Covered	Beach is fully covered by DW pieces

Table 4. Plastic Abundance Index (PAI). This index determines the plastic abundance (presence) on a beach by calculating the existing relationship between the amount of plastic and the total amount of litter items collected. This index can be used to monitor changes in beach state over time.

PAI	Type (Number of plastics per unit area)	Description
0	Very low abundance/Absence	No plastics are seen
0.1 to 1	Low abundance	Some plastics are in the sample unit
1.1 to 4	Moderate abundance	A considerable number of plastics are seen
4.1 to 8	High abundance	A lot of plastics are in the sample unit
More than 8	Very high abundance	Most of the sampling unit is composed of plastics

Magnitudes

Of the eight beaches surveyed, five beaches can be considered as sheltered while the remaining three can be considered as exposed (Table 1). Due to their typology and localization, the eight beaches never have been subjected to clean-up operations (i.e., data from these study locations have not been skewed by removal of litter or DW).

Refuse was found along the eight beaches surveyed (Table 5 and Fig. 2) and specifically, a total amount of 7,292 items (Avg: 18.23 items/m^2) were found, of which 50.8% was driftwood (3,701 items), 44.3% was plastics (3,232 items) and the remaining 4.9% were other litter items (359 items).

In terms of driftwood presence, the average along the study area was 9.2 items/m^2, but this value changed significantly between beaches (Table 5 and Fig. 2). Highest abundances (in terms of amount and density) were observed on A7 Puerto Velero, a remote exposed beach located in the centermost part of the study area, with 780 DW items equivalents to a density of 15.6 items/m^2 (Table 5 and Fig. 2). Conversely, the lowest DW abundance was found at A17 Playa Linda, a sheltered remote beach located in the south of the study area, with 214 DW pieces equivalent to a density of 4.28 items/m^2 (Table 5 and Fig. 2).

Anthropogenic litter was found in all the surveyed beaches (Table 5 and Fig. 2). Specifically, a total amount of 3,591 items (Avg: 8.9 items/m^2) were found, of which 90% was plastic (3,232 items) while the remaining 10% (359 items) were other items (rubber, textile, paper, processed wood, metal, glass, biohazards, organic and miscellaneous items).

In terms of plastic presence, the plastic items' average in the study area was 8.08 plastic items/m^2, but this value varied substantially between beaches (Table 5 and Fig. 2). Highest plastic abundances (in terms of amounts and densities) were observed on A23 Bocatocinos, a rural exposed beach located in the southernmost part of the study area, with 736 plastic items, equivalent to a density of 14.72 items/m^2 (Table 5 and Fig. 2). Conversely, the lowest plastic abundances and densities were found on A20 Salinas del Rey, a village sheltered beach located in the central part of the study area, with 154 plastic pieces equivalent to a density of 3.08 items/m^2 (Table 5 and Fig. 2).

Table 5. Driftwood and plastic amounts and densities along the central Caribbean Coast of Colombia. (Note that DW density averages show considerable variation with higher values on exposed beaches, and lower values on sheltered beaches.)

Site		Plastics		Other Litter		Driftwood	
Code	**Beach**	**Total**	**Density (items/m²)**	**Total**	**Density (items/m²)**	**Total**	**Density (items/m²)**
A7	Puerto Velero - Exposed	278	5.56	56	1.12	780	15.6
A17	Playa linda	266	5.32	21	0.42	214	4.28
A20	Salinas del Rey	154	3.08	23	0.46	244	4.88
A21	Loma de Piedra	293	5.86	27	0.54	308	6.16
A23	Bocatocinos	736	14.72	83	1.66	440	8.8
A24	Punta Astilleros	461	9.22	37	0.74	475	9.5
A25	Salinas de Galerazamba	699	13.98	73	1.46	735	14.7
A26	Galerazamba	345	6.9	39	0.78	505	10.1

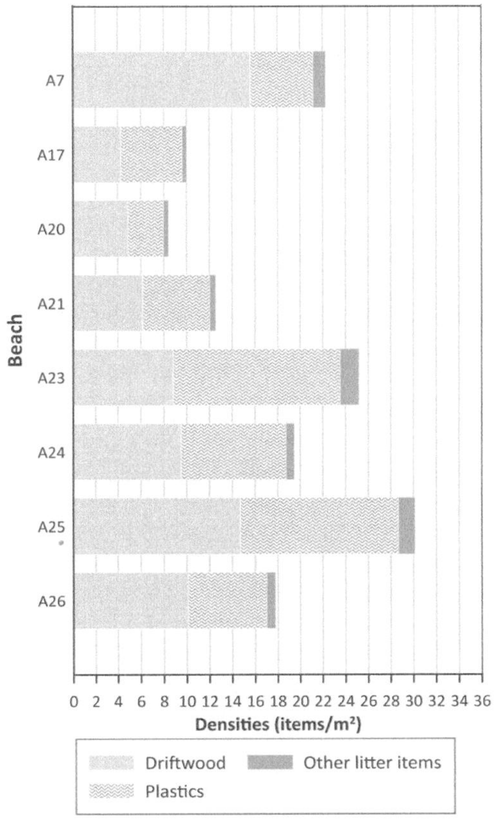

Figure 2. Driftwood, plastics and other items densities.

The 3,232 plastics items collected along the study area belong to 43 of the 54 plastic/polystyrene typologies in the OSPAR (2010) classification. Specifically, ten typologies equivalent to 2,865 items (Avg: 7.16 items/m^2) are responsible for 88.6% of all plastic collected (Table 6 and Fig. 3). These eight typologies are: plastic/polystyrene pieces 0–2.5 cm (32.18%, 1040 items, Avg: 2.6 items/m^2), caps/lids (18.13%, 586 items, Avg: 1.4 items/m^2), other plastic items (15.87%, 513 items, Avg: 1.2 items/m^2), lolly sticks (4.52%, 146 items, Avg: 0.36 items/m^2), drinks - bottles, containers and drums (4.39%, 142 items,

Table 6. Types. Sources. Numerical amounts. Percentages and densities (items/m^2) of plastic observed along the central Caribbean Coast of Colombia.

OSPAR code	Type	Source/Activity	Amount	Percentage	Densities (items/m^2)
2	Bags shopping	Dumping	43	1.33	0.108
2	Bags	Dumping	4	0.12	0.010
3	Small Plastic Bags	Dumping	5	0.15	0.013
112	Plastic Ends	Dumping	5	0.15	0.013
4	Drinks (bottles, containers and drums)	Shoreline and Recreational	142	4.39	0.355
5	Cleaner	Dumping	27	0.84	0.068
6	Food Containers	Shoreline and Recreational	6	0.19	0.015
7	Cosmetics	Medical and Sanitary Waste	87	2.69	0.218
8	Engine Oil < 50 cm	Ocean/waterway	38	1.18	0.095
9	Engine Oil > 50 cm	Ocean/waterway	5	0.15	0.013
10	Jerry Cans	Ocean/waterway	3	0.09	0.008
12	Others bottles	Medical and Sanitary Waste	52	1.61	0.130
19	Lolly sticks	Shoreline and Recreational	146	4.52	0.365
13	Crates	Dumping	0	0.00	0.000
14	Car Parts	Dumping	3	0.09	0.008
15	Caps/lids	Shoreline and Recreational	586	18.13	1.465
16	Cigarette Lighters	Smoking-Related	23	0.71	0.058
17	Pens	Dumping	10	0.31	0.025
18	Combs/Hair brushes	Dumping	23	0.71	0.058
19	Crips/Sweet Packets	Shoreline and Recreational	47	1.45	0.118
20	Toys and party poppers	Dumping	25	0.77	0.063
21	Cups	Shoreline and Recreational	97	3.00	0.243
22	Cutlery/Trays/Straws	Shoreline and Recreational	88	2.72	0.220
23	Fertiliser/Animal feed bags	Dumping	4	0.12	0.010
25	Gloves	Dumping	0	0.00	0.000
31	Rope > 1 cm	Ocean/waterway	5	0.15	0.013
115	Nets and Pieces of net < 50 cm	Ocean/waterway	4	0.12	0.010
116	Nets and Pieces of net > 50 cm	Ocean/waterway	1	0.03	0.003
33	Tangled net/cord	Ocean/waterway	3	0.09	0.008
36	Light Sticks	Ocean/waterway	1	0.03	0.003
37	Floats Buoys	Ocean/waterway	1	0.03	0.003
39	Strapping bands	Ocean/waterway	0	0.00	0.000
42	Hard hats	Dumping	2	0.06	0.005
43	Shotgun Cartridges	Dumping	1	0.03	0.003

Table 6 contd. ...

...Table 6 contd.

OSPAR code	Type	Source/Activity	Amount	Percentage	Densities (items/m²)
44	Shoes/Sandals	Shoreline and Recreational	114	3.53	0.285
45	Foam Sponge	Ocean/waterway	49	1.52	0.123
117	Plastic/polystyrene pieces 0–2.5 cm	Dumping	1040	32.18	2.600
46	Plastic/Polystyre pieces 2.5–50 cm	Dumping	8	0.25	0.020
47	Plastic/Polystyre pieces > 50 cm	Dumping	5	0.15	0.013
48	Other Plastic Items	Dumping	513	15.87	1.283
49	Balloons	Dumping	8	0.25	0.020
52	Tyres and Belts	Dumping	2	0.06	0.005
53	Other pieces	Dumping	6	0.19	0.015
TOTAL			3232	100	8.08

Figure 3. Examples of plastics items found along the central Caribbean Coast of Colombia.

Avg: 0.35 items/m^2), shoes/sandals (3.53%, 114 items, Avg: 0.28 items/m^2), cups (3%, 97 items, Avg: 0.24 items/m^2), cutlery/trays/straws (2.72%, 88 items, Avg: 0.22 items/m^2), cosmetics (2.69%, 87 items, Avg: 0.21 items/m^2), and other bottles (1.61%, 52 items, Avg: 0.13 items/m^2).

Plastic abundance and percentage determined in this work agree with findings obtained in the Moroccan Atlantic coast (Rida et al. 2021), Senegal beaches (Castro Tavares et al. 2020), Gran Canaria

Table 7. Beach classification into one of the categories belonging to the Driftwood Abundance Index (DWAI), Driftwood Coverage Index (DWCI) and Plastic Abundance Index (PAI).

Code	Beach	DWAI	DWAI TYPE	DWCI	DWCI TYPE	PAI	PAI TYPE	GROUP
A7	Puerto Velero - Exposed	51.2	Very High	51.2	Very High	36.5	Very High	A
A17	Playa linda	15.9	Very High	15.9	Very High	39.4	Very High	A
A20	Salinas del Rey	46.3	Very High	46.3	Very High	88.0	Very High	A
A21	Loma de Piedra	18.6	Very High	18.6	Very High	23.5	Very High	A
A23	Bocatocinos	28.4	Very High	28.4	Very High	95.0	Very High	B
A24	Punta Astilleros	31.8	Very High	31.8	Very High	61.7	Very High	A
A25	Salinas de Galerazamba	22.0	Very High	22.0	Very High	41.9	Very High	B
A26	Galerazamba	34.3	Very High	34.3	Very High	46.8	Very High	A

Island beaches (Rapp et al. 2020) and SE Black Sea beaches (Terzi et al. 2020), where more than 85% of litter found were plastic items.

The eight beaches surveyed were assessed and classified into one of the categories belonging to the Driftwood Abundance Index (DWAI), the Driftwood Coverage Index (DWCI) and Plastic Abundance Index (PAI). The DWAI shows all beaches (100%) with a very high abundance of DW (Table 7). The DWCI shows that all beaches (100%) are fully covered with WD pieces (Table 7). The general PAI for the whole area was 54.1, which categorized the central Caribbean coast of Colombia as an area with a very high abundance of plastics. This same analysis highlights that all beaches (100%) have a very high abundance of plastics (Table 7).

A Hierarchical Cluster Analysis (HCA) was developed to show similarities and differences between beaches concerning driftwood and plastics abundances (Fig. 4). The HCA extracts the existing natural groups and allows the visualization of similarity or dissimilarity level existing between surveyed beaches. This analysis highlights that according to driftwood and plastics abundances, the study area can be divided into two groups (Fig. 4):

- **Group A:** This group is composed of six beaches (3 exposed and 5 sheltered) located in remote (3), village (2) and one rural site (1). These beaches encompass 80% and 68% of all driftwood and plastic found in the study area. Driftwood densities inside this group of beaches ranged from 16.6 items/m^2 to 4.28 items/m^2 (Avg: 9.8 items/m^2) while plastic densities ranged from 13.9 items/m^2 to 3.08 items/m^2 (Avg: 7.34 items/m^2).
- **Group B:** This group is composed of two beaches (both exposed) located in remote and rural sites. These beaches encompass 20% and 31% of all driftwood and plastic found in the study area. Driftwood densities inside this group of beaches ranged from 8.8 items/m^2 to 6.1 items/m^2 (Avg: 7.4 items/m^2) while for plastics the densities ranged from 14.7 items/m^2 to 5.8 items/m^2 (Avg: 10.3 items/m^2).

After defining these two groups of beaches, determining the environmental information in terms of the principal components is a useful tool for evaluating the environmental status of the study area. The PCA scatter plot (Fig. 4) shows factors F1 and F2 as responsible for the 47.26% of the total variance existing along the study area. In the same way, there is a trend along with the first component (F1) from sectors with fewer amounts of driftwood and plastics items to the left (Group A) to those sites with more quantities of driftwood and plastic pieces on the right (Group B).

The PCA vector plot (Fig. 4) also shows the 43 plastic categories that reach distinct locations on the principal axes, and hence these determine the corresponding axis positions for the beaches. Thirty-two of the forty-three plastic typologies are located at the positive axis of F1 and are related to the most polluted beaches. This analysis could reflect numerically significant plastic typologies in each of the surveyed beaches (e.g., plastic/polystyrene pieces 0–2.5 cm, caps/lids, lolly sticks). In the same way, the analysis

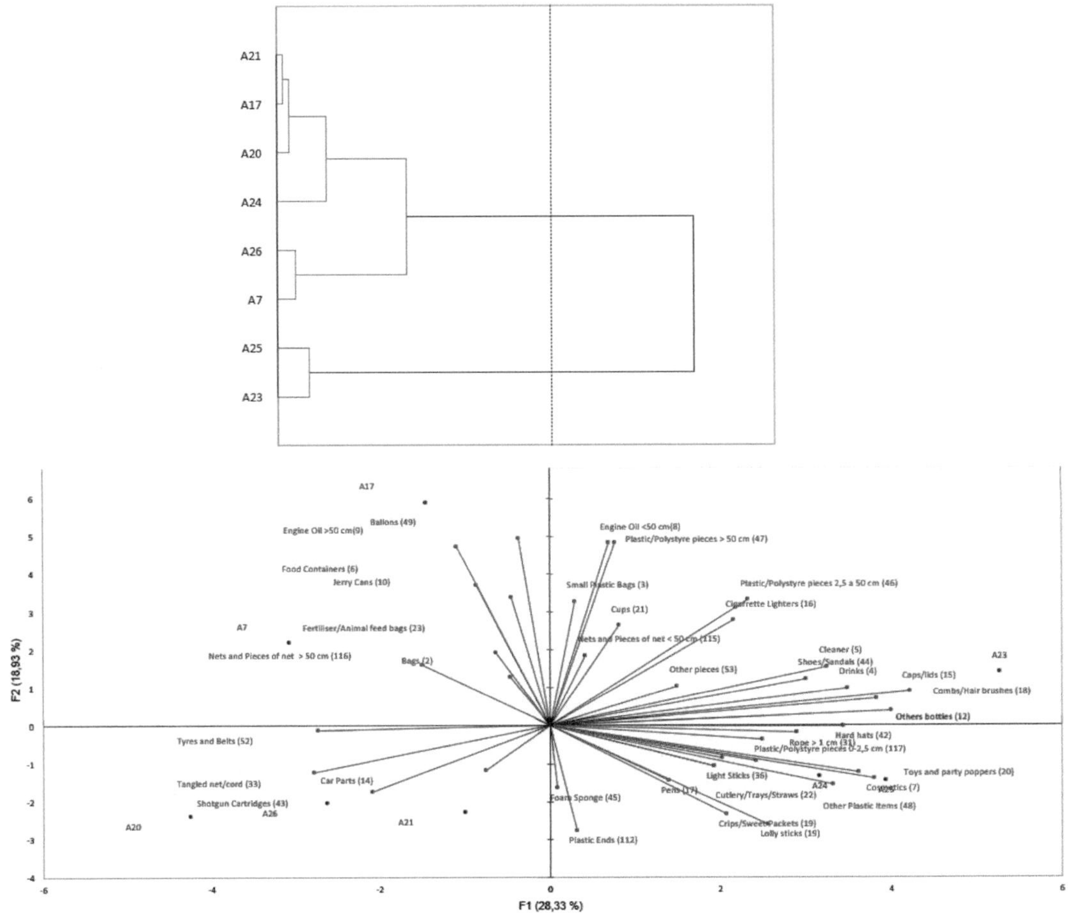

Figure 4. Agglomerative Hierarchical Clustering (AHC) and Principal Component Analysis (PCA) for the study area. The dendrogram allows the clear definition of two groups of beaches while the PCA scatter plot shows factors F1 and F2 as responsible for the 47.26% of the total variance. The PCA vector plot also shows the 43 plastic categories, which reach distinct locations on the principal axes, and hence these determine the corresponding axis positions for the beaches.

also confirms the elevated degree of pollution related to driftwood and plastic presence along the whole study area. In that sense, the clusters formed after the application of AHC make a distinction between:

- **Beaches in bad condition:** Extremely dirty beaches because they are fully covered by driftwood, which traps high amounts of plastics - Group A (e.g., A24 - Punta Astilleros).
- **Beaches in extremely bad condition:** Extremely dirty beaches because they are fully covered by driftwood, which traps extreme amounts of plastics - Group B (e.g., A25 - Salinas de Galerazamba).

Sources and fates

After the identification of magnitudes of driftwood and plastics along the study area, two questions must be addressed: Where do these materials come from? What are their pathways and fates?

Buoyancy is the essential property for both materials, and as such, can be used as an indicator to determine sources and delivery paths. Driftwood and plastics are prevalent buoyant items that can float without sinking over long distances (including river basins, beaches, and oceans). Therefore, their presence on a beach infers a land-based source related to riverine inputs (basins), or beach uses that are followed by longshore current or oceanic transport. In the case of driftwood origin, the process is clear: vegetated areas (e.g., a forest) adjacent to rivers supply most of this material, so drainage basins serve as pathways that collect the wood available to them, followed by transport toward beaches and offshore.

Due to its geographical position, the study area is highly influenced by three river drainage basins (Caribe, Mallorquin, and Magdalena) that discharge wood along the Atlantico and Bolivar Departments' coastlines. The Caribbean Basin drains an area of 965 km^2, encompasses 31 sub-basins, and includes ten municipalities located in the study area. The Mallorquin Basin drains an area of 400 km^2, encompasses three sub-basins, and belongs to four municipalities located in the Atlantico Department. Both basins concentrate 38 dendritic to sub-dendritic small rivers that flow directly to the sea. The Magdalena River Basin is the most important basin of the study area (as well the whole country). This basin encompasses a total area of 268,460 km^2 occupying at least 70% of the country's total area. The Magdalena River Basin encompasses 725 municipalities, and its area has 81% of the total Colombian population.

Along these three basins, processes such as bank undercutting and collapse, tree blowdown, tree collapse, and mass soil movements contribute to the origin of driftwood and its transport. In a big basin such as the Magdalena River Basin, large floods move significant quantities of wood onto the upper flood plain to be subsequently carried to and deposited on the delta, beaches and offshore. Along medium-size basins such as the Caribe and Mallorquin Basins, low frequency, high magnitude floods add much material to related streams. In small coastal streams, wood concentrations may temporarily dam sections of these streams; then, these dams fail, resulting in floods, scouring wood debris from the channel, and flushing it to the closest beach.

Plastics' magnitudes and typologies observed along the study area indicate a combination of five primary sources/activities related to the various usages carried out along this region's coast, beaches, and related basins. Sources/activities responsible for the plastic generation in the study area are presented in Table 8. Our findings highlight that plastics are mainly related to land-based sources (Table 8). The precise primary sources can be found in nearby basins and the same surveyed beaches. The sources/related activities responsible for all plastic found in the study area are:

Plastic items related to dumping activities: Plastics related to dumping activities include 21 typologies corresponding to 53.65% of all plastic observed along the study area (1,734 items, Avg: 4.34 items/m^2). Main dumped items include plastic/polystyrene pieces 0–2.5 cm, other plastic items, shopping bags, cleaner, toys and party poppers. The dumping action is based on tossing, tipping, or depositing plastic items on nearby streets and in river basins, which then end up accumulating along the surveyed beaches. The amount going into the ocean, or moving as flotsam to offshore islands remains unknown, but such latter accumulation has been documented (Rangel-Buitrago et al. 2019).

Plastic items related to activities developed on beaches: Plastics related to these activities include 8 typologies corresponding to 37.9% of all plastic observed along the study area (1226 items, Avg: 3.07 items/m^2). Main plastic items related to activities developed on beaches include caps/lids, lolly sticks, bottles, containers and drums, cups, cutlery/trays/straws, shoes/sandals. These plastic items are mainly single-use plastics that are carried, used, and then discarded by users during/after their beach-use activities.

Plastic items related to medical and sanitary activities: Plastics related to these activities include 2 typologies (pharmaceutical products/cosmetic bottles) corresponding 4.30% of all plastic observed along the study area (139 items, Avg: 0.35 items/m^2). Currently, handling, storage, treatment, transports, and final disposal of these items is still far from well developed. These types of plastic items are not being collected in containers that are leak-proof or sufficiently strong to prevent breakage during handling.

Table 8. Sources and activities with respective number of items, densities and percentages that give rise to plastic found on the central Caribbean Coast of Colombia.

Source/Activity	Types of Plastic	N°	Densities (items/m²)	%
Dumping	Plastic derived from building and construction materials, tires, car parts, drums, household trashes and appliances.	1734	4.34	53.65
Shoreline and Recreational	Plastic, mainly littering by beachgoers, but also from waterside sports and those carried from streets, drains and gutters.	1226	3.07	37.93
Ocean/waterway	Plastics from recreational fishing/boating, recreational/ commercial fishing and shipping, and of oil and gas industry.	110	0.28	3.40
Medical and Sanitary Waste	Plastics discarded into the sewer system, toilets, or left behind by beachgoers.	139	0.35	4.30
Smoking-Related	Smoking-related packaging and materials.	23	0.06	0.71
	TOTAL	3232	8.08	100

Plastic related to ocean and waterway activities: Plastics related to these activities includes 11 typologies corresponding 3.40% of all plastic observed along the study area (110 items, Avg: 0.28 items/m^2). Main plastic items related to ocean and waterway activities are foam sponge, engine oil tanks, ropes, tangled net/cord, and jerry cans.

Plastic items related to smoking-related activities: These items include one typology (cigarette lighters) corresponding 0.71% of all plastic observed in the study area (23 items, Avg: 0.06 items/m^2). Cigarette lighters are carried, used, and then discarded at the same beach by smokers.

No matter their source, driftwood and plastics pieces leave the continental environment to enter into coastal and oceanic waters. In these areas, the transport and fate are determined by waves and currents.

Driftwood and plastics found along the eight surveyed beaches highlight a strong influence from longshore-current transport. The wave-generated current patterns suggest that transport routes along the area are related to diffracted wavefronts approaching from NE (occasionally from NW and SE). Those wave directions give rise to a dominant S-SW oriented longshore current that transports driftwood and plastics along the area from NE/E to SW/W.

Apart from the wave regime which controls transport along beaches, the wave exposition and morphodynamic state of each beach are considered equally important in the ability of the beach to retain or lose driftwood and plastics (Critchell et al. 2016). Exposed and reflective beaches receive and thus accumulate greater amounts because they are under the direct influence of incoming NE-W waves, while sheltered and dissipative beaches can accumulate fewer amounts because they receive the same waves, but these are reflected and dissipated. Longshore current transport combined with extreme runoff can also produce the southward movement of significant driftwood and litter quantities (blooms). When wave energy decreases in sheltered beaches with shoal areas (i.e., A17 Playa Linda), driftwood and plastics accumulate, generating deposits that can reach up to 3 km^2 in area.

In ocean waters, driftwood and plastics are transported by two dominant currents: (i) the Caribbean Current, which is a warm ocean current that moves significant amounts of water and flows west-northwestward through the Caribbean from the east along the coast of South America and into the Gulf of Mexico (Wust 1963, Morrison and Nowlin 1982), and (ii) the Panama-Colombia Gyre (PCG), which is a cyclonic current that flows counter-clockwise to the Caribbean Current (Schott and Molinari 1996, Mooers and Maul 1998).

Existing relations

Driftwood and plastics accumulation in beach environments depends on factors that include inputs (Rangel-Buitrago et al. 2021), environmental conditions (Andriolo and Goncalves 2022), and coastline configuration (Rangel-Buitrago et al. 2019), among other factors. Given that plastics are trapped by driftwood, the driftwood indexes must contribute to or reflect aspects of the accumulation processes.

To understand this, a series of Pearson correlations and Mann-Whitney tests were performed considering plastics densities and the driftwood indexes. In the same way, a multiple linear regression analysis developed between plastic densities and driftwood indexes suggested that approximately 91% of the plastic presence could be related to the driftwood indexes used in this work.

Having established that driftwood has a strong influence on plastics accumulation in beach environments, a set of linear regressions were performed. The equations that govern plastics accumulation in terms of driftwood presence along the study area is:

$$\textbf{Plastic} = 84.105 - 6.813 \text{ DWAI} + 0.9265 \text{ DWCI } (r^2 = 0.90)$$

A second set of Kruskal-Wallis tests were performed to confirm the above-presented equations, obtaining P values of 0.64 for DWAI and 0.58 in DWCI (alpha = 0.050). Because the calculated P-values were more significant than the alpha level of 0.05, it can be considered that plastics respond to driftwood presence during their accumulation in beach environments.

Management considerations

Having established the magnitudes, sources, and fates of driftwood and litter along the study area, the important issue of "How do we manage the problems of driftwood and plastic accumulation?" still remains.

Driftwood plays an essential role in the ecological processes of a beach, and its management should include actions that balance the designated uses of the beach and maintaining the ecological integrity of the system. Its management should consider multiple factors. The scopes and actions of the management should be based on the beach uses, whether they are being managing for wildlife habitat, or natural storm protection versus commercial use, or recreational purposes. To determine management options, frequent assessment of driftwood conditions on the beach is necessary. Management options may include taking no action, removal with general litter, or minimal removal to complete harvesting of the wood.

Williams and Rangel-Buitrago (2022) highlighted the exponential growth of plastics into the beach environment due to lack of management at sources. A cost-effective approach to reducing current plastic litter impacts must address the input processes along the basins (discards), by engaging industry, municipalities, regulatory authorities, and the public in an integrated way. In the same way, specific actions must involve plastic management facilities (disposal/recycling), optimal reception practices and beach user's education.

Along the study area, as well worldwide, the bulk of beach litter is composed of plastics, a material whose production and use has grown exponentially, reaching almost 475 million tons per year in 2021 (UNEP 2021). Plastics' impacts are related to their excessive abundance in all natural environments, their high persistence (because complete decomposition takes hundreds to thousands of years), the formation of toxic substances, and their ability to act as a vector for contaminants, including Persistent Organic Pollutants (POPs) and heavy metals (Williams and Rangel-Buitrago 2019). They have displaced other materials, such as glass, and are used as an essential component in an uncountable variety of products that humanity uses in daily life.

All the above makes clear that plastic litter management is a necessity. This approach must be dynamic, multidisciplinary, and interactive as a strong response to avoid, prevent or mitigate environmental, economic, and social losses derived from lack of management or bad management practices. This management requires a detailed knowledge of magnitudes and sources of plastic litter as well as its interchange within the marine environment (offshore, shore and inland). The cycle of management begins with magnitude quantification and source identification. This basis then guides planning, preparation and decision making as to the appropriate response, considering the time frame of expected results, including monitoring whether the solution is working. Likewise, the DW problem must be addressed at the source—managing forestry practices that generate excessive wood waste, as well as managing erosion associated with water runoff and riverine processes.

Driftwood and plastic management along the study area should be based on strategies that allow the inherent ability of the coast to recuperate from the impacts induced by driftwood presence, while maintaining the functions fulfilled by the coastal system over medium and long terms. A short-term approach does not allow the above to occur. For proper management to take place, the different typologies and their related processes must be considered to define the specific management strategies to be implemented. Proactive action is the key. Flexible strategies, and not reactive measures, should be adopted to strengthen management and thus improve the coastal environmental quality (e.g., driftwood forms natural sand traps and plays a role in natural beach protection, but when beach cleaning removes woody debris, that protective role is lost. However, where plastic is a common to dominant component of litter, then woody debris entrapment makes litter removal difficult.).

Conclusions

Driftwood can be an essential component of beaches, influencing both the biological and physical functions and processes of this ecosystem. However, in some cases, its presence (particularly if excessive) can be a problem that conflicts with regular beach uses.

The analysis of 8 beaches with different typology along the central Caribbean coast of Colombia demonstrated an elevated presence of driftwood that is serving as a trap for accumulating plastic items. The average abundance was 9.2 DW items/m^2 and 8.1 plastic items/m^2, but this value ranged significantly between beaches, reflecting varied exposure to different coastal processes and litter trapping mechanisms. Plastic items were grouped into 43 typologies, and ten typologies were responsible for 88% of all plastic collected.

Driftwood and plastics found along the study area infer land-based sources related primarily to riverine inputs (drainage basins), which are followed by oceanic or longshore current transport. Once wood and plastic litter are transported through the three basins that influence the study area, these materials then leave the continental environment to enter oceanic and coastal waters. In ocean waters, they are transported by two dominant currents: the Caribbean Current and the Panama-Colombia Gyre. Wave-generated current patterns suggest that transport routes along the coast are related to diffracted wavefronts approaching from NE (occasionally from NW and SE). Those wave directions give rise to a dominant S-SW oriented longshore current that transports WD along the central Caribbean coast of Colombia from NE/E to SW/W.

Tri-dimensional structures of driftwood compartments become traps in which plastics accumulate as the result of wind deposition due to 'wind-shadow' effect of driftwood elements, or litter being tossed into this "mesh" by wave overtopping driftwood in storms. Such plastic, buried by beach and aeolian processes, can be re-exhumed to add to the litter total so that the amount of plastic on driftwood-covered beaches is time cumulative.

Driftwood plays an essential role in the ecological processes of a beach, and its management should include actions that balance beach uses, maintaining the ecological integrity of the system. A plastic litter-degraded beach can experience a substantial reduction in its economic potential, and in that sense, stakeholders must work on developing strategies for management based on litter reduction at the sources.

Results of this work are related to parameters geared to Atlantico Department economic aspirations, specifically the tourism industry. Currently, this area is encouraging its coastal tourism industry with a clear aim of increasing its international and national appeal. The primary goal of this effort must focus on solving the driftwood and plastic litter issues, because the current low-cleanliness conditions contribute to lower touristic values, which will not be appealing to potential visitors.

Acknowledgements

This work is a contribution from "Geología, Geofísica y Procesos Marino-Costeros", Universidad del Atlántico (Colombia), "Laboratory of Aquatic Systems, Marine and Continental Environments (AQUAMAR)- Faculty of Sciences", Ibn Zohr University (Morocco), and "Department of Geology," Grand Valley State University (USA).

References

Al-Najjar, T. and Al-Shiyab, A. (2011). Marine litter at Al-Ghandoor Area the Most Northern Part of the Jordanian Coast of the Gulf of Aqaba, Red Sea. Nat. Sci. 3: 921–926.

Andrade, C. (2008). Cambios recientes del nivel del mar en Colombia. pp. 103–122. *In*: Restrepo, J. D. (ed.). Deltas de Colombia: morfodinámica y vulnerabilidad ante el Cambio Global. EAFIT Univ. Press. Medellin.

Andriolo, H. and Goncalves, G. (2022). Is coastal erosion a source of marine litter pollution? Evidence of coastal dunes being a reservoir of plastics. Mar. Pollut. Bull. 174: 113307.

Anfuso, G., Rangel-Buitrago, N. and Correa, I. (2015). Evolution of sandspits along the Caribbean Coast of Colombia: natural and human influences. pp. 1–21. *In*: Randazzo, G., Jackson, D. and Cooper, J. A. (eds.). Sand and Gravel Spits. Springer, New York.

Barnes, D. K., Galgani, F. and Thompson, R. (2009). Accumulation and fragmentation of plastic debris in global environments. Philos. Trans. R. Soc. B. 364(1526): 1985–1998.

Bull, E. L. (2002). The value of coarse woody debris to vertebrates in the Pacific Northwest. Department of Agriculture, U.S.D.A., Forest Service, Tech. Rep. PSW-GTR-181. Seattle, USA.

Cardona, L. and Garcia, M. (2008). Beach-cast seagrass material fertilizes the foredune vegetation of Mediterranean coastal dunes. Acta Ecol. 34: 97–103.

Castro Tavares, D., Moura, J., Ceesay, A. and Merico, A. (2020). Density and composition of surface and buried plastic debris in beaches of Senegal. Sci. Total Environ. 737: 139633.

Critchell, K. and Lambrechts, J. (2016). Modelling accumulation of marine plastics in the coastal zone; what are the dominant physical processes? Estuar. Coast. Shelf Sci. 171: 111–122.

Davidson, S. G., Hesp, P. A. and Silva, G. M. D. (2020). Controls on dune scarping. Prog. Phys. Geogr. Earth Environ. 44(6): 923–947.

Davis, D. and Fitzgerald, D. (2004). Beaches and Coasts. Blackwell Pub. Co. New York.

de Jonge, V., Schuckel, U. and Baird, D. (2019). Subsets of food webs cannot be used as a substitute to assess the functioning of entire ecosystems. MEPS 613: 49–66.

Everett, R.A. and Ruiz, G.M. (1993). Coarse woody debris as a refuge from predation in aquatic communities. Oecologia 93(4): 475–486.

Gracia, A., Rangel-Buitrago, N. and Florez, P. (2018). Beach litter and woody-debris colonizers on the Atlantico department Caribbean coastline, Colombia. Mar. Poll. Bull. 128: 185–196.

Gracia, A., Duran-Fuentes, J., Santodomingo, N. and Rangel-Buitrago, N. (2021). Artificial structures as biological "influencers": Hydrozoa and Anthozoa diversity in a Colombian Caribbean Marina. Mar. Poll. Bull. 173: 113058.

Grilliot, M. J., Walker, I. J. and Bauer, B. O. (2019). Aeolian sand transport and deposition patterns within a large woody debris matrix fronting a foredune. Geomorphology 338: 1–15.

Grilliot, M. J. (2019). The Role of Large Woody Debris on Sandy Beach-dune Morphodynamics. PhD Thesis. University of Victoria, Victoria.

Heathfield, D. K. and Walker, I. J. (2011). Analysis of coastal dune dynamics, shoreline position, and large woody debris at Wickaninnish Bay, Pacific Rim National Park, British Columbia. Can. J. Earth Sci. 48(7): 1185–1198.

Hood, W. G. (2007). Large woody debris influences vegetation zonation in an oligohaline tidal marsh. Estuar. Coasts 30(3): 441–450.

Jackson, D. and Short, A. (2020). Sandy Beach Morphodynamics. Elsevier. Amsterdam.

Kramer, N. (2016). Great River Wood Dynamics in Northern Canada. PhD Thesis. Colorado State University, Fort Collins.

Kwak, J.-H., Chang, S., Naeth, M. and Schaaf, W. (2015). Coarse woody debris increases microbial community functional diversity but not enzyme activities in reclaimed oil sands soils. PLoS One 10(11): e0143857.

Lavers, J. and Bond, A. (2019). Exceptional and rapid accumulation of anthropogenic debris on one of the world's most remote and pristine islands. PNAS 1619818114.

Luijendik, A., Hagenaars, G., Ranasinghe, R., Baart, F., Donchyts, G. and Aarninkhof, S. (2018). The State of the World's Beaches. Sci. Rep. 8: 6641.

Lynch, K. (2022). Hijacking the 3Ss to ground reflective co-management of beach-dune environments. Ocean Coast Manage. 217: 106006.

Macreadie, P. I., Trevathan-Tackett, S. M., Baldock, J. A. and Kelleway, J. J. (2017). Converting beach-cast seagrass wrack into biochar: A climate-friendly solution to a coastal problem. Sci. Total Environ. 574: 90–94.

Maser, C., Tarrant, R., Trappe, J. and Franklin, J. (1988). From the forest to the sea: a story of fallen trees. Gen. Tech. Rep. PNW-GTR-229. Bureau of land management. Portland, Oregon, USA.

Mateo, M. A., Sanchez-Lizaso, J. L. and Romero, J. (2003). Posidonia oceanica "banquettes": a preliminary assessment of the relevance for meadow carbon and nutrient budget. Est Coast Shelf. Sci. 56: 85–90.

Mooers, C. N. K. and Maul, G. A. (1998). Intra-Americas Sea Circulation, in the Sea. John Wiley, New York.

Morrison, J. M. and Nowlin, W. D. (1982). General distributions of water masses within the Eastern Caribbean Sea during the winter of 1972 and fall of 1973. J. Geophys Res. 87: 4207–4229.

Murphy, E., Nistor, I., Cornett, A., Wilson, J. and Pilechie, A. (2021). Fate and transport of coastal driftwood: A critical review. Mar. Poll. Bull. 170: 112649.

Ocean Conservancy. (2010). Trash travels – 2010 annual report. Washington, DC. https://oceanconservancy.org/wp-content/uploads/2017/04/2010-Ocean-Conservancy-ICC-Report.pdf (assessed 14 April 2022).

OSPAR. (2010). Guideline for Monitoring Marine Litter on the Beaches in the OSPAR Maritime Area. OSPAR, London.

Pilkey, O. H., Neal, W. J., Kelley, J. T. and Cooper, J. A. G. (2011). The World's Beaches: A Global Guide to the Science of the Shoreline. University of California Press, Berkeley, CA, 283 p.

Rangel-Buitrago, N., Williams, A. T., Anfuso, G., Arias, M. and Gracia, A. (2017). Magnitudes, sources, and management of beach litter along the Atlantico department coastline, Caribbean coast of Colombia. Ocean Coast Manage. 138: 142–157.

Rangel-Buitrago, N., Williams, A. and Anfuso, G. (2018). Killing the goose with the golden eggs: Litter effects on scenic quality of the Caribbean coast of Colombia. Mar. Poll. Bull. 127: 22–38.

Rangel-Buitrago, N., Gracia, A., Vélez-Mendoza, A., Carvajal-Florián, A., Mojica-Martinez, L. and Neal, W. (2019). Where did this refuse come from? Marine anthropogenic litter on a remote island of the Colombian Caribbean Sea. Mar. Pollut. Bull. 149: 110611.

Rangel-Buitrago, N., Velez-Mendoza, A., Gracia, A. and Neal, W. (2020). The impact of anthropogenic litter on Colombia's central Caribbean beaches. Mar. Pollut. Bull. 152: 110909.

Rangel-Buitrago, N., Arroyo-Olarte, H., Trilleras, J., Arana, V. A., Mantilla-Barbosa, E., Gracia, A. et al. (2021). Microplastics pollution on Colombian Central Caribbean beaches. Mar. Pollut. Bull. 170: 112685.

Rangel-Buitrago, N., Williams, A, Micallef, A. T., Neal, W. and Pilkey, O. (2022). Not all that glitters is gold: Can the real scenic value of the Colombian Caribbean coast be restored? Ocean Coast. Manage. 227: 106292.

Rangel-Buitrago, N. (2019). Coastal Scenery: Evaluation and Management, Springer, Amsterdam.

Rangel-Buitrago, N., Anfuso, G., Correa, I., Ergin, A. and Williams, A. T. (2013). Assessing and managing scenery of the Caribbean Coast of Colombia. Tour Manage. 35: 41–58.

Rangel-Buitrago, N., Anfuso, G. and Williams, A. T. (2015). Coastal erosion along the Caribbean coast of Colombia: magnitudes, causes and management. Ocean Coast. Manage. 114: 129–144.

Rapp, J., Herrera, A., Martinez, I., Raymond, E., Santana, A. and Gomez, M. (2020). Study of plastic pollution and its potential sources on Gran Canaria Island beaches (Canary Islands, Spain). Mar. Pollut. Bull. 153: 110967.

Rida, M., Ben-Haddad, M., Alla, A. and Rangel-Buitrago, N. (2021). Marine litter in the central Atlantic coast of Morocco. Ocean Coast. Manag. 214: 105940.

Sánchez-Páez, H., Alvarez-Leon, A., Guevara-Mancera, O. A. and Zamora, A. (1997). Diagnistico y zonificación preliminar de los manglares del Caribe Colombiano. Ministerio de Medio Ambiente de Colombia, Bogota.

Scheungrab, D., Trettin, C. C., Russ, L. and Jurgensen, M. F. (2000). Woody debris. Gen. Tech. Rep. SRS-38. U.S. Department of Agriculture, Forest Service, Southern Research Station. Asheville, NC, USA.

Schott, F. A. and Molinari, R. L. (1996). The Western Boundary Circulation of the Subtropical Warmwatersphere. Gebrueder Bontraeger, Berlin.

Stevens, V. (1997). The ecological role of coarse woody debris: an overview of the ecological importance of CWD in BC forests. Forest Science Program. British Columbia, Canada.

Storry, K. A., Weldrick, C. K., Mews, M., Zimmer, M. and Jelinski, D. E. (2006). Intertidal coarse woody debris: a spatial subsidy as shelter or feeding habitat for gastropods? Estuar. Coast. Shelf Sci. 66(1–2): 197–203.

Terich, T. (2019). Driftwood. *In*: Finkl, C. W. and Makowski, C. (eds.). Encyclopedia of Coastal Science. Encyclopedia of Earth Sciences Series. Springer, Cham.

Terzi, Y., Eruz, C. and Ozseker, K. (2020). Marine litter composition and sources on coasts of south-eastern Black Sea: A long-term case study. Was. Man. 10: 139–147.

UNEP. (2021). United Nations Environment Programme, UNEP Yearbook: Emerging Issues in Our Global Environment. United Nations Environment Programme (UNEP), Nairobi, Kenya.

Williams, A. T. and Micallef, A. (2009). Beach Management Principles and Practice. London, Earthscan.

Williams, A. T. and Rangel-Buitrago, N. (2019). Marine litter: solutions for a major environmental problem. J. Coast. Res. 35: 648–663.

Williams, A. T. and Rangel-Buitrago, N. (2022). The past, present, and future of plastic pollution. Mar. Poll. Bull. 176: 113429.

Wust, G. (1963). On the stratification and the circulation in the cold-water sphere of the Antillean-Caribbean basins. Deep-Sea Res. 10: 165–187.

Section II
Challenges in Particular Marine Ecosystems

CHAPTER 6

Sandy Shores

Highly Dynamic and Vulnerable Ecosystems

M. Cecilia Carcedo,[1,2,*] *M. Clara Menéndez,*[1]
Gabriela E. Blasina[1,2] and *Sandra M. Fiori*[1,2]

Introduction

Sandy shorelines occupy approximately 31% of the world's ice-free coasts (Luijendijk et al. 2018) and are defined as energetic environments determined by the interaction between sand, waves and tides. They consist of three main compartments along the land-sea axis: dunes, beaches and surf zones, which are connected by the interchange of materials, especially sand, and energy through trophic interactions between biological communities (Brown and McLachlan 2002). Together, these three compartments comprise a single geomorphic system named the "littoral active zone" (LAZ, Fig. 1), with boundaries defined by the effects of the waves (seaward) and aeolian (landward) transportation of sediment from active dunes

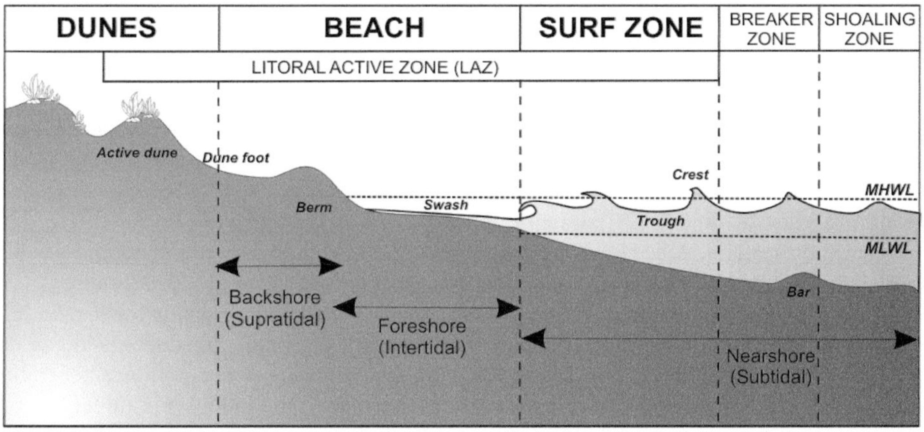

Figure 1. Zonation of a typical sandy shore and its three main components (in bold and capital letters): dunes, beach and surf zone. MHWL: mean high water level; MLWL: mean lower water level.

[1] Instituto Argentino de Oceanografía, Universidad Nacional del Sur, CONICET, IADO (Camino La Carrindanga km 7.5), B8000FWB Bahía Blanca, Argentina.

[2] Departamento de Biología, Bioquímica y Farmacia, Universidad Nacional del Sur, (UNS), Bahía Blanca, Argentina.

* Corresponding author: ccarcedo@iado-conicet.gob.ar

(Tinley 1985). The LAZ supports a wide range of species biodiversity and sustains economic activities such as fisheries and recreation (Schlacher et al. 2016). Across the LAZ, the beach and the surf zone together constitute the main wave-dissipating area, while dunes have a substantial role in releasing sand during meteorological extreme events, acting as a natural coastal defence against storms and sea-level rise. The LAZ is resilient insofar as it remains free to fluctuate within its natural boundaries (McLachlan and Defeo 2018). However, in recent decades these environments have been heavily modified by human activities and are currently considered to be highly vulnerable social-ecological systems (Defeo et al. 2021).

Physical factors influencing sandy shores

Within the LAZ, dunes have a strong terrestrial component being wind-controlled environments, whereas sandy beaches and their adjacent surf zones are mainly marine and wave-driven ecosystems. Given the differential nature of the physical forces acting on the three compartments that make up sandy shores, each one will be treated separately in the next sections.

Dunes

Coastal dunes are aeolian landforms that develop where an abundant supply of loose, sand-size sediment is available to be transported inland by winds. They are defined by the exchange of sand with beaches and by the strong influence of wind forces, representing a transition between continental and marine environments. Thus, coastal dunes are part of the sand-sharing system extending between the landward edge of the beach and the landward limit of aeolian sand transport (Martínez et al. 2004) (Fig. 1). The landward boundary of the dunes may be abrupt, or it may grade gradually into other terrestrial ecosystems. Several recent scientific studies have reported the contribution of coastal dunes to different ecosystem services such as protection against floods and storms, recreation, sediment reservoirs for beach nourishment, salinization prevention, and drinking water production (e.g., Everard et al. 2010, Lithgow et al. 2013, Van der Biest et al. 2017). Coastal dunes also serve as groundwater recharge areas and as a buffer against salt-water intrusion, assisting in the retention of freshwater (Martínez et al. 2004).

Coastal dunes consist mainly of fine to medium sand particles (125–300 μm) of silica and/or carbonate. The driving force of sand particles is the wind (Amir et al. 2014), and particles begin to move at wind speeds near 4 m s^{-1}, the movement increasing significantly above 10 m s^{-1}. The amount of sand moved by winds depends on the moisture content of the sand, the wind speed over the surface of the sand and sand grain size. As a consequence of this movement, dune sand tends to be finer than beach sand (McLachlan and Defeo 2018). The higher the wind energy, the greater the potential for dune development, particularly for those coasts with prevailing onshore winds (Jennings 1957). Considering factors such as prevailing winds, sand supply, climate, vegetation and others, dune coasts can be classified into four main types: (1) foredunes or frontal dunes, (2) parabolic dunes, (3) blowouts and (4) transgressive dunefields. Foredunes are shore-parallel dune ridges formed on the landward boundary of the beach by aeolian sand deposition within vegetation and are the most common type; they can be small or totally absent if the beach is backed by cliffs or hard structures. Parabolic dunes occur where strong winds blow sand inland and trailing ridges are held by vegetation. Blowouts occur wherever the vegetation cover is low or destroyed by strong onshore winds and occasional storm-wave erosion. Transgressive dunefields generally consist of transverse dunes (i.e., with their axes running perpendicular to the main wind direction) and occur where large to huge volumes of sand are transported back from beaches under strong wind and wave regimes (Hesp 2011).

Physical stressors for biological communities in coastal dunes include sand movement, salt spray produced by breaking waves, temperature changes, wind, and disruptive events such as storms. Increasing vegetation cover, often coupled with succession, moderates physical stress, creates stable microenvironments, and is responsible for a general increase in habitat complexity landwards (McLachlan 1991). Fauna responds to this coastal-inland gradient in several ways, which will be detailed in upcoming sections.

Beaches

Beaches are wave-deposited accumulations of sediment at the shoreline, lying between the base of wave activity and the limit of wave run-up or swash (Fig. 1), a turbulent layer of water that washes up on the beach after an incoming wave has broken (Short 1999). As in coastal dunes, the two main sand types on beaches are silica (terrestrial origin) and carbonate (marine) (McLachlan and Defeo 2018). In general, sand grain size varies with the distance from the water. Greater amounts of finer and smaller sand grains (< 0.25 mm) are pushed higher up the beach by waves or wind, whereas larger (> 0.5 mm), coarser grains are deposited closer to the water (Spalding et al. 2009). Beaches require a base to reside on (the bedrock geology), waves to shape them, sediment to form them, and the periodic movement of tides (Short 2012). In this chapter, we refer to "beach" as a unit consisting of both the intertidal zone and the backshore. The intertidal zone is the term used to describe seashore, which is alternately flooded by seawater and exposed to air (and desiccation) according to the tidal rhythm, revealing a unique biome that survives under such fluctuating conditions. Backshore refers to the zone of the beach beyond the high-water line up to the dunefoot, covered by water only during storms of exceptional severity.

Across the intertidal zone, the tide is perceived as a periodic rise and fall of the sea surface, with the highest elevation of water known as high tide and the lowest, low tide. The vertical difference between tidal levels is the tidal range, a very important measure for sandy beach ecology (McLachlan and Dorvlo 2005). Tides are produced by the gravitational pull of the moon and sun acting on a rotating earth, producing a very slight bulge in the ocean. Tidal cycles are influenced by a variety of factors, meaning they can vary across different regions of the world. Most coastal areas experience two high and two low tides a day, with highs and lows of roughly the same height. This pattern is referred to as semi-daily or semi-diurnal tide. If the high and low tides differ in height, the pattern is called a mixed semidiurnal tide. Some areas have only one high and one low tide a day, referred to as a diurnal tide.

When the sun and the moon are aligned (e.g., new moon and full moon), they pull the ocean's surface in the same direction. This causes higher high tides and lower low tides, which are known as 'spring tides'. When the sun and the moon are at right angles to each other, they pull in opposite directions, producing lower high tides and higher low tides than usual. These are called 'neap tides' and occur between the first and third quarter moon. Tides and the currents they generate are responsible for about 50% of the marine energy delivered to the coast (Short 2012). Moreover, the hydrodynamic forces resulting from the action of waves and tides, in association with sand particle size, strongly determine the beach morphodynamics of a sandy shore.

Surf zones

This zone is the shoreline region from the seaward limit of initial wave breaking to the landward limit of still water level, where the wave action is the outstanding factor. Basic waves have a crest and a trough, the highest and lowest points of the wave, respectively. Wave height is the vertical distance between the trough and the crest, and wavelength is the horizontal distance between two identical points on successive waves: crest to crest or trough to trough (Fig. 1). A number of terms are used to describe wave motion: wave period (the time it takes for one complete wave to pass a particular point), wave frequency (the number of waves passing a fixed point in a given amount of time) and wave speed (how fast the wave travels, or the distance travelled per unit of time), which is determined by multiplying wavelength by wave frequency.

As waves move shoreward, the interaction with the bottom alters the wave motion and the wavelength and velocity decrease. The slowing down of the wave and the shortening of the wavelength causes an increase in wave height and a deformation of the wave shape that makes it asymmetrical and 'peaked', a process known as shoaling (Fig. 1). In even shallower waters, waves will become unstable and start breaking (breaker zone). After breaking, a wave propagates as a bore (foam-shaped broken wave) through the surf zone. This zone is characterized by highly turbulent flows and intense sediment transport, being the region where wave energy is dissipated thereby protecting the dry beach. When the wave is near the waterline, it will collapse and a thin layer of water will travel up (run-up) on the beach and back down (run-down) in an area known as the swash zone (Masselink et al. 2011).

Wave breaking is the dominant hydrodynamic process in surf zone waters driving nearshore currents and sediment transport. Three types of wave-induced currents exist, whose intensity increases with increasing incident wave height: (1) bed return flow, undertow/close to bed, strong offshore-directed mean current; (2) rip currents, relatively narrow and fast currents flowing away from the shore; and (3) long-shore currents, produced when waves approach the shore at an oblique angle, generating a current which runs parallel to the shoreline (McLachlan and Defeo 2018). The total energy transferred to the shore from incoming waves will vary under different wave climates, depending on the origin and nature of the wave regime. Locally generated wind waves with periods generally < 8 s determine "sea" conditions, whereas periods generally > 8 s define "swell" conditions. Shorter period waves transfer more energy than longer period waves of the same height, which is relevant for classification (McLachlan et al. 2018).

Sandy shore classification

The classification of sandy shores is based on the characteristics of two compartments of LAZ, beach and surf zone. The primary factors used to characterize a beach are sediment type and size, waves and tides (Masselink and Short 1993). Types of sandy beaches are classified based on the relative contribution of waves and tides, using the relative tide range: TR = TR/H_b, where TR is the tide range, which classifies beaches into the following categories: < 2 m (microtidal), 2–4 m (mesotidal), 4–8 m (macrotidal) and > 8 m (megatidal); and H_b is the significant breaker wave height (defined as the average wave height of the highest third of the waves at breaking). According to RTR, three beach types can be differentiated: RTR values < 3 represent wave-dominated (WD) beaches, RTR in the range of 3–10 is associated with tide-modified (TM) beaches, and RTR values > 10 indicate tide-dominated (TD) beaches.

Another parameter used to classify beaches is the beach morphodynamic state within each type, which can range from higher energy dissipative to lower energy reflective, particularly within the WD and TM beach states. WD dissipative beaches are associated with a combination of high waves, fine sand, gentle slope (< 0.02) and wide surf zones (> 100 m), with at least two lines of breakers. They are favoured by the short-wave periods (< 8 s) of sea environments. In swell conditions, to attain this morphodynamic state, waves need to be higher than in a sea environment, with H_b > 2.5 m. As wave height decreases, the intermediate beach states begin to prevail. WD intermediate beaches are characterized by narrow surf zones (< 100 m) with at least one line of breakers and variable sand size. WD reflective beaches are associated with waves < 1 m, medium to coarse sands, steep slopes (> 0.05) and without surf zones (McLachlan et al. 2018, Short 2020).

Though there is general agreement on the terminology and environments associated with the range of WD beaches, this is not the case regarding the nature and range of TM beaches. Within TM beaches, three states are recognized in sea and swell environments and across a range of increased wave energy: TM reflective, TM intermediate and TM ultradissipative. The three states usually occur under large tides (at least mesotidal), with a different morphology during low and high tides. For more detailed descriptions of each beach state, see McLachlan et al. (2018) and Short (2020), and references therein. The wave energy in TD beaches is very low (H_b < 0.5 m) and the role of tides becomes increasingly important. These beaches only occur in the sea environment and a variety of substrates occur within this state, most having some form of beach on the upper shore, with flats below. The WD, TM and TD continuum ends where tidal flats take over, usually at a boundary RTR ~ 50 (Short 2020).

The morphodynamic state of beaches is an essential factor for determining the volume of sand delivered by aeolian transport across the backshore to dunes. The volume is higher in dissipative beaches, due to the higher potential for aeolian sediment transport, than in reflective beaches, where the potential is minimal. There is therefore a strong link between beach state and foredune height and volume; the highest and largest foredunes occur on dissipative beaches and the smallest are found on reflective beaches. As transgressive dunefields are formed by the breakdown of large established foredunes, this dune type is commonly found on high-energy dissipative beaches. Moderate to higher-energy intermediate beaches are characterized by parabolic dunes. On reflective beaches, the foredunes remain relatively stable over time, resulting in little dune transgression (Hesp 2011).

Major biological components and interactions

Sandy shores provide specialized habitats for many organisms across the environmental heterogeneity of the LAZ. Recognizable sets of species appear and disappear across the marine gradient, making zonation a common feature (Martínez et al. 2004, Schlacher and Thompson 2013, Carcedo et al. 2017). Assemblages of invertebrates such as insects, crustaceans, molluscs and polychaetes play a vital role in the processing of nutrients and organic matter and act as a link between higher trophic consumers. Sandy shores also provide habitats for a diversity of fishes, being prime sites for commercial net fisheries (McLachlan et al. 1996a, Olds et al. 2018), and they are also critical habitats for endangered species such as turtles and birds (Rumbold et al. 2001, Schlacher et al. 2007).

Dunes

The main biological component of sand dunes is vegetation (Fig. 2a). Here, plant species are tolerant to the sandy substrate, high wind velocities, salt spray, sand accretion, and environmental heterogeneity. The various plant species are not randomly distributed; a clear zonation of plant communities has been well documented along seacoasts worldwide, with differences in species composition and dominance (Maun 2009). Four distinct zones of vegetation can usually be distinguished across the sea-inland gradient, related to the ability to withstand prevailing environmental factors (Tinley 1985). Closest to the sea is a zone characterized by dune pioneers such as creeping grasses and succulent herbs with rhizomatous and stoloniferous growth. A second zone is distinguished by a shrub community or dune heath, which

Figure 2. Biological components in sandy shore ecosystems. (a) Dune vegetation; (b) Dense population of a filter-feeder clam in the intertidal zone; (c) Birds foraging in the beach; (d) Insects and fragments of vegetation from the dunes blown seaward by the wind. Photograph (a) by A. M. Domínguez, and photographs (b), (c) and (d) by M. C. Carcedo.

may include a mixture of pioneer plants together with psammophytes from other habitats. A third zone consists of dwarf trees and shrubs, with a compact canopy whose height increases with rainfall. The latter zone presents tall thickets and/or true forest species, only developing in areas of high rainfall, behind the protection of larger dunes. Wind, shelter, and rainfall are key factors for the development of these zones (McLachlan and Defeo 2018).

In areas with high moisture content, dunes support rich interstitial biotas dominated by bacteria, fungi, and meiofaunal forms (< 1 mm) which may be important in the decomposition of organic material (McLachlan 1991, Van der Merwe and McLachlan 1991). The macrofauna (> 1 mm) is composed of vertebrates and arthropods, and on lime-rich soils, molluscs may be abundant. Among arthropods, insects are usually dominant, especially Hymenoptera, Coleoptera, and Diptera. Segregation of the main orders of the insect community across a dune-beach interface was described by Mourglia et al. (2015). Hymenoptera (especially Formicidae) were reported to prefer the dune zone, characterized by high elevation and low sand moisture and compaction levels, whereas Coleoptera prefer the gentle slopes and fine and humid sands of the beach. Arachnids are common in dunes and their diversity increases with the complexity of the habitat structure (Duffey 1986, McLachlan 1991). Crustaceans may be important near the beach, the backshore and dunes being one of the avenues of land colonization by talitrid amphipods and oniscid isopods (Vader and de Wolf 1988); their abundance decreasing landwards owes to desiccation stress caused by the lack of moisture (McLachlan 1991).

Dune systems support a relatively high diversity of vertebrates such as small mammals, birds, reptiles, and amphibians (Van Aarde et al. 2004), few of which are unique or restricted to the dune habitat (McLachlan 1991). The diversity of faunal species responds primarily to changes in vegetation along the gradient perpendicular to the shore, which usually increases away from the sea. Diversity is thus positively related to the structural diversity of soil and vegetation, which usually increases in cover and complexity in the same direction, as salt spray and sand movement rates decline. Small mammals such as rats, mice, gerbils, and moles are found in dunes and may occasionally forage on the backshore (McLachlan and Defeo 2018). Many shorebirds such as terns and oystercatchers (Medeiros et al. 2012) nest at the dune-beach interface above the highest drift line and sometimes into the dunes (Medeiros et al. 2012). They commonly feed on the beach, importing significant quantities of energy into the dunes (McLachlan and Defeo 2018).

Beaches

Beaches support an under-appreciated biodiversity of invertebrates, mainly because they are not visible most of the time, remaining retracted in the sediment. A significant body of knowledge has been accumulated in recent decades relating to the community and population ecology in sandy beach ecosystems, and how morphodynamic states are useful for describing variations in abundance, richness, and diversity of macrofaunal communities (Defeo and McLachlan 2005). In the case of supratidal species, reflective beaches represent more stable environments than dissipative ones (Short 1996), and therefore constitute a safer habitat for the species development, particularly at juvenile stages (Defeo and Gómez 2005). Another contributory factor to the shaping of supratidal communities is the presence of wrack deposits, composed mainly of macrophytes, carrion and other strandings of organic allochthonous materials, which could influence the structure and taxonomic composition of invertebrate assemblages (Rodil et al. 2008, Ruiz Delgado et al. 2014). The distribution of wrack-associated species, including meio- and macrofaunal invertebrates, shorebirds and other secondary consumers such as spiders, scorpions, lizards and rodents, is related to the distribution of their primary food sources (Colombini and Chelazzi 2003).

The intertidal zone is characterized by a diverse and abundant faunal assemblage, with some species attaining exceptionally high densities (Carcedo et al. 2015a; Fig. 2b). Meiofaunal organisms like nematodes, harpacticoid copepods, turbellarians, oligochaetes, mystacocarids, gastrotrichs, archiannelids, ostracods, mites and tardigrades share characteristics such as slender and filiform/vermiform body shape and extreme flexibility, cuticles ornamented with spines and/or scales and specialized organs for adhesion to sand grains (Giere 2009). The macrofauna is characterized by organisms with a high degree of mobility that allows them to relocate within the intertidal area. Crustaceans commonly dominate the most exposed shores, polychaetes the most sheltered, and molluscs on those shores of intermediate exposure (McLachlan 1983). Molluscs are usually abundant in terms of biomass and are composed mainly of filter-feeder bivalves (Fig. 2b). Other groups include nemerteans, anthozoans, platyhelminthes, sipunculids, echiurids, insects and echinoderms. Macrofaunal communities living in the intertidal zone are structured by the independent responses of individual species to the physical environment (Noy-Meir 1979, McLachlan 1990). This would suggest an increment in community descriptors (richness, density, total abundance and biomass) from harsh microtidal reflective beaches to benign macrotidal dissipative beaches through the exclusion of species as a result of both short and turbulent swashes and coarse sands (Barboza and Defeo 2015, Defeo and McLachlan 2013).

Supra and intertidal zones are also important for large vertebrates. In the case of marine mammals, such as seals and other pinnipeds, the availability of suitable coastal habitats is essential for copulation, parturition, and the safe rearing of pups (Rodriguez and Bastida 1998). Regarding marine turtles, there are seven known species belonging to two families, Cheloniidae and Dermochelyidae (Wallace et al. 2010). They live and mate at sea but during the reproductive season, they depend on sandy beaches for nesting and laying their eggs on the backshore zone (Lutz and Musick 1996, McLachlan and Defeo 2018). Amphibians and other terrestrial forms of reptiles may occasionally stray onto the backshore when foraging. Birds also spend time on the beach (Fig. 2c), either for foraging (e.g., shorebirds, waders, some seabirds, waterbirds, crows, and raptors) or for roosting and/or resting (e.g., cormorants, pelicans, penguins) (McLachlan and Defeo 2018).

Surf zones

There are few studies on surf zone biota, mainly because of the turbulent nature of these waters. In general terms, surf zone faunistic assemblages are characterized by numerous species with a low number of individuals and a few dominant species, many of them shared with near coastal ecosystems like estuaries (Fiori et al. 2022). In the case of phytoplankton, a few species of diatoms manage to accumulate and predominate on sandy shores, particularly those where intermediate to dissipative wave energy conditions prevail (Odebrecht et al. 2014). Surf-diatom accumulations can generate exceptionally high primary production rates. Together with the organic matter and detritus, these accumulations sustain macrobenthic filter-feeders in the intertidal area and other surf-water organisms, transferring energy to consumers such as fish, birds and even marine mammals (Bergamino et al. 2016, Garcia et al. 2019).

The zooplankton in surf zones is dominated mainly by crustaceans, especially copepods, with larval stages of benthic adults (bivalves, cirripeds, polychaetes) and gelatinous forms as intermittent visitors, among others (e.g., Costa 2011, Pinheiro et al. 2013, Menéndez et al. 2019, Rodrigues et al. 2019). Resident species (those with permanent occurrence) may be strictly planktonic (copepods, mysids, small prawns) or bentho-planktonic (isopods, amphipods), while non-residents may belong to holoplankton (organisms whose entire life cycle occurs in the water column) or meroplankton (that spend only a portion of their early life in the plankton). Holoplankton includes micro (2–200 μm, mainly tintinnid ciliates and copepod nauplii), meso (0.2–20 mm) and macrozooplankton (20–200 mm) (copepods, cladocerans, chaetognaths, and jellyfishes). Meroplankton is primarily made up of larval stages of non-planktonic animals from nearby coastal areas, such as crabs, bivalves, polychaetes, and fishes. In surf waters, the importance of hydrodynamic factors in determining zooplankton community structure has been demonstrated (Pinheiro et al. 2011, Menéndez et al. 2019, Baleani et al. 2020). The abundance of dominant species is strongly affected by short-term variations in wave height and the associated littoral current velocity (Baleani et al. 2020), whereas the general structure of a zooplankton community (richness, abundance, diversity) is significantly affected by storm events (Baleani et al. 2021). Finally, environmental variables such as water temperature, the concentration of suspended sediments and chlorophyll, and the precipitation regime also help explain seasonal community patterns on sandy beaches (e.g., Costa 2011, Pinheiro et al. 2013, Aboul-Ezz et al. 2014).

Several faunistic groups of invertebrates make up the community inhabiting the bottom layer in surf zones; however, due to the non-standardization of definitions such as suprabenthos (e.g., San Vicente and Sorbe 1999), hyperbenthos (e.g., Beyst et al. 2001), epibenthos (e.g., Carcedo et al. 2015b), bentho-planktonic or benthopelagic species (e.g., McLachlan and Defeo 2018), the information about these organisms remains scarce and disperse. Despite the disparate terminology, some generalizations can be made, such as the high values of species richness, abundance and biomass, emphasizing the importance of this shallow habitat for benthic invertebrates. Aggregations of shrimps, prawns, mysids and other peracarids are major prey items for fishes and important links for energy circulation between benthic and pelagic food webs (McLachlan 1983). The main variables determining the occurrence of these organisms in surf waters are physical factors such as water temperature, wave height and turbidity (Beyst et al. 2001), accumulations of detached macrophytes (Crawley and Hyndes 2007, Marin Jarrin and Shanks 2011) and storms, which ultimately influence their occurrence (Baring et al. 2014).

The surf zone is also intensely used by fishes (e.g., Marin Jarrin and Miller 2016, Beyst et al. 2002). A wide variety of fishes can be encountered in these waters, including different stages (adults, juveniles, and larvae), with different diets (benthic feeders, planktivores, piscivores, and omnivores) and habitats or behaviour (pelagic and demersal species). The body plans of the most common and abundant species in surf zones are likely advantageous for living in turbulent conditions (Tatematsu et al. 2014). Examples of these are streamlined fusiform bodies (e.g., mullets), laterally compressed bodies which increase manoeuvrability (e.g., trevally) or dorsoventrally flattened, which allows fish to bury or remain on the sand surface (e.g., puffers) as waves pass over them (Lauder 2015, Olds et al. 2018). Some fishes are less well adapted to life in the surf, being abundant only when they arrive with mats of drifting algae or are brought by strong winds and waves (Connolly et al. 2002). Surf zones are used by fish species that migrate among different ecosystems in coastal seascapes and as feeding habitats. The high turbulence of exposed surf zones makes these areas hazardous for spawning fishes, possibly limiting their use for this purpose (Hyndes and Potter 1997), while their role as nursery for juvenile fish still needs to be further corroborated (Olds et al. 2018). Birds such as cormorants and terns can be found feeding offshore by diving, and an abundance of these birds can therefore have a considerable impact on fish species. Finally, the passage of cetaceans through these zones is transitory and usually related to their use as feeding areas (McLachlan and Defeo 2018).

Species interactions

Because of the relatively small scale of most sandy shores, much of the exchange of organic materials occurs across the different compartments. This system is thus very open and many of its food webs may start in the sea and end inland, or vice versa. Animal movements give rise to the direct exchange of organic materials across the dune-beach interface: birds and mammals from the dunes may be dependent on intertidal organisms for food and some beach animals, such as supralittoral crustaceans, move into the dunes to feed. The exchange can also be mediated by physical forces: when winds blow offshore, insects are often blown from the dunes toward the sea (Fig. 2d) and many of those landing on the beach are consumed by birds, crabs, and snails (McLachlan and Defeo 2018). This exchange across the dune-beach interface also extends to the surf zone: insects have been recorded from the stomachs of fishes (Wilber et al. 2003, Costa and Zalmon 2017, Vaske-Junior et al. 2018). In addition to insects, organic detritus such as fragments of vegetation from the dunes, blown seaward by the wind, represent a significant input to the beach/surf zone (McLachlan and Defeo 2018). Allochthonous inputs include wrack brought in from the sea, providing a potential food source for animals in the dunes and beach areas. All these temporarily available resources can induce changes in trophic relationships and trophic niche properties of organisms across the entire sandy shore ecosystem (Schlacher et al. 2013, Bessa et al. 2014, Quillien et al. 2016).

The role of trophic interactions in the structure and dynamics of sandy shore communities has received limited attention in dune ecosystems. Some studies are devoted to analyzing the effects of animals on the flora (see Maun 2009), and quantifying energy flow (McLachlan et al. 1996b). Trophic interactions have been more deeply studied in the beach/surf zone ecosystem. The presence of various components (diatoms, water-column microbes, interstitial organisms, meio- and macrofauna) and their relative importance differ between beach states. Dissipative beaches sustain sufficient primary production to be self-supporting, whereas reflective beaches are dependent on food imports from the sea, with intermediate states occurring between these two extremes. A comparison of macroscopic food webs on beaches in northern Uruguay (Lercari et al. 2010, Bergamino et al. 2011, 2016) confirmed the considerable differences between these two beach types in terms of energy flow through macrofauna: dissipative beaches supported greater productivity and a more complex food web (larger number of discrete trophic levels and higher transfer efficiencies) than reflective beaches.

Several studies have described the importance of biological interactions in shaping community structure (see Defeo et al. 1992, Defeo 1996, McLachlan 1998, Van Tomme et al. 2014). The earliest evidence of competition on sandy beaches was provided by Croker and Hatfield (1980), who found horizontal and vertical space partitioning between three sympatric amphipods along with differential survival in coexistence experiments. Defeo et al. (1997) also detected interspecific interactions between two sympatric isopods: *Excirolana braziliensis* is most abundant in fine sands and has a greater niche breadth throughout most of its range than that observed in Uruguay, where it mainly inhabits coarse sands and upper beach levels on reflective beaches, being displaced by *E. armata*. Dugan et al. (2004) investigated the burrowing performance of the hippid crab *Emerita analoga* in three experimental densities of the clam *Mesodesma donacium*, finding burrowing times of large crabs significantly longer in all densities of clams than in controls (without clams). This delayed burrowing increases the exposure of the individuals involved, with potentially negative consequences.

Other examples are related to temporal variations in the abundance of dominant community members. Fishing pressure on the yellow clam *Amarilladesma mactroides* explained temporal variations in recruitment in the non-targeted species, the wedge clam *Donax hanleyanus*. The recruitment density of *D. hanleyanus* was inversely related to the density of juveniles and adults of *A. mactroides* (Defeo 1998), defining an 'interspecific stock-recruitment relationship' (*sensu* Pauly 1980). Fluctuations in species abundance related to specific events of natural mass mortality also explain temporal variations in macrobenthic community structure. A significant decrease in abundance of this community, in particular of the pea crab *Austinixa patagoniensis*, was associated with an increase in yellow clam abundance after a period of very low density. This high density of the yellow clam reduces the available area for the establishment of burrows of the ghost shrimp *Audacallichirus mirim*, in which the crab lives as a commensal (Carcedo et al. 2019). These examples provide evidence that even though communities on sandy beaches are generally structured by independent responses of individuals to the physical environment, in dissipative beaches with high species richness and densities, the potential for biological interactions increases due to possible resource limitations (Defeo and McLachlan 2005).

Effect of Anthropogenic activity

Sandy shores are considered the most common natural habitats used for human activities (Schlacher et al. 2007), supporting a range of ecosystem services such as storm protection, water filtration, erosion control, recreational opportunities, and cultural value, among others (Barbier et al. 2011). A host of anthropogenic stressors associated with these ecosystems has given rise to degradation and a net loss of total area in various regions of the world (Lester et al. 2020), shores with smaller spatial extents being the most vulnerable to these impacts (Knights et al. 2013, Batista et al. 2014, Rocha et al. 2015). The impact of anthropic activity varies depending on the type and intensity of resource use, as well as the life history characteristics and resilience of individual species and populations (Jennings et al. 1998).

Main anthropogenic stressors on sandy shores

The intensity of human stressors and their impact on 21 marine ecosystems around the world were analyzed by Halpern et al. (2019). These authors highlight that the most direct human activities impacting sandy shores are nutrient pollution, artisanal fishing, light pollution, shipping, coastal constructions, and population density. In terms of the latter, sandy shores are highly affected by growing urban populations and mass tourism leading to the rapid development of urbanized areas and consequently to the loss of natural habitats (Costa et al. 2017a, Orlando et al. 2020). Dune suppression by infrastructure and defence constructions, such as groynes, seawalls and sand fences, modify the natural sediment dynamic and geomorphology of beaches (Rangel-Buitrago et al. 2018), with the consequent loss of their sand reservoir and their buffering capacity against storms (Defeo et al. 2021). These constructions fragment the habitat and cause loss of migratory paths for organisms; furthermore, the introduction of artificial substrates could favour the settlement of exotic species, with severe impacts on biodiversity at the community level (Laitano et al. 2022). Urbanized sandy beach biota becomes more vulnerable than non-urbanized beach communities (Machado et al. 2016, 2017, Orlando et al. 2020). Adverse effects on vegetal and faunal organisms can lead to reductions in energy transfer efficiency through trophic levels, altering the trophic functioning of the ecosystems (Costa et al. 2017b).

Climate-driven changes can also impact sandy shore ecosystems through sea level rise and warmer temperatures, with associated storminess, excessive precipitation, and flooding. This intensifies erosion, pushing the LAZ landward and causing the retreat of beaches (Defeo et al. 2009, Bindoff et al. 2019). If the dune component is constrained due to recreational, urban, and industrial development on the land, the beach has no capacity to respond naturally to the sea level rise. This phenomenon, called "coastal squeeze", is considered a major long-term threat to sandy shores worldwide (Defeo et al. 2021): if it persists over time, beaches will continue to diminish until only the surf zone remains. The reduction of the LAZ has been associated with a loss of vegetal and faunal species (Veloso et al. 2006, Schlacher et al. 2016, Costa et al. 2017a, Lester et al. 2020) and its associated ecological consequences, plus loss of the species' main ecosystem services, leading to social-ecological collapse (Defeo et al. 2021).

The usual way to protect sandy shores from erosion processes is nourishment, filling the beach with sediments extracted from another place in the sea, often from harbours where dredging operations are carried out for navigational purposes. When the wrong sediments are chosen for transferral, changes in sand mean grain size, colour and beach slope occur, thus modifying beach morphodynamics. Furthermore, this practice entails the risk of transferring the contaminating substances produced in port zones together with sediments. Negative effects of nourishment procedures on sandy beach biota have been extensively documented, including mortality by burial, crushing and suffocation of intertidal invertebrates (Veloso et al. 2006, Machado et al. 2017), reduced foraging activity by fish and shorebirds due to the decrease in prey density (Peterson et al. 2006, Costa et al. 2017b), and lower nesting and hatching success for sea turtles (Cisneros et al. 2017), among others. Another source of pollution on sandy shores, closely associated with urbanization, is the artificial lighting at night that alters the behavioural and physiological patterns of those organisms dependent on day-night cycles (Luarte et al. 2016).

Since recreational activities and tourism involve the arrival and movement of thousands of millions of people to coastal areas (Mason 2020), they constitute a constant threat to the biodiversity and functioning of the sandy shoreline ecosystem (Schlacher et al. 2007, Laitano et al. 2022). Intensified beach recreation often involves trampling by both human and off-road vehicles, inducing sediment compaction and crushing of resident organisms (Schlacher et al. 2008, Fanini et al. 2014). In addition, litter produced by humans can obstruct the gastrointestinal tract of organisms, lower reproductive output, and increase the probability of death (Li et al. 2016). Benthic macroinvertebrates, insects, fish, reptiles, birds and mammals from beach and surf zones have been negatively affected worldwide (see literature cited in Costa et al. 2022). The main deleterious effect of plastic is associated with its entanglement and consumption by various species; it can also be a vector of metal ions, persistent organic pollutants, and polycyclic aromatic hydrocarbons (Truchet et al. 2021, Ben-Haddad et al. 2022).

Human activities related to resource exploitation, which contribute to the local economy by providing an important source of employment for coastal communities, can also have a negative impact (McLachlan and Defeo 2018). The significant biomass of intertidal filter-feeder clams turns them into target species, extensively collected for artisanal purposes, as part of recreational extraction, or for commercial purposes. Species subjected to elevated harvesting rates can experience density population declines, changes in demographic characteristics such as truncation of size and age composition, reduced size and age at reproduction, and altered growth and mortality rates (Gray 2016). Their exploitation is sometimes amplified by the accumulation of toxins, such as those associated with blooms of toxic algae (Odebrecht et al. 1995), which can cause the mortality of the clams themselves, weakening or decimating stocks. Detrimental effects of resource exploitation over intertidal clam populations can also be exacerbated by the effects of climate change, showing modifications in population features related to increasing sea surface temperatures (Ortega et al. 2016). The case of a clam fishery in an intertidal zone suffering from some of the aforementioned impacts is presented in the next section.

From abundance to collapse: A yellow clam fishery on the southwestern Atlantic coast

The yellow clam *Amarilladesma mactroides* (= *Mesodesma mactroides*) is an intertidal species distributed along the temperate Atlantic coast of South America, from Sao Pablo state (Brazil, 24°S) to the southern end of Buenos Aires province (Argentina, 41°S) (Fiori and Defeo 2006) (Fig. 3a). The species thrives primarily in dissipative sandy beaches with a gentle slope, fine sand and heavy wave action. Its patchy distribution is mainly due to variations in sand granulometry and water physicochemical variables, particularly salinity, owing to the presence of estuaries or freshwater courses (Olivier et al. 1971, Thompson and Sanchez De Bock 2009, Carvalho et al. 2015, Fiori and Carcedo 2015). This large clam (up to 80 mm in length) is considered a fast burrower, lives buried in the sediment and migrates seasonally into the intertidal zone (Coscarón 1959, Fiori et al. 2004a). Due to its trophic position, it is one of the most relevant links between the bento-pelagic and marine-terrestrial communities. The clam diet is mainly composed of phytoplankton, microphytobenthos and detritus. Among its main predators are teleost fish, rays, sharks, marine birds and sublittoral snails, which rise with the tides and capture the clams when they carry out their seasonal migrations (Coscarón 1959, Olivier et al. 1971, Defeo and Scarabino 1990, Bergamino et al. 2011).

The yellow clam has been a food resource for South American coastal populations since ancient times. The exploitation of this species during the Holocene by hunter-gatherer groups is evidenced by the numerous archaeological shell middens found in Brazil (Correa Luz de Souza et al. 2010) and Argentina (Eugenio and Aldazabal 2004, Frontini and Bayon 2017, Martínez 2017, Fiori et al. 2019). In the recent past, commercial, and recreational harvesters (reviewed in Defeo 2003) have also exploited the yellow clam throughout its distribution range. In Uruguay, this clam supported artisanal fishery over several decades, with marked periods of overexploitation leading to the implementation of management strategies such as fishery closure, rotating extraction zones and more recently, an ecosystem approach with the active involvement of fishers in the management process (Castilla and Defeo 2001, Gianelli

Figure 3. (a) Geographic distribution of *Amarilladesma mactroides* (adapted from Olivier et al. 1971); (b) Summer tourism and coastal urbanization in an Argentinian sandy beach; (c) Mass mortality of *A. mactroides* during November 1995; (d) *A. mactroides*. Map (a) by W. D. Melo, photograph (b) by F. Latorre and photographs (c) and (d) by S. M. Fiori.

et al. 2015). In Argentina, commercial fishing activity started in the 1940s, coinciding with the development of the canned food industry. Available at more advantageous market prices, clams became more popular and reached their commercial peak between 1940 and 1957. The growing demand subsequently caused an increase in the number of shellfish collectors, a diversification of extraction tools and hence overexploitation (Coscarón 1959). Though commercial harvesting was prohibited in 1958, stocks continued to gradually decline and the species disappeared completely from many beaches because of tourist harvesting, poaching and the loss of habitat quality due to increasing coastal urbanization (Bastida et al. 1991, Dadon et al. 2001) (Fig. 3b).

In addition to this background of deterioration, large-scale mass mortality events (Fig. 3c,d) progressively decimated all yellow clam populations throughout their geographic range (Fiori and Cazzaniga 1999). Mortalities, which followed a sequential north-south direction, started in Brazil in 1993 (Odebrecht et al. 1995), in Uruguay in 1994 (Méndez 1995), and in Argentina from 1995 to 2002, affecting even the southernmost population of this species (Fiori and Cazzaniga 1999, Fiori et al. 2004b). After these initial events, mortality became recurrent but less noticeable given the slow recovery of the depleted clam populations. Mortality events occurred mainly during late spring and summer, indicating greater vulnerability of the clams to high sea temperatures or a combination of environmental factors favouring the agent causing the mortality (Fiori et al. 2004b). Reinforcing this hypothesis, long-term interannual fluctuations of the yellow clam population on the Uruguayan coast show a positive correlation between periods of low abundance and clam body abnormalities with increasing sea surface temperature, suggesting a link with climate stress (Manta et al. 2016, Ortega et al. 2016). Mortalities did not affect other members of the macrofaunal community including the very abundant suspension feeders *Donax hanleyanus* (Defeo and de Alava 1995) and *Emerita brasiliensis* (Defeo and Cardoso 2002), which co-occurs with *A. mactroides*. The strong host-specific mortality together with histopathological analyses of clam tissues (with several pathogens with mortality potential), suggests that parasites could play a relevant role in the massive mortalities (Fiori et al. 2004b, Cremonte and Figueras 2004, Carvalho et al. 2013, Vázquez et al. 2016). Other proposed hypotheses relate to harmful algae blooms, oxygen depletion (Méndez 1995, Odebrecht et al. 1995) and acute contamination (Thompson and Sánchez de Bock 2009). Causes and mechanisms of mass mortalities in natural populations are usually difficult to ascertain and are probably due to a combination of factors. These catastrophic episodes changed the structure and functioning of the ecosystem, temporarily eliminating one of the main species of the benthic community of the sandy beaches of South America. The yellow clam, an important economic resource in the past, has become a threatened species and likely to achieve the status of "critically endangered" (Fiori and Cazzaniga 1999).

Final remarks

Sandy shorelines are permanent or transitory habitats of several biological communities highly specialized to severe and variable physical conditions. Although it is mainly physical factors that govern the communities and populations features of sandy beaches, biological interactions also play an important role as regulatory agents. In addition to the variability in terms of physical and biological factors, human-induced impacts impose additional changes on organisms living in these ecosystems, as well as on their ecological functions and the ecosystem services that they provide. Within a scenario of expanding human activities, the impact on coastal ecosystems is also expected to increase and be even further exacerbated by global climate change. In the case of urbanization-threatened ecosystems such as sandy shores, where both the ecological and socio-economical aspects coexist on the same physical unit, further studies to assess the specific impacts on each compartment along the LAZ are urgently required. There is a lack of balance in current studies devoted to analyzing variability within these three compartments, with the surf zone often being neglected in favour of the beach, for which in-depth studies on patterns and processes at the community level are available, though even within this zone, the supratidal zone remains less studied than the intertidal zone. Dunes are extensively studied as a terrestrial, wind-controlled system, but are often separated from their adjacent marine beach/surf zone system. In order to determine the precise functioning of the entire LAZ and how it responds to drivers of change, it is therefore necessary to integrate long-term ecological studies to assess biodiversity changes, interactions, and energy exchanges between the different compartments. In particular, studies should include the most sensitive LAZ subsystem, the frontal dunes, to reinforce the idea that it is only by conserving this sand reservoir with its buffering capacity against storms that the sandy shore will have the capacity to deal with approaching threats from land and sea.

Acknowledgments

We gratefully acknowledge Alejandro Vitale and Walter Melo for their valuable comments and suggestions.

References

Aboul Ezz, S. M., Heneash, A. M. H. and Gharib, S. M. (2014). Variability of spatial and temporal distribution of zooplankton communities at Matrouh beaches, southeastern Mediterranean Sea, Egypt. Egypt. J. Aquat. 40: 283–290.

Amir, R., Kinast, S., Tsoar, H., Yizhaq, H., Zaady, E. and Ashkenazy, Y. (2014). The effect of wind and precipitation on vegetation and biogenic crust covers in the Sde-Hallamish sand dunes. J. Geophys. Res. Earth. Surf. 119: 437–450.

Baleani, C. A., Menéndez, M. and Piccolo, M. C. (2020). Surf zone zooplankton temporal dynamics and their environmental regulation in a southwestern Atlantic sandy beach (Pehuen Co, Argentina). J. Sea Res. 156: 101833.

Baleani, C. A., Manstretta, G. M., Menéndez, M. C., Vitale, A. J., Piccolo, M. C. and Perillo, G. M. E. (2021). Influence of storm events on surf zone zooplankton in a Southwestern Atlantic sandy beach: A preliminary methodology using high-frequency physical data. J. Mar. Syst. 220: 103560.

Barbier, E. B., Hacker, S. D., Kennedy, C., Koch, E. W., Stier, A. C. and Silliman, B. R. (2011). The value of estuarine and coastal ecosystem services. Ecol. Monogr. 81: 169–193.

Barboza, F. R. and Defeo, O. (2015). Global diversity patterns in sandy beach macrofauna: A biogeographic analysis. Sci. Rep. 5: 14515.

Baring, R. J., Fairweather, P. G. and Lester, R. E. (2014). Storm versus calm: variation in fauna associated with drifting macrophytes in sandy beach surf zones. J. Exp. Mar. Biol. Ecol. 461: 397–406.

Batista, M. I., Henriques, S., Pais, M. P. and Cabral, H. P. (2014). Assessment of cumulative human pressures on a coastal area: integrating information for MPA planning and management. Ocean Coast. Manage. 102: 248–257.

Bastida, R., Roux, A., Bremec, C., Gerpe, M. and Sorensen, M. (1991). Estructura poblacional de la almeja amarilla (*Mesodesma mactroides*) durante el verano de 1989 en la provincia de Buenos Aires, Argentina. Frente Marítimo 9(A): 83–92.

Ben-Haddad, M., Rida, M., Hajji, S., De-la-torre, G. E., Abou, H., Rangel-buitrago, N. et al. (2022). The Wedge Clam *Donax trunculus* L., 1758 as a bioindicator of microplastic pollution. Mar. Pollut. Bull. 178: 113607.

Bergamino, L., Lercari, D. and Defeo, O. (2011). Food web structure of sandy beaches: temporal and spatial variation using stable isotopes analysis. Estuar. Coast. Shelf Sci. 91: 536–543.

Bergamino, L., Martínez, A., Han, E., Lercari, D. and Defeo, O. (2016). Trophic niche shifts driven by phytoplankton in sandy beach ecosystems. Estuar. Coast. Shelf Sci. 180: 33–40.

Bessa, F., Baeta, A. and Marques, J. C. (2014). Niche segregation amongst sympatric species at exposed sandy shores with contrasting wrack availability illustrated by stable isotopic analysis. Ecol. Indic. 36: 694–702.

Beyst, B., Buysse, D., Dewicke, A. and Mees, J. (2001). Surf zone hyperbenthos of Belgian sandy beaches: seasonal patterns. Estuar. Coast. Shelf Sci. 53(6): 877–896.

Beyst, B., Hostens, K. and Mees, J. (2002). Factors influencing the spatial variation in fish and macrocrustacean communities in the surf zone of sandy beaches in Belgium. J. Mar. Biolog. Assoc. U. K. 82(2): 181–187.

Bindoff, N. L., Cheung, W. W. L., Kairo, J. G., Arístegui, J., Guinder, V. A., Hallberg, R. et al. (2019). Changing ocean, marine ecosystems, and dependent communities. pp. 447–587. *In*: Pörtner, H. O., Roberts, D. C., Masson-Delmotte, V., Zhai, P., Tignor, M., Poloczanska, E. et al. (eds.). IPCC Special Report on the Ocean and Cryosphere in a Changing Climate. Cambridge University Press, New York, NY, USA.

Brown, A. C. and McLachlan, A. (2002). Sandy shore ecosystems and the threats facing them: some predictions for the year 2025. Environ. Conserv. J. 29(1): 62–77.

Carcedo, M. C., Fiori, S. M., Piccolo, M. C., López Abbate, M. C. and Bremec, C. (2015a). Variations in macrobenthic community structure in relation to changing environmental conditions in sandy beaches of Argentina. Estuar. Coast. Shelf Sci. 166: 56–64.

Carcedo, M. C., Fiori, S. M. and Bremec, C. S. (2015b). Macrobenthic surf zone community of temperate sandy beaches: spatial and temporal patterns. Mar. Ecol. 36: 326–336.

Carcedo, M. C., Fiori, S. M. and Bremec, C. S. (2017). Zonation of macrobenthos across a mesotidal sandy beach: Variability based on physical factors. J. Sea Res. 121: 1–10.

Carcedo, M. C., Fiori, S. M., Scotti, M., Ito, M., Dutto, M. S. and Carbone, M. E. (2019). Dominant bivalve in an exposed sandy beach regulates community structure through spatial competition. Estuar. Coast. 42(7): 1912–1923.

Carvalho, Y. B. M., Poersch, L. H. and Romano, L. A. (2013). Rickettsia-associated mortality of the yellow clam Mesodesma mactroides (Bivalvia: Mesodesmatidae) in southern Brazil. Malacologia 56: 301–307.

Carvalho, Y. B. M., Romano, L. A. and Poersch, L. H. S. (2015). Effect of low salinity on the yellow clam Mesodesma mactroides. Braz. J. Biol. 75(1): 8–12.

Castilla, J. C. and Defeo, O. (2001). Latin American benthic shellfisheries: emphasis on comanagement and experimental practices. Rev. Fish Biol. Fish. 11: 1–30.

Cisneros, J. A., Briggs, T. R. and Martin, K. (2017). Placed sediment characteristics compared to sea turtle nesting and hatching patterns: a case study from Palm Beach County, FL. Shore Beach. 85(2): 35–40.

Colombini, I. and Chelazzi, L. (2003). Influence of marine allochthonous input on sandy beach communities. Oceanogr. Mar. Biol. Annu. Rev. 41: 123–127.

Connolly, R. M., Melville, A. J. and Keesing, J. K. (2002). Abundance, movement and individual identification of leafy seadragons, Phycodurus eques (Pisces: Syngnathidae). Mar. Freshw. Res. 53: 777–780.

Correa Luz de Souza, R. C., Andrade Lima, T. and Pereira Silva, E. (2010). Holocene molluscs from Rio de Janeiro state coast, Brazil. Checklist 6(2): 301–308.

Coscarón, S. (1959). La almeja amarilla (*Mesodesma (T.) mactroides* Deshayes) de la costa de la provincia de Buenos Aires. Agro, Publicación Técnica 1(3): 1–66.

Costa, K. G. (2011). Effects of seasonality on zooplankton community dynamics in the macrotidal coastal zone of the Amazon region. J. Coast. Res. 64: 364–368.

Costa, L. L. and Zalmon, I. R. (2017). Surf zone fish diet as an indicator of environmental and anthropogenic influences. J. Sea Res. 128: 61–75.

Costa L. L., Landmann, J. G., Gaelzer, L. R. and Zalmon, I. R. (2017a). Does human pressure affect the community structure of surf zone fish in sandy beaches? Cont. Shelf Res. 132: 1–10.

Costa, L. L., Tavares, D. C., Suciu, M. C., Rangel, D. F. and Zalmon, I. R. (2017b). Human-induced changes in the trophic functioning of sandy beaches. Ecol. Ind. 82: 304–315.

Costa, L. L., Fanini, L., Ben-Haddad, M., Pinna, M. and Zalmon, I. R. (2022). Marine litter impact on sandy beach fauna: a review to obtain an indication of where research should contribute more. Microplastics 1: 554–571.

Crawley, K. R. and Hyndes, G. A. (2007). The role of different types of detached macrophytes in the food and habitat choice of a surf-zone inhabiting amphipod. Mar. Biol. 151(4): 1433–1443.

Cremonte, F. and Figueras, A. (2004). Parasites as possible cause of mass mortalities of the critically endangered clam Mesodesma mactroides on the Atlantic coast of Argentina. Bull. Eur. Ass. Fish. Pathol. 24: 166–171.

Croker, R. A. and Hatfield, E. B. (1980). Space partitioning and interactions in an intertidal sand burrowing amphipod guild. Mar. Biol. 61: 79–88.

Dadon, J. R., Castaños, C., García, R. E. P., Chiappini, G. M. T. and Cruses, J. M. (2001). Efectos a largo plazo de las pesquerías de almeja amarilla (*Mesodesma mactroides*) y berberecho (*Donax hanleyanus*), y de la urbanización sobre las comunidades intermareales. pp. 703–715. *In*: Alveal, K. and Antezana, T. (eds.). Sustentabilidad de la Biodiversidad, un Problema Actual. Bases Científico-Técnicas, Teorizaciones y Proyecciones. Universidad de Concepción, Chile.

Defeo, O. (1996). Experimental management of an exploited sandy beach bivalve population. Rev. Chil. Hist. Nat. 69: 605–614.

Defeo, O. (1998). Testing hypotheses on recruitment, growth and mortality in exploited bivalves: an experimental perspective. Can. Spec. Publ. Fish. Aquat. Sci. 125: 257–264.

Defeo, O. (2003). Marine invertebrate fisheries in sandy beaches:an overview. J. Coast. Res. (Spec. Iss.) 35: 56–65.

Defeo, O. and Scarabino, V. (1990). Ecological significance of a possible deposit-feeding strategy in Mesodesma mactroides (Deshayes, 1854) (Mollusca: Pelecyoda). Atlantica 12(1): 55–65.

Defeo, O., Jaramillo, E. and Lyonnet, A. (1992). Community structure and intertidal zonation of the macroinfauna in the Atlantic coast of Uruguay. J. Coast. Res. 8: 830–839.

Defeo, O. and de Alava, A. (1995). Effects of human activities on long-term trends in sandy beach populations: The wedge clam *Donax hanleyanus* in Uruguay. Mar. Ecol. Prog. Ser. 123: 73–82.

Defeo, O., Brazeiro, A., de Alava, A. and Riestra, G. (1997). Is Sandy beach macroinfauna only physically controlled? Role of substrate and competition in isopods. Estuar. Coast. Shelf. Sci. 45: 453–462.

Defeo, O. and Cardoso, R. S. (2002). Macroecology of population dynamics and life history traits of the mole crab *Emerita brasiliensis* in Atlantic sandy beaches of South America. Mar. Ecol. Prog. Ser. 239: 169–179.

Defeo, O. and Gómez, J. (2005). Morphodynamics and habitat safety in sandy beaches: life-history adaptations in a supralittoral amphipod. Mar. Ecol. Prog. Ser. 293: 143–153.

Defeo, O. and McLachlan, A. (2005). Patterns, processes and regulatory mechanisms in sandy beach macrofauna: a multi-scale analysis. Mar. Ecol. Prog. Ser. 295: 1–20.

Defeo, O., McLachlan, A., Schoeman, D. S., Thomas, A., Schlacher, J. D., Jones, A. et al. (2009). Threats to sandy beach ecosystems: A review. Estuar. Coast. Shelf Sci. 81(1): 1–12.

Defeo, O. and McLachlan, A. (2013). Global patterns in sandy beach macrofauna: Species richness, abundance, biomass and body size. Geomorphology 199: 106–114.

Defeo, O., McLachlan, A., Armitage, D., Elliot, M. and Pittman, J. (2021). Sandy beach social-ecological systems at risk: regime shifts, collapses, and governance challenges. Front. Ecol. Environ. 19(10): 564–573.

Duffey, E. (1968). An ecological analysis of the spider fauna of sand dunes. J. Anim. Ecol. 37: 641–674.

Dugan, J. E., Jaramillo, E., Hubbard, D. M., Contreras, H. and Duarte, C. (2004). Competitive interactions in macroinfaunal animals of exposed sandy beaches. Oecologia 139: 630–640.

Eugenio, E. and Aldazabal, V. B. (2004). Los cazadores recolectores del litoral marıtimo del area de Bahıa de San Blas, Pcia. de Buenos Aires. pp. 687–700. *In*: Civalero, M. T., Fernandez, P. M. and Guraieb, A. G. (eds.). Contra viento y Marea. Arqueologıa de Patagonia. Instituto Nacional de Antropología y Pensamiento Latinoamericano, Buenos Aires, Argentina.

Everard, M., Jones, L. and Watts, B. (2010). Have we neglected the societal importance of sand dunes? An ecosystem services perspective. Aquat. Conserv.: Mar. Freshw. Ecosyst. 20: 476–487.

Fanini, L., Zampicinini, G. and Pafilis, E. (2014). Beach parties: a case study on recreational human use of the beach and its effects on mobile arthropod fauna. Ethol. Ecol. Evol. 26: 69–79.

Fiori, S. M. and Cazzaniga, N. J. (1999). Mass mortality of Yellow clam *Mesodesma mactroides* (Bivalvia: Mactracea) in Monte Hermoso beach Argentina. Biol. Conserv. 89(3): 305–309.

Fiori, S. M., Cazzaniga, N. J. and Estebenet, A. L. (2004a). Winter distribution, density and size of *Mesodesma mactroides* (Bivalvia, Mactracea) in Monte Hermoso beach (Argentina). Braz. J. Oceanogr. 52(1): 1–9.

Fiori, S. M., Vidal-Martınez, V., Sima-Alvarez, R., Rodrıguez-Canul, R., Aguirre-Macedo, M. and Defeo, O. (2004b). Field and laboratory observations of the mass mortality of the yellow clam *Mesodesma mactroides* in South America: the case of Isla del Jabalí, Argentina. J. Shellfish Res. 23(2): 451–455.

Fiori, S. M. and Defeo, O. (2006). Biogeographic patterns in life history traits of the yellow clam, *Mesodesma mactroides*, in sandy beaches of South America. J. Coast. Res. 22(4): 872–880.

Fiori, S. M. and Carcedo, M. C. (2015). Influence of grain size on burrowing and alongshore distribution of the yellow clam (*Amarilladesma mactroides*). J. Shellfish Res. 34(3): 785–789.

Fiori, S. M., Frontini, R. and Bayon, C. (2019). Morphometric analysis of modern yellow clam (*Amarilladesma mactroides*) as a tool for estimating shell length and clam biomass of fragmented fossil clams from the southern coast of Buenos Aires, Argentina. J. Island Coast. Archaeol. 14(2): 291–300.

Fiori, S. M., Lopez Cazorla, A., Martínez, A., Carcedo, M. C., Blasina, G., Molina, J. M. et al. (2022). Life in the surf-zone: Variations of faunal assemblage structure in temperate sandy beaches of the Southwestern Atlantic coast. Cont. Shelf Res. 244: 104781.

Frontini, R. and Bayon, C. (2017). Use of marine resources (fauna and tool stones) in the southwest of Buenos Aires Province (Argentina) during the Middle and Late Holocene. pp. 25–46. *In*: Mondini, M., Muñoz, S. and Fernández, P. M. (eds.). Zooarchaeology in the Neotropics: Environmental Diversity and Human-animal Interactions. Springer, New York, NY, USA.

Garcia, A., Oliveira, M. C. L. M., Odebrecht, C., Colling, J. L. A., Vieira, J. P., Rodrigues, F. L. et al. (2019). Allochthonous versus autochthonous organic matter sustaining macroconsumers in a subtropical sandy beach revealed by stable isotopes. Mar. Biol. Res. 15(3): 241–258.

Gianelli, I., Martínez, G. and Defeo, O. (2015). An ecosystem approach to small-scale co-managed fisheries: The yellow clam fishery in Uruguay. Mar. Policy 62: 196–202.

Giere, O. (2009). Meiobenthology. The Microscopic Motile Fauna of Aquatic Sediments. Springer-Verlag, Berlin, Heidelberg.

Gray, C. A. (2016). Effects of fishing and fishing closures on beach clams: experimental evaluation across commercially fished and non-fished beaches before and during harvesting. PLoS One 11(1): e0146122.

Halpern, B. S., Frazier, M., Afflerbach, J., Lowndes, J. S., Micheli, F., O'Hara, C. et al. (2019). Recent pace of change in human impact on the world's ocean. Sci. Rep. 9: 11609.

Hesp, P. A. (2011). Dune coasts. pp. 193–221. *In*: Wolanski, E. and McLusky, D. S. (eds.). Treatise on Estuarine and Coastal Science, vol 3. Academic Press, Waltham.

Hyndes, G. A. and Potter, I. C. (1997). Age, growth and reproduction of *Sillago schomburgkii* in south-western Australian, nearshore waters and comparisons of life history styles of a suite of Sillago species. Environ. Biol. Fishes 49: 435–447.

Jennings, J. N. (1957). On the orientation of parabolic or U-dunes. Geogr. J. 123(4): 474–480.

Jennings, S., Reynolds, J. D. and Mills, S. C. (1998). Life history correlates of responses to fisheries exploitation. Proc. R. Soc. B: Biol. Sci. 265: 333–339.

Knights, A. M., Koss, R. S. and Robinson, L. A. (2013). Identifying common pressure pathways from a complex network of human activities to support ecosystem-based management. Ecol. Appl. 23: 755–765.

Laitano, M. V., Chiaradia, N. M. and Nuñez, J. D. (2022). Human Impacts over Sandy Beaches. pp. 54–88. *In*: Gonçalves, S. C. and Ferreira, S. M. F. (eds.). Sandy Beaches as Endangered Ecosystems: Environmental Problems and Possible Assessment and Management Solutions. CRC Press, Boca Raton, Florida.

Lauder, G. V. (2015). Fish locomotion: Recent advances and new directions. Annu. Rev. Mar. Sci. 7: 521–545.

Lercari, D., Bergamino, L. and Defeo, O. (2010). Trophic models in sandy beaches with contrasting morphodynamics: comparing ecosystem structure and biomass flow. Ecol. Model 221: 2751–2759.

Lester, S. E., Dubel, A. K., Hernán, G., McHenry, J. and Rassweiler, A. (2020). Spatial planning principles for marine ecosystem restoration. Front. Mar. Sci. 7: 328.

Li, W. C., Tse, H. F. and Fok, L. (2016). Plastic waste in the marine environment: a review of sources, occurrence and effects. Sci. Total Environ. 566: 333–349.

Lithgow, D., Martínez, M. L., Gallego-Fernande, J. B., Hesp, P. A., Flores, P., Gachuz, S. et al. (2013). Linking restoration ecology with coastal dune restoration. Geomorphology 199: 214–224.

Luarte, T., Bonta, C. C., Silva-Rodriguez, E. A., Quijón, P. A., Miranda, C., Faria, A. A. et al. (2016). Light pollution reduces activity, food consumption and growth rates in a sandy beach invertebrate. Environ. Pollut. 218: 1147–1153.

Luijendijk, A., Hagenaars, G., Ranasinghe, R., Baart, F., Donchyts, G. and Aarninkhof, S. (2018). The State of the World's beaches. Sci. Rep. 8: 6641.

Lutz, P. L. and Musick, J. A. (1996). The Biology of Sea Turtles, vol. 1. CRC Press, Boca Raton, Florida.

Machado, P. M., Costa, L. L., Suciu, M. C., Tavares, D. C. and Zalmon, I. R. (2016). Extreme storm wave influence on sandy beach macrofauna with distinct human pressures. Mar. Poll. Bull. 107: 125–135.

Machado, P. M., Suciu, M. C., Costa, L. L., Tavares, D. C. and Zalmon, I. R. (2017). Tourism impacts on benthic communities of sandy beaches. Mar. Ecol. 38: e12440.

Manta, G., Barreiro, M., Ortega, L. and Defeo, O. (2016). The effect of climate variability on the abundance of the sandy beach clam (*Mesodesma mactroides*) in the southwestern Atlantic. J. Coast. Res. 33(3): 531–536.

Marin Jarrin, J. R. and Shanks, A. L. (2011). Spatio-temporal dynamics of the surf-zone faunal assemblages at a southern Oregon sandy beach. Mar. Ecol. 32(2): 232–242.

Marin Jarrin, J. R. and Miller, J. A. (2016). Spatial variability of the surf zone fish and macroinvertebrate community within dissipative sandy beaches in Oregon, USA. Mar. Ecol. 37(5): 1027–1035.

Martínez, G. (2017). Arqueología de cazadores-recolectores del curso inferior del Río Colorado (Provincia de Buenos Aires, Argentina). Aportes al conocimiento de las ocupaciones humanas Pampeano Patagónicas, no. 6. INCUAPA-CONICET UNICEN, Olavarría, Argentina.

Martínez, M. L., Psuty, N. P. and Lubke, R. A. (2004). A perspective on coastal dunes. pp. 3–10. *In*: Martínez, M. L. and Psuty, N. P. (eds.). Coastal Dunes. Ecological Studies, vol. 171. Springer, Berlin, Heidelberg.

Mason, P. (2020). Tourism Impacts, Planning and Management (4th ed.). Routledge, London.

Masselink, G. and Short, A. D. (1993). The effect of tide range on beach morphodynamics and morphology: a conceptual model. J. Coast Res. 9: 785–800.

Masselink, G., Hughes, M. and Knight, J. (2011). Introduction to Coastal Processes and Geomorphology (2nd ed.). Routledge. https://doi.org/10.4324/9780203785461.

Maun, A. (2009). The Biology of Coastal Sand Dunes. Oxford University Press, Oxford.

McLachlan, A. (1983). Sandy beach ecology—a review. pp. 321–380. *In*: McLachlan, A. and Erasmus, T. (eds.). Sandy Beaches as Ecosystems. W Junk, The Hague.

McLachlan, A. (1990). Dissipative beaches and macrofauna communities on exposed intertidal sands. J. Coast. Res. 6: 57–71.

McLachlan, A. (1991). Ecology of Coastal dune fauna. J. Arid Environ. 21: 229–243.

McLachlan, A. (1998). Interactions between two species of *Donax* on a high energy beach: an experimental approach. J. Molluscan Stud. 64: 492–49.
McLachlan, A., Dugan, J. E., Defeo, O., Ansell, A. D., Hubbard, D. M., Jaramillo, E. et al. (1996a). Beach clam fisheries. Oceanogr. Mar. Biol. Annu. Rev. 34: 163–232.
McLachlan, A., Kerley, G. and Richard, C. (1996b). Ecology and energetics of slacks in the Alexandria coastal dunefield. Landsc. Urban Plan. 34: 267–276.
McLachlan, A. and Dorvlo, A. (2005). Global patterns in sandy beach macrobenthic communities. J. Coast. Res. 21(4): 674–687.
McLachlan, A. and Defeo, O. (2018). The Ecology of Sandy Shores, 3rd edition. Academic Press, California, United States.
McLachlan, A., Defeo, O. and Short, A. D. (2018). Characterizing sandy beaches into major types and states: implications for ecologists and managers. Estuar. Coast Shelf Sci. 215: 152–160.
Medeiros, R., Ramos, J. A., Pedro, P. and Thomas, R. J. (2012). Reproductive consequences of nest site selection by little terns breeding on sandy beaches. Waterbirds 35: 512–524.
Méndez, S. (1995). Bivalve mortality on southwest Atlantic shores. HAN 10-11: 12.
Menéndez, M. C., Baleani, C. A., Amodeo, M. R., Acha, E. M. and Piccolo, M. C. (2019). Assessment of surf zone zooplankton dynamics in a Southwestern Atlantic sandy beach: Seasonal cycle and tidal height influence. Estuar. Coast. Shelf Sci. 7: 106307.
Mourglia, V., González-Vainer, P. and Defeo, O. (2015). Distributional patterns in an insect community inhabiting a sandy beach of Uruguay. Estuar. Coast. Shelf Sci. 166: 65–73.
Noy-Meir, I. (1979). Structure and function of desert ecosystems. Israel J. Bot. 28: 1–19.
Odebrecht, C., Roig, L., García, V. T. and Abreu, P. C. (1995). Shellfish mortality and a red tide in southern Brazil. pp. 213–218. *In*: Lassus, P. (ed.). Harmful Marine Algal Blooms. Springer-Verlag, New York.
Odebrecht, C., Du Preez, D. R., Abreu, P. C. and Campbell, E. E. (2014). Surf zone diatoms: a review of the drivers, patterns and role in sandy beaches food chains. Estuar. Coast. Shelf Sci. 150: 24–35.
Olds, A. D., Vargas-Fonseca, E., Connolly, R. M., Gilby, B. L., Huijbers, C. M., Hyndes, G. A. et al. (2018). The ecology of fish in the surf zones of ocean beaches: A global review. Fish Fish. 19(1): 78–89.
Olivier, S. R., Capezzani, D. A. A., Carreto, J. I., Christiansen, H. E., Moreno, V. J., Moreno, J. E. A. et al. (1971). Estructura de la comunidad, dinámica de la población y biología de la almeja amarilla (*Mesodesma mactroides* Desh., 1854) en Mar Azul (Pdo. de Gral. Madariaga, Bs. As., Argentina). Proy. Desarr. Pesq. FAO, Ser. Inf. Tecn. Publ. 27: 1–90.
Orlando, L., Ortega, L. and Defeo, O. (2020). Urbanization effects on sandy beach macrofauna along an estuarine gradient. Ecol. Ind. 111: 106036.
Ortega, L., Celentano, E., Delgado, E. and Defeo, O. (2016). Climate change influences on abundance, individual size and body abnormalities in a sandy beach clam. Mar. Ecol. Prog. Ser. 545: 203–213.
Pauly, D. (1980). A new methodology for rapidly acquiring basic information on tropical fish stocks: growth, mortality and stock-recruitment relationship. pp. 154–172. *In*: Saila, S. B. and Roedel, P. M. (eds.). Stock Assessment for Tropical Small-scale Fisheries. Proc. Int. Workshop Univ. Rhode Island, Narragansett.
Peterson, C. H., Bishop, M. J., Johnson, G. A., D'Anna, L. M. and Manning, L. M. (2006). Exploiting beach filling as an unaffordable experiment: benthic intertidal impacts propagating upwards to shorebirds. J. Exp. Mar. Biol. Ecol. 338: 205–221.
Pinheiro, S. C. C., Leite, N. R., Costa, V. B., Costa, K. G. L., Pereira, C. C. and Costa, R. M. (2011). Spatial-temporal influence of hydrological variables on the diversity and abundance of copepods on an equatorial macrotidal beach in the Brazilian Amazon region. J. Coast. Res. 64: 425–429.
Pinheiro, S. C. C., Magalhaes, A., Costa, V. B., Pereira, C. C. and Costa, R. M. (2013). Temporal variation of zooplankton on a tropical Amazonian beach. J. Coast. Res. 65: 1838–1843.
Quillien, N., Nordström, M. C., Schaal, G., Bonsdorff, E. and Grall, J. (2016). Opportunistic basal resource simplifies food web structure and functioning of a highly dynamic marine environment. J. Exp. Mar. Biol. Ecol. 477: 92–102.
Rangel-Buitrago, N., Williams, A. T. and Anfuso, G. (2018). Hard protection structures as a principal coastal erosion management strategy along the Caribbean coast of Colombia. A chronicle of pitfalls. Ocean Coast. Manag. 156: 58–75.
Rocha, J., Yletyinen, J., Biggs, R., Blenckner, T. and Peterson, G. (2015). Marine regime shifts: drivers and impacts on ecosystem services. Phil. Trans. R. Soc. B 370: 20130273.
Rodil, I. F., Olabarria, C., Lastra, M. and Lopez, J. (2008). Differential effects of native and invasive algal wrack on macrofaunal assemblages inhabiting exposed sandy beaches. J. Exp. Mar. Biol. Ecol. 358: 1e13.

susceptible to desiccation tend to occur in the higher levels and the more vulnerable ones in the lower levels (Connell 1972). However, desiccation is not the only factor that may affect zonation as other variables such as water energy (currents) and wave action, geology and geomorphology, tidal amplitude and others can also determine the configuration of local communities depending on how they all interact. Finally, biotic interactions have also been shown to be key aspects that shape intertidal zonation, particularly related to the lower limits of species distributions often being determined by predation or herbivory (Connell 1972, Lubchenco 1980). Intertidal communities have been shown to respond not only to local features, as mentioned above, but also to variables that are modeled at an oceanographic scale such as nutrient, propagule and particulate subsidies (Menge et al. 2003).

During the 1950s and early 60s, manipulative experimental field studies redirected rocky intertidal ecology. Inter- and intraspecific interactions were shown to have a fundamental role in species distribution patterns on rocky shores (e.g., Connell 1961, Paine 1966, Lubchenco 1980). These studies showed how, for example, predation and herbivory can limit species distribution or control competitive dominants increasing local biodiversity. Interactions between organisms were classified as negative when one species benefits in detriment of the other (e.g., predation, competition and herbivory) or positive when there is a benefit to a species without harm to the other (e.g., facilitation). These forces along with physical environmental conditions and recruitment were later complemented with studies that demonstrated the important role that bottom-up processes also play on rocky intertidal communities (i.e., Bustamante et al. 1995, Menge et al. 2003). Some generalities have been suggested and described for these processes; however, it is important to stress the context dependency of these concepts. Complex interactions that operate at different scales can often lead to contrasting and even opposite results, which can be explained by complementing the observed interactions with context dependent variables (Rilov and Schiel 2011). It is clear that interspecific interactions on intertidal rocky shores are broad ranging, convoluted and difficult to predict and can operate on a range of different scales. Hence, the idea that general models and mechanisms that occur in certain parts of the world can be extrapolated must be considered with care. Models should be tested *in situ* in order to ensure that assumptions and predictions are correct in order to avoid reaching inappropriate conclusions. Natural environmental conditions and local community structure and composition must be accounted for along with a seascape context. For example, previously we mentioned that low intertidal assemblages are often determined by biotic factors; in particular, predation has been shown to have this effect on many well studied shores, such as the NE Pacific, NW and NE Atlantic and New Zealand (e.g., Paine 1966, 1974, Lubchenco and Menge 1978, Rilov and Schiel 2011). However, what has been known as "supply-side ecology" introduced another important aspect in species interactions by accounting for how productivity or juvenile settlement and growth can also affect assemblage structure and interact with predation to obtain the observed results. Thus, the predation paradigm that focused on this feature as the main interaction occurring on intertidal rocky shores was complemented by supply-side and bottom-up dynamics that are context dependent (Menge et al. 2002).

The natural environmental conditions in which communities develop determine, at least in part, which species can successfully establish, grow and reproduce. Environmental stressors can have more or less weight in terms of community structure and composition depending on how species respond to them. Furthermore, many of the natural stressors that normally affect intertidal communities are being exacerbated by the effects of climate change.

Two of the more relevant stressors that determine community composition and structure on rocky intertidal shores are wave action (McQuiad and Branch 1995, Blamey and Branch 2009) and temperature, both ambient and water (Leonard 2000, Harley and Helmtuh 2003). Storms that create large waves and surges are predicted to become more extreme and frequent in the near future (Hawkins et al. 2008, Bruno et al. 2014). Wave action can have different effects on intertidal organisms, they can dislodge organisms attached to rocks and create gaps or bare patches within an otherwise homogeneous community which can enhance local biodiversity. Large waves and the spray they create can also extend the "inhabitable" area by keeping moisture in areas that would be dry. On the other hand, strong water motion may limit access to mobile organisms or prevent the establishment of sessile ones due the high energy. Heatwaves are also predicted to increase in frequency and intensity (Hobday et al. 2018, Stillman 2021). Both ambient and water temperatures are stressors that influence species distribution on a biogeographically scale as well as survival within microhabitats. Survival rates of adults or pelagic larval stages of many marine intertidal organisms are often determined by the temperature ranges they are exposed to. Concurrently, the temperature individuals experience is determined not only by ambient temperature but by other factors that exacerbate or mitigate it. Radiation, for example, that is whether they are in a shaded or un-shaded area directly influences individual temperature. The nature and color of the surface on which organisms are on is another example; sedimentary rocks that retain water can reduce temperature by evaporation and provide humidity, whilst granite rocks retain heat (not water) and elevate the organisms' temperature well above ambient (Helmuth et al. 2006, Lathlean and Michinton 2012).

Population level responses from intertidal organisms to natural stressors can be species specific and interact with other stressors in additive, synergistic or antagonistic ways (Crain et al. 2008, Manriquez et al. 2016, Kunze et al. 2021). Furthermore, interspecific interactions may also be an intrinsic part of community level changes and often vary from those predicted from individual species response (Gestoso et al. 2016). Hence, the importance of considering these environmental variables on scales pertinent to organisms when studying larger scale responses (Helmuth et al. 2006).

Anthropogenic stressors on rocky shores

Coastal areas, where human populations are markedly concentrated in a narrow strip of land, are mainly subject to disturbances from anthropogenic activities. Human land reclamation and activities directly and indirectly influence the adjacent coastal ecosystems. In rocky intertidal areas, anthropogenic disturbances (i.e., any potential source of stress to natural populations and assemblages due to human activities, Benedetti-Cechi et al. 2001) are major processes that can determine the distribution and abundance of the species that form their communities (Thompson et al. 2002, Claudet and Fraschetti 2010). Regrettably, rocky shores communities around the world are greatly different from what they were several decades ago. Attempting to classify the various and diverse forms of human activities that can affect rocky intertidal areas is a truly complicated task, due to the scale at which they can occur, their intensity, frequency and the level, or even the time in the life cycle of a species, that they can affect (Halpern et al. 2007, Kunze et al. 2021). Traditionally, animal and plant extraction and habitat degradation were the main human impacts on marine systems (Worm and Lenihan 2013). Over the past decades, many other stressors were added to this list, including pollution, invasive species, and more recently, climate change. In an exhaustive review, Halpern and collaborators (2007) documented nearly forty threats for rocky shores.

In a simplified manner, we can first distinguish between global anthropogenic stressors (e.g., climate change and its many consequences) and local anthropogenic stressors (e.g., invasive species and pollution). This first step will only help with the analysis of potential isolated effects of a given stressor. However, if an attempt to properly understand the real effect of a stressor is expected, interaction among global and local-scale impacts should be considered since they are not independent of one another (Crain et al. 2008, Kunze et al. 2021). Likewise, some anthropogenic disturbances occur at scales where it is possible to distinguish from other natural processes and in these cases, it is also possible to experimentally evaluate their impacts (Benedetti-Cechi et al. 2001). The intrinsic characteristics and properties of rocky intertidal shores make them an exceptional test system for experimental ecology in general and have been historically used to comprehend community structure and functioning (Hawkins et al. 2008, 2020). Given the location of rocky shores at the land-sea interface and the number of potential stressors acting on them, biodiversity changes in this ecosystem are often strong. In this context, rocky shores are appropriate systems for exploring impacts of anthropogenic disturbance and their consequences on different aspects of the affected species (Schiel et al. 2004).

As are the variety of potential sources, the impacts generated by anthropogenic stressors on rocky intertidal communities are very diverse. Species respond to these disturbances both directly and indirectly. Human actions usually modify the physical and biotic characteristics of the benthic habitat but also of the water column, and can thus affect larvae supply, recruitment processes and mortality rates, consequently affecting the entire dynamics of these habitats. Changes in the distribution and abundance of invertebrates and algae, alterations in their physiology or behavior, and even local extinctions have been documented.

The stressors, in addition to modifying the abundance and distribution of a single species, are usually responsible for physiological changes that derive in sub-lethal effects on organisms. Such modifications in individual physiology can scale up to the community level through species interactions (Bulleri et al. 2018). It is not easy *a priori* to determine which human activities will have the greatest impact on communities. Even activities that can be thought of as "simple" can generate drastic changes in the abundance of certain species. Similar effects can be expected by different environmental stressors that have a larger or smaller impact depending on the local context. Organism's responses to stressors may also be context dependent, being very rigid for some cases and plastic in others. Predictability on how species will respond is extremely difficult as trait mediated indirect interactions (TMIIs) can be expected to occur. This is when one species has a plastic response and changes its phenotype, but not necessarily its abundance, and it is this plastic change that causes a cascading effect throughout the community (Werner and Peacor 2003). Hence, TMIIs add a yet another layer of complexity to marine intertidal community ecology, particularly in this global change scenario, which adds anthropogenic stressors to the already variable natural ones. As mentioned at the beginning of the chapter, a small number of species tend to dominate rocky intertidal shores: habitat-forming species that harbor great diversity. It is evident, then, that when stressors are able to affect these species, the impacts on the existing communities are greater.

In the following paragraphs, we will examine anthropogenic stressors that have been highlighted in the literature during recent decades as the most influential drivers acting on rocky intertidal systems. This does not intend to be an exhaustive list, but attempts to synthesize a variety of stressors with both global and local influence. After introducing the main aspects of each stressor, an example is given to illustrate its effects on benthic species and communities. In selecting examples, special care was taken to include different taxonomic groups, including invertebrates and algae, and also to represent areas around the world.

Climate change: global warming and ocean acidification

Climate change emerges as the major stressor affecting natural communities (Hawkins et al. 2008, Bruno et al. 2013). The changes associated with this global stressor are unprecedented and the scientific community is dedicating special efforts to assess its impacts on populations and ecosystems. Anthropogenic greenhouse gas emissions have completely altered Earth's climate, changing the physics and chemistry of the ocean. As a result, nowadays we detected increments in temperatures and UVB, rises in sea level, acidification, changes in salinity gradients, altered biochemical cycles, changes in wind, waves and storm frequencies, changes in costal upwellings, reduction in oxygen concentration, and sea ice loss (Hawkins et al. 2008, Bruno et al. 2013). Given that structural processes on rocky shores are well studied and the ecologies of many species are well known, they are an adequate system for assessing the consequences of climate change on the whole community and their ecosystem processes (Hawkins et al. 2008). Of all of the mentioned consequences of climate change, most attention was given to study the effect of global warming and ocean acidification. We will now go into these two aspects in more detail.

The global warming forecasts under the "business as usual" scenario SSP5-8.5 (where emissions will continue to increase until 2080) predict gradual increases of the atmospheric and marine water temperatures and more frequent and intense atmospheric and marine heatwaves by the end of the century (IPCC 2019, Carvalho et al. 2021). The gradual increase in seawater temperatures and the marine heatwaves decreases the physiological performance and the reproductive success of cold-affinity species, whereas warm-affinity species can thrive and occupy their space, which results in gradual changes in the composition of the communities and geographic shifts in their distribution (Mieszkowska 2009). This process has already been reported for the macroalgal assemblages of the northern and northwestern Iberian Peninsula (Fernández 2011, 2016, Martínez et al. 2012, Voerman et al. 2013, Méndez-Sandín and Fernández 2016, Monteiro et al. 2022) or for sessile invertebrate's communities in the Gulf of California (Favoretto et al. 2022).

Importantly, the more frequent and intense atmospheric heatwaves will be particularly harmful for the rocky intertidal ecosystems (Helmuth et al. 2006, Stillman 2019, Gosling 2015). Unlike what can occur under gradual temperature increases, the acuteness of heatwave events may not allow intertidal organism to acclimate. Although Hobday et al. (2016) attempted to homogenize the nomenclature in the marine realm, several definitions of "heatwave" exist in the literature. Studies about the impact of heatwaves on marine species have used alternative definitions, with close but not identical number of days or percentiles to those in the cited study (Hereward et al. 2019, Domínguez et al. 2021). Generally, they are defined as periods of consecutive days (more than three) where the temperature registers are above a percentile (usually higher than 90th) of the long-term registers (years to decades) at the site.

Whilst the majority of research about warming effects on intertidal ecosystems has focused on the effects of marine heatwaves, the atmospheric heatwaves during low tides are especially concerning. The macroalgae and sessile invertebrates that are subjected to heat stress during emersion experience exposure to solar radiation, dessication and body heating (Ferreira et al. 2014, Chapperon et al. 2016, Herewards et al. 2019, Hesketh and Harley 2022). In such conditions, they invest great amounts of energy in maintaining homeostasis (Shick et al. 1988) and have less energy available for growth and reproduction, so their physiological performance decreases (Shonbeck and Norton 1980, Gosling 2015). Eventually, if the survival mechanisms are not enough to overcome the physiological stress, mortality occurs (Olabarria et al. 2016, Román et al. 2020).

The impact of climate change driven heatwaves on rocky intertidal ecosystems can be portrayed at the Western English Channel in the summer of 2018, where biological and temperature surveys revealed that the massive mussel mortalities reported in the region were linked with the occurrence of consecutive atmospheric heatwaves. Concurrently, the exposure to solar radiation and the absence of the microhabitats created by mussel beds were highly influential for the body heating and mortality, rather than the sole exposure to air temperatures (Seuront et al. 2019). Another relevant case of the effects of atmospheric heatwaves was the 2021 Western North America Heat Dome, during which temperatures close to 40ºC in the intertidal zone paired with spring low tides caused mass mortalities of shellfish and a general deterioration of the condition of organisms (Raymond et al. 2022) and whose longer-term ecological, economic and social implications are still to be quantified.

The acidification of the ocean is another thread derived from the climate change to relevant functional groups of invertebrates in rocky intertidal ecosystems, especially those with calcareous shells or skeletons such as mollusks, echinoderms and calcareous macroalgae. The atmospheric CO_2 concentrations have oscillated throughout Earth's history. However, the increase of human population, the changes in land use and the burning of fossil fuels since the middle of the nineteenth century have changed the global biogeochemical cycles. One of these changes is the increase of the CO_2 emissions towards the atmosphere, which derives in the oceans absorbing more CO_2 (Rockström et al. 2009). The surface waters of the oceans and the atmosphere are continuously interchanging this CO_2 through diffusion and its uptake by phytoplankton or photosynthetic microalgae, who incorporate it to the ocean carbon cycle. Intertidal calcifiers are highly sensitive to ocean acidification because some parts of their bodies are composed of calcium carbonate ($CaCO_3$), mainly in the forms of aragonite and calcite, which are synthetized through biocalcification reactions (1). These reactions involve the transformation of dissolved CO_2 into carbonic acid (H_2CO_3), which dissociates into bicarbonate (HCO_3^-) and carbonate (CO_3^{2-}) ions and protons (H^+), and finally the CO_3^{2-} reacts with (Ca^{2+}) calcium ions precipitating the $CaCO_3$. But under increasing inputs of CO_2, the equilibrium reaction (1) is altered and the generation of HCO_3^- is enhanced at the expense of the generation of $CaCO_3$. This process increases the concentration of H^+, which causes a decrease of the pH and subsequent acidification, and also retires the CO_3^{2-} ions necessary for the formation of $CaCO_3$, which ends up in an increase of the dissolution rates of calcium carbonate (Turley and Findlay 2000).

$$2\ HCO_3^- + Ca^{2+} \longleftrightarrow CaCO_3 + CO_2 + H_2O$$

The major reported impacts of ocean acidification on calcifying fauna at the rocky intertidal are the weakening of their shells and increases of the respiration rates (Harvey et al. 2016), decreases of their growth and lower reproductive success (Turley and Findlay 2000), or altered behavioral responses against predation (Bibby et al. 2007). The ocean acidification can alter the community structure by favoring the invasive species, as for example, decreases of pH reduced the growth and survivorship of the native mussel *M. galloprovincialis*, but not of the invader *Xenostrobus securis* at NW Iberian Peninsula (Gestoso et al. 2016). Interestingly, macroalgae assemblages composed of mixtures of calcifying and of non-calcifying species showed general decreases in biomass and photosynthetic efficiency under high CO_2 concentrations (Olabarria et al. 2012), which highlights that the negative impacts of ocean acidification on intertidal organisms go beyond the calcifying species.

Example: Weakening of the mussel byssus of Mytilus trossulus under high pCO_2

The mussel byssus is crucial for the attachment of the individuals to the substrate, especially under high-energy environments, and enhances their ability to form clumps, modify their environment and enhances survivorship and reproductive success. Many studies have demonstrated the negative effects of acidification on shell strength or physiology on intertidal invertebrates, yet fewer have focused on the effects on other structures without calcium carbonate. O'Donnell and collaborators (2013) tested the effect of increasing pCO_2 in seawater on the attachment properties of the byssus of the mussel *M. trossulus* from the Pacific coast of North America in a mesocosm experiment, by cutting the byssus of the specimens and letting them regrow under different pCO_2 increases during 20 days. After this period, the thread strength and the maximum strain of the byssus showed decreasing trends with increasing CO_2 pressures, but the growth and condition of the individuals was not altered. This implied that a direct influence of the water chemistry, and not a decreased physiological performance, was probably the cause of the byssus weakening, and suggested concerning effects for the sustainability of this ecologically and economically relevant species under future perspectives of ocean acidification.

Aquaculture and fisheries

The use of intertidal habitats by mankind to obtain food resources can be dated to prehistoric times. Different organisms like mollusks, crabs and algae were easily collected whilst larger animals such as seals and sea lions resting upon intertidal habitats were also hunted. In more recent times, collection and aquaculture of mainly bivalves and algae of commercial interests that go from edible resources to industrial products such as fertilizers, construction material, pharmaceuticals and nutraceuticals have transformed the use of intertidal habitats (Naylor et al. 2020). In this context, there are many examples in which fisheries and aquaculture of commercial species has led to the deterioration of habitats and the services they provide. Over-exploitation of the target resource is the most obvious and direct impact that has been widely recorded (see Moreno et al. 1986, Davies 1995, Airoldi and Beck 2007, Marra et al. 2018 for examples). However, eutrophication, pollution, diseases, introduction of non-indigenous species, over exploitation of non-target species (either as by catch or as food supply for commercial species), among others can lead to other ways in which the ecosystems may be affected (Naylor et al. 2001, Primavera 2006, Airoldi and Beck 2007, Naylor et al. 2020). These impacts are often the consequence of fish aquaculture done in coastal areas, which have important effects on the benthic assemblages including intertidal rocky shores (Primavera 2006, Ottinger et al. 2016). In addition to environmental issues, aquaculture may interfere with other ecosystem services such as restriction for recreational use of land, alteration of the landscape and even human demographic changes through industrial developments (Banta and Gibbs 2015). Interests between different sectors of society often clash and a holistic perspective that encompasses all the ecosystem services these habitats provide should be incorporated in the resolution of the problem.

On the other hand, some aquaculture practices can generate changes that do not have an obvious negative affect on the surrounding ecosystem and may even have some effects that can be considered positive. Oyster farming on intertidal shores has been suggested to benefit avian predator species that feed on bivalves (Kirk et al. 2007, Maslo et al. 2020) and in some occasions, does not affect biodiversity around trestles, although impacts on surrounding habitats due to heavy vehicles transit were observed (Forde et al. 2015). In France, the introduction of oysters (*Magallana* (*Crassostreas*) *gigas*) for aquaculture has led to the creation of wild oyster reefs on both soft and hard bottoms where they previously did not exist, which increases local biodiversity (Lejart and Hily 2010) and improves water quality (Lindahl et al. 2005). Introduction of *M. gigas*, in areas previously occupied by other oyster species may be considered as positive change in which ecosystem services are restored under new climate change conditions and a new economic opportunity may arise (McAfee and Connell 2020).

The Chilean or false abalone, *C. concholepas*, known in Chile as *loco* and in Peru as *chanque* is a large carnivorous gastropod with a fleshy foot that is target for human consumption since prehistoric times (Santoro et al. 2005, Rivadeneria et al. 2009). It also has an important ecological role in the intertidal and shallow subtidal community as a keystone predator. The removal of this predator from the foodweb through overexploitation (Geaghan and Castilla 1986, 1988) had an effect on the remaining community because it strongly limits the density of otherwise space dominant species such as mussels. When *loco* populations were diminished, predation pressure was released and the cascading effect led to the spatial dominance of the mussel *Perumytilus purpuratus* as a monoculture in the mid intertidal area (Moreno et al. 1986, Durán and Castilla 1989).

Co-management strategies that included a 4 year ban, land use rights to artisanal fishers, along with capture sizes and reproductive fishery bans resulted in an improvement of the conservation status of *loco* (Castilla 1994, González 1996, Castilla and Fernández 1998). The history of *loco* in Chile highlights some of the aforementioned impacts of fisheries on intertidal communities. *Loco* collection, fishery and promising aquaculture is a good example of the many aspects that need to be addressed to sustainably exploit a natural resource. These activities need management, which should include the continued interaction at different scales of all actors of society, starting with local fishermen and women and ending with national regulation bodies. The creation of Territorial Use Rights for Fisheries, along with basic research, has provided Chile with a good basis. The need for mission-orientated research policy in Chile has been proposed (Castilla 2015). These policies should be promoted in any developing country with research capabilities in order to gain in sovereignty over local resources and exploit them in a sustainable fashion that considers economic as well as social, cultural and ecological aspects of the endeavor.

Assessment and monitoring management of aquaculture in general and mariculture sites in particular are needed if sustainable practices are to be obtained (Ottinger et al. 2016). A step in this direction may be the promotion of polyculture systems that integrate the rearing of several commercial species, with the ensuing impact reduction on the environment (Primavera 2006). Practices for farmed marine protein through economically viable and ecologically acceptable processes may be possible and if correctly implemented, may be a compromise solution to human harvesting with less socio-ecological impact.

Example: Crab collection for bait from rocky shores in Kuwait

Intertidal gleaning in rocky shores can have several purposes: from providing a direct food source to collecting bait for other fisheries. The disturbances caused by such activity and the effects on the intertidal habitat and its associated biota were studied by Al-Wazzan and collaborators (2020) on the coasts of Kuwait. By using combined approaches from structured online and field surveys to anglers and bait collectors, as well as data from direct observation and photographic records, they provided a comprehensive picture of how bait collection, directly, and recreational fisheries, indirectly, impacted rocky shore habitat. Live *Leptodius exaratus* crabs are collected from local rocky shores and used as bait for target species of sea bream, which are highly valued by anglers. The collection methods include overturning and, sometimes, breaking rocks in order to expose and trap the crabs. The anglers perceived that this gleaning activity caused declines in crab abundance and habitat degradation, and size class comparison between crabs collected by gleaners and those sampled randomly suggested a selective collection of larger sizes by gleaners. Although the mortality estimated from direct collection of crabs seemed relatively low (8% of the population per year), the removal of larger size classes may affect reproductive potential and its ensuing recovery. Furthermore, habitat degradation by overturning stones and leaving them unturned and them being broken into smaller pieces not only affects the biota that lives on and under rocks and boulders, it also systematically reduces the available habitat for the actual target species. This coincided with a drop in the perceived and described abundances of *L. exaratus* in the highly visited shores. As recreational fishing is an important aspect of Kuwait's culture and tradition, Al-Wazzan et al. suggest that there is a need to develop management strategies, before the ecological impacts on the crab population as well as on the wider intertidal community are irreversible.

Pollution

Pollution (i.e., the presence of unnatural elements) is a widespread and complex problem in all natural system. Industrial and urban activities are severely contaminating coastal ecosystems. Anthropogenic pollution, or the effects of pollutants introduced to the environment through human activities, embraces different process and events, including eutrophication, oil spills, chemicals associated to maritime traffic or urban runoff and litter (Cloern 2001, Bishop et al. 2002, Castège et al. 2014). All of these are not only linked to changes in the biodiversity of coastal communities and effects on their species, but also generate problems for human health. Hydrocarbons, persistent organic pollutants (POPs), heavy metals, trace elements and pathogens are just some of the agents that are part of the varied list of polluting agents. Pollution may substantially increase the amount of various nutrients, organic matter and sediments. Although most pollutants are commonly found associated with water circulation or concentrated in sediments, in the rocky intertidal they can be directly transferred to certain species, thus affecting the composition of benthic communities. In fact, historical declines in rocky intertidal biodiversity have largely been attributed to chronic and persistent disturbances such as sewage and industrial effluents discharges (Fairweather 1990, Bishop et al. 2002). Most recently, microplastics (< 5 mm) are emerging pollutants in all the oceans and coasts (Thompson et al. 2004, Thompson 2015, Doyle et al. 2019). These small-size elements are incorporated by animals through ingestion and studies on the effects of such process in rocky intertidal organisms are limited but indicate that they may produce several negative impacts. For example, reductions in energy reserves, obstruction of organs, changes in behaviors and incidence in mortality rate have been reported for other marine organisms (Green 2016, Seuront 2018, Doyle et al. 2019).

Contamination by sewage outfall is among the most studied types of pollution impacts on rocky shores (López Gappa et al. 1990, Bishop et al. 2002, Arévalo et al. 2007, O'Connor 2013, Cabral-Oliveira et al. 2014). This process directly results in organic and nutrient enrichment of the environment, thus favoring eutrophication processes and decreasing water or sediment quality. Overall, it has been observed that communities disturbed by sewage (or eutrophication derived from a different source) exhibit a shift in their composition, with sensitive species being replaced by opportunistic ones (mostly ephemeral algae). Because of this, mussel or perennial macroalgae-dominated communities shift to completely different communities dominated by short-lived opportunistic algae (Fairweather 1990, Bishop et al. 2002, Pinedo et al. 2015, Becherucci et al. 2016). Additionally, hydrocarbons are also well-studied contamination agents. Petroleum and its derived products have been responsible for massive coastal pollution and its effects on the benthic communities were significant, causing direct mortalities of most of the species present whilst producing long-lasting effects (Carls et al. 2004, Commendatore and Esteves 2007, Castège et al. 2014). Likewise, these products (and many others containing POPs) are highly resistant to degradation and are transferred through the food web, specially affecting predator species. For example, much attention was given to the tributyltin (TBT), a non-fouling agent in boat paints, effects on benthic species (Bauer et al. 1995, Bigatti et al. 2009, Langston et al. 2015). TBT causes imposex (masculinization of female gastropods) and exposure to very high concentrations of TBT can alter population dynamics by reduction or elimination of wild populations. As is the case of numerous pollutant compounds, there is still a lack of knowledge about the long-term action and impact of TBT on marine organisms.

Example: Trophic transfer of pollutants in South Africa

Erasmus and collaborators (2020) used stable isotope analysis to study trophic transfer of pollutants to determine the effects of this type of anthropogenic pressure at two sites of different biogeographical regions on the African coast. They generated data on the current levels and distribution of 15 metals and organochlorine pesticides (OCPs) in intertidal organisms and assessed the trophic transfer of these elements within intertidal food webs. They found no correlation between metal and OCPs accumulation with dietary source or trophic level but they found biomagnification of several metals along a short benthic food web. Additionally, they registered biodilution of OCPs at both sites. With this, they suggested that food web composition rather than biogeography influenced the transfer of metals and OCPs. They highlighted the importance of identifying the dynamics of pollutants within marine ecosystems and the influence that associated characteristic (e.g., species composition) could have in this regard. Concerning biogeography, they showed higher concentration of some metals and OCPs at the subtropical site and they suggested a possible correlation of elements diffusion with temperature. Lastly, the authors stated that profounder knowledge on physiological mechanisms and processes related with metal and organic metabolism are needed to fully comprehend the transfer of contaminants to and among intertidal organisms.

Invasive species

The rate at which human travel and goods transport has increased over the last two centuries is astonishing (Hulme 2009). This globalization, which involves many means of transport, has not been without consequences to the environment. In this particular case, we are referring to the transport of marine organisms from their places of origin to distant areas where it is unlikely they would have reached without human assistance. Mainly through travel in ballast waters of ships and the fouling of their hulls, marine species have been able to overcome natural barriers and arrive at different parts of the world where they may have advantages over the species already living there (Bailey 2015). These physical, biological or ecological advantages allow for some species to become invasive and alter the original communities resulting in impacts on the entire ecosystem and the services they provide (McNeely 2006).

Intertidal rocky shores are not an exception with many invasions reported in these marine habitats. Some invasions are successful because the alien species can out-compete local species for space or resources and displace them in the community (Shinen and Morgan 2009). Other times, the lack of predators or resistance to local ones allows for invasive species to be established and suffer less pressure than their local counterparts (Battini et al. 2021). In other cases, they may occupy niches that are locally unoccupied; such is the case of some invasive barnacles (Orensanz et al. 2002). Habitat forming species such as oysters, mussels, algae and ascidians have successfully invaded rocky shores around the world causing very visible changes to the original environment and displacing native species of commercial and cultural interest (Byers et al. 2010, Troost 2010, Davis et al. 2018). Not all introduced species become invasive and hence may have minimal or no impacts; however, it is widely agreed that invasive species are a risk to biodiversity and can have negative ecological impacts (Boltoskoy et al. 2022). For example, the relatively recent introduction of the invasive shore crab, *Carcinus maenas*, along parts of the Atlantic Patagonian coastline and its rapid rate of expansion, may change the role predation plays on these intertidal communities. Until now, predation in Atlantic Patagonia was a very weak ecologic interaction on intertidal rocky shores. *Carcinus maenas* has been shown to have important impacts on bivalve populations (Leignel et al. 2014); however, the fact that it is a generalist predator with a broad diet in the area (Cordone et al. 2022) may, in turn, ameliorate the consequences at a community level.

Example: When the invader is also an ecosystem engineer, a case study from New Zealand

Ecosystem engineers on rocky intertidal shores tend to have a key role for associated species that depend on them to provide suitable habitat with conditions that allow their existence. On the shores of New Zealand, the green-lipped mussel *Perna canaliculus* commonly dominates the rocky intertidal habitat and provides substrate and shelter to an entire community. Recently, the ascidian *Pyura doppelgangera* was observed on the northern shores of the North Island. Davis and collaborators (2018) studied the community-wide shift in species association from one habitat forming species (the native green-lipped mussel) to another habitat-forming species of different characteristics (the invasive ascidian). They compared the communities associated to each species and predicted that organisms that lived upon each habitat-forming species would differ due to the nature of the substratum each species provides, namely, hard substratum by the mussel and "rubbery" substratum by the ascidian. They also predicted that organisms that used the interstices of the matrix of the habitat-forming species would not differ among communities. To assess these potential shifts in communities, they sampled two different habitat types (shallow rock pools and exposed substrate) at two reef locations in Northland where both ascidian and mussel clumps could be found. All fauna and flora associated to each habitat-forming species was counted and identified. Overall, no differences were found in the abundance of invertebrates. As predicted, they found that fouling organisms, such as barnacles, that depend on the hard substratum of the mussel were not present in the ascidian clumps. However, they found that motile fauna that uses the matrix spaces changed dramatically from a mollusk and crab dominated assemblage in mussel clumps to errant polychaetes that dominated ascidian clumps. These changes in community composition can have ecosystem level impacts through alteration in processes, for example, nutrient cycling. The ascidian invasion in northern New Zealand is expected to keep extending its way further south. The foreseen consequences are not limited to their associated assemblages; they may also have important cultural and economic impacts. Green-lipped mussels are collected by Maori culture, they are part of a recreational harvest, and some areas where they dominate are used to provide wild-caught mussel-spat for the mussel farming industry.

Tourism and recreation

The tourism industry has grown exponentially in the last century and coastal cities have become favorite destinations mainly due to the attractiveness of their beaches (Miller 1993, Davenport and Davenport 2006). Massive tourism is a relatively modern phenomenon that has grown significantly since the development and improvement of transportation systems (Dahl 1993, Miller 1993, Davenport and Davenport 2006). Although tourism brings economic benefits, there are usually substantial environmental costs associated with its development (Davenport and Davenport 2006). Furthermore, the intense use of natural environments may lead to a deterioration of the original attractions. Over the years, other activities such as nature contemplation have become increasingly popular (Miller 1993, Davenport and Davenport 2006). Initially, the main threat from tourism activity is related to the development of the infrastructure required to accommodate visitors, usually leading to irreversible environmental degradation (Davenport and Davenport 2006). Later, simpler activities occur on a daily basis that significantly impact ecosystems (Addessi 1994, Davenport and Davenport 2006, Mendez et al. 2017).

In rocky intertidal areas, the activities carried out by tourists are diverse and can produce different levels of impact on the communities (Addessi 1994, Smith et al. 2008). The simple act of walking is often combined with curiosity about flora and fauna, their subsequent manipulation or extraction, and the physical disturbance of the intertidal (e.g., turning over loose rocks to observe hidden organisms). In addition, informal exploitation of the species present is commonly observed, which are used as food, bait for recreational fishing (as mentioned above) or even souvenirs (Addessi 1994, Airoldi et al. 2005, Minchinton and Fels 2013). Thus, the activities carried out represent an important source of disturbance that directly or indirectly affects marine organisms (Addessi 1994, Milazzo et al. 2002, 2004, Smith et al. 2008). Recreational activities have been reported to reduce species richness of benthic communities, and abundance, biomass and body size of intertidal macroinvertebrates and algae. With the exception of trampling, most studies report the effects of recreational activities by comparing community diversity and other variables between sites with different tourism uses. In this sense, there is a need to develop manipulative experiments to confirm the observed patterns and rule out other potential confounding effects.

Among the activities performed on rocky shores, human trampling has been, by far, the main focus of the studies (Brosnan and Crumrine 1994, Brown and Taylor 1999, Milazzo et al. 2004, Araujo et al. 2009, Micheli et al. 2016, Nicastro et al. 2019, among many others). Trampling influences the benthic assemblage directly removing individuals and can drastically change the abundance of many species, from bivalves to algae. As a consequence, ecological interactions between species are disrupted by these changes. In addition, since it is a simple-replication activity, its effects have been evaluated experimentally in a great variety of environmental conditions around the world and involving different taxonomic groups.

Example: Tourism disturbance and long-term changes on United States benthic communities

Although Addessi's (1994) work was published almost 20 years ago, few of the studies that have followed had this degree of detail. This paper describes the effects of tourism activities on the rocky shores of San

Diego, United States. The work begins describing the activities that people performed on the coast. After that, echinoderms and mollusks were compared along a gradient of beach use and, at the most visited site, densities of different species were compared after 20 years (1971 and 1991). Addessi was able to obtain patterns of use of the rocky shores and to distinguish the main activities that people do when visiting the coast. The author also described how the abundance of macroalgae increased as a response to reduced herbivory power. The work reported that the densities of most of the species increased with distance from the public access to the coast. With the caution of comparing two surveys separated by 20 years without considering natural fluctuations that may have occurred, the work highlights that almost all species decreased in abundance in the time period analyzed and that several of them were not even recorded in the 1991 survey. Although the example described may seem simple at first look, it is extremely enriching to have this type of work as a starting point to describe the general patterns of occurrence of activities and the changes generated on the coastal biodiversity. Experiments can then be developed for the most affected species or work can be concentrated on the most frequent activities.

Local cases

Effects of recreational activities on Atlantic Patagonian mussel beds

Among the harshest rocky shore environments described, we find Patagonian intertidal habitats along the Argentinean coast (Bertness et al. 2006). Extending from ~ 40°S to 54°S, the rocky coasts are primarily made of sedimentary rocks (e.g., mudstone, limestone, etc.) with occasional volcanic outcrops (Haller et al. 2005). These intertidal rocky shores experience extreme physical conditions with persistent strong dry winds, little rainfall, large temperature amplitude and high solar radiation, which tend to modulate the benthic assemblage structure (Bertness et al. 2006). At the highest latitudes, rainfall tends to be slightly higher due to the decreasing altitude of the Andes in that area. Mean temperatures in the coastal areas show a decreasing trend with increasing latitudes and larger amplitude on the northern Patagonian shores (Pessacg et al. 2022). Shores physiognomy and high amplitude tides give place to exceptionally wide platforms (more than 500 m) with little slope.

Despite the severe environmental conditions of intertidal rocky shores in Atlantic Patagonia, these habitats are home to a specific and rich biodiversity. Much of this diversity is dependent on the above mentioned habitat forming species such as scorched mussels, *Brachidontes rodriguezii* and *Perumytilus purpuratus* that form extended mono- or multi-layered mussel-beds; or macroalgae such as *Corallina officinalis*, which provide protection from desiccation and predation (Bertness et al. 2006, Raffo et al. 2014, Livore et al. 2021). Algal assemblages, mainly occurring in the low intertidal areas, also add to intertidal biodiversity with more than 30 different algal species reported (Rechimont et al. 2013, Raffo et al. 2014). On these rocky shores, a general vertical zonation pattern is relatively consistent along the entire coast with three distinct levels. Barnacles (the invasive *Balanus glandula*), pulmonate limpet species and seasonal algae are present in the high intertidal level (Bertness et al. 2006, Raffo et al. 2014, Livore et al. 2021). The mid intertidal level is dominated by scorched mussels, with extremely high levels of cover (> 85%) and few bare patches (Livore et al. 2021). The low intertidal level is generally occupied by stands of algae that have *Corallina officinalis* as a common species that occurs along the entire latitudinal gradient in combination with other species, some of which can be very abundant seasonally such as *Ulva* spp. and *Ceramium* sp. (Bertness et al. 2006, Raffo et al. 2014, Livore et al. 2021).

Environmental stressors such as desiccation are believed to have a strong influence on species distribution along the shore. Physical stress is the main modeling force on Patagonian intertidal assemblages. Hence, many species are entirely dependent on facilitation processes (Bertness et al. 2006, Silliman et al. 2011). For example, scorched mussels have been shown to recruit into interstitial spaces between barnacles and byssal threads of other conspecifics due to the retention of humidity in these spaces (Silliman et al. 2011). Furthermore, the microhabitat created by scorched mussel matrices along the mid intertidal rocky shores provide a benign environment for the vast majority of the > 40 species that have to live within it to avoid the extreme desiccation conditions outside the matrix (Hidalgo et al. 2007, Liuzzi and Lopez-Gappa 2008). Most of them are small (> 2 cm) and either sessile or with movement at a very small scale (Bertness et al. 2006, Silliman et al. 2011, Bagur et al. 2016). Accordingly, *Corallina* turfs also harbor rich assemblages of species with some overlap to those found in mussel beds, but with many that are specific to these microhabitats (Liuzzi and López Gappa 2008, Kelaher et al. 2007, Rechimont et al. 2013). In certain cases, species interactions such as competition, predation, and herbivory may also play a role. The pulmonate limpet, *Siphonaria lessonii*, in its northern distribution where ambient conditions are not as harsh as the ones in Patagonia, was shown to remove fast growing algae from a surface, which then allowed for other organisms such as barnacles to colonize the cleared area (Bastida et al. 1971). However, this same species does not seem to have such an effect even at relatively high densities on Patagonian shores due to the physical and environmental stress they are subject to (Bazterrica et al. 2007). During the first decade of the 21st century, a new and very active predator, *Carcinus maenas* the European shore crab, was reported on Atlantic Patagonian rocky shores, at Bahía Camarones (44°46' S), and has since extended its range north. Although predation has never been described as a strong species interaction on these shores, the recent addition of this invasive species to Patagonian assemblages may well change that fact.

Species interactions along Patagonian rocky intertidal shores are a good example of how strong role context-dependent variables can play. As mentioned before, scorched mussel beds generally dominated mid intertidal rocky shores. Under these circumstances, in other parts of the world, vertical distribution is determined by predation pressure. Even on the Pacific Patagonian coast in Chile, where scorched mussels also dominate the mid intertidal habitat, predation strongly affects mussel recruitment (Wieters 2005). However, on Atlantic Patagonian intertidal shores predation has, for the moment, very little to no influence (Bertness et al. 2006, Hidalgo et al. 2007).

Coastal tourism represents a growing and large-scale industry in Argentina. Traditionally, the sandy beaches in the northern part of the country have been the most visited. In the last decades, tourism has been growing exponentially along the entire coastline, developing in "pristine" areas such as the Patagonian region. Natural sites as the Valdés Peninsula, a UNESCO World Heritage Site, have become highly attractive tourist destinations, receiving more than 250 thousand tourists a year. Rocky shores in this region have been studied primarily from a biological and ecological point of view though, in the last years, studies of how and to what extent tourism and recreation could be affecting their communities were developed (Mendez et al. 2017, 2018, 2019). Local people and visitors intensively use rocky intertidal areas, carrying out different activities such as walking, snorkeling and collecting organisms. Furthermore, given the topography, vehicle traffic over the intertidal rocky shores (including bikes and cars) is sadly common to observe (Fig. 2).

The work published in 2017 by Mendez and collaborators is the first precedent in the country where the effect of tourism activities on rocky intertidal shores is described. The study reported the different activities performed on the coast and showed that human trampling is the most frequent activity carried out on the intertidal, with the ability to drastically modify the natural cover and abundance of algae and invertebrate species that inhabit the area. As we introduced earlier, mussel beds characterized the middle intertidal level in Atlantic Patagonia and cover close to 90% of the rock surface. Mussel beds have a fundamental ecological and structural role in the region. Trampling occurred mostly in this part of the intertidal that remains uncovered by seawater several hours a day. Therefore, a series of subsequent works were aimed at experimentally assessing how trampling affects mussel beds abundance (Mendez et al. 2018, 2019). It was observed that the effect of trampling is proportional to its intensity. However, there is an intensity of trampling (50 tramples on 3 consecutive weekends) after which the mussel cover is reduced in its entirety. Subsequent recovery is very slow, with only a few species being observed in the first 2 years. In a second field experiment, the effect of trampling on mussel beds exposed to contrasting natural disturbance conditions (wave exposure) was evaluated (Mendez et al. 2019). The work shows that the susceptibility of mussel beds to trampling is closely associated with the wave exposure of each site, with a greater loss of cover being recorded at the wave protected site (Fig. 3). Authors found that mussel resistance to trampling increased in exposed sites due to greater byssus attachment and valve compressive strength.

Together, the results obtained from the studies have served as input to suggest to local authorities the maximum number of people that should visit a recreational site (or the carrying capacity), the potential establishment of temporary exclusion zones and to discuss the idea of implementing a rotation in site use as potential palliatives to recreation impacts.

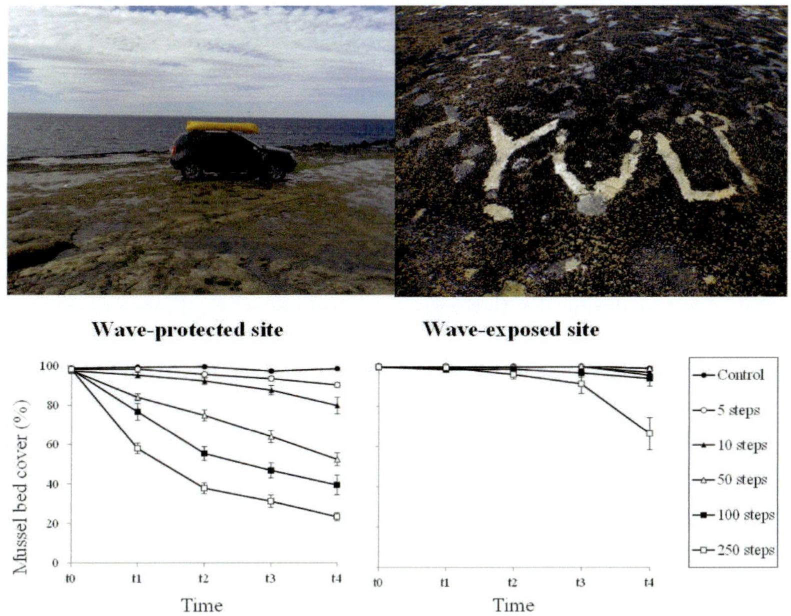

Figure 2. Top: Photographs illustrating the impacts of recreational activities carried out in the coast of Golfo Nuevo, Patagonia, Argentina. Bottom: Effects of different trampling intensities on mussel bed cover in sites with different wave force (modified from Mendez et al. 2019).

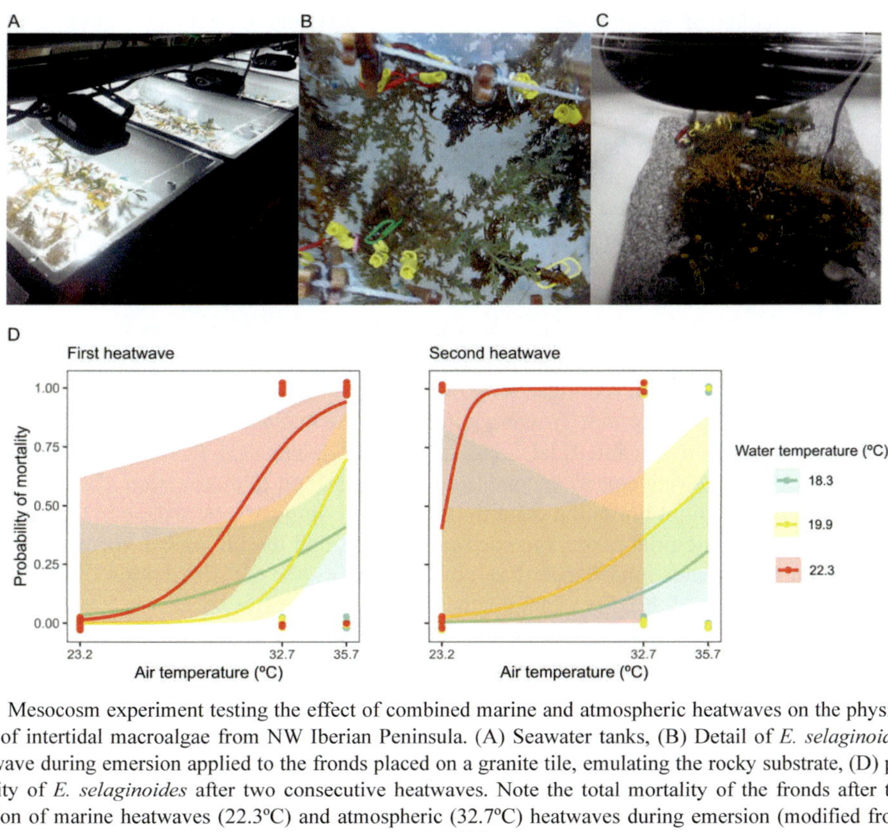

Figure 3. Mesocosm experiment testing the effect of combined marine and atmospheric heatwaves on the physiology and mortality of intertidal macroalgae from NW Iberian Peninsula. (A) Seawater tanks, (B) Detail of *E. selaginoides* fronds, (C) Heatwave during emersion applied to the fronds placed on a granite tile, emulating the rocky substrate, (D) probability of mortality of *E. selaginoides* after two consecutive heatwaves. Note the total mortality of the fronds after the second combination of marine heatwaves (22.3°C) and atmospheric (32.7°C) heatwaves during emersion (modified from Román et al. 2020).

Incidence of heatwaves on the algae species on the North-Western Iberian Peninsula

The NW Iberian coast can be geographically delimited as the coastline from Estaca de Bares Cape, the most septentrional geographic point of the Iberian Peninsula (43°7' N, 7°7' W) to the outflow of the Miño River (41°8' N, 8°9' W), in the natural frontier between Spain and Portugal. The mean annual temperature is 12.5–15°C and the annual precipitation ranges between 1400–2000 mm year^{-1} (AEMET 2011), so the meteorology of the region can be defined as temperate and humid. The high pluviometry results in the outflow of several rivers into the shore, which occasionally causes low salinity events in the winter (Díez et al. 2000). This coastline is composed of exposed shores combined with narrow estuaries or *rias*, which were formed by either the influence of fluvial systems, the falling of emerged tectonic blocks, the flooding of fluvial basins during the Tertiary, or a combination of all these processes (Méndez and Vilas 2005). From the inner and less energetic sector of the *rias* towards the more exposed sectors, one can find tidal flats, beaches intermingled with formations of large boulders and rocky cliffs. A seasonal upwelling occurs in this region, more intensely from the Spanish Cape Finisterre (42°8' N, 9°2' W) to the south. The predominant equatorward winds occurring from April to October are rotated rightwards, causing an offshore transport of surface water masses that are replaced in the coast by deeper, colder and nutrient rich waters' masses through vertical advection (Wooster et al. 1976, Fraga 1981). The high concentration of nutrients promote photosynthesis and support the basis of a diverse and complex trophic web, and the colder temperature favors the development of septentrional flora associated with cool waters (Díez et al. 2000, Sibaja-Cordero and Troncoso 2011). Indeed, in spite of their geographical closeness, its biogeography can be differentiated from the adjacent coasts of northern Spain (Ramos et al. 2015) and Portugal (Monteiro et al. 2022).

The rocky intertidal has a very heterogeneous morphology and is often irregular, with a mixture of flat and inclined surfaces, crevices and boulders. The tidal range can reach 4 m (mesotidal), and the critical levels of intertidal zonation sometimes cannot be clearly delimited; rather, one can observe a smooth change in the distribution of sessile species along the intertidal range (Rafaelli and Hawkins 1996). The horizontal zonation fringes are often patchy and fragmented, driven by the variable morphology, orientation and/or exposure of the rocks to the wave splash. The upper section of the intertidal is dominated by barnacles as *Chthamalus montagui*, *Chthamalus stellatus* and *Semibalanus balanoides, Littorina* gastropods, and *Patella* limpets. In the upper sector of the middle intertidal clumps of the mussels, *Mytilus galloprovincialis* thrive and can coexist with the stalked barnacle *Pollicipes pollicipes* in crevices and inclined rocks at exposed shores. In this zone, macroalgae canopies of *Pelvetia canaliculata, Ascophyllum nodosum*, and *Fucus* spp. appear. Towards the lower intertidal, the cover of macroalgae canopies increases, mainly of the species *Bifucaria bifurcata*, *Codium* spp., *Mastocarpus stellatus*, *Chondrus crispus*, *Gelidium sesquipedale, Ericaria selaginoides* and *Treptacantha baccata*, the invasive *Sargassum muticum* or the ephimerous *Ulva* spp. The most frequent invertebrates in this sector are the sea urchin *Paracentrotus lividus,* the anemone *Anemonia viridis* and the star *Marthasterias glaciaris.* Towards the subtidal, the invasive red algae *Asparagopsis armata* can be found, and the predominant macroalgae are *Saccorhiza polyschides,* growing jointly with the invasive kelp *Undaria pinnatifida.*

Generally, as in other rocky intertidal areas of the world, physical factors as the resistance to desiccation and heat stress during emersion conditions the upper limit of distribution of sessile species in the upper intertidal and the light limitation conditions the lower limits of macroalgae in the subtidal. Yet, in intermediate sectors of the tidal range, the species diversity, and the biological interactions increase, and therefore acquire a greater importance than physical factors for the distribution of species on the shore (Rafaelli and Hawkins 1996). For example, the presence of invasive species as *S. muticum* reduced the number of native species and increased the abundance of filamentous and foliose algae, although the composition of the macroalgae assemblages were not altered (Olabarria et al. 2009); which demonstrates that competition for space is a relevant interaction in the lower intertidal at the NW Iberian Peninsula.

The effect of heatwaves on intertidal seaweeds is of high importance because their habitat and food provisioning function makes them crucial for the whole ecosystem well-being (Umanzor et al. 2018), and their sessile nature makes them particularly vulnerable to heat stress or desiccation. The studies addressing heat stress effects in macroalgae from the NW Iberian Peninsula rocky intertidal have shown different responses depending on the temperature affinity and physiological thresholds of the species. These responses generally indicate a worse performance and a greater mortality in cold-affinity than in warm-affinity species, which imply range contractions and expansions, respectively, of their geographical distribution.

For example, two consecutive experimental cycles of simmulated heatwaves during low tide were applied to macroalgae assemblages. The heatwave treatments consisted of 4 days with: (1) temperature set at the 95th percentile of maximum daily temperatures in summer, and 2)temperature set at 3° above this 95th percentile. This increase was forecasted for the period 2081–2100 (IPCC 2014).

The applied heatwaves resulted in cumulative decreases of the physiological performance of predominant macroalgae species from the NW Iberian Peninsula. Specifically, a greater probability of mortality in *E. selaginoides* (formerly *Cystoseira tamariscifolia*) than in *B. bifurcata* and *Codium tomentosum* was found, and the two latter species were more resistant and resilient. The obtained results suggested that more frequent and more intense heatwaves during emersion in the region could result in changes of the relative abundances of the studied species and in changes in the community structure. Specifically, increases of *Codium tomentosum* and *B. bifurcata*, and decreases of *E. selaginoides* would be expected (Fig. 3; Román et al. 2020).

Other studies, that did not apply heatwaves *per se*, but that subjected macroalgae from NW Iberia to anomalously high seawater temperatures, revealed their physiological responses and thresholds and can give clues about their expected responses under the more frequent and intense marine heatwaves expected under global warming. For instance, eight macroalgal species showed a consistent decrease of photosynthesis rates after being incubated in seawater with temperatures beyond the historical registers (from 24ºC to 30º) (Piñeiro-Corbeira et al. 2018). Also, *H. elongata* showed less tolerance to experimentally warming seawater and a lower physiological temperature threshold than *B. bifurcata* (18°C vs. 24.7°C), and prediction models based on these thresholds indicated a local extinction of the former at its southern distribution limit in Spain in future warming scenarios (Martínez et al. 2014).

References

Addessi, L. (1994). Human disturbance and long-term changes on a rocky intertidal community. Ecology 4: 786–797.

Airoldi, L., Bacchiocchi, F., Cagliola, C., Bulleri, F. and Abbiati, M. (2005). Impact of recreational harvesting on assemblages in artificial rocky habitats. Mar. Ecol. Prog. Ser. 299: 55–66.

Airoldi, L. and Beck, M. (2007). Status of Euromediterranean coastal marine habitats: loss and trends. Rapp. Comm. Int. Mer. Médit. 38: 410.

Araújo, R., Vaselli, S., Almeida, M., Serrão, E. and Sousa-Pinto, I. (2009). Effects of disturbance on marginal populations: human trampling on *Ascophyllum nodosum* assemblages at its southern distribution limit. Mar. Ecol. Prog. Ser. 378: 81–92.

Arévalo, R., Pinedo, S. and Ballesteros, E. (2007). Changes in the composition and structure of Mediterranean rocky-shore communities following a gradient of nutrient enrichment: descriptive study and test of proposed methods to assess water quality regarding macroalgae. Mar. Poll. Bull. 55: 104–113.

Bagur, M., Gutiérrez, J. L., Arribas, L. P. and Palomo, M. G. (2016). Complementary influences of co-occurring physical ecosystem engineers on species richness: insights from a Patagonian rocky shore. Biod. Conserv. 25: 2787–2802.

Bailey, S. A. (2015). An overview of thirty years of research on ballast water as a vector for aquatic invasive species to freshwater and marine environments. Aq. Ecosyst. Health 18: 261–268.

Bastida, R., Capezzani, A. and Torti, M. R. (1971). Fouling organisms in the port of Mar del Plata (Argentina). I. *Siphonaria lessoni*: ecological and biometric aspects. Mar. Biol. 10: 297–307.

Battini, N., Giachetti, C. B., Castro, K. L., Bortolus, A. and Schwindt, E. (2021). Predator–prey interactions as key drivers for the invasion success of a potentially neurotoxic sea slug. Biol. Invasions 23: 1207–1229.

Bauer, B., Fioroni, P., Ide, I., Liebe, S., Oehlmann, J., Stroben, E. et al. (1995). TBT effects on the female genital system of *Littorina littorea*: a possible indicator of tributyltin pollution. Hydrobiologia 309: 15–27.

Bazterrica, M. C., Silliman, B. R., Hidalgo, F. J., Crain, C. M. and Bertness, M. D. (2007). Limpet grazing on a physically stressful Patagonian rocky shore. J. Exp. Mar. Biol. Ecol. 353: 22–34.

Becherucci, M. E., Llanos, E. N., Garaffo, G. V. and Vallarino, E. A. (2016). Succession in an intertidal benthic community affected by untreated sewage effluent: A case of study in the SW Atlantic shore. Mar. Poll. Bull. 109: 95–103.

Benedetti-Cecchi, L., Pannacciulli, F., Bulleri, F., Moschella, P. S., Airoldi, L., Relini, G. et al. (2001). Predicting the consequences of anthropogenic disturbance: large-scale effects of loss of canopy algae on rocky shores. Mar. Ecol. Prog. Ser. 214: 137–150.

Bertness, M. D., Crain, C. M., Silliman, B. R., Bazterrica, M. C., Reyna, M. V., Hildago, F. et al. (2006). The community structure of Western Atlantic Patagonian rocky shores. Ecol. Monogr. 76: 439–460.

Bibby, R., Cleall-Harding, P., Rundle, S., Widdicombe, S. and Spicer, J. (2007). Ocean acidification disrupts induced defences in the intertidal gastropod *Littorina littorea*. Biol. Lett. 3: 699–701.

Bigatti, G., Primost, M. A., Cledón, M., Averbuj, A., Theobald, N., Gerwinski, W. E. et al. (2009). Biomonitoring of TBT contamination and imposex incidence along 4700 km of Argentinean shoreline (SW Atlantic: From 38S to 54S). Mar. Poll. Bull. 58: 695–701.

Bishop, M. J., Underwood, A. J. and Archambault, P. (2002). Sewage and environmental impacts on rocky shores: necessity of identifying relevant spatial scales. Mar. Ecol. Prog. Ser. 236: 121–128.

Blamey, L. K. and Branch, G. M. (2009). Habitat diversity relative to wave action on rocky shores: implications for the selection of marine protected areas. Aq. Conserv.: Mar. Fresh. Ecosyst. 19: 645–657.

Boltovskoy, D., Sylvester, F. and Paolucci, E. M. (2018). Invasive species denialism: sorting out facts, beliefs, and definitions. Ecol. Evol. 8: 11190–11198.

Brosnan, D. M. and Crumrine, L. L. (1994). Effects of human trampling on marine rocky shore communities. J. Exp. Mar. Biol. 177: 79–97.

Brown, P. J. and Taylor, R. B. (1999). Effects of trampling by humans on animals inhabiting coralline algal turf in the rocky intertidal. J. Exp. Mar. Biol. Ecol. 235: 45–53.

Bruno, J. F., Harley, C. D. and Burrows, M. T. (2014). Climate change in marine communities. pp. 427–447. *In*: Bertness, M. D., Bruno, J. F., Silliman, B. R. and Stachowicz, J. J. (eds.). Marine Community Ecology and Conservation. Sinauer Associates, Inc., Massachusetts, USA.

Bulleri, F., Eriksson, B. K., Queirós, A., Airoldi, L., Arenas, F., Arvanitidis, C. et al. (2018). Harnessing positive species interactions as a tool against climate-driven loss of coastal biodiversity. PLoS Biol. 16: e2006852.

Byers, J. E., Wright, J. T. and Gribben, P. E. (2010). Variable direct and indirect effects of a habitat-modifying invasive species on mortality of native fauna. Ecology 91: 1787–1798.

Cabral-Oliveira, J., Mendes, S., Maranhão, P. and Pardal, M. A. (2014). Effects of sewage pollution on the structure of rocky shore macroinvertebrate assemblages. Hydrobiologia 726: 271–283.

Carls, M. G., Harris, P. M. and Marine, R. (2004). Restoration of oiled mussel beds in Prince William Sound, Alaska. Environ. Res. 57: 359–376.

Chapperon, C., Volkenborn, N., Clavier, J., Séité, S., Seabra, R. and Lima, F. P. (2016). Exposure to solar radiation drives organismal vulnerability to climate: Evidence from an intertidal limpet. J. Thermal Biol. 57: 92–100.

Claudet, J. and Fraschetti, S. (2010). Human-driven impacts on marine habitats: a regional meta-analysis in the Mediterranean Sea. Biol. Conserv. 143: 2195–2206.

Cloern, J. E. (2001). Our evolving conceptual model of the coastal eutrophication problem. Mar. Ecol. Prog. Ser. 210: 223–253.

Commendatore, M. G. and Esteves, J. L. (2007). An assessment of oil pollution in the coastal zone of Patagonia, Argentina. Environ. Manag. 40: 814–821.

Connell, J. H. (1961). The influence of interspecific competition and other factors on the distribution of the barnacle *Chthamalus stellatus*. Ecology 42: 710–723.

Connell, J. H. (1972). Community interactions on marine rocky intertidal shores. Ann. Rev. Ecol. Syst. 3: 169–192.

Cordone, G., Lozada, M., Vilacoba, E., Thalinger, B., Bigatti, G., Lijtmaer, D. A. et al. (2022). Metabarcoding, direct stomach observation and stable isotope analysis reveal a highly diverse diet for the invasive green crab in Atlantic Patagonia. Biol. Invasions 24: 505–526.

Crain, C. M., Kroeker, K. and Halpern, B. S. (2008). Interactive and cumulative effects of multiple human stressors in marine systems. Ecol. Lett. 11: 1304–1315.

Dahl, C. (1993). Tourism development on the Island of Pohnpei (Federated States of Micronesia): Sacredness, control and autonomy. Oc. Coastal Manag. 20: 241–265.

Davenport, J. and Davenport, J. L. (2006). The impact of tourism and personal leisure transport on coastal environments: A review. Estuar. Coast. Shelf. Sci. 67: 280–292.

Davis, A. R., Walls, K. and Jeffs, A. (2018). Biotic consequences of a shift in invertebrate ecosystem engineers: Invasion of New Zealand rocky shores by a zone-forming ascidian. Mar. Ecol. 39: e12502.

Díez, I., Secilla, J. A., Santolaria, A. and Gorostiaga, J. M. (2000). The North coast of Spain. pp. 135–150. *In*: Sheppard, C. (ed.). Seas at the Millennium. An Environmental Evaluation. Vol I. Regional Chapters: Europe, The Americas and West Africa. Pergamon, Oxford, UK.

Domínguez, R., Olabarria, C., Woodin, S. A., Wethey, D. S., Peteiro, L. G., Macho, G. et al. (2021). Contrasting responsiveness of four ecologically and economically important bivalves to simulated heat waves. Mar. Environ. Res. 164: 105–229.

Doyle, D., Gammell, M., Frias, J., Griffin, G. and Nash, R. (2019). Low levels of microplastics recorded from the common periwinkle, *Littorina littorea* on the west coast of Ireland. Mar. Poll. Bull. 149: 110645.

Ellis, E. C. (2011). Anthropogenic transformation of the terrestrial biosphere. Philos. Trans. R. Soc. A 369: 1010–1035.

Erasmus, A., Ikenaka, Y., Nakayama, S. M., Ishizuka, M., Smit, N. J. and Wepener, V. (2020). Trophic transfer of pollutants within two intertidal rocky shore ecosystems in different biogeographic regions of South Africa. Mar. Poll. Bull. 157: 111309.

Fairweather, P. G. (1990). Sewage and the biota on seashores: Assessment of impact in relation to natural variability. Environ. Monit. Assess. 14: 197–210.

Favoretto, F., Sánchez, C. and Aburto-Oropeza, O. (2022). Warming and marine heatwaves tropicalize rocky reefs communities in the Gulf of California. Progr. Oceanogr. 206: 102838.

Fernández, C. (2011). The retreat of large brown seaweeds on the north coast of Spain: The case of *Saccorhiza polyschides*. Europ. J. Phycol. 46: 352–360.

Fernández, C. (2016). Current status and multidecadal biogeographical changes in rocky intertidal algal assemblages: The northern Spanish coast. Estuar. Coast. Shelf Sci. 171: 35–40.

Ferreira, J. G., Arenas, F., Martínez, B., Hawkins, S. J. and Jenkins, S. R. (2014). Physiological response of fucoid algae to environmental stress: comparing range centre and southern populations. New Phytologist 202: 1157–1172.

Gestoso, I., Arenas, F. and Olabarria, C. (2016). Ecological interactions modulate responses of two intertidal mussel species to changes in temperature and pH. J. Exp. Mar. Biol. Ecol. 474: 116–125.

Gosling, E. (2015). Marine Bivalve Molluscs. Second edition. Wiley Blackwell, United Kingdom.

Green, D. S. (2016). Effects of microplastics on European flat oysters, *Ostrea edulis* and their associated benthic communities. Environ. Pollut. 216: 95–103.

Haller, M. J. (2000). Regional geological setting. *In*: Isla, F. I., Schnack, E. J. and Weiler, N. E. (eds.). Geological setting and quaternary environments of Peninsula Valdés, Chubut, Argentina. Field Meeting Guidebook, IGCP, Project 437.

Halpern, B. S., Selkoe, K. A., Micheli, F. and Kappel, C. V. (2007). Evaluating and ranking the vulnerability of global marine ecosystems to anthropogenic threats. Conserv. Biol. 21: 1301–1315.

Harley, C. D. G. and Helmuth, B. S. T. (2003). Local- and regional-scale effects of wave exposure, thermal stress, and absolute vs. effective shore level on patterns of intertidal zonation. Limnol. Oceanogr. 48: 1498–1508.

Harvey, B. P., McKeown, N. J., Rastrick, S. P. S., Bertolini, C., Foggo, A,, Graham, H. et al. (2016). Individual and population-level responses to ocean acidification. Scient. Rep. 6: 20194.

Hawkins, S. J., Moore, P. J., Burrows, M. T., Poloczanska, E., Mieszkowska, N., Herbert, R. J. H. et al. (2008). Complex interactions in a rapidly changing world: responses of rocky shore communities to recent climate change. Clim. Res. 37: 123–133.

Hawkins, S. J., Pack, K. E., Hyder, K., Benedetti-Cecchi, L. and Jenkins, S. R. (2020). Rocky shores as tractable test systems for experimental ecology. J. Mar. Biol. Assoc. UK 100: 1017–1041.

Helmuth, B., Mieszkowska, N., Moore, P. and Hawkins, S. J. (2006). Living on the edge of two changing worlds: forecasting the responses of rocky intertidal ecosystems to climate change. Annu. Rev. Ecol. Evol. Syst. 37: 373–404.

Hereward, H. F. R., King, N. G. and Smale, D. A. (2019). Intra-annual variability in responses of a canopy forming kelp to cumulative low tide heat stress: implications for populations at the trailing range edge. J. Phycol. 56: 146–158.

Hesketh, A. V. and Harley, C. D. G. (2022). Extreme heatwave drives topography-dependent patterns of mortality in a bed-forming intertidal barnacle, with implications for associated community structure. Glob. Change Biol. 00: 1–14.

Hidalgo, F. J., Silliman, B. R., Bazterrica, M. C. and Bertness, M. D. (2007). Predation on the rocky shores of Patagonia, Argentina. Estuar. Coasts 30: 886–894.

Hobday, A. J., Alexander, L. V., Perkins, S. E., Smale, D. A., Straub, S. C., Oliver, E. C. J. et al. (2016). A hierarchical approach to defining marine heatwaves. Prog. Oceanogr. 141: 227–238.

Hulme, P. E. (2009). Trade, transport and trouble: managing invasive species pathways in an era of globalization. J. Appl. Ecol. 46: 10–18.

Kelaher, B. P., Castilla, J. C., Prado, L., York, P., Schwindt, E. and Bortolus, A. (2007). Spatial variation in molluscan assemblages from coralline turfs of Argentinean Patagonia. J. Molluscan Stud. 73: 139–146.

Kunze, C., Wölfelschneider, M. and Rölfer, L. (2021). Multiple driver impacts on rocky intertidal systems: The need for an integrated approach. Front. Mar. Sci. 8: 667168.

Langston, W. J., Pope, N. D., Davey, M., Langston, K. M., O'Hara, S. C. M., Gibbs, P. E. et al. (2015). Recovery from TBT pollution in English Channel environments: a problem solved?. Mar. Poll. Bull. 95: 551–564.

Lathlean, J. A. and Minchinton, T. E. (2012). Manipulating thermal stress on rocky shores to predict patterns of recruitment of marine invertebrates under a changing climate. Mar. Ecol. Prog. Ser. 467: 121–136.

Leignel, V., Stillman, J. H., Baringou, S., Thabet, R. and Metais, I. (2014). Overview on the European green crab *Carcinus* spp. (Portunidae, Decapoda), one of the most famous marine invaders and ecotoxicological models. Environ. Sci. Poll. Res. 21: 9129–9144.

Leonard, G. H. (2000). Latitudinal variation in species interactions: a test in the New England rocky intertidal zone. Ecology 81: 1015–1030.

Liuzzi, M. G. and López Gappa, J. (2008). Macrofaunal assemblages associated with coralline turf: species turnover and changes in structure at different spatial scales. Mar. Ecol. Prog. Ser. 363: 147–156.

Livore, J. P., Mendez, M. M., Miloslavich, P., Rilov, G. and Bigatti, G. (2021). Biodiversity monitoring in rocky shores: Challenges of devising a globally applicable and cost-effective protocol. Ocean Coast. Manag. 205: 105548.

López Gappa, J. J. and Tablado, N. M. (1990). Influence of sewage pollution on a rocky intertidal community dominated by the mytilid Brachidontes. Mar. Ecol. Prog. Ser. 63: 163–175.

Lubchenco, J. (1980). Algal zonation in the New England rocky intertidal community: an experimental analysis. Ecology 61: 333–344.

Manríquez, P. H., Jara, M. E., Seguel, M. E., Torres, R., Alarcon, E. and Lee, M. R. (2016). Ocean acidification and increased temperature have both positive and negative effects on early ontogenetic traits of a rocky shore keystone predator species. PLoS ONE 11: e0151920.

Martínez, B., Viejo, R. M., Carreño, F. and Aranda, S. C. (2012). Habitat distribution models for intertidal seaweeds: Responses to climatic and non-climatic drivers. J. Biogeogr. 39: 1877–1890.

McNeely, J. A. (2006). As the world gets smaller, the chances of invasion grow. Euphytica 148: 5–15.

McQuaid, C. and G. M. Branch. (1985). Trophic structure of rocky intertidal communities: Response to wave action and implications for energy flow. Mar. Ecol. Prog. Ser. 22: 153–161.

Mendez, M. M., Livore, J. P. and Bigatt, G. (2018). Effects of trampling on mussel beds: importance of disturbance intensity. Mar. Ecol. Prog. Ser. 606: 231–235.

Mendez, M. M., Livore, J. P. and Bigatti, G. (2019). Interaction of natural and anthropogenic stressors on rocky shores: community resistance to trampling. Mar. Ecol. Prog. Ser. 631: 117–126.

Mendez, M. M., Livore, J. P., Calcagno, J. and Bigatti, G. (2017). Effects of recreational activities on Patagonian rocky shores. Mar. Environ. Res. 130: 213–220.

Méndez, G. and Vilas, F. (2005). Geological antecedents of the Rias Baixas (Galicia, Northwest Iberian Peninsula). J. Mar. Syst. 54: 195–207.

Menge, B. A., Lubchenco, J., Bracken, M. E. S., Chan, F., Foley, M. M. and Freidenburg, T. L. (2003). Coastal oceanography sets the pace of rocky intertidal community dynamics. Proc. Natl. Acad. Sci. U.S.A. 100: 12229–12234.

Menge, B. A., Olson, A. M. and Dahlhoff, E. P. (2002). Environmental stress, bottom-up effects, and community dynamics: integrating molecular-physiological and ecological approaches. Integr. Comp. Biol. 42: 892–908.

Micheli, F., Heiman, K. W., Kappel, C. V., Martone, R. G., Sethi, S. A., Osio, G. C. et al. (2016). Combined impacts of natural and human disturbances on rocky shore communities. Ocean Coastal Manag. 126: 42–50.

Mieszkowska, N. (2009). Intertidal indicators of climate and global change. pp. 281–296. *In*: Letcher, T. M. (ed.). Climate Change. Observed Impact on Planet Earth, Elsevier. Amsterdam, The Netherlands.

Milazzo, M., Badalamenti, F., Riggio, S. and Chemello, R. (2004). Patterns of algal recovery and small-scale effects of canopy removal as a result of human trampling on a Mediterranean rocky shallow community. Biol. Conserv. 117: 191–202.

Milazzo, M., Chemello, R., Badalementi, F. and Riggio, S. (2002). Short-term effect of human trampling on the upper infralittoral macroalgae of Ustica Island MPA (western Mediterranean, Italy). J. Mar. Biol. Ass. U.K. 82: 745–748.

Miller, M. L. (1993). The rise of coastal and marine tourism. Ocean Coastal Manag. 20: 181–199.

Minchinton, T. E. and Fels, K. J. (2013). Sediment disturbance associated with trampling by humans alters species assemblages on a rocky intertidal seashore. Mar. Ecol. Prog. Ser. 472: 129–140.

Monteiro, C., Pereira, J., Seabra, R. and Lima, F. P. (2022). Fine-scale survey of intertidal macroalgae reveals recent changes in a cold-water biogeographic stronghold. Front. Mar. Sci. 9: 880074.

Nagelkerken, I. and Munday, P. L. (2016). Animal behaviour shapes the ecological effects of ocean acidification and warming: Moving from individual to community-level responses. Glob. Change Biol. 22: 974–989.

Nicastro, K. R., McQuaid, C. D. and Zardi, G. I. (2019). Between a rock and a hard place: combined effect of trampling and phototrophic shell-degrading endoliths in marine intertidal mussels. Mar. Biodivers. 49: 1581–1586.

O'Connor, N. E. (2013). Impacts of sewage outfalls on rocky shores: Incorporating scale, biotic assemblage structure and variability into monitoring tools. Ecol. Indic. 29: 501–509.

O'Donnell, M. J., George, M. N. and Carrington, E. (2013). Mussel byssus attachment weakened by ocean acidification. Nat. Clim. Change 3: 587–590.

Olabarria, C., Arenas, F., Viejo, R. M., Gestoso, I., Vaz-Pinto, F., Incera, M. et al. (2012). Response of macroalgal assemblages from rockpools to climate change: effects of persistent increase in temperature and CO_2. Oikos 122: 1065–1079.

Olabarria, C., Gestoso, I., Lima, F. P., Vázquez, E., Comeau, L. A., Gomes, F. et al. (2016). Response of two mytilids to a heatwave: the complex interplay of physiology, behaviour and ecological interactions. PLoS ONE 11: e0164330.

Olabarria, C., Rodil, I. F., Incera, M. and Troncoso, J. S. (2009). Limited impact of *Sargassum muticum* on native algal assemblages from rocky intertidal shores. Mar. Env. Res. 67: 153–158.

Orensanz, J., Schwindt, E., Pastorino, G., Bortolus, A., Casas, G., Darrigran, G. et al. (2002). No longer the pristine confines of the world ocean: a survey of exotic marine species in the southwestern Atlantic. Biol. Invasions 4: 115–143.

Paine, R. T. (1966). Food web complexity and species diversity. Am. Nat. 100: 65–75.

Pessacg, N., Blázquez, J., Lancelotti, J. and Solman, S. (2021). Climate changes in coastal areas of Patagonia: observed trends and future projections. pp. 13–42. *In*: Helbling, E. W., Narvarte, M. A., González, R. A. and Villafañe, V. E. (eds.). Global Change in Atlantic Coastal Patagonian Ecosystems. Springer, New York, USA.

Pinedo, S., Arévalo, R. and Ballesteros, E. (2015). Seasonal dynamics of upper sublittoral assemblages on Mediterranean rocky shores along a eutrophication gradient. Estuar. Coast. Shelf Sci. 161: 93–101.

Piñeiro-Corbeira, C., Barreiro, R., Cremades, J. and Arenas, F. (2018). Seaweed assemblages under a climate change scenario: Functional responses to temperature of eight intertidal seaweeds match recent abundance shifts. Sci. Rep. 8: 1–9.

Rafaelli, D. and Hawkins, S. (1996). Human impact on the shore. *In*: Rafaelli, D. and Hawkins, S. (eds.). Intertidal Ecology. Kluwer Academic Publishers, Dordretch, The Netherlands.

Raffo, M. P., Lo Russo, V. and Schwindt, E. (2014). Introduced and native species on rocky shore macroalgal assemblages: Zonation patterns, composition and diversity. Aquat. Bot. 112: 57–65.

Raymond, W. W., Barber, J. S., Dethier, M. N., Hayford, H. A., Harley, C. D. G., King, T. L. et al. (2022). Assessment of the impacts of an unprecedented heatwave on intertidal shellfish of the Salish Sea. Ecology 103: e3798.

Rechimont, M. E., Galván, D. E., Sueiro, M. C., Casas, G., Piriz, M. L., Diez, M. E. et al. (2013). Benthic diversity and assemblage structure of a north Patagonian rocky shore: a monitoring legacy of the NaGISA project. J. Mar. Biol. Assoc. U.K. 93: 2049–2058.

Rilov, G. and Schiel, D. R. (2011). Community regulation: the relative importance of recruitment and predation intensity of an intertidal community dominant in a seascape context. PLoS ONE 6(8): e23958.

Rockström, J., Steffen, W., Noone, K., Persson, A., Chapin, F. S., Lambin, E. et al. (2009). Planetary boundaries: Exploring the safe operating space for humanity. Ecol. Soc. 14: 32.

Schiel, D. R., Steinbeck, J. R. and Foster, M. S. (2004). Ten years of induced ocean warming causes comprehensive changes in marine benthic communities. Ecology 85: 1833–1839.

Schonbeck, M. W. and Norton, T. A. (1980). The effects on intertidal fucoid algae of exposure to air under various conditions. Bot. Mar. 23: 141–148.

Seuront, L. (2018). Microplastic leachates impair behavioural vigilance and predator avoidance in a temperate intertidal gastropod. Biol. Lett. 14: 20180453.

Seuront, L., Nicastro, K. R., Zardi, G. I. and Goberville, E. (2019). Decreased thermal tolerance under recurrent heat stress conditions explains summer mass mortality of the blue mussel *Mytilus edulis*. Sci. Rep. 9: 1–14.

Shinen, J. S. and Morgan, S. G. (2009). Mechanisms of invasion resistance: competition among intertidal mussels promotes establishment of invasive species and displacement of native species. Mar. Ecol. Prog. Ser. 383: 187–197.

Sibaja-Cordero, J. and Souza Troncoso, J. (2011). Upper and lower limits of rocky shore organisms at different spatial scales and wave exposure. Thalassas 27: 81–100.

Silliman, B. R., Bertness, M. D., Altieri, A. H., Griffin, J. N., Bazterrica, M. C., Hidalgo, F. J. et al. (2011). Whole-community facilitation regulates biodiversity on Patagonian rocky shores. PLoS ONE 6: e24502.

Smith, J. R., Fong, P. and Ambrose, R. F. (2008). The impacts of human visitation on mussel bed communities along the California coast: are regulatory Marine Reserves effective in protecting these communities? Environ. Manag. 41: 599–612.

Steffen, W., Grinevald, J., Crutzen, P. and McNeill, J. (2011). The Anthropocene: conceptual and historical perspectives. Philos. T. R. Soc. 369: 842–867.

Steffen, W., Leinfelder, R., Zalasiewicz, J., Waters, C. N., Williams, M., Summerhayes, C. et al. (2016). Stratigraphic and Earth System approaches to defining the Anthropocene. Earth's Future 4: 324–345.

Stillman, J. H. (2019). Heat waves, the new normal: Summertime temperature extremes will impact animals, ecosystems, and human communities. Physiology 34: 86–100.

Thompson, R. C., Crowe, T. P. and Hawkins, S. J. (2002). Rocky intertidal communities: past environmental changes, present status and predictions for the next 25 years. Env. Cons. 29: 168–191.

Thompson, R. C., Olsen, Y., Mitchell, R. P., Davis, A., Rowland, S. J., John, A. W. et al. (2004). Lost at sea: where is all the plastic? Science 304: 838–838.

Thompson, R. C. (2015). Microplastics in the marine environment: sources, consequences and solutions. pp. 185–200 *In*: Bergmann, M., Gutow, L. and Klages, M. (eds.). Marine Anthropogenic Litter. Springer International Publishing AG Switzerland.

Troost, K. (2010). Causes and effects of a highly successful marine invasion: case-study of 0the introduced Pacific oyster *Crassostrea gigas* in continental NW European estuaries. J. Sea Res. 64: 145–165

Turley, C. and Findlay, H. S. (2000). Ocean acidification as an indicator of climate change. pp. 367–390. *In*: Letcher, T. M. (eds.). Climate Change. Observed Impact on Planet Earth. Elsevier, Amsterdam, The Netherlands.

Umanzor, S., Ladah, L. and Calderon-Aguilera, L. E. (2018). Testing the relative importance of intertidal seaweeds as ecosystem engineers across tidal heights. J. Exp. Mar. Biol. Ecol. 511: 100–107.

Voerman, S. E., Llera, E. and Rico, J. M. (2013). Climate driven changes in subtidal kelp forest communities in NW Spain. Mar. Environ. Res. 90: 119–127.

Werner, E. E. and Peacor, S. D. (2003). A review of trait-mediated indirect interactions in ecological communities. Ecology 84: 1083–1100.

Wieters, E. A. (2005). Upwelling control of positive interactions over mesoscales: a new link between bottom-up and top-down processes on rocky shores. Mar. Ecol. Prog. Ser. 301: 43–54.

Wooster, W. S., Bakun, A. and Mclain, D. R. (1976). The seasonal upwelling cycle along the eastern boundary of the North Atlantic. J. Mar. Res. 34: 131–141.

Worm, B. and Lenihan, H. S. (2013). Threats to marine ecosystems: overfishing and habitat degradation. pp. 449–476. *In*: Bertness, M. D., Bruno, J. F., Silliman, B. R. and Stachowicz, J. J. (eds.). Marine Community Ecology and Conservation. Sinauer Associates, Inc., Massachusetts, USA.

Zalasiewicz, J., Waters, C. N., Summerhayes, C. P., Wolfe, A. P., Barnosky, A. D., Cearreta, A. et al. (2017). The Working Group on the Anthropocene: Summary of evidence and interim recommendations. Anthropocene 19: 55–60.

CHAPTER 8

Estuaries

Patterns and Trends on Ecosystem Structure and Functioning under Global Changes

Henrique N. Cabral,[1,*] *Elorri Arevalo*[2] and *Laure Carassou*[1]

The estuarine ecosystem: Patterns, processes and functioning of estuarine systems

Estuaries are coastal features where river water mixes with sea water, but their definition has been changing over time without consensus (syntheses in Whitfield and Elliott 2011, Kennish 2024). Among the most used definitions are those attributed to Pritchard (1952), i.e., semi-enclosed coastal body of water, which has a free connection with the open sea and within which seawater is measurably diluted with freshwater derived from land drainage, or to Day et al. (2012), i.e., coastal indentation that has a restricted connection to the ocean and remains open at least intermittently. Some of the most recent definitions are quite complex, integrating biological aspects or even using expert judgement (e.g., Elliott and McLusky 2002).

From an ecological point of view, estuaries have also generated some ambiguity, being considered a transition system, and therefore an ecotone, by some authors, while for others they are considered a distinct ecosystem in their own right. Regarding a geological perspective, estuaries are systems with a relatively short life span (in the order of tens of thousands of years) essentially dictated by the cycles of glaciation and the rise or fall of sea level, isostatic adjustment of continents, and sedimentation rates (Kennish 1976, Whitfield and Elliott 2011). Estuaries that exist today originated during the Holocene epoch (about 12 000 years BP), an interglacial period when sea level raised and inundated fluvial valleys (Kennish 1976, 2024, Whitfield and Elliott 2011, Wolanski and Elliott 2015).

Despite this ephemeral nature, estuaries have been present on the planet for more than 200 million years and the biggest challenges to their continued existence in the future are linked to the greater or lesser difficulty in transposing estuarine environments to areas further upstream in the hydrographic basins as a result of sea level rise and human occupation and associated artificialisation of the territory (e.g., Whitfield and Elliott 2011, Cabral 2024). The types and classification of estuaries may be quite diverse according to the criteria used (e.g., geomorphology, physiography, vertical salinity structure, salinity ranges, water balance, hydrodynamics, tidal ranges, sediment infilling, anthropogenic effects; see Kennish 2024 for a review). Combinations of two of the most important factors structuring estuaries, i.e., geomorphology

[1] INRAE, EABX, 50 Avenue de Verdun, 33610 Cestas, France.
[2] AZTI Marine Research, Basque Research and Technology Alliance (BRTA), Txatxarramendi Ugartea z/g, 48395 Sukarrieta, Spain.
* Corresponding author: henrique.cabral at inrae.fr

Figure 1. Main types of estuarine ecosystems according to their physiography (adapted from Whitfield and Elliott 2011).

and hydrology, generates a wide diversity of physiographic types, which in turn have a strong influence on the habitats and biological communities that will occur in these systems. Whitfield and Elliott (2011) have proposed three main physiographic categories, for which different sub-categories may be recognised: valley estuarine ecosystems, estuarine lagoon/lakes and river-mouth ecosystems (Fig. 1).

Indeed, physiography, freshwater input through river(s) flow(s) and tidal range dictate, to a large extent, the hydrodynamics in estuaries, and consequently the abiotic conditions and biological communities, according to different temporal scales (circadian, weekly, monthly, seasonal, interannual, decadal). The relevance of the four main salinity range zones usually referred for estuaries, i.e., the oligohaline zone (typically with a salinity from 0.5 up to 5), the mesohaline zone (from 5 to 18), the polyhaline zone (from 18 to near 30), and the euhaline zone (salinity higher than 30 up to 40), deeply depends on the factors indicated above. In estuaries with a high freshwater input, the oligohaline zone tends to prevail, while for estuaries receiving a small river flow, marine conditions found in the polyhaline zone are expected to represent most of the estuarine area. Nonetheless, the combination of these structuring forces in space and time originates from rather specific and case-dependent conditions, making each estuary a rather unique system. Two additional characteristics that are common in certain geographical areas are the existence of hyperhaline (salinity higher than 40), estuaries, usually found in shallow systems of arid zones with a low freshwater input and a high evaporation rate, and non-permanent-systems, i.e., estuaries that are closed to the sea in a period of the year, due to a high sedimentation rate or sediment transport into the estuarine mouth, and/or to a low freshwater flow (Whitfield 1992, Whitfield and Elliott 2011).

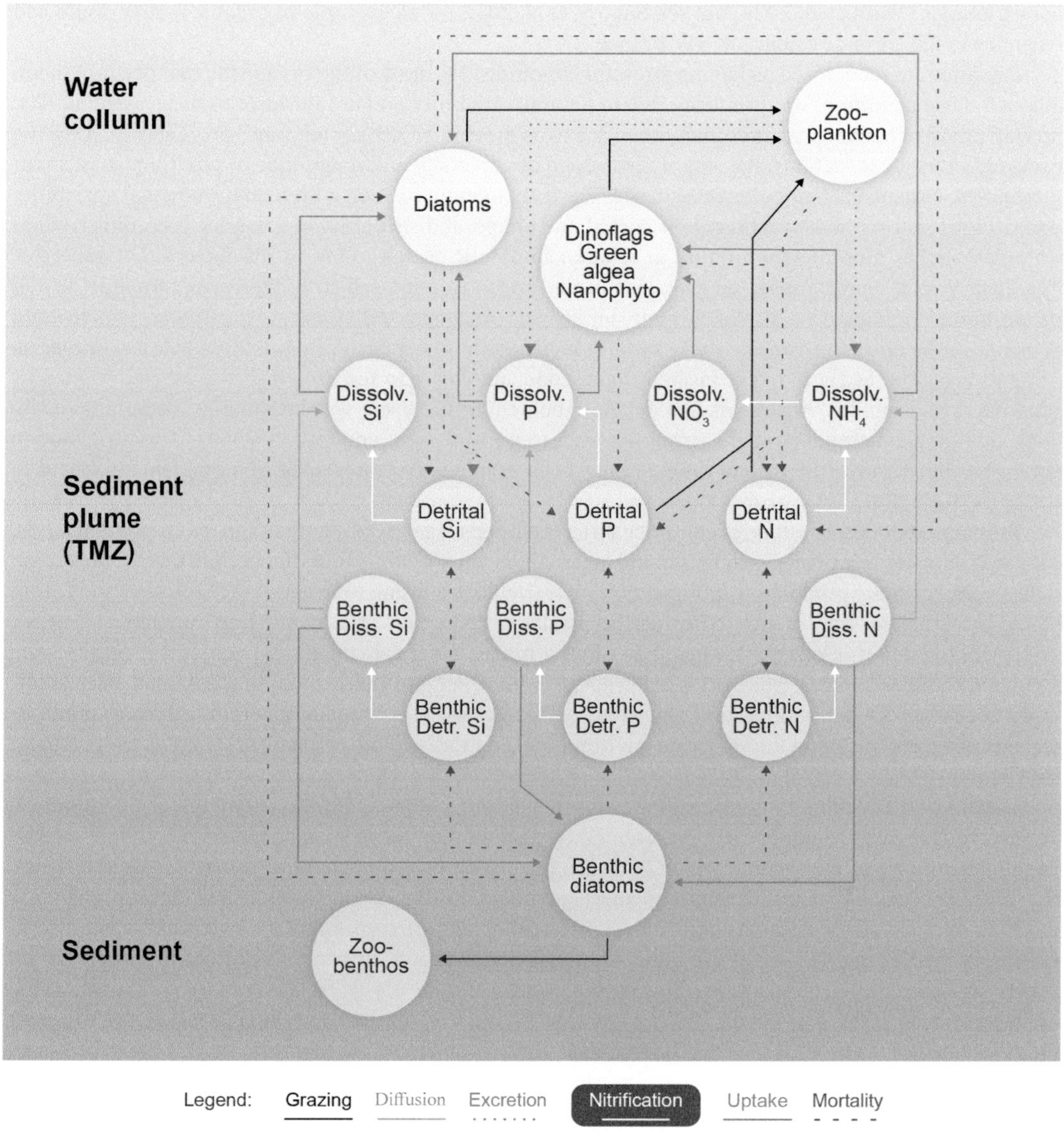

Figure 2. Schematic representation of the main biogeochemical processes in the water column and sediment compartments, and the exchanges between them, described by Romero et al. 2019, for the Seine estuary (France). **Acronyms; nanophyto = nanophytoplankton; dissolv. = dissolved; diss. = dissolved; detr. = detrital.**

The hydrological balance mentioned above and the sources of organic matter largely determine the biogeochemical cycles of estuaries, including their spatial and temporal dynamics. Sources of organic matter in estuaries can be of allochthonous or autochthonous origin. Their relative importance depends greatly on the system. The allochthonous sources are mainly discharged by rivers or by outwelling from coastal areas and include organic matter, nutrients, trace metals and sediments (Wolanski and Elliott 2015, Santos et al. 2021, Ward et al. 2024). Autochthonous sources are mainly due to internal recirculation and mobilisation of these elements, the organic matter being essentially due to estuarine primary producers. Biogeochemical processes are mainly driven by primary producers affecting the entire estuarine food webs (as an example, Romero et al. 2019 presented a schematic representation of these processes for the Seine estuary, France; Fig. 2). The organic matter storage and dynamics in estuaries are influenced by a large number of factors, depend on sedimentation/burial rates, but are still poorly understood for many estuaries (Ward et al. 2024, but see Savoye et al. 2012 for an example of organic matter origin and dynamics in the Gironde estuary in SW France).

Estuaries are recognised as having stressful conditions for most of the organisms that occur in them, although these conditions are primarily due to natural variability and are intrinsic to these systems. The intensification of human pressures that took place from the mid XIXth century until present, which will be discussed in the next sections, has largely increased the diversity and magnitude of environmental stress in estuaries. Organisms that colonise estuaries are truly tolerant to such a high environmental variability (euryoecious species) and/or adjust their spatial and temporal distribution in estuaries according to their ecological niches and tolerance limits in relation to abiotic factors. One of the factors that generates a gradient with a major impact on estuarine organisms is salinity, which is known to structure spatial distribution of biological assemblages. Quite intuitively, organisms with freshwater affinities will be more abundant in the upstream zones, where salinity is generally lower than 5, while organisms with marine affinities are more likely to occur in the middle zone and closer to the mouth of the estuary. Typically, estuarine organisms, i.e., organisms that develop their entire life cycle within estuaries, occur practically along this gradient throughout the estuarine area (e.g., McLusky and Elliott 2004). This distribution pattern has been known for a long time and has been graphically represented in a diagram attributed to Remane (1934) (Fig. 3).

Primary producers in estuaries comprise different groups, namely phytoplankton, microphytobenthos, saltmarsh or mangrove plants, seagrasses and macroalgae, their relative importance, and their contribution to the overall primary production, being extremely variable according to the system (McLusky and Elliott 2004). Nonetheless, the fraction of estuarine primary production originating from microalgae, either phytoplankton and microphytobenthos, has often a higher biogeochemical and ecological significance compared to other sources, because it is enriched in nitrogen and lipids including essential fatty acids, easily accessible to consumers, and also due to the fact that this production is mostly consumed or decomposed locally (within the estuarine environment, contrarily to other fractions) (Furnas 1990, Duarte and Cebrian 1996, Cloern et al. 2014).

Detritivores and primary consumers are also major components in estuarine biological communities. Detritivorous organisms are usually represented by a small number of species but with a high relevance for the functioning of the system, which has been mainly described for benthic invertebrates and fishes. The diversity, functional guilds and processes involved in the decomposition and detritic chains is poorly known for estuarine microorganisms. Different groups of consumers, taxonomically, but also in relation to their estuarine habitat use patterns, may be recognised. In general, benthic invertebrates are diverse and extremely abundant, feeding up the highest trophic levels in estuaries (e.g., Kalejta 1993, Granadeiro et al. 2006, Rosa et al. 2008). Fishes occurring in estuaries are often classified into different guilds or categories according to their habitat use. There is an extensive literature on this topic, but as an example, Potter et al. (2015) identified the marine (i.e., species that spawn at sea), estuarine (i.e., species with populations in which the individuals complete their life cycles within the estuary), diadromous (i.e., species that migrate between the sea and fresh water) and freshwater categories (i.e., species that spawn in freshwater), each comprising several sub-categories. Estuaries are usually important nursery areas for fish, and their function in the life cycle of such species is well documented (Beck et al. 2001, Nagelkerken et al. 2015, Cabral 2022). Many birds, especially migratory waders, depend on estuarine areas for wintering or just for resting during their long distance journeys, where they occur in large numbers and benefit from the abundant prey resources (e.g., van de Kam et al. 2004).

Estuaries provide a wide range of valuable ecosystem services to human communities (Costanza et al. 1997, Barbier et al. 2011, Jacobs et al. 2013, Boerema and Meire 2017, Booi et al. 2022; Table 1), which may be broadly categorized into provisioning, regulating, and cultural services. These ecosystems have long been used for fisheries, providing food and income for coastal communities. They also supply freshwater, which can be used for drinking, agriculture, and industry. Additionally, estuaries contain valuable mineral and energy resources, such as oil, gas, and sand and gravel deposits. Estuaries also provide key regulating services, playing a crucial role in regulating water quality by filtering and processing nutrients and pollutants, and helping attenuate waves and storm surges, therefore providing protection against coastal flooding and erosion. They also contribute to climate regulation through

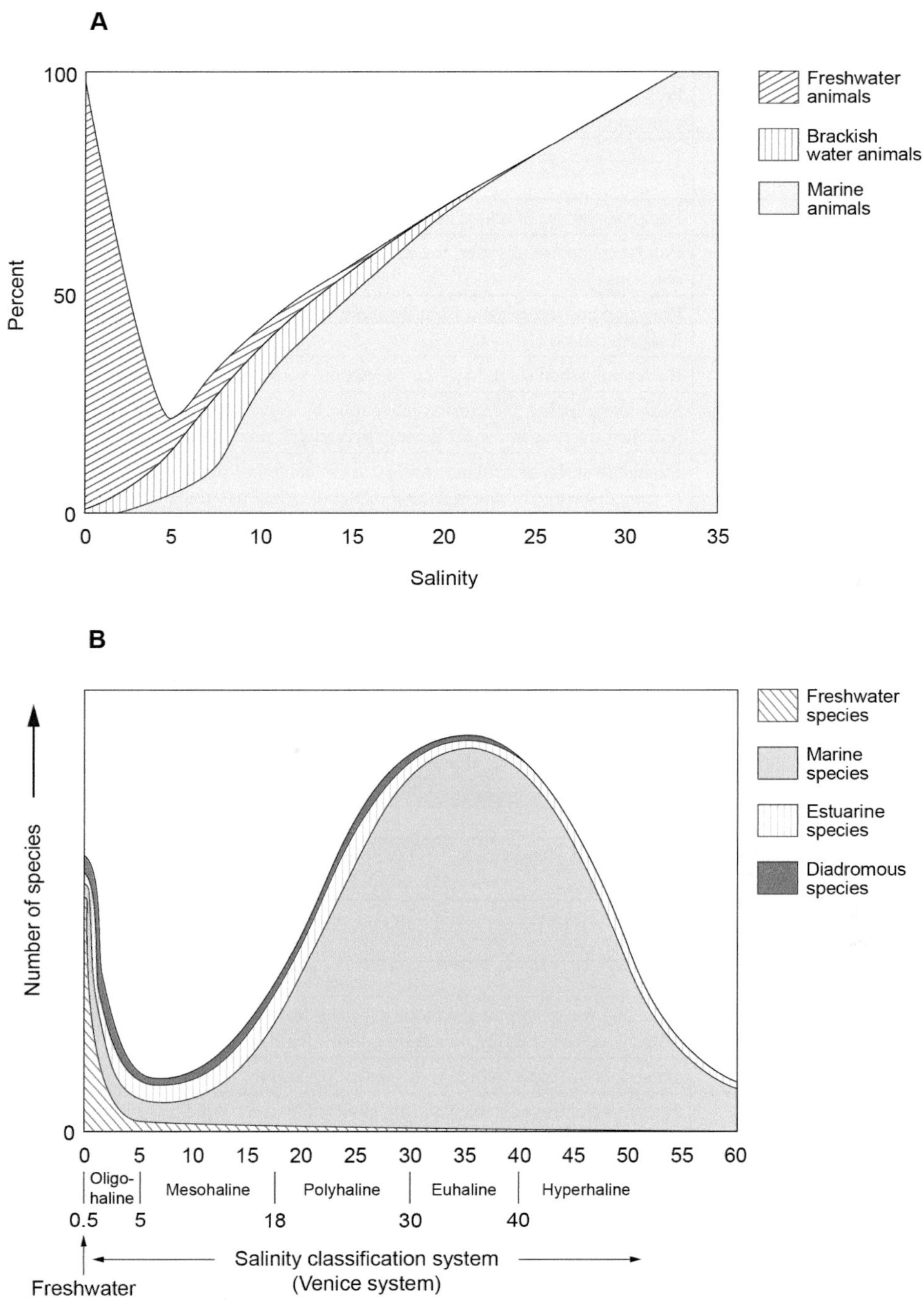

Figure 3. Conceptual diagram of the number or representativeness of species occurring in estuaries along the salinity gradient, as proposed by Remane and Schlieper (1958) (a) and Whitfield et al. (2012) (b).

carbon sequestration in their sediments and vegetation. Cultural services encompass the non-material benefits that people derive from estuaries, which are often sites of significant cultural and historical importance, with many indigenous communities having strong spiritual and traditional connections to these ecosystems. Estuarine systems also provide recreational opportunities, such as fishing, boating, and birdwatching, which can contribute to human well-being and local economies, and are often important areas for nature conservation.

Table 1. Main ecosystem services provided by estuaries (from Boerema and Meire 2017).

Ecosystem services		Description
Provisioning	Food: crops	Presence and use of edible plants, including agricultural production for direct food consumption
	Food: animals (cattle, sheep)	Presence and use of edible animals, including livestock growth and fodder production
	Food: animals (fish)	Presence and use of edible animals, including aquatic species (fish, shellfish)
	Water for industrial use	Provision and use of water, for e.g., cooling water, rinsing water, water for chemical reactions
	Materials (e.g., sand)	Provision and use of sand from dynamic environments which are renewed within a few generations (100 y)
Regulating	Climate regulation	Buffering carbon stock in living vegetation, burial of organic matter in soils
	Erosion and sedimentation regulation	Sediment trapping and erosion prevention by vegetation, effects of bioturbation; sediment trapping and gully erosion by variable water currents and topography
	Flood protection	Storage of storm or extreme spring tides in natural or flood control habitats; storage of peak discharge floods in natural or flood control habitats
	Water current reduction	Reduction of water current by physical features or vegetation
	Wave reduction	Reduction of wave height by physical features or vegetation
	Water quality regulation	Transport of pollutants from source, dilution; binding of N, P in sediments and pelagic food web
	Drainage of river water	Drainage of the catchment by the river
	Water quantity regulation for transportation	Discharge and tidal input for shipping, including water use for canals and docks
Cultural	Aesthetic information	Appreciation of beauty of organisms, landscapes
	Information for cognitive development	Use of organisms, landscapes for (self-) educational purposes
	Inspiration for culture, art and design	Appreciation of organisms, landscapes, as inspiration for culture, art and design
	Opportunities for recreation and tourism	Opportunities and exploitation for recreation & tourism
Additional categories:		
Other	Biodiversity	Total amount of abiotic and biotic diversity at all levels (gene-landscape), regardless of rarity or vulnerability. Vegetation, birds, fish
	Habitat	Presence of suitable habitat for estuarine organisms
	Infrastructure (houses, roads…)	Presence and use of stable and safe environments for building of infrastructure: housing, roads

Anthropogenic pressures and their effects on estuarine structure and functioning

Estuaries are highly productive environments and provide important ecosystem services, such as flood protection and providing recreational space, so they are economically and socially important (Barbier et al. 2011). With more than 60% of Earth's population living in coastal areas, estuarine ecosystems worldwide have been extensively altered by human activities for many decades, thereby inducing profound changes in their structure and functioning (Ray 2006, Elliott and Whitfield 2011, Halpern et al. 2008). Estuaries are being physically transformed and organisms are facing a reduction in the quantity of available habitats. Concurrently, the quality of habitats is also being degraded. Direct habitat loss, reductions in freshwater inputs, pollution, overfishing and the introduction of non-indigenous species are the most significant threats to biological populations in estuaries (Cabral et al. 2019). Although difficult to quantify, these impacts on estuarine ecosystems are at least partly linked to climate change (Robins et al. 2016). All these increasing perturbations occur in a cumulative way, interacting with each other and generating effects that may be additive, antagonistic or synergistic (Cabral et al. 2019), impairing the sustainable use of estuarine resources and services (Whitfield and Elliott 2002, Santos et al. 2017). In the following sections, we present briefly the magnitude and impacts of these human pressures, illustrated with a few case-studies.

Direct habitat loss

Increasing human activity on a large geographical scale over the past 300 years has worsen the vulnerability of estuarine and coastal marine ecosystems to degradation or total loss of habitats (McLusky and Elliott 2004, Lotze et al. 2006), resulting in economic losses associated with, for example, declining recreational and commercial fisheries (Kennish 2002). Estuarine wetlands have been massively destroyed by coastal development (e.g., construction of marinas, widening and dredging of channels for navigation, development of touristic infrastructures, aquaculture, waste disposal, water extraction and diversion, oil and gas production, electric power generation and transmission, expansion of agricultural areas on polders) and defence (e.g., breakwaters, seawalls, jetties, dykes; Airoldi and Beck 2007, Elliott et al. 2016, Wolanski and Elliott 2015). The modification of shorelines and the introduction of large amounts of man-made structures and materials adversely alter the dynamics of the tides, waves and river discharge, changing hydrodynamic regimes in estuaries (Schettini et al. 2017). As a result, sediment erosion and deposition are modified, as well as the hydromorphology of estuaries.

For example, approximately 50% of the original tidal saltmarsh habitat in the USA has been destroyed by human activities (Kennish 2001); this is particularly noticeable around San Francisco where up to 95% of the wetland habitats have been lost since 1950 (Pinet 2000). In the Dutch Wadden Sea, natural saltmarshes were extensively diked and drained (i.e., converted to polders) in the past centuries and almost completely disappeared (Kennish 2002). This diminution in habitat availability and complexity leads to changes in biological communities (Beck et al. 2001, Boesch and Turner 1984, Dorenbosch et al. 2004, Robertson and Duke 1987). For example, eelgrass beds are an important habitat for juvenile cod (Peterson and Boysen-Jensen 1911, Peterson 1918). Tupper and Butilier (1995) and Pihl et al. (2006) recorded a 96% loss of 0+ juvenile cod *Gadus morhua* at sites in the Swedish Skagerrak where eelgrass *Zostera marina* had disappeared. In Buzzards Bay (USA), the loss of *Z. marina* was accompanied by significant declines in fish abundance, biomass, species richness and functional diversity (Hughes et al. 2002). In Western Port Bay (Australia), a 70% loss of seagrass directly induced a 40% loss in total commercial fish catches, particularly in species specifically adapted to seagrass habitat (Jenkins et al. 1993). Lastly, Rochette et al. (2010) used a habitat suitability model based on bathymetry and sediment structure to hindcast the effect of nursery habitat degradation of the Seine estuary on *Solea solea* population. Channel dredging, constructions of dikes and harbour extensions substantially reduced the Seine estuary area and subsequently suitable juvenile fish habitats, such as shallow muddy areas (> 75% decrease). Consequently, the authors suggest that, since 1850, the estuary nursery capacity has decreased by 42% due to habitat loss. This reduction of the nursery capacity (i.e., size and quality) alone led to a loss of nearly 23% (between 8 and 36%) of the total juvenile production as well as a reduction of flatfish growth performances (Gilliers et al. 2006). Loss of estuarine wetlands, loss of connectivity and habitat fragmentation generally compromise the ability of organisms to reach favourable habitats and have negative effects on biological populations (Peterson 2003). This loss of habitat is practically permanent, as it requires massive and large-scale engineering works to reverse (Amorim et al. 2017).

River flow regulation and water abstraction

Freshwater flow has a strong impact on physical, chemical and biological characteristics of estuarine environments (Sklar and Browler 1998). River discharge influences geomorphology, salinity, dissolved oxygen and turbidity in estuaries (Costa et al. 2007). Overall, nutrients supplied by rivers are vital to estuarine productivity (Schlacher and Wooldridge 1996, Livingston 1997). However, more than half the world's large river systems are dammed (Nilsson et al. 2005) and the cumulative impact of dams has led to a 60% decline in discharge of midlatitude rivers since 1950 (Cloern et al. 2016). River discharges are regulated mainly for agriculture, although other significant uses include industries, such as hydroelectric power generation or water supply to urban settlements (Postel et al. 1996, Vasconcelos et al. 2007, Zhang et al. 2012). Growing demand for freshwater has led to excessive abstraction of surface and underground water resources (Whitfield 2005) and the storage capacity of some reservoirs far exceeds the annual runoff from rivers feeding them (Schlacher and Wooldridge 1996).

Freshwater discharge into estuarine environments varies on predictable timescales showing seasonal and annual patterns, to which stochastic variability is commonly added (Gillanders and Kingsford 2002). Typically, discharge directly increases in winter during high rainfall or snow melting periods. This natural variability does not necessarily coincide with the freshwater requirements for human activities, which means that water resources is almost completely controlled by dams, weirs and levee banks in the long term (González-Ortegón and Drake 2012). Small floods are totally removed from the river discharge spectrum and the frequency and duration of large floods are also attenuated (Whitfield 2005). Flood reduction has a deep impact on the physical structure of estuaries, namely morphometry, sedimentary processes, depth profiles, mouth configuration and tidal influence (Cooper 2002). For example, the middle and upper estuarine reaches may be subjected to an excessive mud accumulation because floods no longer flush sediment out of the estuary (Reddering 1988), while the downstream reach of the estuary as well as the coast could be eroded (Cloern et al. 2016). A direct consequence of the removal of extreme events is the disappearance of species that are well adapted to environmental fluctuations, in favour of species that colonise more constant environments (Whitfield 2005). In contrast, water storage during summer periods magnifies the effect of low freshwaters inputs and droughts (López-Moreno et al. 2009, Rodrigues et al. 2019).

One of the most notorious negative impacts of a change in the pattern of freshwater inputs in estuaries is the modification of the salinity gradient. The salinity gradient depends on the balance between the freshwater inputs and the ingress of tides. Estuaries are composed of oligohaline upstream reaches (with low salinity, between 0.5 and 4.9); salinity then increases gradually until the estuary mouth, which is a euhaline reach (with high salinity, between 30 and 39.9) (Whitfield 2015). This salinity gradient directly affects the spatial distribution of organisms (Barletta et al. 2005, Elliott and Hemingway 2002, Vilas et al. 2009). The ability of organisms to cope with changes in salinity, called osmoregulation, is stage- and species-specific (Arevalo et al. 2023). As a result, freshwater species settle in the upstream reaches, while marine species colonise the downstream reaches, connected to sea. Species with a very narrow salinity preferendum are stenohaline species, occurring only in specific estuarine reaches. Species with a broad salinity preferendum, such as estuarine species, can face salinity variations and colonise all the estuary (Whitfield 2015). When freshwater inputs are significantly reduced due to water abstraction in catchment, the estuary becomes tide-dominated, i.e., the tidal range expands and salt penetration rises. This phenomenon of "marinization" has been described extensively in the Gironde estuary (France), where the salinity of the estuary is sometimes higher than the surrounding seawater (David et al. 2005). This situation is all the more pronounced during summer months when precipitation levels are typically low. High salinities result in a loss of freshwater species, decline in estuarine species and the expansion of marine species, from zooplankton (Chaalali et al. 2013) to fishes (Chevillot et al. 2016, Pasquaud et al. 2012).

Reductions in freshwater inputs therefore modify species assemblage and interspecific interactions, including prey-predator relations (Durant et al. 2007). Primary production is directly impacted as phytoplankton biomass is reduced (Gillanders and Kingsford 2002). In a West Australian estuary, 90% of freshwater inputs are concentrated over a short period of 3 winter months. The nutrients supplied by freshwater inputs stimulate primary production and induce phytoplankton blooms (Lucas et al. 1999). Then, this rainy season systematically induces diatom blooms. In dry years, freshwater and nutrient inputs are reduced and there are no significant blooms of phytoplankton (Gillanders and Kingsford 2002). Low freshwater inputs also have deleterious effects on recruitment, growth, movement, mortality and fecundity of zooplankton, including copepods (Chaalali et al. 2013, Devreker et al. 2009, Sautour and Castel 1995). Concerning fish, abundance is negatively correlated with salinity (Mann et al. 2002, Whitfield 2005). Consequently, the food web structure is totally disrupted: the reduction in nutrients, and in particular in terrestrial organic matter, reduces the diversity of prey available to consumers (Liu et al. 2020). This bottom-up effect shortens and simplifies the trophic chain (Costa et al. 2007, Sabo et al. 2018), which has implications in terms of ecosystem resilience in a global change context (Vinagre et al. 2010).

Freshwater inputs contribute greatly to the extension of the estuarine plume, which spreads olfactory cues to trigger the spawning of many species (Strydom et al. 2002) and attracts migrating organisms into estuaries (Costa et al. 2007, O'Connor et al. 2017, Teodósio et al. 2016). Estuarine plume areas are sites of intense biological activity with high densities of phytoplankton, zooplankton, larval fishes and nekton (Grimes and Kingsford 1996). The dynamics of these areas may be deeply affected by changes in water use patterns and water abstraction.

Pollution

Estuaries are generally believed to be exposed to the highest levels of contamination of any marine environment due to their proximity to human settlements and their position directly downstream of agricultural and industrial activities (Kennish 2002, Lotze et al. 2006). Human activities have introduced thousands of substances and/or materials in aquatic environments which, when reaching concentrations above certain tresholds, might present negative effects on biological components and, therefore, become pollutants (Cabral et al. 2019). Pollutants originate from a large number of human activities (e.g., agriculture, industrial, urban and port development, transportation, fisheries and aquaculture, recreation) and may be of a different nature – chemical (e.g., nutrients, biocides, metals, oil, pharmaceuticals), physical (e.g., plastic debris, large hard structures, temperature, radiation, noise, artificial light), or biological (e.g., introduced non-indigenous invasive species). Pollutants incorporate biogeochemical cycles, therefore introducing a high complexity to their quantification and assessment of their impacts. Nutrient load inputs to estuaries, and especially nitrogen (N) and phosphorus (P) inputs, have dramatically increased in the past decades due to rapid intensive agricultural development. Nutrients within estuaries and eutrophication are intimately linked. Eutrophication is characterised by high primary production and an increased activity of heterotrophic bacteria. Nutrient enrichment leads to deep changes in the estuarine and coastal environments: occurrences of dense monospecific phytoplankton blooms, decreased water clarity and loss of macrophytes, proliferation of jellyfish and harmful algal blooms, highly variable oxygen levels (from hypoxia to supersaturated DO according to photosynthetic activity) and loss of a large proportion of the macrobenthos community (Wolanski et al. 2006, Statham 2012, Cloern et al. 2016). Manifestations of nutrient enrichment are evident along the world coastlines and include 10-fold chlorophyll increase in regions of Chesapeake Bay since the 1950s (Harding and Perry 1997), doubling of primary production in Danish coastal waters from the 1950s to the late 1970s (Ærtebjerg Nielsen and Gargas 1984), fourfold expansion of hypoxic zones in the Baltic Sea from 1950 to 1970 (Carstensen et al. 2014) and recurrent massive green tides in coastal waters of the southern Yellow Sea since 2007 (Li et al. 2014).

The occurrence and concentration of metals in estuarine ecosystems has increased drastically in recent years, with industrial and mining activities being the main anthropogenic sources of metallic contamination (Lu et al. 2018). Potentially toxic metals such as copper, zinc, lead, cadmium and mercury typically enter waters through industrial and domestic waste, atmospheric deposition and stormwater runoff. Recreational boating, shipping, fisheries and aquaculture activities can all be sources of metal-based antifoulants (Macleod and Coughanowr 2019). The toxicity of heavy metals in marine sediments and waters depends upon their bioavailability, and this is principally controlled by the geochemistry of sediments (Peng et al. 2004, Gillan et al. 2012, Hoffmann et al. 2012) or waters (Money et al. 2011, Sinoir et al. 2012). One of the greatest problems with metals in the environment is that they are highly persistent, with even some metals such as copper and zinc that cannot be degraded by biotic or abiotic processes. When exposed to metals, organisms need to allocate energy from other functions to face chemical stress (e.g., metabolic cost of detoxification), generally resulting in decreased performances (growth and survival; Handy et al. 1999, Vieira et al. 2009, Kerambrun et al. 2013). Metals generally produce non-viable larvae due to developmental retardation, morphological and functional deformities

(Jezierska et al. 2009, Sfakianakis et al. 2015) or the production of poor-quality offspring (Barbee et al. 2014). Behaviour of organisms can also be affected due to damage to the chemosensory organs, including olfactory bulbs (Carreau and Pyle 2005, Bettini et al. 2006), neuromast cells of the lateral line (Linbo et al. 2006, Froehlicher et al. 2009), and the eye components of fish (Bodammer 1985, Osman et al. 2007, Chow et al. 2009). While most studies of metal toxicity on aquatic organisms have looked at effects of single metals, those that have used metal mixtures have found similar synergistic effects on fish (Vosylienė et al. 2003, Zhu et al. 2011). More globally, ecosystem effects of metals, such as shifts in community structure and associated changes in functional groups, have also been observed (Morrisey et al. 1996) as well as bioaccumulation and magnification in estuarine food webs (Fonseca et al. 2019). Organisms accumulate contaminants by absorbing them from the environment (bioaccumulation) or by accumulating contaminants assimilated by their prey (biomagnification). Potential for bioaccumulation in organisms and biomagnification in food webs differ depending on the investigated contaminants, environmental contexts, and physiological characteristics of species (Bodiguel et al. 2009, Connolly and Glaser 2002, Gobas 1993). For example, Chen et al. (2016) have shown that mercury (Hg), methylmercury (MeHg) and selenium (Se) accumulate particularly well in organisms in the New England estuary (USA).

Persistent Organic Pollutents (POPs) are synthetic chemicals that have been widely used since the latter part of the 20th century (Pintado-Herrera et al. 2017). Most of them present a high structural stability, giving them a high bioaccumulation potential and a long-range transport (Barletta et al. 2019). These pollutants originate from a wide diversity of human activities, namely the ones related to pest control in agriculture, pharmaceuticals, detergents, flame retardants, combustion by-products, among many others (Muir and Howard 2006). Polycyclic aromatic hydrocarbons (PAH), polychlorinated biphenyls (PCB), polybromodiphenyl ethers (PBDE), organochlorine pesticides (OP), hexabromocyclododecanes (HBCD), Dechlorane Plus (DP) and perfluoralkyl substances (PFAS) are some examples of these pollutants (Han and Currell 2017). Many of these compounds are highly toxic to biota (from phytoplankton, crustaceans, molluscs to fish). Emerging pollutants, such as pharmaceuticals and other personal care compounds (PPCP), have also gained attention in the last decade. Personal care products are used for improvement of the quality of life and include products such as cosmetics, UV blocker agents (sunscreens) or fragrances (Daughton 2016). PPCP residues are continuously released into the aquatic environment, leading to chronic exposure even for relatively degradable compounds. Exposure of organisms to these xenobiotics induce bioaccumulation (Barletta et al. 2019). Early stages of fish die or exhibit growth reductions, abnormal behaviours, yolk utilization, brain and liver development, neurological and cellular dysfunctions, hormonal disturbances and cardiovascular and skeletal deformities when exposed to these compounds (Matthiessen and Law 2002, Ostrach et al. 2008, Faria et al. 2020, Kibria et al. 2021). Multigenerational (i.e., on the F1 generation, exposed as germ cells in the F0 parents) and transgenerational effects (i.e., on the F2 generation) have been observed (Schwindt et al. 2014, Decourten et al. 2020) and could have major consequences on long-term population viability.

Microplastics are also emerging contaminants for which a large number of studies have been recently published (Guo and Wang 2019, Rodrigues et al. 2019, 2020). Plastic production has increased drastically along with their accumulation and contamination in the environment (Lima et al. 2014, Correa-Herrera et al. 2017). The latest estimations pointed out between 1.1 and 8.8 MT of mismanaged plastic waste being generated annually by land-based human activities at each country (Jambeck et al. 2015). Characteristics such as durability, buoyancy, and resistance allow plastic being extremely durable and long-lasting. Over time, plastics fragment into smaller particles and become microplastics (MPs; defined as particles between 1 μm and 5 mm; Arthur et al. 2009). Due to their size, MPs can be easily mistaken for food, and their ingestion may cause false sense of satiety (Eerkes-Medrano et al. 2015) or physical injuries such as damage in the gastrointestinal tract with inflammatory responses (Von Moos et al. 2012, Eerkes-Medrano et al. 2015, Rezania et al. 2018). Also, MPs have potential capacity to adsorb metals or pollutants from the environment and when ingested, may eventually increase the risk of toxic effects on organisms (Fendall and Sewell 2009, Cole et al. 2011, Frias et al. 2010, 2014). The impact of these pollutions (i.e., eutrophication, metals, POPs, or emerging pollutants) considered independently is not yet clearly identified. Consequently, the synergetic actions of several stressors are understudied (Cabral et al. 2019, Kibria et al. 2021).

Fisheries and aquaculture

Fisheries, and especially over exploitation of fishery resources, is the oldest pervasive human disturbance to coastal and estuarine ecosystems (Cloern et al. 2016). Estuarine fisheries impact both finfish and shellfish (Wilson 2002). Finfish fisheries tend to be seasonal, depending on the migration and ingress of species into estuaries, while shellfish exploitation (both crustaceans and molluscs) can be constant all year round. Fish and many of the invertebrates of estuaries and coastal waters are a valuable and usually readily accessible human food resource (Blaber et al. 2000). These resources constitute the main source of food for people living along the coast and have led to the development of a great variety of fisheries (Blaber 1997). In addition, recreational fishing is also an essential component of estuarine fisheries.

The main effects of fishing on target organisms are decreases in their abundance, changes in age structure, and size composition and changes in species composition and diversity. The largest fish and the oldest age classes are generally targeted by fisheries, which tends to concern long-lived and slow growing species (K-strategy species; Breitburg et al. 2009, Ferguson et al. 2013). Intensive fishing pressure leads to a drastic drop in abundance of K-strategy species worldwide (Bauchbaum and Powell 2008, Ecoutin et al. 2010, Lotze et al. 2006, Musick et al. 2000). Species with an extended lifespan reach larger maximum sizes but they also mature later. Among a diverse taxonomic range of species, older or larger females have higher fecundity both on absolute (total eggs per batch, number of batches per season) and relative (eggs g^{-1} female body weight) scales, produce higher quality larvae (e.g., egg size, lipid content, larval performance) and adopt different spawning phenology (greater spread of reproduction across the season for batch spawners or different timing of spawning for total spawners) (Hixon et al. 2014). These "bet-hedging" strategies allow older and larger fish to integrate over environmental fluctuations and increase their fitness by maximising the survival of the offspring (Hutchings and Myers 1993, Berkeley et al. 2004, Anderson et al. 2008). The removal of large/old, highly fecund individuals from the population can reduce the capacity to produce strong year classes when environmental conditions are favourable (Beamish et al. 2006, Jennings and Kaiser 1998, Laë 1995, Ferguson et al. 2013), the productivity and the sustainability of the species. It is tricky to disentangle the effect of fishing from other pressures (Cabral et al. 2001). Also, biological data does not necessarily exist prior to the implementation of fisheries, some of which have been established for centuries (Blaber et al. 2000, Cloern et al. 2016). However, truncation of the age structures of a species likely reflects removal of older, larger individuals by fishing (Hilborn and Walters 1992, Planque et al. 2010, Walsh et al. 2010).

Fishing tends to favour fast-growing, short-lived species (r-strategy species; Llopiz et al. 2014). These r-strategy species are characterized by small body size, early maturation, small eggs and continuous spawning (Winemiller and Rose 1992). These ecological traits enable them to be highly resilient due to high turnover rates and to persist in altered environments (Jennings and Kaiser 1998). Additionally, the larger species usually predate the smaller ones, thus regulating their abundance. If the larger species are harvested, the smaller ones can proliferate (Jennings and Kaiser 1998, Pauly et al. 1998). The overexploitation of certain species has led to the complete depletion of some stocks. This is the case of cod (*Gadus morhua*) in the Baltic Sea (Casini et al. 2008), striped bass (*Morone saxatilis*) in US Atlantic coastal waters (Richards and Rago 1999), and anchovy kilka (*Clupeonella engrauliformis*) in the Caspian Sea (Daskalov and Mamedov 2007). Management measures that allow fishing and stock renewal at the same time are therefore invaluable tools for ensuring the sustainability of this human activity (Elliott et al. 2022). There are also examples of trophic cascades due to the collapse of apical predators that have been overexploited (Casini et al. 2008). Loss of cod from the Baltic Sea in the early 1980s reduced predation mortality on spratt (*Sprattus sprattus*), allowing its population to grow while biomass of its zooplankton prey fell and phytoplankton biomass grew (Cloern et al. 2016).

Fishing also directly affects the abundance of non-target species as well as non-commercial sizes of target species through accidental catches. For example, catches of marine mammals (e.g., dolphins, porpoises, dugongs, and manatees) or turtles are regularly reported (Blaber et al. 2000). In fisheries using small mesh sizes, such as in shrimp fisheries, small fish species and young stages (larvae or juveniles) of large species represent the bulk of discards (Liggins and Kennelly 1996, Sobrino et al. 2005). Basically, the composition of the discards is representative of the estuarine community both spatially and temporally, i.e., all the species present in estuaries are likely to be caught. This aspect is very detrimental to the nursery function that estuaries can have for young stages of fish, especially as young individuals caught accidentally die inside the net because of water pressure (Sobrino et al. 2005). The survival of any fish returned after capture varies with species. While gadids, smelt and small flatfish rarely survive release after capture, the larger flatfish has a good survival rate (Robin 1992). Other indirect effects of fishing can impact estuarine organisms, such as the physical disturbance of the habitat by dredging or trawling, or the decrease of the water quality by releasing wastewater and organic pollutants (Blaber et al. 2000).

Aquaculture is the farming or cultivation of aquatic organisms, in particular fishes, molluscs, crustaceans and aquatic plants. Concerning fish production, aquaculture provided 43% of the world's total fish supply in 2012 (FAO 2014) with the goal of reaching 60% by 2030, corresponding to 181 million tonnes (Lem et al. 2014). Mariculture is predominately practiced in coastal waters, including estuaries where conditions are calmer and shore-based infrastructures and markets are more accessible. As a result, mariculture will increase considerably over the coming years to meet the production goals. Marine-based aquaculture that produce finfish favours open cages which allow dissolved and particulate nutrient enriched waste to deposit in the surrounding water (Hadley et al. 2016). It may result in increased primary production and lead to algal blooms (Buschmann et al. 2009, Bonsdorff et al. 1997) and in worst case, result in eutrophication (Anderson et al. 2002, Carmona et al. 2006). Similarly, shrimp aquaculture has several detrimental environmental impacts, including destruction of mangrove forests and poor water quality in adjacent estuarine waters. Another impact is the capture of wild fish species as post-larvae to supply the ponds, which once again reduces the nursery capacity of estuaries (Blaber et al. 2000).

Intertidal oyster cultivation is one of the most important aquaculture industries globally (Forrest et al. 2009). The most widely described environmental effects associated with this activity are changes to sediments and associated infaunal assemblages beneath cultivation areas (Forrest and Creese 2006, Dubois et al. 2007, Bouchet and Sauriau 2008). The set up of oyster reefs (as well as mussel beds) creates a complex habitat that can support a considerably greater biomass, richness and density of organisms than adjacent natural habitats (Dealteris et al. 2004, Murray et al. 2007). These news habitats can provide a range of ecosystem services including an enhanced food supply, a refuge from predation, a settlement surface, and protection from physical (e.g., water movement) and physiological (e.g., dessication) stress (McKindsey et al. 2006). However, they can be detrimental to other types of habitats that are of greater ecological value. In Willapa Bay or in San Diego Bay (USA), the density and growth of eelgrass (*Zostera marina*) have been drastically reduced due to high oyster cover or mussel beds (Tallis et al. 2009). Nutrient cycles in the surroundings areas are also deeply affected. Suspended particulate matter (especially phytoplankton) are removed, transformed into biomass and released as dissolved waste products (Dame and Dankers 1988, Dame et al. 1989). There are also noticeable qualitative changes, for instance in the carbon, nitrogen and phosphorus ratios (Wilson 2002) and seabed sediments beneath oyster cultures can become anoxic (Forrest and Creese 2006). Changes in physico-chemical characteristics beneath oyster cultures can lead to a displacement of large-bodied macrofauna (e.g., heart urchins, brittle stars, large bivalves) and the proliferation of small-bodied disturbance-tolerant opportunistic species (e.g., capitellid polychaetes and other marine worms) (Forrest et al. 2009). The introduction of pests (fouling pests, toxic/noxious microalgae, non-indigenous species, disease) related to aquaculture can have implication for the whole estuarine ecosystem, including fish, seabirds and marine mammals (Forrest et al. 2009).

Non-indigenous invasive species

The introduction of non-indigenous invasive species (NIS) has increased dramatically in recent decades, causing significant changes in ecosystem structure and functioning (Cabral et al. 2020, Novais et al. 2015, Sousa et al. 2008). Anthropogenic activities are responsible for the translocation of hundreds of species beyond their natural range and dispersal limits. Only a small fraction establishes successfully and permanently in their new environment. An even smaller fraction reaches high population densities and/or successfully disperses over wider ranges (Sakai et al. 2001, Colautti and MacIsaac 2004). Additionally, it generally takes several years for the few individuals introduced to become established and proliferate.

Estuaries support a broad spectrum of human activities that modify thermal environment and habitat characteristics, which can improve opportunities to invade (Ruiz et al. 1999, 2000, 2009). Additionally, habitats or communities submitted to severe and/or frequent disturbances such as estuaries, which are highly dynamic habitats, are often more susceptible to invasion (Simberloff 1986). Central and upper estuarine reaches generally support low species diversity in direct relation with the salinity gradient and the associated osmotic stress (Remane 1934). However, communities with reduced biological diversity are more susceptible to invasions when compared to those with high biodiversity (Essink and Dekker 2002, Afonso et al. 2020). NIS generally originate from regions with high biodiversity (Tsiamis et al. 2018) and are, therefore, well adapted to strong competition. Competition is less intense when species richness is low and NIS can invade more easily a wide range of unoccupied niches (the "enemy-release-hypothesis"; Bax et al. 2001, Brockerhoff and McLay 2011, Paavola et al. 2005). Pressures already exerted on estuarine systems weaken the native species, making them less resistant to the introduction of NIS (Williams and Grosholz 2008).

Commercial trade by ship can facilitate the introduction of NIS in two ways. First, their hulls constitute a habitat for fouling communities of sessile species, which are transported between ports and may be released into new environments (Ruiz et al. 1997, Gollasch 2002). Second, the exchange of huge amounts of ballast water holds the potential for all species with planktonic stages or swimming lifestyle to be taken up in one and be released in another port (Geburzi and McCarthy 2018). In the past 20–30 years, invasion rates have increased due to changes in ballast water transport and the technological advances made (Carlton and Geller 1993). More ships with larger volumes of ballast water arrive rapidly from more regions and increase the overall abundance, density and survival of species transferred in ballast water (Ruiz et al. 1997). Aquaculture is another important vector for NIS and the rapid expansion of this industry in recent years has led to an increase in the number of NIS introduced (Naylor et al. 2001). For example, aquaculture directly caused the invasion of the Pacific oyster *Magallana gigas* along the southeastern coast of the European North Sea and the northern coasts and estuaries of Patagonia. Larval planktonic stage, associated with favourable environmental conditions (e.g., particularly warm summers), greatly increased the success of the species invasion (Diederich et al. 2005, Smaal et al. 2009, Croce and Parodi 2012). Recently, microplastics have emerged as a vector for transporting a variety of species from different taxonomic groups (including bryozoans, barnacles and mollusks) over large distances (Barnes and Milner 2005, Gregory 2009, Gil and Pfaller 2016). The release of ornamental and aquarium-kept organisms or organisms associated with aquarium decorations (especially epibionts on seaweeds and live rock used, such as macro- and microalgae, aquatic plants, (hemi)sessile cnidarians, crustaceans, polychaetes or mollusks) can be a significant pathway for the introduction of NIS (Padilla and Williams 2004). Many natural barriers controlling dispersal, such as distance or currents, are removed by human activities, increasing the number of species likely to invade an ecosystem and the number of source regions from which invasions can occur (Ruiz et al. 1997).

Some ecological traits facilitate the establishment of NIS. Invaders are often characterized by a r-strategy with early maturity, short generation time, high fecundity, rapid growth rates (Sakai et al. 2001). For example, these traits enabled the Asian clam *Corbicula fluminea* to recover rapidly from heatwaves in the River Minho estuary (Portugal), while native bivalve species disappeared (Sousa et al. 2008). An extended spawning season also enables offspring to hatch when environmental conditions are favourable (Anderson and Epifanio 2010, van den Brink et al. 2013). Other traits are an advantage for colonising new environments, such as a planktonic larval phase that enables organisms to travel long distances, or omnivory that confer plasticity to modify their diet according to the resources available (Geburzi and McCarthy 2018).

Newly arrived species develop interactions with both native and other non-native species in their new environment. Zoobenthos comprise the predominant group of NIS in marine ecosystems (including estuaries) representing approximately 57% of the introduced species (Streftaris et al. 2005). Bivalves are one of the most invasive groups because certain species rapidly attain very high densities and account for most of the benthic faunal biomass (Carlton et al. 1990, Sousa et al. 2009). Bivalves possess key attributes such as the production of durable shells, bioturbating and filter-feeding behaviour that control the availability of resources to other species (Sousa et al. 2009, Novais et al. 2015). They can generate positive interactions with other species as the structure provided by their shells might serve as a refuge from biotic and abiotic stress, predation and competition (Gutiérrez et al. 2003, Sousa et al. 2009, Novais et al. 2015). Similarly, the invasion of the submerged plant, *Hydrilla verticillate*, in the Chesapeake region (USA) increased populations of largemouth bass (*Micropterus salmoides*), and other predatory fish, by providing a refuge for high densities of early stages (Killgore et al. 1989, Phelps 1994, Serafy et al. 1994). However, the NIS are strong competitors for food and shelter and have generally negative effects on native species occupying the same niche (Griffen et al. 2011). The Manila clam, *Ruditapes philippinarum*, was intentionally introduced for massive aquaculture production and became established in several regions worldwide including the Pacific coast of North America and along the shores of Europe (Jensen et al. 2004, Melià et al. 2004, Melià and Gatto 2005, Moura et al. 2017). The colonisation of several European estuaries by *R. philippinarum* coincided with a sharp and global decrease in abundance of native bivalves, possibly due to interspecific competition for the same resources (Chainho 2014, Chainho et al. 2015, Ramajal et al. 2016). Several studies also reported direct predation pressure by NIS on native species (Ruiz et al. 1999). If the invasive species is introduced at the upper levels of the trophic web, "top-down" control of the food web is expected to promote changes in the abundance and biomass of lower trophic levels (Pace et al. 1999). In contrast, changes in either primary producers or the input of limiting nutrients by invasive species affect the total energy available to the ecosystem with "bottom-up" control (Thomsen et al. 2014). Finally, invasions may alter genetic diversity both within and among populations, resulting from increased gene flow and differential success of particular genotypes (Ruiz et al. 1997).

Climate change

As a major component of climate change, sea surface temperature has been increasing globally since the end of the 19th century, at an annual rate of about 0.11°C from 1970 onwards (IPCC 2013, Robins et al. 2016, Hallett et al. 2018). At the same time, river water temperature also increases significantly (Van Vliet et al. 2013, Arevalo et al. 2020). These two phenomena lead to a warming of estuaries and influence estuarine fish community structure and composition (de Souza and Dos Santos 2023). Studies conducted in the Bay of Biscay and in several temperate European estuaries detected a decrease in the occurrence/abundance of species with northern affinities (deborealization) and an increase of southern ones (tropicalization; Poulard and Blanchard 2005, Désaunay et al. 2006, Hermant et al. 2010, Cabral et al. 2001, Henderson 2007, Nicolas et al. 2011, McLean et al. 2021). A shift in the species distribution area generally results in a shift in spawning grounds and timing, locations, and environments in which fish eggs and larvae occur (Hsieh et al. 2009, Shoji et al. 2011). These phenological shifts potentially caused by climate change may threaten the synchronisation of peak production of offspring and their main prey in both time and space (the "Match-Mismatch Hypothesis"; Cushing 1990). Global warming also has indirect effects on ecological processes such as oxygen loss, amplifying hypoxia episodes (Claret et al. 2018, Levin 2018, Schmidtko et al. 2017). Among the most relevant deoxygenation drivers, warming reduces ventilation of deeper waters due to stronger stratification, and decreases oxygen solubility (Keeling et al. 2010) while raising microbial metabolic rates and oxygen consumption (Deutsch et al. 2011).

Models from the Intergovernmental Panel on Climate Change (IPCC) estimate that sea levels will rise by around 40–75 cm by 2100 (IPCC 2013). As a direct consequence, the rise in sea level could impact estuarine tidal range. Along the coast of the Bay of Biscay, sea levels have increased by up to 3 cm decade^{-1} since 1993 (Chust et al. 2022).

Extreme weather events (heatwaves, storms, hurricanes, floods and droughts) are expected to become more frequent, intense and unpredictable (IPCC 2013). Forecasts for these events vary greatly depending on the geographical area and the season. For example, precipitation levels are predicted to increase significantly in northern Europe, while they are expected to decrease significantly in southern Europe and the Mediterranean region, regardless of the season in both cases (Blöschl et al. 2017, 2019). Results between these two edges are more contrasted. Climate change has already led to a modification in annual river discharge (e.g., in large French rivers; Arevalo et al. 2020) as well as in the supply of freshwaters to estuaries, and this trend is expected to intensify in the future (Dayon et al. 2018). The cumulative effects of climate changes (e.g., precipitation deficits combined with intense evaporation) and human activities, especially at river basin scale (land use changes, dams, among many others), have caused drastic changes in natural river flow regimes in the short and long term (Palmer et al. 2008, Van Vliet et al. 2013, Wada et al. 2013). For these reasons, river discharge has decreased, leading to an increase in seawater intrusion and to the "marinisation" of estuaries (Martinho et al. 2007). As a result, Pasquaud et al. (2012) observed a predominance of marine species in the fish assemblage of the Gironde estuary (France) in years of severe drought. The decrease in river discharge could be responsible for the lower concentration of estuarine chemical cues reaching coastal waters, and thus limit estuarine ingress of marine fish larvae and juveniles, e.g., *Solea solea*, *Dicentrarchus labrax* or *Engraulis encrasicolus* (Le Pape et al. 2003, Darnaude et al. 2004, Vinagre et al. 2007, Martinho et al. 2009).

Marine Heat Waves (MHW) are discrete, prolonged, anomalously warm water events affecting marine ecosystems, including coastal and estuarine systems (Hobday et al. 2016). MHWs have become longer, more frequent and intense over the past century and this trend is projected to increase with future climate change (Fragkopoulou et al. 2023, Laufkötter et al. 2020, Oliver et al. 2018). Biological impacts related to MHW have been thoroughly reported worldwide, including algal blooms, coral bleaching, mass mortality and re-structuring of entire ecosystems (Izquierdo et al. 2022, Pinsky et al. 2020, Ummenhofer and Meehl 2017, Wernberg et al. 2016). Beyond the immediate effect, MHW lead to serious, long-lasting impacts on the structure and function of marine ecosystems (O'Leary et al. 2017, Smale et al. 2019). However, the effects of MHW could be difficult to disentangle from natural and sampling variability, particularly on fish community, which limit our current understanding of their consequences (Fredston et al. 2023). By 2050, the intensification of the occurrence and magnitude of MHW is expected to reduce fish biomass and accelerate shifts in biogeography (Cheung and Frolicher 2020). These changes are expected to be at least four times faster and bigger in magnitude than to average longer-term trends throughout the 21st century. Consequently, MHW will be a major challenge for fisheries and their management in the near future.

The ocean absorbs approximately 30% of the carbon dioxide emitted from human activities (Sabine et al. 2004). An increasing concentration of CO_2 in the atmosphere leads to more CO_2 being absorbed into the ocean, which lowers seawater pH and reduces the availability of carbonate ions (Doney et al. 2009, Feely et al. 2009). Along the Australian coastline including 166 estuaries, the water in estuaries is acidifying by 0.09 pH units year^{-1} (Scanes et al. 2020). Elevated CO_2 affects the activity of neurotransmitters (Nilsson et al. 2012, Chivers et al. 2014, Hamilton et al. 2014), which adversely impacts olfactory (Munday et al. 2009a, 2014, Dixson et al. 2010) and auditory preferences (Simpson et al. 2011), visual reactivity (Chung et al. 2014), behavioural lateralization (Domenici et al. 2012), activity levels (Munday et al. 2010, 2013, 2014, Pimentel et al. 2014), and learning (Ferrari et al. 2012) of nekton. These modifications to the sensory system alter the behaviour of fishes (Briffa et al. 2012, Leduc et al. 2013), including decreased predator avoidance (Nagelkerken et al. 2016), decreased response to food stimuli (Caprio et al. 2014, Pistevos et al. 2015), unsuitable settlement behaviour (Munday et al. 2009b, Rossi et al. 2015), and changes in sexual selection preferences (Leduc et al. 2013).

Prospective trajectories of estuaries under global changes

Human pressures negatively affect the productivity and functionality of estuaries (Adams et al. 2020), and are magnified when acting synergistically to reduce resistance and resilience of ecosystems, sustainable use of estuarine resources and, ultimately, ecosystem goods and services (Defeo and Elliott 2021, Kennish 2022). The causes and consequences of the aforementioned endogenic pressures (i.e., pressures occurring inside the estuarine area) need to be managed directly at the estuarine scale. However, estuaries are also subjected to exogenic pressures acting at larger spatial scale, i.e., at the upstream catchment scale (e.g., water abstraction), or at regional and global scales (e.g., climate change) (Elliott 2011, Elliott et al. 2019, Lonsdale et al. 2022). Although the cause may be located outside the estuary, the consequences of both endogenic and exogenic pressures must still be managed within the estuary, but requires coordination with larger-scale management efforts (regional, international). To do so, it is essential to identify the various pressures acting at both local and global scales, quantify their

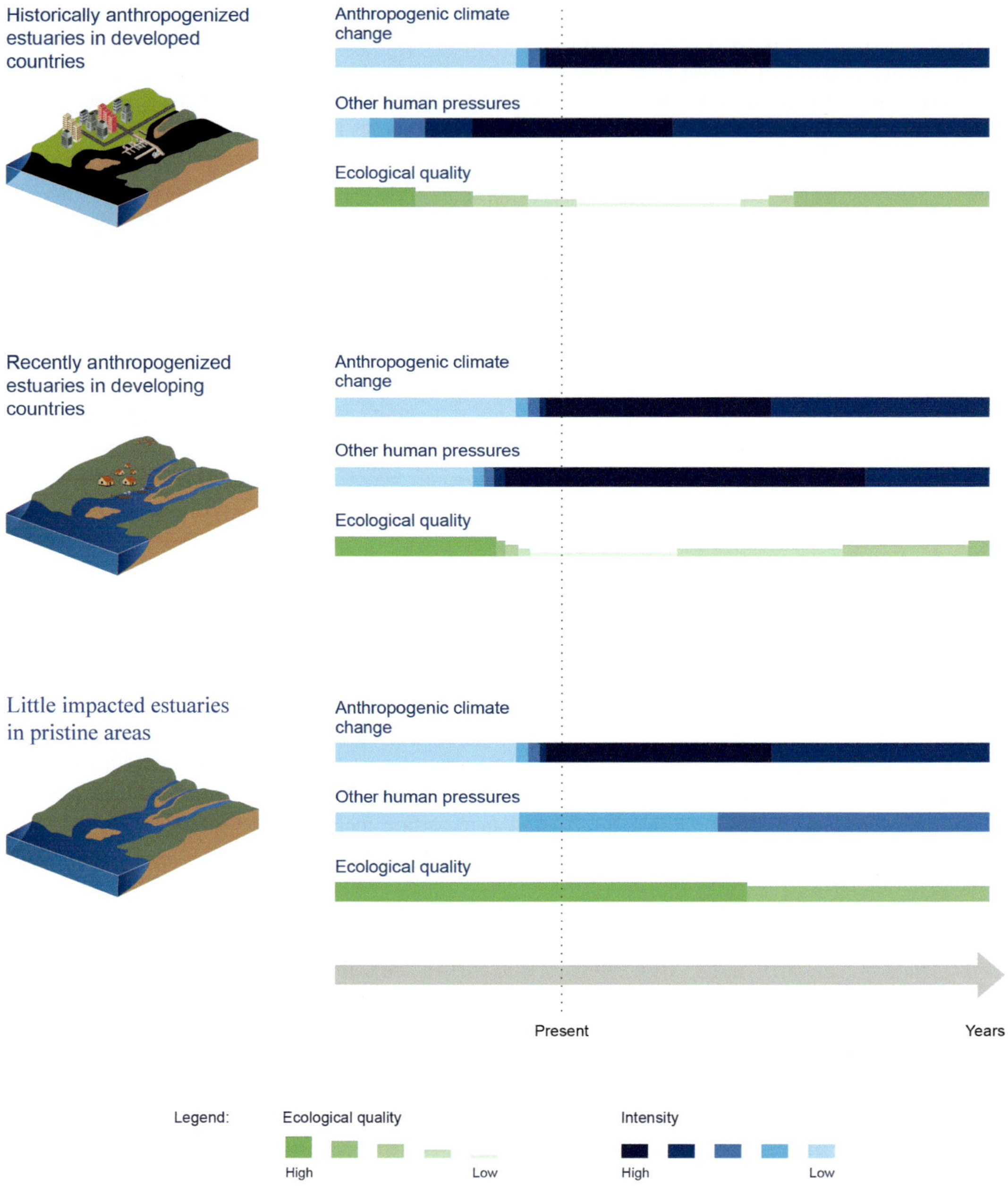

Figure 4. Archetypes of different groups of estuaries taking into account human pressures and their ecological quality status throughout.

singular and combined effects on multiple processes contributing to estuarine structure and functioning, and coordinate the mobilization of all the parties involved (scientists, managers, politicians, society). Those steps are needed to apply effective, suitable, and sustainable management measures and secure the valuable ecosystem capital of estuaries, before it is lost to society (Borja et al. 2009, 2010, 2016a, b). It is also important to evaluate scenarios of estuarine response to management actions undertaken in order to be able to anticipate estuarine recovery trajectories and adapt management schemes accordingly. However, predictive scenarios of estuarine recovery are difficult to establish, first because of the complex and multiple processes at play as highlighted earlier, and second because of the variability in estuarine responses expected from their very diverse contexts in terms of types and severity of local pressures, historical and socio-economic frameworks (governance, management philosophies, political stability, etc.) and climatic risks (as all regions in the world are not affected similarly by climate change). Therefore, we propose herein an approach based on feedbacks from multiple case studies informed in the literature, that we summarize using three typologies of estuarine responses to local and global pressures and associated management actions: (1) estuaries in developed countries that were severely degraded by local human pressures over historical times, where those pressures are currently being tackled thanks to improvements in environmental policies and regulation, but still suffer from direct or indirect impacts of global pressures (examples from North America, Europe, Australia); (2) estuaries in developing countries that are suffering more recent local human pressures and decrease in ecological quality and concomitantly with higher pressure of climate change (examples from South America, China, South Africa); (3) estuaries in less occupied, more pristine, areas, that are also changing but mostly due to global pressures such as climate change (examples from Australia and the Arctic) (Fig. 4).

This synthesis therefore seeks at highlighting how the respective influences of local vs. global stressors are handled differently throughout the world with variable effects on estuarine recovery trajectories. The rise of ecological restoration projects for improving estuarine ecosystem health is also highlighted.

Historically anthropogenized estuaries in developed countries

Several estuaries in developed countries (e.g., Western Europe, North America, Australia) typically have accumulated a large number and high magnitude of human pressures over the past centuries, but progresses in environmental policies and regulations try to invert a decreasing quality trend, through the implementation of well-developed management schemes with restoration of some of the natural features (Day et al. 2019). In addition to management policies, ecological restoration of estuarine habitats indeed quickly developed in those areas, stimulated by the United Nations 2021–2030 decade on ecosystem restoration. In fact, North America, Europe, and Australia have been involved in estuarine and coastal marine ecosystem management since the 1970s and have begun to introduce a series of environmental policies since then. Kennish (2022) provided a detailed timeline of the legislation put in place. Examples of relevant federal laws in the USA are the National Environmental Policy Act (1970), Clean Air Act (1970) and its amendments (1977 and 1990), Clean Water Act (1972) and its amendments (1977 and 1987), Coastal Zone Management Act (1973), and Endangered Species Act (1973). In the European Union (EU), directives of significance include the Habitats and Species Directive (1992), Water Framework Directive (WFD 2000), Strategic Environmental Assessment Directive (2001), Marine Strategy Framework Directive (2008), Birds Directive (2009), Single-Use-Plastic Directive (2019), Maritime Spatial Planning Directive (2014), and the very recent European law on the restoration of nature (2024). Also worth mentioning is the NATURA 2000 network of EU protected areas designated under the Birds and Habitats Directives. Similar legislations were also enacted in Australia (Environmental Protection and Biodiversity Conservation Act 1999).

The idea behind these laws was initially focused more on water quality and pollution control, but gradually broadened to protect biotic communities and habitats from adverse effects of human activities and natural events (Ballinger and Stojanovic 2010). For example, on the Severn Estuary (United Kingdom), the introduction of legislation, and in particular the Water Framework Directive (2000/60), has led to an improvement in water quality (Ballinger and Stojanovic 2010). Concentrations of heavy metals still exceed environmental standards but have declined over the last few decades. Improving wastewater treatment processes also contributed to improve the water quality in the highly impacted Seine system in France (Azimi and Rocher 2016). The efficiency of implementing such regulations is however competing with climate change impacts, making the improvement of water quality in estuaries challenging. For example, on the east coast of the United-States, despite the implementation of Total Maximum Daily Loads (TMDL) of nutrients arising from growing urban or agricultural activities in the watersheds together with stream restoration efforts, the overall eutrophication and hypoxia exposure risks of the Chesapeake Bay estuary remains concerning (Williams et al. 2017, Bhatt et al. 2023) (Fig. 5). On

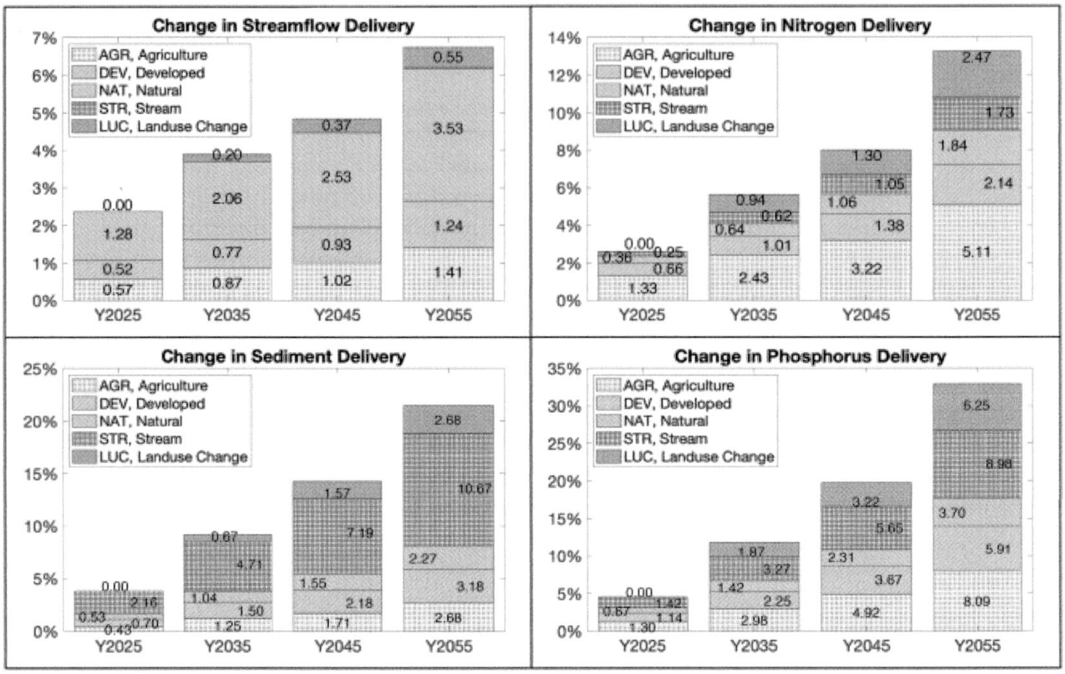

Figure 5. Simulated changes in the delivery of streamflow, nutrients, and sediment loads to the Chesapeake Bay for 2025, 2035, 2045, and 2055 climate change, population growth, and land use change (LUC) scenarios. Plots show that change in flow and loads due to climate change from AGR, Agriculture (dots), DEV, Developed (upward stripes), NAT, Natural (Downward stripes), STR, Streams (grid), and LUC is the compounded effect of population growth, land use, and climate change (diamond grid) (from Bhatt et al. 2023).

the West coast, despite a constant interest from multiple management parties as soon as the 1970s (Ackerly et al. 2012, Eichenberg 2013), the San Francisco Bay estuary remains subjected to pollution (Davis et al. 2004, Greenfield et al. 2005). Another notable effect of human activities in this area is the dramatic shrink of estuarine wetlands, with, for example in Suisun Bay, 5488 hectares of wetlands remaining in 1988 among the 26,449 ha historically present (circa 1800), among which only 850 ha are subjected to specific restoration measures (Callaway et al. 2011). All those endogenic pressures act concurrently with climate change (Stern et al. 2020). Fish populations in the bay also declined in recent decades, including populations of native and non-native but long-established species supporting historical fisheries (Stompe et al. 2020).

In Europe, the WFD directive has stimulated a lot of research on methods to assess the ecological quality of estuaries, and multiple indicators are used to inform management with the objective of maintaining or improving the status of the monitored water masses, depending on evaluation results. In case of a bad evaluation, member states are obliged to put into place some management or regulation efforts in order to inverse the tendency and recover good ecological status. Despite those efforts, French estuaries remain in a concerning situation although some estuaries cope better than others. For example, the Seine and Gironde estuaries are still considered as highly stressed through water quality issues and intense recalibration of physical habitats for navigation, associated with sediment dynamics, which is itself influenced by climate change (Dauvin et al. 2009). The Loire estuary is considered less impacted, possibly due to a larger and less urbanized watershed, more efforts towards restoration of the river and estuarine habitats, morphology, especially longitudinal and lateral connectivity, leading to a larger intertidal area availability (Selleslagh et al. 2012, 2015).

In Australia, most estuaries are subjected to eutrophication risks, especially in the south. This is due to their constricted or barred ocean entrances, which limits tidal exchanges with the ocean, together with shallow estuarine basins with extensive agricultural clearing and draining in catchments, and large inland catchments with limited vegetation cover (Deeley and Paling 1998). For example, the permanently open and extensively modified Swan-Canning estuary, one of the largest estuary in western Australia, displays signs of a general decline in quality in ecosystem health with for example frequent events of massive fish mortalities during the 2000s (Hallett et al. 2018). This degraded estuarine health can be related to the concomitant effects of both climate and anthropogenic pressures acting at embedded spatial scales, making the tackling of pressure sources challenging for local managers.

Those examples illustrate the complexity of the management of estuarine health, even in countries where environmental regulations were enforced and multiple management parties are involved. This is because management and restoration efforts necessarily focus on time and group-specific issues, so that it is difficult to generalize upon localized actions that may efficiently initiate estuarine dynamics recovery at large scale (Verdonschot et al. 2013). The time needed for estuarine recovery to happen is also highly variable, depending on initial causes and duration of estuarine degradation. Borja et al. (2010) for instance, based on 51 examples, suggested that recovery trajectories may take less than 5 years for rare cases where management and restoration efforts targeted short-lived and high turn-over biological components in estuaries, to a minimum of 15–25 years when the whole assets of estuarine structure and functioning are being tackled.

Recently anthropogenized estuaries in developing countries

Among developing countries characterized by a large diversity of estuarine ecosystems, South Africa stands out as an interesting case study of how recent human development interacts with climate change effects to affect estuarine health. South Africa also initiated a number of environmental regulations to tackle coastal and estuarine management issues. Examples of such regulations include the National Water Act (1998), the Integrated Coastal Management Act (1998), and the Marine Living Resources Act (2008). Van Niekerk et al. (2013) determined the state of health of estuaries in South Africa and the authors highlighted that small estuaries are in a better state of health because there were fewer pressures on them, at least at the time of the study. This is because small systems are mainly located in rural areas. However, small systems are not as resilient to change as large estuaries, due to their small size and "self-purification" capacities lowered by limited tidal exchanges. Larger estuaries, which are important as fish nursery grounds and of higher economic and ecological importance, are more heavily affected by catchment activities and direct development pressure. However, despite a higher resilience capacity related to tidal exchanges, larger systems are more degraded and in a poor health status. This illustrates that all South African estuaries are different, but all share a similar major threat related to the provision of freshwater, as, like in most developing countries, there is a high demand for catchment water for irrigation and hydroelectric schemes (Niang et al. 2019). The National Water Act voted in 1998 regulates this demand, based on an estuarine national ranking system built upon ecological criteria (Ecological Reserve Categories) and river runoff scenarios taking into account perspectives related to climate change (Adams et al. 2002). This method was applied for example to regulate freshwater use along the Mtata estuary's watershed in the Eastern Cape Province. Besides freshwater input, eutrophication issues concern most South African estuaries. For example, the Sundays and Swartkops estuaries are subjected to phytoplankton blooms and bottom-water hypoxia; the Knysna estuary is considered eutrophic, and the Wildevoëlvlei and Zeekee estuaries are regularly subjected to toxic algal blooms (Adams et al. 2020). The Saint Lucia estuarine reserve, classified as a UNESCO World Heritage site, is also under threat from agricultural pressure. Management of excessive wastewater effluents is in fact a national concern, threatening all aquatic environments including rivers, estuaries and coastal systems in South Africa (Adams et al. 2020).

As human population increases, pressures on natural systems are exacerbated, while awareness towards environmental issues rise, but implementation of appropriate measures to conserve natural assets and ecosystem services relying on them is slow. Li and Chen (2019) demonstrated that freshwater will not be available each decade after 2040 during the annual low-flow periods in the Changjiang River estuary (China) due to an increase in population combined with large existing catchment pressures. Similarly, population growth in the Río de la Plata catchment, one of the largest estuaries in the world and the second largest in South-America (and with drainage basin crossing five countries: Argentina, Bolivia, Brazil, Paraguay and Uruguay), is threatening cultural ecosystem services for recreation and tourism supported by the estuary (García-Alonso et al. 2019). Moreover, in arid regions or affected by complex geopolitical contexts, there is a significant potential for conflicts caused by resource scarcity (e.g., water) (Kidwai et al. 2019). This emphasizes the need for early planning for management and remedial measures, possibly starting decades in advance. However, a lack of data and weak socio-political will, as well as geopolitical uncertainties, create impediments to integrated management (Elliott et al. 2019, Ramesh et al. 2019).

Low impacted estuaries in pristine areas

Australia's tropical coast and the Arctic are the only places worldwide with numerous pristine or nearly pristine estuarine ecosystems (Wolanski and Ducrotoy 2014, Elliott et al. 2019). These areas indeed remain sparsely populated and include little permanent anthropic infrastructure, but remain subjected to exogenic pressures such as global warming, which cause severe environmental changes. Much of the human population of Australia is concentrated along the temperate coast, but the tropical coasts, where rainfall is abundant, are much less populated and this is where some estuarine systems remain preserved from most endogenic anthropogenic pressures (Wolanski and Ducrotoy 2014). However, sea level rise due to climate change is a significant threat to those systems, mostly dominated by mangrove forests, and where sea level rise may reach up to 9 mm $year^{-1}$ (5-year running average 1996–2013; CoastAdapt 2018, Webb et al. 2013). The sustainability of those otherwise pristine tropical Australia wetlands is therefore uncertain, as rising sea levels will lead to increasingly regular flooding. If wetland plant populations are unable to adapt, the increase in inundation time will likely lead to stand mortality and an associated loss of ecosystem services (Rogers et al. 2017, Munksgaard et al. 2019). Australia also still has several slightly affected and even pristine rivers and estuaries in far north Western Australia and Queensland, western Tasmania, and the Northern Territory. However, these regions represent vast tracts of potentially arable land. To develop their economy, the States and Territories proposed large-scale irrigation projects and construction of up to 100 dams. As a result, most estuaries throughout Australia, including those located within currently relatively pristine areas, are now threatened by new projects of urbanisation, irrigation, mining, or industrialisation, which threatens the pristine nature of these systems. In contrast, the relatively lesser-studied conditions of the Pan-Arctic coasts will increase the challenges against impending climate change. The scientific literature regarding the structure, functioning and potential threats affecting those coastal systems however remain very scarce.

Perspectives on the recovery and ecological restoration of estuarine systems

In response to the degradation of estuarine ecosystems, restoration and rehabilitation projects are designed to recover and increase resiliency of degraded ecosystems (Buitenhuis and Dieperink 2019). Restoration refers to the return of a damaged, degraded, or destroyed ecosystem to a previous condition, before the occurrence of stress (Abelson et al. 2020). Instead, rehabilitation is more about improving an ecosystem function, rather than one that has returned to its original state. Capderrey et al. (2016) reviewed the literature about estuarine restoration and reported that 58% of the articles published were from North America (114 articles), particularly from the United States (112 articles). Europe accounted for 28% of publications (mainly from the United Kingdom, Belgium and the Netherlands). Only 5% of articles come from other regions of the world (e.g., Australia, China or Arab countries) and 9% of articles concern more general feedback or opinions. The first major restoration projects took place in the 1970s in the United States (e.g., San Francisco Estuary Project, Chesapeake Bay Program, Everglades Comprehensive Restoration Plan). In Europe, the first projects were conducted in the 1980s on the Scheldt and the Wadden Sea coast. It is therefore possible to benefit from solid experience feedbacks, but it is also necessary to take into account the fact that the time needed to notice significant beneficial effects for restored ecosystem can be very long, between 10 and 25 years (Borja et al. 2010, Jones and Schmitz 2009). The lack of information about the initial state of systems also limits our ability to assess the effectiveness of restoration projects. Sufficient pre-restoration data is needed to separate the effects of restoration from the natural variability of the system. Pre- and post-restoration monitoring is therefore a fundamental component of future restoration projects. Additionally, studying recovery trajectories need to rely on some reference systems from which rehabilitated or restored systems can be compared with (Palmer et al. 2016). As most estuarine systems throughout the world are highly modified by anthropogenic pressures, finding such reference conditions is a challenge. Studies looking at estuarine structure and functioning from poorly studied pristine areas from tropical Australia or Pan-Arctic areas may provide insights on such reference conditions on a global perspective.

References

Abelson, A., Reed, D. C., Edgar, G. J., Smith, C. S., Kendrick, G. A., Orth, R. J. et al. (2020). Challenges for restoration of coastal marine ecosystems in the Anthropocene. Frontiers in Marine Science 7. https://doi.org/10.3389/fmars.2020.544105.

Ackerly, D. D., Ryals, R. A., Cornwell, W. K., Loarie, S. R., Veloz, S., Higgasen, K. D. et al. (2012) Potential impacts of climate change on biodiversity and ecosystem services in the San Francisco Bay area. California Energy Commission. CEC-500-2012-037.

Adams, J. B., Whitfield, A. K. and Van Niekerk, L. (2020). A socio-ecological systems approach towards future research for the restoration, conservation and management of southern African estuaries. African Journal of Aquatic Science 45(1-2): 231–241.

Adams, J. B., Bate, G. C., Harrison, T. D., Huizinga, P., Taljaard, S., van Niekerk, L. et al. (2002). A methoed to assess the freshwater inflow requirements of estuaries and application to the Mtata Estuary, South AFrica. Estuaries 25: 1382–1393.

Ærtebjerg Nielsen, G. and Gargas, E. (1984). Oxygen, nutrients and primary production in the open Danish waters. Limnologica (Berlin) 15: 303–310.

Afonso, I., Berecibar, E., Castro, N., Costa, J. L., Frias, P., Henriques, F. et al. (2020). Assessment of the colonization and dispersal success of non-indigenous species introduced in recreational marinas along the estuarine gradient. Ecological Indicators 113. https://doi.org/10.1016/j.ecolind.2020.106147.

Airoldi, L. and Beck, M. (2007). Status of Euromediterranean coastal marine habitats: loss and trends. Rapp. Comm. Int. Mer Médit. 38: 410.

Amorim, E., Ramos, S., Elliott, M., Franco, A. and Bordalo, A. A. (2017). Habitat loss and gain: Influence on habitat attractiveness for estuarine fish communities. Estuarine, Coastal and Shelf Science 197: 244–257.

Anderson, D. M., Glibert, P. M. and Burkholder, J. M. (2002). Harmful algal blooms and eutrophication: nutrient sources, composition, and consequences. Estuaries 25: 704–726.

Anderson, C. N., Hsieh, C. H., Sandin, S. A., Hewitt, R., Hollowed, A., Beddington, J. et al. (2008). Why fishing magnifies fluctuations in fish abundance. Nature 452(7189): 835–839.

Anderson, J. A. and Epifanio, C. E. (2010). Mating and sperm storage of the Asian shore crab Hemigrapsus sanguineus. J. Shellfish Res. 29: 497–501. https://doi.org/10.2983/035.029.0228.

Arevalo, E., Lassalle, G., Tétard, S., Maire, A., Sauquet, E., Lambert, P. et al. (2020). An innovative bivariate approach to detect joint temporal trends in environmental conditions: Application to large French rivers and diadromous fish. Science of the Total Environment 748: 141260.

Arevalo, E., Cabral, H. N., Villeneuve, B., Possémé, C. and Lepage, M. (2023). Fish larvae dynamics in temperate estuaries: A review on processes, patterns and factors that determine recruitment. Fish and Fisheries 24(3): 466–487. https://doi.org/10.1111/faf.12740.

Azimi, S. and Rocher, V. (2016). Influence of the water quality improvement on fish population in the Seine River (Paris, France) over the 1990–2013 period. Sci. Total Environ. 542(Pt A): 955–964.

Ballinger, R. and Stojanovic, T. (2010). Policy development and the estuary environment: a Severn Estuary case study. Mar Pollut Bull 61(1-3): 132–145.

Barbee, N. C., Ganio, K. and Swearer, S. E. (2014). Integrating multiple bioassays to detect and assess impacts of sublethal exposure to metal mixtures in an estuarine fish. Aquatic Toxicology 152: 244–255.

Barbier, E. B., Hacker, S. D., Kennedy, C. J., Koch, E. W., Stier, A. C. and Silliman, B. R. (2011). The value of estuarine and coastal ecosystem services. Ecological Monographs 81: 169–193.

Barletta, M., Barletta-Bergan, A., Saint-Paul, U. S. G. H. and Hubold, G. (2005). The role of salinity in structuring the fish assemblages in a tropical estuary. Journal of Fish Biology 66(1): 45–72.

Barletta, M., Lima, A. R. A. and Costa, M. F. (2019). Distribution, sources and consequences of nutrients, persistent organic pollutants, metals and microplastics in South American estuaries. Sci Total Environ. 651(Pt 1): 1199–1218. https://doi.org/10.1016/j.scitotenv.2018.09.276.

Barnes, D. K. A. and Milner, P. (2005). Drifting plastic and its consequences for sessile organism dispersal in the Atlantic Ocean. Marine Biology 146: 815–825.

Bax, N. C. J. T., Carlton, J. T., Mathews-Amos, A., Haedrich, R. L., Howarth, F. G., Purcell, J. E. et al. (2001). The control of biological invasions in the world's oceans. Conservation Biology 15(5): 1234–1246.

Beamish, R. J., McFarlane, G. A. and Benson, A. (2006). Longevity overfishing. Progress in Oceanography 68: 289–302.

Beck, M. W., Heck, K. L., Able, K. W., Childers, D. L., Eggleston, D. B., Gillanders, B. M. et al. (2001). The identification, conservation, and management of estuarine and marine nurseries for fish and invertebrates. BioScience 51: 633–641.

Berkeley, S. A., Hixon, M. A., Larson, R. J. and Love, M. S. (2004). Fisheries sustainability via protection of age structure and spatial distribution of fish populations. Fisheries 29(8): 23–32.

Bettini, S., Ciani, F. and Franceschini, V. (2006). Recovery of the olfactory receptor neurons in the African Tilapia mariae following exposure to low copper level. Aquatic Toxicology 76(3-4): 321–328.

Bhatt, G., Linker, L., Shenk, G., Bertani, I., Tian, R., Rigerlman, J. et al. (2023). Water quality impacts of climate change, land use, and population growth in the Chesapeake Bay watershed. J. Am. Water Resour. Assoc. 59: 1313–1341.

Blaber, S. J. (1997). Fish and Fisheries in Tropical Estuaries (Vol. 22). Springer Science & Business Media.

Blaber, S. J., Cyrus, D., Albaret, J.-J., Ching, C. V., Day, J., Elliott, M. et al. (2000). Effects of fishing on the structure and functioning of estuarine and nearshore ecosystems. ICES Journal of Marine Science 57(3): 590–602.

Blöschl, G., Hall, J., Parajka, J., Perdigão, R. A., Merz, B., Arheimer, B. et al. (2017). Changing climate shifts timing of European floods. Science 357(6351): 588–590.

Blöschl, G., Hall, J., Viglione, A., Perdigão, R. A., Parajka, J., Merz, B. et al. (2019). Changing climate both increases and decreases European river floods. Nature 573(7772): 108–111.

Bodammer, J. E. (1985). Corneal damage in larvae of striped bass Morone saxatilis exposed to copper. Transactions of the American Fisheries Society 114(4): 577–583.

Bodiguel, X., Maury, O., Mellon-Duval, C., Roupsard, F., Le Guellec, A. M. and Loizeau, V. (2009). A dynamic and mechanistic model of PCB bioaccumulation in the European hake (Merluccius merluccius). Journal of Sea Research 62(2-3): 124–134.

Boerema, A. and Meire, P. (2017). Management for estuarine ecosystem services: A review. Ecological Engineering 98: 172–182.

Boesch, D. F. and Turner, R. E. (1984). Dependence of fishery species on salt marshes: the role of food and refuge. Estuaries 7: 460–468.

Bonsdorff, E., Blomqvist, E. M., Mattila, J. and Norkko, A. (1997). Coastal eutrophication: causes, consequences and perspectives in the archipelago areas of the northern Baltic Sea. Estuarine, Coastal and Shelf Science 44: 63–72.

Booi, S., Mishi, S. and Andersen, O. (2022). Ecosystem services: a systematic review of provisioning and cultural ecosystem services in estuaries. Sustainability 14: 7252.

Borja, A., Bald, J., Franco, J., Larreta, J., Muxika, I., Revilla, M. et al. (2009). Using multiple ecosystem components, in assessing ecological status in Spanish (Basque Country) Atlantic marine waters. Marine Pollution Bulletin 59(1-3): 54–64.

Borja, Á., Dauer, D. M., Elliott, M. and Simenstad, C. A. (2010). Medium- and long-term recovery of estuarine and coastal ecosystems: patterns, rates and restoration effectiveness. Estuaries and Coasts 33(6): 1249–1260. https://doi.org/10.1007/s12237-010-9347-5.

Borja, A., Elliott, M., Snelgrove, P. V., Austen, M. C., Berg, T., Cochrane, S. et al. (2016a). Bridging the gap between policy and science in assessing the health status of marine ecosystems. Frontiers in Marine Science 3: 175.

Borja, A., Elliott, M., Andersen, J. H., Berg, T., Carstensen, J., Halpern, B. S. et al. (2016b). Overview of integrative assessment of marine systems: the ecosystem approach in practice. Frontiers in Marine Science 3: 20.

Bouchet, V. M. and Sauriau, P. G. (2008). Influence of oyster culture practices and environmental conditions on the ecological status of intertidal mudflats in the Pertuis Charentais (SW France): A multi-index approach. Marine Pollution Bulletin 56(11): 1898–1912.

Breitburg, D. L., Craig, J. K., Fulford, R. S., Rose, K. A., Boynton, W. R., Brady, D. C. et al. (2009). Nutrient enrichment and fisheries exploitation: interactive effects on estuarine living resources and their management. Hydrobiologia 629(1): 31–47. https://doi.org/10.1007/s10750-009-9762-4.

Briffa, M., de la Haye, K. and Munday, P. L. (2012). High CO_2 and marine animal behaviour: potential mechanisms and ecological consequences. Marine Pollution Bulletin 64(8): 1519–1528.

Brockerhoff, A. and McLay, C. (2011). Human-mediated spread of alien crabs. pp. 27–106. *In*: In the Wrong Place-Alien Marine Crustaceans: Distribution, Biology and Impacts. Dordrecht: Springer Netherlands.

Buchsbaum, R. and Powell, J. C. (2008). Symposium review: long-term shifts in faunal assemblages in eastern North American estuaries: a review of a workshop held at the biennial meeting of the Coastal and Estuarine Research Federation (CERF), November 2007, Providence, Rhode Island. Reviews in Fish Biology and Fisheries 18: 447–450.

Buitenhuis, Y. and Dieperink, C. (2019). Governance conditions for successful ecological restoration of estuaries: lessons from the Dutch Haringvliet case. Journal of Environmental Planning and Management 62(11): 1990–2009. https://doi.org/10.1080/09640568.2018.1529556.

Buschmann, A. H., Cabello, F., Young, K., Carvajal, J., Varela, D. A. and Henríquez, L. (2009). Salmon aquaculture and coastal ecosystem health in Chile: analysis of regulations, environmental impacts and bioremediation systems. Ocean & Coastal Management 52(5): 243–249.

Cabral, H. N. (2022). The role of estuaries and coastal areas as nurseries for fish: concepts, methodological challenges and future perspectives. *In*: Humphreys, J. and Little, S. (eds.). Challenges in Estuarine and Coastal Science. Pelagic Publishing.

Cabral, H. N. (2024). Climate change impacts on the nursery function for fish of coastal zones. *In*: Treatise on Estuarine and Coastal Science, 2nd Edition.

Cabral, H. N., Fonseca, V. F., Sousa, T. J. and Leal, M. C. (2019). Synergistic effects of climate change and marine pollution: an overlooked interaction in coastal and estuarine areas. International Journal of Environmental Research and Public Health 16: 27–37.

Cabral, H. N., Costa, M. J. and Salgado, J. P. (2001). Does the Tagus estuary fish community reflect environmental changes? Climate Research 18: 119–126.

Cabral, S., Carvalho, F., Gaspar, M., Ramajal, J., Sá, E., Santos, C. et al. (2020). Non-indigenous species in soft-sediments: Are some estuaries more invaded than others? Ecological Indicators 110. https://doi.org/10.1016/j.ecolind.2019.105640.

Callaway, J. C., Parker, V. T., Vasey, M. C., Schile, L. M. and Herbert, E. R. (2011). Tidal wetland restoration in San Francisco Bay: history and current issues. San Francisco Estuary and Watershed Science 9: 1–12.

Capderrey, C., Olivier, J.-M., Moussard, S., Foussard, V. and Bacq, N. (2016). Retours d'expériences de restauration écologique en milieu estuarien. Analyse de la littérature scientifique publiée (Rapport final). GIP Seine Aval—Onema, Rouen.

Caprio, J., Shimohara, M., Marui, T., Harada, S. and Kiyohara, S. (2014). Marine teleost locates live prey through pH sensing. Science 344(6188): 1154–1156.

Cariton, J. T. and Geller, J. B. (1993). Ecological roulette: the global transport of nonindigenous marine organisms. Science 261(5117): 78–82.

Carlton, J. T., Thompson, J. K., Schemel, L. E. and Nichols, F. H. (1990). Remarkable invasion of San Francisco Bay (California, USA) by the Asian clam Potamocorbula amurensis. I. Introduction and dispersal. Marine Ecology Progress Series 81–94.

Carmona, R., Kraemer, G. P. and Yarish, C. (2006). Exploring Northeast American and Asian species of Porphyra for use in an integrated finfish–algal aquaculture system. Aquaculture 252(1): 54–65.

Carreau, N. D. and Pyle, G. G. (2005). Effect of copper exposure during embryonic development on chemosensory function of juvenile fathead minnows (Pimephales promelas). Ecotoxicology and Environmental Safety 61(1): 1–6.

Carstensen, J., Andersen, J. H., Gustafsson, B. G. and Conley, D. J. (2014). Deoxygenation of the Baltic Sea during the last century. Proceedings of the National Academy of Sciences 111(15): 5628–5633.

Casini, M., Lövgren, J., Hjelm, J., Cardinale, M., Molinero, J. C. and Kornilovs, G. (2008). Multi-level trophic cascades in a heavily exploited open marine ecosystem. Proceedings of the Royal Society B: Biological Sciences 275(1644): 1793–1801.

Chaalali, A., Chevillot, X., Beaugrand, G., David, V., Luczak, C., Boët, P. et al. (2013). Changes in the distribution of copepods in the Gironde estuary: A warming and marinisation consequence? Estuarine, Coastal and Shelf Science 134: 150–161. https://doi.org/10.1016/j.ecss.2012.12.004.

Chainho, P. (2014). Report of the Working Group on Introduction and Transfers of Marine Organisms (WGITMO). ICES CM.

Chainho, P., Fernandes, A., Amorim, A., Ávila, S. P., Canning-Clode, J., Castro, J. J. et al. (2015). Non-indigenous species in Portuguese coastal areas, coastal lagoons, estuaries and islands. Estuarine, Coastal and Shelf Science 167: 199–211.

Chen, C. Y., Ward, D. M., Williams, J. J. and Fisher, N. S. (2016). Metal bioaccumulation by estuarine food webs in New England, USA. J. Mar. Sci. Eng. 4(2). https://doi.org/10.3390/jmse4020041.

Cheung, W. W. L. and Frolicher, T. L. (2020). Marine heatwaves exacerbate climate change impacts for fisheries in the northeast Pacific. Sci. Rep. 10(1): 6678. https://doi.org/10.1038/s41598-020-63650-z.

Chevillot, X., Pierre, M., Rigaud, A., Drouineau, H., Chaalali, A., Sautour, B. et al. (2016). Abrupt shifts in the Gironde fish community: an indicator of ecological changes in an estuarine ecosystem. Marine Ecology Progress Series 549: 137–151. https://doi.org/10.3354/meps11681.

Chivers, D. P., McCormick, M. I., Nilsson, G. E., Munday, P. L., Watson, S. A., Meekan, M. G. et al. (2014). Impaired learning of predators and lower prey survival under elevated CO_2: a consequence of neurotransmitter interference. Global Change Biology 20(2): 515–522.

Chow, E. S. H., Hui, M. N. Y., Cheng, C. W. and Cheng, S. H. (2009). Cadmium affects retinogenesis during zebrafish embryonic development. Toxicology and Applied Pharmacology 235(1): 68–76.

Chust, G., Gonzalez, M., Fontan, A., Revilla, M., Alvarez, P., Santos, M. et al. (2022). Climate regime shifts and biodiversity redistribution in the Bay of Biscay. Sci. Total Environ. 803: 149622. https://doi.org/10.1016/j.scitotenv.2021.149622.

Claret, M., Galbraith, E. D., Palter, J. B., Bianchi, D., Fennel, K., Gilbert, D. et al. (2018). Rapid coastal deoxygenation due to ocean circulation shift in the northwest Atlantic. Nature Climate Change 8(10): 868–872.

Cloern, J. E., Abreu, P. C., Carstensen, J., Chauvaud, L., Elmgren, R., Grall, J. et al. (2016). Human activities and climate variability drive fast-paced change across the world's estuarine-coastal ecosystems. Glob Chang Biol. 22(2): 513–529. https://doi.org/10.1111/gcb.13059.

Colautti, R. I. and MacIsaac, H. J. (2004). A neutral terminology to define 'invasive'species. Diversity and Distributions 10(2): 135–141.

Cole, M., Lindeque, P., Halsband, C. and Galloway, T. S. (2011). Microplastics as contaminants in the marine environment: a review. Marine Pollution Bulletin 62(12): 2588–2597.

Connolly, J. P. and Glaser, D. (2002). p, p′-DDE bioaccumulation in female sea lions of the California Channel Islands. Continental Shelf Research 22(6-7): 1059–1078.

Cooper, J. A. G. (2002). The role of extreme floods in estuary-coastal behaviour: contrasts between river- and tide-dominated microtidal estuaries. Sedimentary Geology 150(1-2): 123–137. https://doi.org/10.1016/s0037-0738(01)00271-8.

Correa-Herrera, T., Barletta, M., Lima, A. R. A., Jiménez-Segura, L. F. and Arango-Sánchez, L. B. (2017). Spatial distribution and seasonality of ichthyoplankton and anthropogenic debris in a river delta in the Caribbean Sea. Journal of Fish Biology 90(4): 1356–1387.

Costa, M. J., Vasconcelos, R., Costa, J. L. and Cabral, H. N. (2007). River flow influence on the fish community of the Tagus estuary (Portugal). Hydrobiologia 587(1): 113–123. https://doi.org/10.1007/s10750-007-0690-x.

Costanza, R., d'Arge, R., de Groot, R., Farber, S., Grasso, M., Hannon, B. et al. (1997). The value of the world's ecosystem services and natural capital. Nature 387: 253–260.

Croce, M. E. and Parodi, E. R. (2014). The Japanese alga Polysiphonia morrowii (Rhodomelaceae, Rhodophyta) on the South Atlantic Ocean: first report of an invasive macroalga inhabiting oyster reefs. Helgoland Marine Research 68: 241–252.

Cushing, D. H. (1990). Plankton production and year-class strength in fish populations: an update of the match/mismatch hypothesis. Advances in Marine Biology 26: 249–293.

Dame, R. F. and Dankers, N. (1988). Uptake and release of materials by a Wadden Sea mussel bed. Journal of Experimental Marine Biology and Ecology 118(3): 207–216.

Dame, R. F., Spurrier, J. D. and Wolaver, T. G. (1989). Carbon, nitrogen and phosphorus processing by an oyster reef. Marine Ecology Progress Series 249–256.

Darnaude, A. M., Salen-Picard, C., Polunin, N. V. and Harmelin-Vivien, M. L. (2004). Trophodynamic linkage between river runoff and coastal fishery yield elucidated by stable isotope data in the Gulf of Lions (NW Mediterranean). Oecologia 138: 325–332.

Daskalov, G. M. and Mamedov, E. V. (2007). Integrated fisheries assessment and possible causes for the collapse of anchovy kilka in the Caspian Sea. ICES Journal of Marine Science 64(3): 503–511.

Daughton, C. G. (2016). Pharmaceuticals and the Environment (PiE): Evolution and impact of the published literature revealed by bibliometric analysis. Science of the Total Environment 562: 391–426.

Dauvin, J.-C., Bachelet, G., Barillé, A.-L., Blanchet, H., De Montaudoin, X., Lavesque, N. et al. (2009). Benthic indicators and index approaches in the three main estuaries along the French Atlantic coast (Seine, Loire and Gironde). Marine Ecology 30: 228–240.

David, V., Sautour, B., Chardy, P. and Leconte, M. (2005). Long-term changes of the zooplankton variability in a turbid environment: the Gironde estuary (France). Estuarine, Coastal and Shelf Science 64(2-3): 171–184.

Davis, J. A. (2004). The long-term fate of polychlorinated biphenyls in San Francisco Bay (USA). Environ. Toxicol. Chem. 23: 2396–2409.

Day, J. W., Crump, B. C., Kemp, W. M. and Yáñez-Arancibia, A. (2012). Introduction to estuarine ecology. pp. 1–18. *In*: Day, J. W., Crump, B. C., Kemp, W. M. and Yáñez-Arancibia, A. (eds.). Estuarine Ecology, Second Edition. John Wiley & Sons, Inc, Hoboken, NJ, USA.

Day, J. W., Ibáñez, C., Pont, D. and Scarton, F. (2019). Status and sustainability of Mediterranean deltas: The case of the Ebro, Rhône, and Po Deltas and Venice Lagoon. pp. 237–249. *In*: Coasts and Estuaries. Elsevier.

Dayon, G., Boe, J., Martin, E. and Gailhard, J. (2018). Impacts of climate change on the hydrological cycle over France and associated uncertainties. Comptes Rendus. Géoscience 350(4): 141–153.

de Souza, J. S. and Dos Santos, L. N. (2023). Resident species, not immigrants, drive reorganization of estuarine fish assemblages in response to warming. Ecology e3987.

DeAlteris, J., Kilpatrick, B. D. and Rheault, R. B. (2004). A comparative evaluation of the habitat value of shellfish aquaculture gear, submerged aquatic vegetation and a non-vegetated seabed.

DeCourten, B. M., Forbes, J. P., Roark, H. K., Burns, N. P., Major, K. M., White, J. W. et al. (2020). Multigenerational and transgenerational effects of environmentally relevant concentrations of endocrine disruptors in an estuarine fish model. Environmental Science & Technology 54(21): 13849–13860.

Deeley, D. M. and Paling, E. I. (1998). Assessing the ecological health of estuaries in southwest Australia. pp. 257–271. *In*: McComb, A. J. (ed.). Wetlands for the Future. Gleneagles Publishing.

Defeo, O. and Elliott, M. (2021). The 'triple whammy' of coasts under threat - Why we should be worried! Mar. Pollut. Bull. 163: 111832.

Désaunay, Y., Guérault, D., Le Pape, O. and Poulard, J. C. (2006). Changes in occurrence and abundance of northern/southern flatfishes over a 20-year period in a coastal nursery area (Bay of Vilaine) and on the eastern continental shelf of the Bay of Biscay. Scientia Marina 70: 193–200.

Deutsch, C., Brix, H., Ito, T., Frenzel, H. and Thompson, L. (2011). Climate-forced variability of ocean hypoxia. Science 333(6040): 336–339.

Devreker, D., Souissi, S., Winkler, G., Forget-Leray, J. and Leboulenger, F. (2009). Effects of salinity, temperature and individual variability on the reproduction of Eurytemora affinis (Copepoda; Calanoida) from the Seine estuary: A laboratory study. Journal of Experimental Marine Biology and Ecology 368(2): 113–123. https://doi.org/10.1016/j.jembe.2008.10.015.

Diederich, S., Nehls, G. J. E. E., Van Beusekom, J. E. and Reise, K. (2005). Introduced Pacific oysters (Crassostrea gigas) in the northern Wadden Sea: invasion accelerated by warm summers? Helgoland Marine Research 59: 97–106.

Dixson, D. L., Munday, P. L. and Jones, G. P. (2010). Ocean acidification disrupts the innate ability of fish to detect predator olfactory cues. Ecology Letters 13(1): 68–75.

Domenici, P., Allan, B., McCormick, M. I. and Munday, P. L. (2012). Elevated carbon dioxide affects behavioural lateralization in a coral reef fish. Biology Letters 8(1): 78–81.

Dorenbosch, M., Van Riel, M. C., Nagelkerken, I. and Van der Velde, G. (2004). The relationship of reef fish densities to the proximity of mangrove and seagrass nurseries. Estuarine, Coastal and Shelf Science 60(1): 37–48.

Duarte, C. M. and Cebrian, J. (1996). The fate of marine autotrophic production, Limnol. Oceanogr. 41: 1758–1766.

Dubois, S., Marin-Léal, J. C., Ropert, M. and Lefebvre, S. (2007). Effects of oyster farming on macrofaunal assemblages associated with Lanice conchilega tubeworm populations: a trophic analysis using natural stable isotopes. Aquaculture 271(1-4): 336–349.

Durant, J. M., Hjermann, D. Ø., Ottersen, G. and Stenseth, N. C. (2007). Climate and the match or mismatch between predator requirements and resource availability. Climate Research 33(3): 271–283.

Ecoutin, J. M., Simier, M., Albaret, J. J., Laë, R. and de Morais, L. T. (2010). Changes over a decade in fish assemblages exposed to both environmental and fishing constraints in the Sine Saloum estuary (Senegal). Estuarine, Coastal and Shelf Science 87(2): 284–292.

Eerkes-Medrano, D., Thompson, R. C. and Aldridge, D. C. (2015). Microplastics in freshwater systems: a review of the emerging threats, identification of knowledge gaps and prioritisation of research needs. Water Research 75: 63–82.

Eichenberg, T. (2013). The challenges of adapting to climate change in San Francisco Bay. Hastings Environmental Law Journal 19(article 18): 393.

Elliott, M. (2011). Marine science and management means tackling exogenic unmanaged pressures and endogenic managed pressures—a numbered guide. Mar. Pollut. Bull. 62(4): 651–655.

Elliott, M. and Hemingway, K. L. (2002). Fishes in Estuaries. Blackwell Science, Oxford.

Elliott, M. and McLusky, D. S. (2002). The need for definitions in understanding estuaries. Estuarine, Coastal and Shelf Science 55: 815–827.

Elliott, M. and Whitfield, A. K. (2011). Challenging paradigms in estuarine ecology and management. Estuarine, Coastal and Shelf Science 94(4): 306–314.

Elliott, M., Day, J. W., Ramachandran, R. and Wolanski, E. (2019). A synthesis: what is the future for coasts, estuaries, deltas and other transitional habitats in 2050 and beyond? pp. 1–28. *In*: Coasts and Estuaries. Elsevier.

Elliott, M., Mander, L., Mazik, K., Simenstad, C., Valesini, F., Whitfield, A. et al. (2016). Ecoengineering with ecohydrology: successes and failures in estuarine restoration. Estuarine, Coastal and Shelf Science 176: 12–35.

Elliott, M., Houde, E. D., Lamberth, S. J., Lonsdale, J. A. and Tweedley, J. R. (2022). Management of fishes and fisheries in estuaries. Fish and Fisheries in Estuaries: A Global Perspective 2: 706–797.

Essink, K. and Dekker, R. (2002). General patterns in invasion ecology tested in the Dutch Wadden Sea: the case of a brackish-marine polychaetous worm. Biological Invasions 4: 359–368.

FAO. (2014). ftp://ftp.fao.org/fi/stat/Overviews/AquacultureStatistics2012.

Faria, M., Wu, X., Luja-Mondragón, M., Prats, E., Gómez-Oliván, L. M., Piña, B. et al. (2020). Screening anti-predator behaviour in fish larvae exposed to environmental pollutants. Science of the Total Environment 714: 136759.

Fendall, L. S. and Sewell, M. A. (2009). Contributing to marine pollution by washing your face: microplastics in facial cleansers. Marine Pollution Bulletin 58(8): 1225–1228.

Ferguson, G. J., Ward, T. M., Ye, Q., Geddes, M. C. and Gillanders, B. M. (2013). Impacts of drought, flow regime, and fishing on the fish assemblage in southern australia's largest temperate estuary. Estuaries and Coasts 36(4): 737–753. https://doi.org/10.1007/s12237-012-9582-z.

Ferrari, M. C., McCormick, M. I., Munday, P. L., Meekan, M. G., Dixson, D. L., Lönnstedt, O. et al. (2012). Effects of ocean acidification on visual risk assessment in coral reef fishes. Functional Ecology 26(3): 553–558.

Fonseca, V. F., França, S., Duarte, B., Caçador, I., Cabral, H. N., Mieiro, C. L. et al. (2019). Spatial variation in mercury bioaccumulation and magnification in a temperate estuarine food web. Frontiers in Marine Science 6: 117.

Forrest, B. M. and Creese, R. G. (2006). Benthic impacts of intertidal oyster culture, with consideration of taxonomic sufficiency. Environmental Monitoring and Assessment 112: 159–176.

Forrest, B. M., Keeley, N. B., Hopkins, G. A., Webb, S. C. and Clement, D. M. (2009). Bivalve aquaculture in estuaries: Review and synthesis of oyster cultivation effects. Aquaculture 298(1-2): 1–15. https://doi.org/10.1016/j.aquaculture.2009.09.032.

Fragkopoulou, E., Sen Gupta, A., Costello, M. J., Wernberg, T., Araújo, M. B., Serrão, E. A. et al. (2023). Marine biodiversity exposed to prolonged and intense subsurface heatwaves. Nature Climate Change. https://doi.org/10.1038/s41558-023-01790-6.

Fredston, A. L., Cheung, W. W. L., Frolicher, T. L., Kitchel, Z. J., Maureaud, A. A., Thorson, J. T. et al. (2023). Marine heatwaves are not a dominant driver of change in demersal fishes. Nature 621: 324–329. https://doi.org/10.1038/s41586-023-06449-y.

Frias, J. P. G. L., Sobral, P. and Ferreira, A. M. (2010). Organic pollutants in microplastics from two beaches of the Portuguese coast. Marine Pollution Bulletin 60(11): 1988–1992.

Frias, J. P., Otero, V. and Sobral, P. (2014). Evidence of microplastics in samples of zooplankton from Portuguese coastal waters. Marine Environmental Research 95: 89–95.

Froehlicher, M., Liedtke, A., Groh, K. J., Neuhauss, S. C., Segner, H. and Eggen, R. I. (2009). Zebrafish (Danio rerio) neuromast: promising biological endpoint linking developmental and toxicological studies. Aquatic Toxicology 95(4): 307–319.

Furnas, M. J. (1990). *In situ* growth rates of marine phytoplankton—approaches to measurement, community and species growth rates. J. Plankton Res. 12: 1117–1151.

García-Alonso, J., Lercari, D. and Defeo, O. (2019). Río de la Plata: a neotropical estuarine system. pp. 45–56. *In*: Coasts and Estuaries. Elsevier.

Geburzi, J. C. and McCarthy, M. L. (2018). How do they do it?–Understanding the success of marine invasive species. YOUMARES 8–Oceans Across Boundaries: Learning from each other: Proceedings of the 2017 Conference for YOUng MARine RESearchers in Kiel, Germany.

Gil, M. A. and Pfaller, J. B. (2016). Oceanic barnacles act as foundation species on plastic debris: implications for marine dispersal. Scientific Reports 6(1): 19987.

Gillan, D. C., Pede, A., Sabbe, K., Gao, Y., Leermakers, M., Baeyens, W. et al. (2012). Effect of bacterial mineralization of phytoplankton-derived phytodetritus on the release of arsenic, cobalt and manganese from muddy sediments in the Southern North Sea. A microcosm study. Science of the Total Environment 419: 98–108.

Gillanders, B. M. and Kingsford, M. J. (1996). Elements in otoliths may elucidate the contribution of estuarine recruitment to sustaining coastal reef populations of a temperate reef fish. Marine Ecology Progress Series 141: 13–20.

Gillanders, B. M. and Kingsford, M. J. (2002). Impact of changes in flow of freshwater on estuarine and open coastal habitats and the associated organisms. In Oceanography and Marine Biology an Annual Review (Vol. 40, pp. 233).

Gilliers, C., Le Pape, O., Désaunay, Y., Morin, J., Guérault, D. and Amara, R. (2006). Are growth and density quantitative indicators of essential fish habitat quality? An application to the common sole Solea solea nursery grounds. Estuarine, Coastal and Shelf Science 69(1-2): 96–106.

Gobas, F. A. (1993). A model for predicting the bioaccumulation of hydrophobic organic chemicals in aquatic food-webs: application to Lake Ontario. Ecological Modelling 69(1-2): 1–17.

Gollasch, S. (2002) The importance of Ship Hull Fouling as a vector of species introductions into the North Sea. Biofouling 18: 105–121.

González-Ortegón, E. and Drake, P. (2012). Effects of freshwater inputs on the lower trophic levels of a temperate estuary: physical, physiological or trophic forcing? Aquatic Sciences 74(3): 455–469. https://doi.org/10.1007/s00027-011-0240-5.

Granadeiro, J. P., Dias, M. P., Martins, R. C. and Palmeirim, J. M. (2006). Variation in numbers and behaviour of waders during the tidal cycle: implications for the use of estuarine sediment flats. Acta Oecologica 29: 293–300.

Greenfield, B. K., Davis, J. A., Fairey, R., Roberts, C., Crane, D. and Ichikawa, G. (2005). Seasonal, interannual, and long-term variation in sport fish contamination, San Francisco Bay. Science of the Total Environment 336: 25–43.

Gregory, M. R. (2009). Environmental implications of plastic debris in marine settings—entanglement, ingestion, smothering, hangers-on, hitch-hiking and alien invasions. Philosophical Transactions of the Royal Society B: Biological Sciences 364(1526): 2013–2025.

Griffen, B. D., Altman, I., Hurley, J. and Mosblack, H. (2011). Reduced fecundity by one invader in the presence of another: a potential mechanism leading to species replacement. Journal of Experimental Marine Biology and Ecology 406(1-2): 6–13.

Guo, X. and Wang, J. (2019). The chemical behaviors of microplastics in marine environment: A review. Marine Pollution Bulletin 142: 1–14.

Gutiérrez, J. L., Jones, C. G., Strayer, D. L. and Iribarne, O. O. (2003). Mollusks as ecosystem engineers: the role of shell production in aquatic habitats. Oikos 101(1): 79–90.

Hadley, S., Wild-Allen, K., Johnson, C. and Macleod, C. (2016). Quantification of the impacts of finfish aquaculture and bioremediation capacity of integrated multi-trophic aquaculture using a 3D estuary model. Journal of Applied Phycology 28(3): 1875–1889. https://doi.org/10.1007/s10811-015-0714-2.

Hallett, C., Hobday, A. J., Tweedley, J., Thompson, P., Mcmahon, K. and Valesini, F. (2018). Observed and predicted impacts of climate change on the estuaries of south-western Australia, a Mediterranean climate region. Regional Environmental Change 18: 1357–1373.

Halpern, B. S., McLeod, K. L., Rosenberg, A. A. and Crowder, L. B. (2008). Managing for cumulative impacts in ecosystem-based management through ocean zoning. Ocean & Coastal Management 51(3): 203–211.

Hamilton, T. J., Holcombe, A. and Tresguerres, M. (2014). CO_2-induced ocean acidification increases anxiety in rockfish via alteration of GABAA receptor functioning. Proceedings of the Royal Society B: Biological Sciences 281(1775): 20132509.

Han, D. and Currell, M. J. (2017). Persistent organic pollutants in China's surface water systems. Science of the Total Environment 580: 602–625.

Handy, R. D., Sims, D. W., Giles, A., Campbell, H. A. and Musonda, M. M. (1999). Metabolic trade-off between locomotion and detoxification for maintenance of blood chemistry and growth parameters by rainbow trout (Oncorhynchus mykiss) during chronic dietary exposure to copper. Aquatic Toxicology 47(1): 23–41.

Harding Jr, L. W. and Perry, E. S. (1997). Long-term increase of phytoplankton biomass in Chesapeake Bay, 1950–1994. Marine Ecology Progress Series 157: 39–52.

Henderson, P. A. (2007). Discrete and continuous change in the fish community of the Bristol Channel in response to climate change. Journal of the Marine Biological Association of the United Kingdom 87(2): 589–598.

Hermant, M., Lobry, J., Bonhommeau, S., Poulard, J. C. and Le Pape, O. (2010). Impact of warming on abundance and occurrence of flatfish populations in the Bay of Biscay (France). Journal of Sea Research 64(1-2): 45–53.

Hilborn, R., and Walters, C. J. (1992). Quantitative Fisheries Stock Assessment: Choice, Dynamics, and Uncertainty. London: Chapman and Hall.

Hixon, M. A., Johnson, D. W. and Sogard, S. M. (2014). BOFFFFs: on the importance of conserving old-growth age structure in fishery populations. ICES Journal of Marine Science 71(8): 2171–2185.

Hobday, A. J., Alexander, L. V., Perkins, S. E., Smale, D. A., Straub, S. C., Oliver, E. C. J. et al. (2016). A hierarchical approach to defining marine heatwaves. Progress in Oceanography 141: 227–238. https://doi.org/10.1016/j.pocean.2015.12.014.

Hoffmann, L. J., Breitbarth, E., Boyd, P. W. and Hunter, K. A. (2012). Influence of ocean warming and acidification on trace metal biogeochemistry. Marine Ecology Progress Series 470: 191–205.

Hsieh, C. H., Kim, H. J., Watson, W., Di Lorenzo, E. and Sugihara, G. (2009). Climate-driven changes in abundance and distribution of larvae of oceanic fishes in the southern California region. Global Change Biology 15(9): 2137–2152.

Hughes, J. E., Deegan, L. A., Wyda, J. C., Weaver, M. J. and Wright, A. (2002). The effects of eelgrass habitat loss on estuarine fish communities of southern New England. Estuaries 25: 235–249.

Hutchings, J. A. and Myers, R. A. (1993). Effect of age on the seasonality of maturation and spawning of Atlantic cod, Gadus morhua, in the Northwest Atlantic. Canadian Journal of Fisheries and Aquatic Sciences 50(11): 2468–2474.

IPCC. (2013). Climate change 2013: the physical science basis. *In*: Stocker, T. F., Qin, D., Plattner, G.-K., Tignor, M., Allen, S. K. and Boschung, J. et al. (eds.). Contribution of Working Group I to the Fifth Assessment Report of the Intergovernmental Panel on Climate Change, p. 1265. IPCC, Cambridge, UK and New York, NY.

Izquierdo, P., Rico, J. M., Taboada, F. G., González-Gil, R. and Arrontes, J. (2022). Characterization of marine heatwaves in the Cantabrian Sea, SW Bay of Biscay. Estuarine, Coastal and Shelf Science 274. https://doi.org/10.1016/j.ecss.2022.107923.

Jacobs, S., Vandenbruwaene, W., Vrebos, D., Beauchard, O., Boerema, A., Wolfstein, K. et al. (2013). Ecosystem service assessment of TIDEestuaries. *In*: Study Report in the Framework of the Interreg IVB Project TIDE,ECOBE, UA, Antwerp, Belgium.

Jambeck, J. R., Geyer, R., Wilcox, C., Siegler, T. R., Perryman, M., Andrady, A. et al. (2015). Plastic waste inputs from land into the ocean. Science 347(6223): 768–771.

Jenkins, G. P., Edgar, G. J., May, H. M. A. and Shaw, C. (1993, August). Ecological basis for parallel declines in seagrass habitat and catches of commercial fish in Western Port Bay, Victoria. pp. 124–136. *In*: Sustainable Fisheries Through Sustaining Fish Habitat. Australian Society of Fish Biology Workshop, Victor Harbour, South Australia. AGPS.

Jennings, S. and Kaiser, M. J. (1998). The effects of fishing on marine ecosystems. pp. 201–352. *In*: Advances in Marine Biology (Vol. 34). Academic Press.

Jensen, A. C., Humphreys, J., Caldow, R. W. G., Grisley, C. and Dyrynda, P. E. J. (2004). Naturalization of the Manila clam (Tapes philippinarum), an alien species, and establishment of a clam fishery within Poole Harbour, Dorset. Journal of the Marine Biological Association of the United Kingdom 84(5): 1069–1073.

Jezierska, B., Ługowska, K. and Witeska, M. (2009). The effects of heavy metals on embryonic development of fish (a review). Fish Physiology and Biochemistry 35: 625–640.

Jones, H. P. and Schmitz, O. J. (2009). Rapid recovery of damaged ecosystems. PLoS One 4(5): e5653. https://doi.org/10.1371/journal.pone.0005653.

Kalejta, B. (1993). Diets of shorebirds at the Berg River Estuary, South Africa - spatial and temporal variation. Ostrich 64: 123–133.

Keeling, R. F., Körtzinger, A. and Gruber, N. (2010). Ocean deoxygenation in a warming world. Annual Review of Marine Science 2: 199–229.

Kennish, M. J. (1986). Ecology of Estuaries. Volume 1: Physical and Chemical Aspects. CRC Press, Boca Raton, FL, p. 391.

Kennish, M. J. (2001). Coastal salt marsh systems in the US: a review of anthropogenic impacts. Journal of Coastal Research 17(3): 731–748.

Kennish, M. J. (2002). Environmental threats and environmental future of estuaries. Environmental Conservation 29(1): 78–107. https://doi.org/10.1017/s0376892902000061.

Kennish, M. J. (2022). Management strategies to mitigate anthropogenic impacts in estuarine and coastal marine environments: a review. Open Journal of Ecology 12(10): 667–688.

Kennish, M. J. (2024). Estuaries: Origin, historical development, and classifications. pp. 43–54. *In*: Kennish, M. J., Paerl, H. W. and Crosswell, J. R. (eds.). Climate Change and Estuaries. CRC Press, Boca Raton.

Kerambrun, E., Henry, F., Cornille, V., Courcot, L. and Amara, R. (2013). A combined measurement of metal bioaccumulation and condition indices in juvenile European flounder, Platichthys flesus, from European estuaries. Chemosphere 91(4): 498–505.

Kibria, G., Nugegoda, D., Rose, G. and Haroon, A. Y. (2021). Climate change impacts on pollutants mobilization and interactive effects of climate change and pollutants on toxicity and bioaccumulation of pollutants in estuarine and marine biota and linkage to seafood security. Marine Pollution Bulletin 167: 112364.

Kidwai, S., Ahmed, W., Tabrez, S. M., Zhang, J., Giosan, L., Clift, P. et al. (2019). The Indus delta—Catchment, river, coast, and people. pp. 213–232. *In*: Coasts and Estuaries. Elsevier.

Killgore, K. J., Morgan, R. P. and Rybicki, N. B. (1989). Distribution and abundance of fishes associated with submersed aquatic plants in the Potomac River. North American Journal of Fisheries Management 9(1): 101–111.

Laë, R. (1995). Climatic and anthropogenic effects on fish diversity and fish yields in the Central Delta of the Niger River. Aquatic Living Resources 8(1): 43–58.

Laufkötter, C., Zscheischler, J. and Frolicher, T. L. (2020). High-impact marine heatwaves attributable to human-induced global warming. Science 369(6511): 1621–1625. https://doi.org/10.1126/science.aba0690.

Le Pape, O., Chauvet, F., Désaunay, Y. and Guérault, D. (2003). Relationship between interannual variations of the river plume and the extent of nursery grounds for the common sole (Solea solea, L.) in Vilaine Bay. Effects on recruitment variability. Journal of Sea Research 50(2-3): 177–185.

Leduc, A. O., Munday, P. L., Brown, G. E. and Ferrari, M. C. (2013). Effects of acidification on olfactory-mediated behaviour in freshwater and marine ecosystems: a synthesis. Philosophical Transactions of the Royal Society B: Biological Sciences 368(1627): 20120447.

Lem, A., Bjorndal, T. and Lappo, A. (2014). Economic analysis of supply and demand for food up to 2030 – Special focus on fish and fishery products. FAO Fisheries and Aquaculture Circular No. 1089. FAO, Rome, p 106.

Levin, S. G., Sirota, N. P., Nenov, M. N., Savina, T. A. and Godukhin, O. V. (2018). Interleukin-10 and PD150606 modulate expression of AMPA receptor GluA1 and GluA2 subunits under hypoxic conditions. Neuroreport 29(2): 84–91.

Li, H. M., Zhang, C. S., Han, X. R. and Shi, X. Y. (2015). Changes in concentrations of oxygen, dissolved nitrogen, phosphate, and silicate in the southern Yellow Sea, 1980–2012: sources and seaward gradients. Estuarine, Coastal and Shelf Science 163: 44–55.

Li, M. and Chen, Z. (2019). An assessment of saltwater intrusion in the Changjiang (Yangtze) River estuary, China. pp. 31–43. *In*: Coasts and Estuaries. Elsevier.

Liggins, G. W. and Kennelly, S. J. (1996). By-catch from prawn trawling in the Clarence River estuary, New South Wales, Australia. Fisheries Research 25(3-4): 347–367.

Lima, A. R. A., Costa, M. F. and Barletta, M. (2014). Distribution patterns of microplastics within the plankton of a tropical estuary. Environmental Research 132: 146–155.

Linbo, T. L., Stehr, C. M., Incardona, J. P. and Scholz, N. L. (2006). Dissolved copper triggers cell death in the peripheral mechanosensory system of larval fish. Environmental Toxicology and Chemistry: An International Journal 25(2): 597–603.

Liu, Q., Yi, Y., Hou, C., Wu, X. and Song, J. (2020). Response of trophic structure and isotopic niches of the food web to flow regime in the Yellow River estuary. Marine Geology 430. https://doi.org/10.1016/j.margeo.2020.106329.

Livingston, R. J., Niu, X., Lewis III, F. G. and Woodsum, G. C. (1997). Freshwater input to a gulf estuary: Long-term control of trophic organization. Ecological Applications 7(1): 277–299.

Llopiz, J. K., Cowen, R. K., Hauff, M. J., Ji, R., Munday, P. L., Muhling, B. A. et al. (2014). Early life history and fisheries oceanography: new questions in a changing world. Oceanography 27(4): 26–41.

Lonsdale, J.-A., Leach, C., Parsons, D., Barkwith, A., Manson, S. and Elliott, M. (2022). Managing estuaries under a changing climate: A case study of the Humber Estuary, UK. Environmental Science & Policy 134: 75–84.

López-Moreno, J. I., Vicente-Serrano, S. M., Beguería, S., García-Ruiz, J. M., Portela, M. M. and Almeida, A. B. (2009). Dam effects on droughts magnitude and duration in a transboundary basin: The Lower River Tagus, Spain and Portugal. Water Resources Research 45(2). https://doi.org/10.1029/2008wr007198.

Lotze, H. K., Lenihan, H. S., Bourque, B. J., Bradbury, R. H., Cooke, R. G., Kay, M. C. et al. (2006). Depletion, degradation, and recovery potential of estuaries and coastal seas. Science 312(5781): 1806–1809.

Lu, Z., Gan, J., Dai, M., Liu, H. and Zhao, X. (2018). Joint effects of extrinsic biophysical fluxes and intrinsic hydrodynamics on the formation of hypoxia west off the Pearl River Estuary. Journal of Geophysical Research: Oceans 123(9): 6241–6259.

Lucas, L. V., Koseff, J. R., Cloern, J. F., Monismith, S. G. and Thompson, J. K. (1999a). Processes governing phytoplankton blooms in estuaries. I: the local production-loss balance. Marine Ecology Progress Series 187: 1–15.

Macleod, C. and Coughanowr, C. (2019). Heavy metal pollution in the derwent estuary: history, science and management. Regional Studies in Marine Science 32: 100866.

Mann, B., James, N. and Beckley, L. (2002). An assessment of the recreational fishery in the St Lucia estuarine system, KwaZulu-Natal, South Africa. African Journal of Marine Science 24: 263–279.

Martinho, F., Leitão, R., Viegas, I., Dolbeth, M., Neto, J. M., Cabral, H. N. et al. (2007). The influence of an extreme drought event in the fish community of a southern Europe temperate estuary. Estuarine, Coastal and Shelf Science 75(4): 537–546.

Martinho, F., Dolbeth, M., Viegas, I., Teixeira, C. M., Cabral, H. N. and Pardal, M. A. (2009). Environmental effects on the recruitment variability of nursery species. Estuarine, Coastal and Shelf Science 83(4): 460–468.

Matthiessen, P. and Law, R. J. (2002). Contaminants and their effects on estuarine and coastal organisms in the United Kingdom in the late twentieth century. Environmental Pollution 120(3): 739–757.
McKindsey, C. W., Anderson, M. R., Barnes, P., Courtenay, S., Landry, T. and Skinner, M. (2006). Effects of shellfish aquaculture on fish habitat. Canadian Science Advisory Secretariat Research Document 2006/011. Fisheries and Oceans, Canada. 84 pp.
McLean, M., Mouillot, D., Maureaud, A. A., Hattab, T., MacNeil, M. A., Goberville, E. et al. (2021). Disentangling tropicalization and deborealization in marine ecosystems under climate change. Curr. Biol. 31(21): 4817–4823 e4815. https://doi.org/10.1016/j.cub.2021.08.034.
McLusky, D. S. and Elliott, M. (2004). The Estuarine Ecosystem: Ecology, Threats and Management. Oxford University Press, New York.
Melià, P. A. C. O., De Leo, G. A. and Gatto, M. (2004). Density and temperature-dependence of vital rates in the Manila clam Tapes philippinarum: a stochastic demographic model. Marine Ecology Progress Series 272: 153–164.
Melia, P. and Gatto, M. (2005). A stochastic bioeconomic model for the management of clam farming. Ecological Modelling 184(1): 163–174.
Money, C., Braungardt, C. B., Jha, A. N., Worsfold, P. J. and Achterberg, E. P. (2011). Metal speciation and toxicity of Tamar Estuary water to larvae of the Pacific oyster, Crassostrea gigas. Marine Environmental Research 72(1-2): 3–12.
Morrisey, D. J., Underwood, A. J. and Howitt, L. (1996). Effects of copper on the faunas of marine soft-sediments: an experimental field study. Marine Biology 125: 199–213.
Moura, P., Garaulet, L., Vasconcelos, P., Chainho, P., Costa, J. and Gaspar, M. (2017). Age and growth of a highly successful invasive species: the Manila clam Ruditapes philippinarum (Adams & Reeve, 1850) in the Tagus Estuary (Portugal). Aquatic Invasions 12(2): 133–146. https://doi.org/10.3391/ai.2017.12.2.02.
Munday, P. L., Cheal, A. J., Dixson, D. L., Rummer, J. L. and Fabricius, K. E. (2014). Behavioural impairment in reef fishes caused by ocean acidification at CO_2 seeps. Nature Climate Change 4(6): 487–492.
Munday, P. L., Crawley, N. E. and Nilsson, G. E. (2009a). Interacting effects of elevated temperature and ocean acidification on the aerobic performance of coral reef fishes. Marine Ecology Progress Series 388: 235–242.
Munday, P. L., Dixson, D. L., Donelson, J. M., Jones, G. P., Pratchett, M. S., Devitsina, G. V. et al. (2009b). Ocean acidification impairs olfactory discrimination and homing ability of a marine fish. Proceedings of the National Academy of Sciences 106(6): 1848–1852.
Munday, P. L., Dixson, D. L., McCormick, M. I., Meekan, M., Ferrari, M. C. and Chivers, D. P. (2010). Replenishment of fish populations is threatened by ocean acidification. Proceedings of the National Academy of Sciences 107(29): 12930–12934.
Munday, P. L., Pratchett, M. S., Dixson, D. L., Donelson, J. M., Endo, G. G., Reynolds, A. D. et al. (2013). Elevated CO_2 affects the behavior of an ecologically and economically important coral reef fish. Marine Biology 160: 2137–2144.
Munksgaard, N. C., Hutley, L. B., Metcalfe, K. N., Padovan, A. C., Palmer, C. and Gibb, K. S. (2019). Environmental challenges in a near-pristine mangrove estuary facing rapid urban and industrial development: Darwin Harbour, Northern Australia. Regional Studies in Marine Science 25.
Murray, L. G., Newell, C. R. and Seed, R. (2007). Changes in the biodiversity of mussel assemblages induced by two methods of cultivation. Journal of Shellfish Research 26(1): 153–162.
Musick, J. A., Harbin, M. M., Berkeley, S. A., Burgess, G. H., Eklund, A. M., Findley, L. et al. (2000). Marine, estuarine, and diadromous fish stocks at risk of extinction in North America (exclusive of Pacific salmonids). Fisheries 25(11): 6–30.
Nagelkerken, I. and Munday, P. L. (2016). Animal behaviour shapes the ecological effects of ocean acidification and warming: moving from individual to community-level responses. Global Change Biology 22(3): 974–989.
Nagelkerken, I., Sheaves, M., Baker, R. and Connolly, R. M. (2015). The seascape nursery: a novel spatial approach to identify and manage nurseries for coastal marine fauna. Fish and Fisheries 16: 362–371.
Naylor, R. L., Williams, S. L. and Strong, D. R. (2001). Aquaculture—A gateway for exotic species. Science 294(5547): 1655–1656.
Neuhaus, E. B., Antonetti, D. A. and Schulz, U. H. (2016). The invasive fish Acestrorhynchus pantaneiro Menezes, 1992 in Southern Brazil: Occurrence and food niche overlap with two native species. Journal of Applied Ichthyology 32(6): 1107–1112.
Niang, A., Scheren, P., Diop, S., Kane, C. and Koulibaly, C. T. (2019). The Senegal and Pangani rivers: examples of over-used river systems within water-stressed environments in Africa. pp. 311–320. *In*: Coasts and Estuaries. Elsevier.

Nicolas, D., Chaalali, A., Drouineau, H., Lobry, J., Uriarte, A., Borja, A. et al. (2011). Impact of global warming on European tidal estuaries: some evidence of northward migration of estuarine fish species. Regional Environmental Change 11(3): 639–649.

Nilsson, C., Reidy, C. A., Dynesius, M. and Revenga, C. (2005). Fragmentation and flow regulation of the world's large river systems. Science 308(5720): 405–408.

Nilsson, G. E., Dixson, D. L., Domenici, P., McCormick, M. I., Sørensen, C., Watson, S. A. et al. (2012). Near-future carbon dioxide levels alter fish behaviour by interfering with neurotransmitter function. Nature Climate Change 2(3): 201–204.

Novais, A., Souza, A. T., Ilarri, M., Pascoal, C. and Sousa, R. (2015). Facilitation in the low intertidal: effects of an invasive species on the structure of an estuarine macrozoobenthic assemblage. Marine Ecology Progress Series 522: 157–167. https://doi.org/10.3354/meps11168.

O'Connor, J. J., Booth, D. J., Swearer, S. E., Fielder, D. S. and Leis, J. M. (2017). Ontogenetic milestones of chemotactic behaviour reflect innate species-specific response to habitat cues in larval fish. Animal Behaviour 132: 61–71.

O'Leary, J. K., Micheli, F., Airoldi, L., Boch, C., De Leo, G., Elahi, R. et al. (2017). The resilience of marine ecosystems to climatic disturbances. BioScience 67(3): 208–220. https://doi.org/10.1093/biosci/biw161.

Oliver, E. C. J., Donat, M. G., Burrows, M. T., Moore, P. J., Smale, D. A., Alexander, L. V. et al. (2018). Longer and more frequent marine heatwaves over the past century. Nature Communications 9(1). https://doi.org/10.1038/s41467-018-03732-9.

Osman, A. G., Wuertz, S., Mekkawy, I. A., Exner, H. J. and Kirschbaum, F. (2007). Lead induced malformations in embryos of the African catfish Clarias gariepinus (Burchell, 1822). Environmental Toxicology: An International Journal 22(4): 375–389.

Ostrach, D. J., Low-Marchelli, J. M., Eder, K. J., Whiteman, S. J. and Zinkl, J. G. (2008). Maternal transfer of xenobiotics and effects on larval striped bass in the San Francisco Estuary. Proceedings of the National Academy of Sciences 105(49): 19354–19359.

Paavola, M., Olenin, S. and Leppäkoski, E. (2005). Are invasive species most successful in habitats of low native species richness across European brackish water seas? Estuarine, Coastal and Shelf Science 64(4): 738–750. https://doi.org/10.1016/j.ecss.2005.03.021.

Pace, M. L., Cole, J. J., Carpenter, S. R. and Kitchell, J. F. (1999). Trophic cascades revealed in diverse ecosystems. Trends in Ecology & Evolution 14(12): 483–488.

Padilla, D. K. and Williams, S. L. (2004). Beyond ballast water: aquarium and ornamental trades as sources of invasive species in aquatic ecosystems. Frontiers in Ecology and the Environment 2(3): 131–138.

Palmer, M. A., Zedler, J. B. and Falk, D. A. (2016). Ecological theory and restoration ecology. Chapter 1, pp. 3–25. *In*: Palmer, M. A., Zedler, J. B. and Falk, D. A. (eds.). Foundations of Restoration Ecology, 2nd edition, Society for Ecological Restoration, EBSCO Publishing, University of Arizona.

Palmer, M. A., Reidy Liermann, C. A., Nilsson, C., Flörke, M., Alcamo, J., Lake, P. S. et al. (2008). Climate change and the world's river basins: anticipating management options. Frontiers in Ecology and the Environment 6(2): 81–89.

Pasquaud, S., Béguer, M., Larsen, M. H., Chaalali, A., Cabral, H. and Lobry, J. (2012). Increase of marine juvenile fish abundances in the middle Gironde estuary related to warmer and more saline waters, due to global changes. Estuarine, Coastal and Shelf Science 104-105: 46–53. https://doi.org/10.1016/j.ecss.2012.03.021.

Pauly, D., Christensen, V., Dalsgaard, J., Froese, R. and Torres Jr, F. (1998). Fishing down marine food webs. Science 279(5352): 860–863.

Peng, S. H., Wang, W. X., Li, X. and Yen, Y. F. (2004). Metal partitioning in river sediments measured by sequential extraction and biomimetic approaches. Chemosphere 57(8): 839–851.

Petersen, C. G. J. (1918). The sea bottom and its production of fish-food. Rep. Dan. Biol. Stat. 25: 1–62.

Petersen, C. G. J. and Boysen Jensen, P. (1911). Valuation of the sea, I. Animal life of the sea bottom, its food and quantity. Rep. Dan. Biol. Stat. 20: 1–81.

Peterson, M. S. (2003). A conceptual view of environment-habitat-production linkages in tidal river estuaries. Reviews in Fisheries Science 11(4): 291–313.

Phelps, H. L. (1994). The Asiatic clam (Corbicula fluminea) invasion and system-level ecological change in the Potomac River estuary near Washington, DC. Estuaries 17: 614–621.

Pihl, L., Baden, S., Kautsky, N., Rönnbäck, P., Söderqvist, T., Troell, M. et al. (2006). Shift in fish assemblage structure due to loss of seagrass Zostera marina habitats in Sweden. Estuarine, Coastal and Shelf Science 67(1-2): 123–132.

Pimentel, M., Pegado, M., Repolho, T. and Rosa, R. (2014). Impact of ocean acidification in the metabolism and swimming behavior of the dolphinfish (Coryphaena hippurus) early larvae. Marine Biology 161: 725–729.

Pinet, P. R. (2000). Invitation to Oceanography, Second Edition. Boston, USA: Jones and Bartlett Publishers: 555 pp.

Pinsky, M. L., Selden, R. L. and Kitchel, Z. J. (2020). Climate-driven shifts in marine species ranges: scaling from organisms to communities. Ann. Rev. Mar. Sci, 12: 153–179. https://doi.org/10.1146/annurev-marine-010419-010916.

Pintado-Herrera, M. G., Wang, C., Lu, J., Chang, Y. P., Chen, W., Li, X. et al. (2017). Distribution, mass inventories, and ecological risk assessment of legacy and emerging contaminants in sediments from the Pearl River Estuary in China. Journal of Hazardous Materials 323: 128–138.

Pistevos, J. C., Nagelkerken, I., Rossi, T., Olmos, M. and Connell, S. D. (2015). Ocean acidification and global warming impair shark hunting behaviour and growth. Scientific Reports 5(1): 1–10.

Planque, B., Fromentin, J. M., Cury, P., Drinkwater, K. F., Jennings, S., Perry, R. I. et al. (2010). How does fishing alter marine populations and ecosystems sensitivity to climate? Journal of Marine Systems 79(3-4): 403–417.

Postel, S. L., Daily, G. C. and Ehrlich, P. R. (1996). Human appropriation of renewable fresh water. Science 271(5250): 785–788.

Potter, I. C., Tweedley, J. R., Elliott, M. and Whitfield, A. K. (2015). The ways in whichfish use estuaries: a refinementand expansion of the guild approach. Fish and Fisheries 16: 230–239.

Poulard, J. C. and Blanchard, F. (2005). The impact of climate change on the fish community structure of the eastern continental shelf of the Bay of Biscay. ICES Journal of Marine Science 62(7): 1436–1443.

Pritchard, D. W. (1952). Estuarine hydrography. Advances in Geophysics 1: 243–280.

Ramajal, J., Picard, D., Costa, J. L., Carvalho, F. B., Gaspar, M. B. and Chainho, P. (2016). Amêijoa-japonesa, uma nova realidade no estuário do Rio Tejo: Pesca e pressão social e impacto socio-económico. Entre Rios e Mares: um Património de Ambientes, Histórias e Saberes 5: 17–30.

Ramesh, R., Lakshmi, A., Sappal, S. M., Bonthu, S., Suganya, M. D., Ganguly, D. et al. (2019). Integrated management of the Ganges delta, India. pp. 187–211. *In*: Coasts and Estuaries. Elsevier.

Ray, A. K., Tripathy, S. C., Patra, S. and Sarma, V. V. (2006). Assessment of Godavari estuarine mangrove ecosystem through trace metal studies. Environment International 32(2): 219–223.

Reddering, J. V. (1988). Prediction of the effects of reduced river discharge on the estuaries of the south-eastern Cape Province, South Africa. South African Journal of Science 84(9): 726–730.

Remane, A. (1934). Die Brackwasserfauna. Verhandlungen Der Deutschen Zoologischen Gesellschaft 36: 34–74.

Remane, A. and Schlieper, C. (1958). Die Biologie des Brackwassers. Schweizerbart'sche Verlagsbuchhandlung, Stuttgart: 348 pp.

Rezania, S., Park, J., Din, M. F. M., Taib, S. M., Talaiekhozani, A., Yadav, K. K. et al. (2018). Microplastics pollution in different aquatic environments and biota: A review of recent studies. Marine Pollution Bulletin 133: 191–208.

Richards, R. A. and Rago, P. J. (1999). A case history of effective fishery management: Chesapeake Bay striped bass. North American Journal of Fisheries Management 19(2): 356–375.

Robertson, A. I. and Duke, N. C. (1987). Mangroves as nursery sites: comparisons of the abundance and species composition of fish and crustaceans in mangroves and other nearshore habitats in tropical Australia. Marine Biology 96: 193–205.

Robin, J. P. (1992). The brown shrimp fishery of the Loire Estuary: production and by-catch of juvenile fish. Fisheries Research 13(2): 153–172.

Robins, P. E., Skov, M. W., Lewis, M. J., Giménez, L., Davies, A. G., Malham, S. K. et al. (2016). Impact of climate change on UK estuaries: A review of past trends and potential projections. Estuarine, Coastal and Shelf Science 169: 119–135. https://doi.org/10.1016/j.ecss.2015.12.016.

Rodrigues, D., Antunes, J., Otero, V., Sobral, P. and Costa, M. H. (2020). Distribution patterns of microplastics in seawater surface at a Portuguese estuary and marine park. Frontiers in Environmental Science 8: 582217.

Rodrigues, M., Fortunato, A. B. and Freire, P. (2019). Saltwater Intrusion in the Upper Tagus Estuary during Droughts. Geosciences 9(9). https://doi.org/10.3390/geosciences9090400.

Rogers, K., Lymburner, L., Salum, R., Brooke, B. P. and Woodroffe, C. D. (2017). Mapping of mangrove extent and zonation using high and low tide composites of Landsat data. Hydrobiologia 803: 49–68.

Romero, E., Garnier, J., Billen, G., Ramarson, A., Riou, P. and Cugier, P. (2019). Modeling the biogeochemical functioning of the Seine estuary and its coastal zone: Export, retention, and transformations. Limnology and Oceanography 64: 818–833.

Rosa, S., Granadeiro, J. P., Vinagre, C., França, S., Cabral, H. N. and Palmeirim, J. M. (2008). Impact of predation on the polychaete Hediste diversicolor in estuarine intertidal flats. Estuarine, Coastal and Shelf Science 78: 655–664.

Rossi, T., Nagelkerken, I., Simpson, S. D., Pistevos, J. C., Watson, S. A., Merillet, L. et al. (2015). Ocean acidification boosts larval fish development but reduces the window of opportunity for successful settlement. Proceedings of the Royal Society B: Biological Sciences 282(1821): 20151954.

Ruiz, G. M., Carlton, J. T., Grosholz, E. D. and Hines, A. H. (1997). Global invasions of marine and estuarine habitats by non-indigenous species: mechanisms, extent, and consequences. American Zoologist 37(6): 621–632. https://doi.org/10.1093/icb/37.6.621.
Ruiz, G. M., Fofonoff, P., Hines, A. H. and Grosholz, E. D. (1999). Non-indigenous species as stressors in estuarine and marine communities: assessing invasion impacts and interactions. Limnology and Oceanography 44(3part2): 950–972.
Ruiz, G. M., Fofonoff, P. W., Carlton, J. T., Wonham, M. J. and Hines, A. H. (2000). Invasion of coastal marine communities in North America: apparent patterns, processes, and biases. Annual Review of Ecology and Systematics 31(1): 481–531.
Sabatés, A. and Masó, M. (1990). Effect of a shelf-slope front on the spatial distribution of mesopelagic fish larvae in the western Mediterranean. Deep Sea Research Part A. Oceanographic Research Papers 37(7): 1085–1098.
Sabine, C. L., Feely, R. A., Gruber, N., Key, R. M., Lee, K., Bullister, J. L. et al. (2004). The oceanic sink for anthropogenic CO_2. Science 305: 367–371.
Sabo, J. L., Caron, M., Doucett, R., Dibble, K. L., Ruhi, A., Marks, J. C. et al. (2018). Pulsed flows, tributary inputs and food-web structure in a highly regulated river. Journal of Applied Ecology 55(4): 1884–1895. https://doi.org/10.1111/1365-2664.13109.
Sakai, A. K., Allendorf, F. W., Holt, J. S., Lodge, D. M., Molofsky, J., With, K. A. et al. (2001). The population biology of invasive species. Annual Review of Ecology and Systematics 32(1): 305–332.
Santos, I. R., Burdige, D. J., Jennerjahn, T. C., Bouillon, S., Cabral, A., Serrano, O. et al. (2021). The renaissance of Odum's outwelling hypothesis in 'Blue Carbon' science. Estuarine, Coastal and Shelf Science 255: 107361.
Santos, L. C. M., Gasalla, M. A., Dahdouh-Guebas, F. and Bitencourt, M. D. (2017). Socio-ecological assessment for environmental planning in coastal fishery areas: A case study in Brazilian mangroves. Ocean & Coastal Management 138: 60–69.
Sautour, B. and Castel, J. (1995). Comparative spring distribution of zooplankton in three macrotidal European estuaries. Hydrobiologia 311(1-3): 139–151. https://doi.org/10.1007/bf00008577.
Savoye, N., David, V., Morisseau, F., Etcheber, H., Abril, G. et al. (2012). Origin and composition of particulate organic matter in a macrotidal turbid estuary: The Gironde Estuary, France. Estuarine, Coastal and Shelf Science 106: 16–28.
Scanes, E., Scanes, P. R. and Ross, P. M. (2020). Climate change rapidly warms and acidifies Australian estuaries. Nat Commun. 11(1): 1803. https://doi.org/10.1038/s41467-020-15550-z.
Schettini, C. A. F., Valle-Levinson, A. and Truccolo, E. C. (2017). Circulation and transport in short, low-inflow estuaries under anthropogenic stresses. Regional Studies in Marine Science 10: 52–64. https://doi.org/10.1016/j.rsma.2017.01.004.
Schlacher, T. A. and Wooldridge, T. H. (1996). Ecological responses to reductions in freshwater supply and quality in South Africa's estuaries: lessons for management and conservation. Journal of Coastal Conservation 2: 115–130.
Schmidtko, S., Stramma, L. and Visbeck, M. (2017). Decline in global oceanic oxygen content during the past five decades. Nature 542(7641): 335–339.
Schwindt, A. R., Winkelman, D. L., Keteles, K., Murphy, M. and Vajda, A. M. (2014). An environmental oestrogen disrupts fish population dynamics through direct and transgenerational effects on survival and fecundity. Journal of Applied Ecology 51(3): 582–591.
Selleslagh. J., Lobry, J., N'Zigou, A. R., Bachelet, G., Blanchet, H. et al. (2012). Seasonal succession of estuarine fish, shrimps, macrozoobenthos and plankton: physico-cheamical and trophic influence. The Gironde estuary as a case study. Estuarine Coastal and Shelf Science 112: 243–254.
Selleslagh, J., Blanchet, H., Bachelet, G. and Lobry, J. (2015). Feeding habitats, connectivity and origin of organic matter supporting fish populations in an estuary with a reduced intertidal area assessed by stable isotope analysis. Estuaries and Coasts 38: 1431–1447.
Serafy, J. E., Harrell, R. M. and Hurley, L. M. (1994). Mechanical removal of hydrilla in the Potomac River, Maryland: Local impacts on vegetation and associated fishes. Journal of Freshwater Ecology 9(2): 135–143.
Sfakianakis, D. G., Renieri, E., Kentouri, M. and Tsatsakis, A. M. (2015). Effect of heavy metals on fish larvae deformities: a review. Environmental Research 137: 246–255.
Shoji, J., Toshito, S. I., Mizuno, K. I., Kamimura, Y., Hori, M. and Hirakawa, K. (2011). Possible effects of global warming on fish recruitment: shifts in spawning season and latitudinal distribution can alter growth of fish early life stages through changes in daylength. ICES Journal of Marine Science 68(6): 1165–1169.
Simberloff, D. (1986). Introduced insects: A biogeographie and systematic perspective. *In*: Mooney, H. A. and Drake, J. A. (eds.), Ecology of Biological Invasions of North America and Hawaii, pp. 326. Springer-Verlag, New York.
Simpson, S. D., Munday, P. L., Wittenrich, M. L., Manassa, R., Dixson, D. L., Gagliano, M. et al. (2011). Ocean acidification erodes crucial auditory behaviour in a marine fish. Biology Letters 7(6): 917–920.

Sinoir, M., Butler, E. C., Bowie, A. R., Mongin, M., Nesterenko, P. N. and Hassler, C. S. (2012). Zinc marine biogeochemistry in seawater: a review. Marine and Freshwater Research 63(7): 644–657.

Sklar, F. H. and Browder, J. A. (1998). Coastal environmental impacts brought about by alterations to freshwater flow in the Gulf of Mexico. Environmental Management 22: 547–562.

Smaal, A. C., Kater, B. J. and Wijsman, J. (2009). Introduction, establishment and expansion of the Pacific oyster Crassostrea gigas in the Oosterschelde (SW Netherlands). Helgoland Marine Research 63: 75–83.

Smale, D. A., Wernberg, T., Oliver, E. C. J., Thomsen, M., Harvey, B. P., Straub, S. C. et al. (2019). Marine heatwaves threaten global biodiversity and the provision of ecosystem services. Nature Climate Change 9(4): 306–312. https://doi.org/10.1038/s41558-019-0412-1.

Sobrino, I., Baldó, F., García-González, D., Cuesta, J. A., Silva-García, A., Fernández-Delgado, C. et al. (2005). The effect of estuarine fisheries on juvenile fish observed within the Guadalquivir Estuary (SW Spain). Fisheries Research 76(2): 229–242. https://doi.org/10.1016/j.fishres.2005.06.016.

Sousa, R., Gutiérrez, J. L. and Aldridge, D. C. (2009). Non-indigenous invasive bivalves as ecosystem engineers. Biological Invasions 11: 2367–2385.

Sousa, R., Nogueira, A. J. A., Gaspar, M. B., Antunes, C. and Guilhermino, L. (2008). Growth and extremely high production of the non-indigenous invasive species Corbicula fluminea (Müller, 1774): Possible implications for ecosystem functioning. Estuarine, Coastal and Shelf Science 80(2): 289–295. https://doi.org/10.1016/j.ecss.2008.08.006.

Statham, P. J. (2012). Nutrients in estuaries—an overview and the potential impacts of climate change. Science of the Total Environment 434: 213–227.

Stern, M. A., Flint, L. E., Flint, A. L., Knowles, N. and Wright, S. A. (2020). The future of sediment transport and streamflow under a changing climate and the implications for long-term resilience of the San Francisco Bay-delta. Water Resour Research 56: e2019WR026245.

Stompe, D. K., Moyle, P. B., Kruger, A. and Durand, J. R. (2020). Comparing and integrating fish surveys in the San Francisco estuary: why diverse long-term monitoring programs are important. San Francisco Estuary and Watershed Science 18: 1–16.

Streftaris, N., Zenetos, A. and Papathanassiou, E. (2005). Globalisation in marine ecosystems: the story of non-indigenous marine species across European seas. pp. 429–464. *In*: Oceanography and Marine Biology. CRC Press.

Strydom, N. A., Whitfield, A. K. and Paterson, A. W. (2002). Influence of altered freshwater flow regimes on abundance of larval and juvenile Gilchristella aestuaria (Pisces: Clupeidae) in the upper reaches of two South African estuaries. Marine and Freshwater Research 53(2): 431–438.

Tallis, H. M., Ruesink, J. L., Dumbauld, B., Hacker, S. and Wisehart, L. M. (2009). Oysters and aquaculture practices affect eelgrass density and productivity in a Pacific Northwest estuary. Journal of Shellfish Research 28(2): 251–261.

Teodosio, M. A., Paris, C. B., Wolanski, E. and Morais, P. (2016). Biophysical processes leading to the ingress of temperate fish larvae into estuarine nursery areas: A review. Estuarine, Coastal and Shelf Science 183: 187–202.

Thomsen, M. S., Byers, J. E., Schiel, D. R., Bruno, J. F., Olden, J. D., Wernberg, T. et al. (2014). Impacts of marine invaders on biodiversity depend on trophic position and functional similarity. Marine Ecology Progress Series 495: 39–47.

Tsiamis, K., Zenetos, A., Deriu, I., Gervasini, E. and Cardoso, A. C. (2018). The native distribution range of the European marine non-indigenous species. Aquatic Invasions 13(2).

Tupper, M. and Boutilier, R. G. (1995). Size and priority at settlement determine growth and competitive success of newly settled Atlantic cod. Marine Ecology Progress Series 295–300.

Ummenhofer, C. C. and Meehl, G. A. (2017). Extreme weather and climate events with ecological relevance: a review. Philos Trans R Soc Lond B Biol Sci. 372(1723). https://doi.org/10.1098/rstb.2016.0135.

van de Kam, J., Ens, B., Piersma, T. and Zwarts, L. (2004). Shorebirds. An Illustrated Behavioural Ecology. KNNV Publishers, Utrecht.

van den Brink, A., Godschalk, M., Smaal, A., Lindeboom, H. and McLay, C. (2013). Some like it hot: the effect of temperature on brood development in the invasive crab Hemigrapsus takanoi (Decapoda: Brachyura: Varunidae). Journal of the Marine Biological Association of the United Kingdom 93(1): 189–196.

Van Niekerk, L., Adams, J. B., Bate, G. C., Forbes, A. T., Forbes, N. T., Huizinga, P. et al. (2013). Country-wide assessment of estuary health: An approach for integrating pressures and ecosystem response in a data limited environment. Estuarine, Coastal and Shelf Science 130: 239–251.

Van Vliet, M. T., Franssen, W. H., Yearsley, J. R., Ludwig, F., Haddeland, I., Lettenmaier, D. P. et al. (2013). Global river discharge and water temperature under climate change. Global Environmental Change 23(2): 450–464.

Vasconcelos, R. P., Reis-Santos, P., Fonseca, V., Maia, A., Ruano, M., França, S. et al. (2007). Assessing anthropogenic pressures on estuarine fish nurseries along the Portuguese coast: a multi-metric index and conceptual approach. Science of the Total Environment 374(2-3): 199–215.

Verdonschot, P. F. M., Spears, B. M., Feld, K. D., Brucet, S., Keizer-Vlek, H., Borja, A. et al. (2013) A comparative review of recovery processess in rivers, lakes, estuarine and coastal waters. Hydrobiologia 704: 453–474.

Vieira, L. R., Gravato, C., Soares, A. M. V. M., Morgado, F. and Guilhermino, L. (2009). Acute effects of copper and mercury on the estuarine fish Pomatoschistus microps: linking biomarkers to behaviour. Chemosphere 76(10): 1416–1427.

Vilas, C., Drake, P. and Pascual, E. (2009). Inter- and intra-specific differences in euryhalinity determine the spatial distribution of mysids in a temperate European estuary. Journal of Experimental Marine Biology and Ecology 369(2): 165–176.

Vinagre, C., Costa, M. J. and Cabral, H. N. (2007). Impact of climate and hydrodynamics on sole larval immigration towards the Tagus estuary, Portugal. Estuarine, Coastal and Shelf Science 75(4): 516–524.

Vinagre, C., Salgado, J., Cabral, H. N. and Costa, M. J. (2010). Food web structure and habitat connectivity in fish estuarine nurseries—impact of river flow. Estuaries and Coasts 34(4): 663–674. https://doi.org/10.1007/s12237-010-9315-0.

Von Moos, N., Burkhardt-Holm, P. and Köhler, A. (2012). Uptake and effects of microplastics on cells and tissue of the blue mussel Mytilus edulis L. after an experimental exposure. Environmental Science & Technology 46(20): 11327–11335.

Vosylienė, M. Z., Kazlauskienė, N. and Svecevičius, G. (2003). Effect of a heavy metal model mixture on biological parameters of rainbow trout Oncorhynchus mykiss. Environmental Science and Pollution Research 10: 103–107.

Wada, Y., Van Beek, L. P., Wanders, N. and Bierkens, M. F. (2013). Human water consumption intensifies hydrological drought worldwide. Environmental Research Letters 8(3): 034036.

Walsh, C. T., Gray, C. A., West, R. J., van der Meulen, D. E. and Williams, L. F. (2010). Growth, episodic recruitment and age truncation in populations of a catadromous percichthyid, Macquaria colonorum. Marine and Freshwater Research 61(4): 397–407.

Ward, N. D., Bianchi, T. S., Osburn, C. L. and Myers-Pigg, A. (2024). Biogeochemical changes in estuaries. pp. 113–142. *In*: Kennish, M. J., Paerl, H. W. and Crosswell, J. R. (eds.). Climate Change and Estuaries. CRC Press, Boca Raton.

Webb, E. L., Friess, D. A., Krauss, K. W., Cahoon, D. R., Guntenspergen, G. R. and Phelps, J. (2013). A global standard for monitoring coastal wetland vulnerability to accelerated sea-level rise. Nature Climate Change 3(5): 458–465.

Wernberg, T., Bennett, S., Babcock, R. C., de Bettignies, T., Cure, K., Depczynski, M. et al. (2016). Climate-driven regime shift of a temperate marine ecosystem. Science 353(6295): 169–172. https://doi.org/10.1126/science.aad8745.

Whitfield, A.K. (1992). A characterization of southern African estuarine systems. Southern African Journal of Aquatic Sciences 18: 89–103.

Whitfield, A. K. (2005). Fishes and freshwater in southern African estuaries—A review. Aquatic Living Resources 18(3): 275–289. https://doi.org/10.1051/alr:2005032.

Whitfield, A. K. (2015). Why are there so few freshwater fish species in most estuaries? J. Fish Biol. 86(4): 1227–1250. https://doi.org/10.1111/jfb.12641.

Whitfield, A. K. and Elliott, M. (2002). Fishes as indicators of environmental and ecological changes within estuaries: a review of progress and some suggestions for the future. Journal of Fish Biology 61: 229–250.

Whitfield, A. K. and Elliott, M. (2011). Ecosystem and biotic classifications of estuaries and coasts. pp. 99–124. *In*: Wolanski, E. and McLusky, D. S. (eds.). Treatise on Estuarine and Coastal Science. Waltham: Academic Press.

Williams, M. R., Bhatt, G., Filoso, S. and Yactayo, G. (2017). Stream restoration performance and its contribution to the Chesapeake Bay TMDL: challenges posed by climate change in urban areas. Estuaries and Coasts 40: 1227–1246.

Williams, S. L. and Grosholz, E. D. (2008). The invasive species challenge in estuarine and coastal environments: marrying management and science. Estuaries and Coasts 31(1): 3–20. https://doi.org/10.1007/s12237-007-9031-6.

Wilson, J. G. (2002). Productivity, fisheries and aquaculture in temperate estuaries. Estuarine, Coastal and Shelf Science 55(6): 953–967. https://doi.org/10.1006/ecss.2002.1038.

Winemiller, K. O. and Rose, K. A. (1992). Patterns of life-history diversification in North American fishes: implications for population regulation. Canadian Journal of Fisheries and Aquatic Sciences 49(10): 2196–2218.
Wolanski, E. and Ducrotoy, J.-P. (2014). Estuaries of Australia in 2050 and beyond—A synthesis. Estuaries of Australia in 2050 and Beyond 1–13.
Wolanski, E., Chicharo, L., Chicharo, M. A. and Morais, P. (2006). An ecohydrology model of the Guadiana estuary (South Portugal). Estuarine, Coastal and Shelf Science 70(1-2): 132–143.
Wolanski, E. and Elliott, M. (2015). Estuarine Ecohydrology: An Introduction. Elsevier, Amsterdam.
Zhang, E., Savenije, H. H. G., Chen, S. and Chen, J. (2012). Water abstraction along the lower Yangtze River, China, and its impact on water discharge into the estuary. Physics and Chemistry of the Earth, Parts A/B/C 47-48: 76–85. https://doi.org/10.1016/j.pce.2011.05.002.
Zhu, B., Wu, Z. F., Li, J. and Wang, G. X. (2011). Single and joint action toxicity of heavy metals on early developmental stages of Chinese rare minnow (Gobiocypris rarus). Ecotoxicology and Environmental Safety 74(8): 2193–2202.

CHAPTER 9

An Ecosystem View of Anthropogenic Impacts in the Deep Ocean

Eva Ramirez-Llodra[1,*] and *Alex David Rogers*[2]

The deep sea: The largest biome on Earth

Nearly 66% of the ocean has experienced increased human impacts in the last decade, with 5% heavily impacted and in need of management measures (Halpern et al. 2015). Deep-sea ecosystems are no exception, and although it was long considered monotonous and with scarce life, exploration and investigations in the last 2 centuries continue to show a highly diverse and dynamic environment that is essential for the health of the planet. The deep sea is the largest biome on Earth, with the deep pelagic environment comprising 73% of water in the ocean, with 50% of this at depths greater than 3000 m, an average depth of 3800 m and the deep seafloor comprising 90% of the total seafloor (Costello et al. 2010, Ramirez-Llodra et al. 2010).

Although a somewhat arbitrary definition, it is generally accepted to consider the deep sea from 200–250 m depth (Ramirez-Llodra et al. 2010, Rogers 2015). This is the depth where the shelf break usually occurs, transitioning from the relatively shallow shelf to the steeper slopes of the continental margins. It is also the depth where, in most regions, the amount of sunlight penetrating in the water column becomes insufficient to support photosynthesis. This is where the ecosystems transition into heterotrophic systems, where the fauna is supported by organic matter sinking through the water column or advected along the slopes from the euphotic zone (Gage 2003). This transition can vary, however, for example, in the tropics, the rariphotic zone has been identified between 150 m to 300 m deep, where fish communities are dominated by species belonging to shallow-water families (Baldwin et al. 2017, Stefanoudis et al. 2019). Marine macroalgae have also been observed at depths greater than 300 m (Friedlander et al. 2014). In contrast, the continental shelves of the Antarctic are depressed by the weight of ice on the continent and so are much deeper than elsewhere.

The deep sea can be divided into different zones, both for the pelagic and benthic systems, based on depth and ecological characteristics (Fig. 1). The deep-water pelagic zone comprises the mesopelagic (200–1000 m), bathypelagic (1000–4000 m), abyssopelagic (4000–6000 m) and hadopelagic (6000–11000 m). The bathypelagic is sometimes also considered to start at 4000 m depth, but here we use the definition based on the 4C isobath, which is usually found at 2000 m depth (Rogers et al. 2015). The benthic deep sea can be divided into the upper bathyal zone (250–1000 m), middle bathyal zone (1000–2000 m), lower bathyal zone (2000–4000 m), abyssal zone (4000–6000 m) and hadal zone (6000–11000 m). The deep sea is the last frontier for exploration and investigation on Earth, with most

[1] REV Ocean, Norway.
[2] Ocean Census, Begbroke, Oxfordshire, U.K. and REV Ocean, Norway.
* Corresponding author: eva.ramirez-llodra@revocean.org

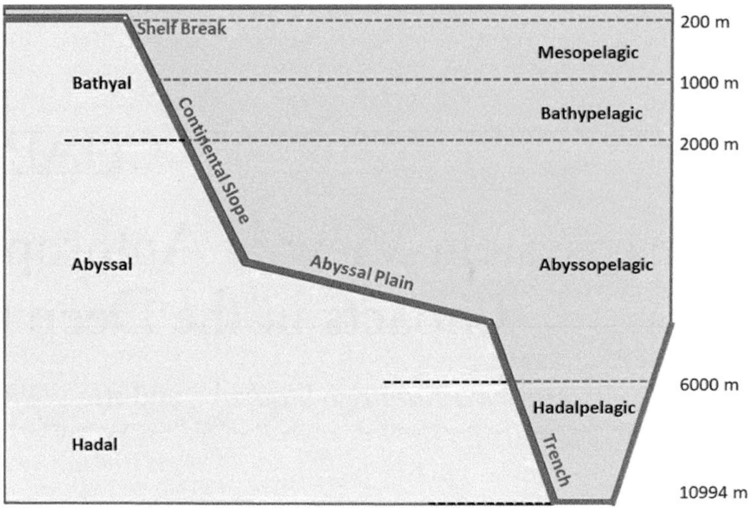

Figure 1. Pelagic and benthic zonation of the ocean.

of its vast ecosystems not sampled to date. However, current knowledge provides evidence of high deep-sea biodiversity, with an estimated 1–2.2 M species found in the entire ocean, of which 75–90% are still undescribed, particularly of small macro- and meiofauna and mainly from the deep sea (Mora et al. 2011, Appeltans et al. 2012, Paulus 2021, Bouchet et al. 2023). Long considered a monotonous and desert-like environment, the deep sea is in fact a highly heterogeneous biome, with a variety of habitats characterised by specific physico-geochemical drivers that shape faunal communities.

The abyssal plains are the largest benthic habitat, covering 75% of the ocean floor (Fig. 2A). These large expanses of mostly inorganic fine sediment are usually exposed to cold water temperatures (0–4°C), excluding in the Mediterranean Sea (13°C) and Red Sea (21°C), high pressures and complete darkness, excluding bioluminescence (Ramirez-Llodra et al. 2010). The abyss is characterized by low food availability, with the quantity and quality of the detrital food sinking from the water column modulating the structure and function of the ecosystem (Gage 2003, Smith et al. 2008). Although this food limitation results in sparse abundance of megafauna, abyssal biodiversity, particularly of macro- and meiofauna, is very high (Snelgrove and Smith 2002, Rex and Etter 2010, Rabone et al. 2023). Physical variation in habitat including sediment characteristics, the presence of hard substrata such as polymetallic nodules and the presence of larger scale abyssal hills can all contribute to variation in biological communities (e.g., Simon-Lledo et al. 2019). Understanding of the connectivity of these communities is relatively poor (Taylor and Roterman 2017), although investigations using environmental (e)DNA and molecular population genetic approaches are showing promise in resolving patterns of species distribution and gene flow across large ocean basins (e.g., Lejzerowicz et al. 2021, Cordier et al. 2022, Stewart et al. 2023). The functioning of the communities in and on the abyssal sediments support important ecosystem services, such as nutrient regeneration, carbon burial and dissolution of calcium carbonate, and because of the large size of abyssal plains, changes in ecosystem functions can result in significant modifications in the ecosystem services provided at a global scale (Snelgrove and Smith 2002). Although the great depths of abyssal plains have protected their ecosystems to some degree, certain stressors such as pollution and warmer temperatures reach abyssal depths. In addition, the planned start of seabed mining for polymetallic nodules in abyssal ecosystems risks biodiversity loss, including species extinction, and habitat destruction at very large geographic scales (Jones et al. 2018, Niner et al. 2018).

Continental margins cover 11% of the ocean floor and are characterized by high habitat diversity shaped by different combinations of geological, geochemical and physical variables (Levin et al. 2010, Menot et al. 2010). The sedimentary slopes can be intersected by submarine canyons (Fig. 2B) that modify currents and act as conduits for particles from the shelf to the deep seafloor, providing enhanced food supply as well as refuge and nursery areas for fauna (Fernandez-Arcaya et al. 2017). Cold-water

Figure 2. Deep-sea ecosystems. (A) Zoroaster asteroid on the Porcupine Abyssal Plain, Atlantic Ocean (David Billett, NOCS); (B) Crinoids on a canyon wall, NE Atlantic (Paul Tyler, NOCS); (C) Cold-water coral reef from Coral Seamount, SW Indian Ridge at 1000 m depth (NERC-IUCN Seamounts Project, A. Rogers, JC066); (D) Cold seeps community in the Gulf of Mexico (Erik Cordes); (E) Biological community on the Atlantis Seamount, Indian Ocean (Alex Rogers); (F) the Enceladus black smoker on the Aurora Vent Field, Gakkel Ridge, Arctic Ocean (HACON, REV Ocean).

coral gardens and reefs (Fig. 2C), sponge gardens and other biogenic ecosystems are also found on continental margins, providing habitat for a myriad of associated organisms that use these habitats for essential activities such as feeding, refuge or nurseries (Rogers 1999, Roberts et al. 2009). Cold seeps are found in active and inactive margins, where fluid charged with methane and other reduced chemicals flows through the sediment (Fig. 2D). These reduced chemicals fuel chemoautotrophy by specialized microorganisms to produce organic matter and support abundant communities of specialized fauna (Suess 2014). This diversity of continental margin habitats supports a variety of ecosystem functions that result in important ecosystem services (Armstrong et al. 2012, Thurber et al. 2014). For example, canyons contribute to carbon sequestration through the enhancement of particle transport from the shelf to the deep seafloor (Fernandez-Arcaya et al. 2017); canyons and cold-water corals can provide nursery areas for other species, including species with commercial value (Rogers 1999, Fernandez-Araya et al. 2017); and sedimentary slopes on continental margins sustain communities of biological resources such as stocks of fish and crustaceans that are regularly fished. Because of their relative accessibility, particularly at the upper bathyal depths, continental margins are subject to intense and increasing anthropogenic impacts from fisheries, hydrocarbon exploitation and associated spill risks, pollution and stressors related to climate change, such as acidification, increased temperatures and expansion of oxygen minimum zones (Menot et al. 2010, Levin and Le Bris 2015, Fernandez-Arcaya et al. 2017).

Seamounts are topographically isolated elevations rising at least 1000 m from the surrounding seabed, with smaller structures called knolls and hills (Schlacher et al. 2010, Rogers 2018). The particular topography of seamounts results in a variety of habitats, from sedimented tops or flanks with low inclinations, to rocky outcrops that provide support to filter feeders such as corals and sponges and their associated fauna (Fig. 2E; Rogers 2018). These communities fulfill the characteristics of Vulnerable Marine Ecosystems (VMEs) as described by the UN Food and Agriculture Organisation (FAO) in the guidelines for management of fisheries in the High Seas (FAO 2009). In addition, seamounts can trap diurnally vertically migrating micronekton such as crustaceans, fish (e.g., lantern fish) and cephalopods, which can attract dense aggregations of fish, including species of commercial value (Rogers 2018). To date, the main impact on seamounts has been caused by commercial fishing (Rogers 2018, 2019). However, the potential development of a deep-sea mining industry targeting cobalt crusts that have formed on certain seamounts over millions of years has the potential to destroy large areas of seamount communities (Schlacher et al. 2014, Rogers 2019).

Mid-ocean ridges are a semi-continuous volcanic mountain chain that run across all oceans, where two tectonic plates are being pulled/pushed apart and new seafloor is formed. In these regions, water percolates through cracks in the oceanic crust and, as the water circulates close to the magma chamber, the fluid is charged with sulphide, hydrogen, methane, manganese and other metals and depleted of magnesium and oxygen. As this superheated fluid (commonly 300–350°C) rises back to the seafloor and comes into contact with the oxygenated cold seawater, the minerals precipitate forming the characteristic black smokers of hydrothermal vents (Fig. 2F). To date, 722 high-temperature vent sites have been confirmed, and an additional 720 high-temperature vents have been inferred from water column data (Beaulieu and Szafranski 2020). Active hydrothermal vents are tiny islands covering a global area of only 50 km^2, forming oases of exotic life in the deep ocean (Van Dover 2000). The vent communities are fueled by high primary productivity from chemoautotrophic microorganisms, both free living and in symbiosis with the fauna, that use the reduced chemicals from the vent fluids as a source of energy to produce organic matter (Van Dover 2000). The vent fauna is highly adapted to a dynamic ecosystem characterized by steep temperature gradients, high concentrations of toxic chemicals and high pressure, resulting in a high degree of species endemism (Van Dover 2000). Because of their physiological adaptations, hydrothermal vent organisms are valued for their genetic diversity and the potential for discovery of marine genetic resources of use in biomedicine, biotechnology, cosmetics and biofuels (Van Dover et al. 2018). In addition, interest in the potential exploitation of accumulated minerals in hydrothermal mounds has increased greatly in recent decades, and exploration licenses have been granted both in national and international waters (Jones et al. 2020). The low number of hydrothermal vent sites occupied by some vent-endemic species, which cover a very small spatial area, means that mining such ecosystems will pose a significant risk of extinction (e.g., Sigwart et al. 2019).

Impacts from extractive industries

Fishing

Fish are a critical component of global food security and provide essential nutrients in the diets of people globally, but especially in developing countries (Béné et al. 2015). Fisheries and aquaculture provide 3.0 billion people with 20% of their annual protein and a further 1.3 billion with 15% (HLPE 2014). They are also critical to food security indirectly through supporting the livelihoods of fishers, aquaculturists, aquatic food processing and trading. Overall, these activities support 660–820 million people including both workers and their families (HLPE 2014, Béné et al. 2015). Fisheries, when not properly managed, can have several impacts on fish populations and ecosystems. The main ones are overfishing, by-catch and damage to the environment.

Overfishing

Global marine fish catches peaked in 1996 at a level reported by FAO as 86.4 million tonnes declining to 79.3 million tonnes by 2016 (FAO 2018, 2022). However, these figures have been disputed and are likely to have peaked at a higher maximum value (130 million tonnes) and declined more rapidly (Pauly and Zeller 2016). This global picture of declining fish catches is reflected in estimated trends in overfishing. The number of fish stocks at biologically sustainable population levels declined from 90% in 1974 to 66.9% in 2015 (FAO 2018, 2022). This means that nearly a third of fish stocks are overfished, whilst 59.9% are fully exploited and 7.0% underexploited (FAO 2018). Many deep-sea fish species have a slow growth, delayed maturity and are long lived, which makes them particularly vulnerable to poorly managed fisheries (Cheung et al. 2007, Simfendorfer and Kyne 2009, Norse et al. 2012). For example, the orange roughy (*Hoplostethus atlanticus*) does not reproduce until individuals reach 30–40 years of age and can live to 150 years (Rogers 2019). They often form large aggregations in specific areas such as over seamounts, making them easy targets for fisheries (Doonan et al. 2015). The extraction of large numbers of mature fish resulted in quick collapse of several populations in New Zealand, Australia and Namibia (Clark et al. 2007). Overfishing can also have cascading effects in the pelagic ecosystem, through the removal of key species of predators such as sharks or cod (Baum et al. 2003, Myers et al. 2007), which can lead to regime shifts. Removal of zooplanktivorous forage fish species can also result in significant regime shifts at lower trophic levels, leading to the "jellification" of marine ecosystems (Daskalov et al. 2007, Roux et al. 2013).

Bycatch

Bycatch has been defined as the catch that is unused or unmanaged (Davies et al. 2009). In addition to non-commercial fish species, bycatch includes marine turtles, seals, seabirds, cetaceans, sharks, rays and chimaeras (Read et al. 2006, Anderson et al. 2011, Worm et al. 2013, Bourjea et al. 2014). Populations of non-targeted fish, sharks and rays have been depleted by deep-sea fisheries, some to the point where they have become critically endangered at a regional level (e.g., Devine et al. 2006, Priede et al. 2011, Neat et al. 2015). Orange roughy fisheries on the Chatham Rise have been associated with the decline of Plunket's shark (*Centroscymnus plunketi*), the black cardinal fish (*Epigonus telescopus*) and the basketwork eel (*Diastobranchus capensis*), likely as a result of bycatch (Clark et al. 2000). Chilean orange roughy fisheries have also shown evidence of decline of bycatch species including the shark *Etmopterus granulosus*, *Gadella obscurus* and several species of macruorids (Niklitschek et al. 2010). In the northeastern Atlantic, monitoring of catches in fisheries down to 1,500 m depth showed a decline in biomass of both targeted and bycatch species down to 2,500 m (Bailey et al. 2009). This was thought to be caused by fisheries removals of these species in the shallow part of their range and demonstrates that deep-sea fisheries can have an impact over a wider area than their apparent footprint (Bailey et al. 2009, Priede et al. 2011).

Environmental damage

Environmental damage of fishing has been extensively documented. Fishing gear deployed on the seabed, especially mobile fishing gear such as trawls but to a lesser extent static gear such as long lines, traps and gill nets, can damage seafloor ecosystems. The level of damage varies according to the gear (Collie et al. 2000), the frequency of exposure to fishing (Clark et al. 2019), the stability or structure of the physical habitat affected (Collie et al. 2000, Buhl-Mortensen et al. 2016) and the morphology, lifestyle and longevity of the species present (Pitcher et al. 2010, Clark et al. 2016, Buhl-Mortensen et al. 2016, Hiddink et al. 2018). Impacts of trawling or dredging include the removal of physical or biogenic habitat structures; mortality of animals; decreased abundance, biomass, species richness and diversity of seabed biota, especially large organisms and changes in community structure (Clark and Rowden 2009, Buhl-Mortensen et al. 2016, Clark et al. 2016, 2019, Hiddinck et al. 2018). Of particular importance in deep-sea ecosystems is the physical damage caused to habitat-forming colonial species, such as cold-water coral reefs or sponge beds (Rogers 1999, 2019, Hall-Spencer et al. 2002) (Fig. 3A). Individual colonies of deep-sea corals, for example, have been found to live for more than 4,000 years (Roark et al. 2009), whilst one species of deep-sea sponge is estimated to live for up to 11,000 years, making it one of the most long-lived organisms on Earth (Jochum et al. 2012). For stony corals, although the longevity of individual colonies may be in the range of decades to > 100 years (Fallon et al. 2014), the reef structures they form can be up to 50,000 years old (Fallon et al. 2014). For such long-lived species and communities, destruction by bottom trawling targeted at deep-sea ecosystems such as seamounts means that recovery may only take place over centuries to millennia (Clark et al. 2016). Examples of impacted cold-water corals from bottom trawling include the Darwin mounds in the Rockall Trough west of Scotland, mostly composed of *Lophelia* and *Madrepora* (e.g., Masson et al. 2003, Wheeler et al. 2005), *Lophelia* reefs off

Figure 3. Examples of anthropogenic impacts on deep-sea ecosystems. (A) Coral rubble and trawl marks at bathyal depths, New Zealand (A. Rowden, NIWA); (B) Entangled fauna on a lost longline in the Whittard Canyon, Atlantic Ocean (P. Tyler, NOCS); (C) cold water corals and ophiuroids covered in oil from the Deep Water Horizon spill, Gulf of Mexico (C. Fisher, PSU); (D) A lander used in seabed mining studies amongst the tracks left by the nodule mining prototype vehicle *Patania II*, Clarion-Clipperton Zone, Pacific Ocean/JPIO-Mining Impact 2 cruise SO295 (GEOMAR, ROV team Kiel 6000); (E) marine litter in the NE Atlantic (P. Tyler, NOCS); (F) ghost fishing example, showing dead/moribund crabs on a lost net collected from 2000 m depth in the Mediterranean Sea (E. Ramirez-Llodra, ICM-CSIC).

western Norway (e.g., Fosså et al. 2002), *Oculina* reefs offshore Florida (e.g., Reed et al. 2007), or stony coral habitats on seamounts offshore New Zealand and Australia (e.g., Williams et al. 2010).

In terms of physical impact on the habitat, trawling can have the effects of homogenizing the seabed, obliterating structures such as sand waves and depressions and bulldozing rocks (Auster et al. 1996, Buhl-Mortensen et al. 2016). Trawls can also leave significant cuts or scouring of soft substrates (Buhl-Mortensen et al. 2016, Mortensen and Mortensen 2018) or physically destroy hard substrata such as manganese crusts on seamounts (Waller et al. 2007). Sediment resuspension and seafloor remodelling by trawling can be so large that it can alter sediment fluxes and submarine landscapes over large spatial scales (Puig et al. 2012). Sediments on the continental shelf and slope have important roles in the transformation and exchange of organic material and nutrients as well as remineralisation (Thrush and Dayton 2002). The impacts of trawling on the physical structure of seabed sediments as well as biogenic structures such as burrows, tubes and tests are known to impact sediment organic content, community structure, including diversity and functional diversity, likely impacting important biogeochemical processes at least partially mediated by organisms (Pusceddu et al. 2014, Ashford et al. 2018).

Whilst the level of impacts of individual longlines are small compared to bottom trawls, bottom long lining is increasingly recognised as also capable of damaging vulnerable marine ecosystems, particularly branching corals including scleractinians and octocorals (Pham et al. 2014a, Stone et al. 2015) (Fig. 3B). Damage occurs from the weights used to hold lines to the seabed, lateral movement of longline sets during retrieval, or as a result of currents at the seabed and also through the hooking and entanglement of organisms (Pham et al. 2014a, Stone et al. 2015).

Illegal, unregulated and unreported fishing

Illegal, unregulated and unreported (IUU) fishing has been reported to capture 11–26 million tonnes of fish per annum worth an estimated \$10–\$23.5 billion USD (Agnew et al. 2009). IUU leads to significant economic losses to the countries most affected (de Coning and Witbooi 2015, Doumbouya et al. 2017) and undermines efforts to move fisheries to sustainability, both in terms of catches of target species and the environmental impacts of fishing activity (Agnew et al. 2009, Richardson et al. 2018). As such, IUU fishing is a direct threat to the conservation status of species and ecosystems (Agnew et al. 2009) and can contribute considerably to extinction risk on a regional and global basis (Riskas et al. 2018). Apart from impacting shallow-water fisheries, IUU fishing is also a threat to deep-sea species and ecosystems especially where such species are valuable (e.g., Patagonian toothfish; Sumaila et al. 2006, Collins et al. 2010). Unreported and unregulated fisheries are also likely to be a significant issue in deep-sea ecosystems in areas beyond national jurisdiction. In these waters, only certain fish stocks are managed by Regional Fisheries Management Organisation (RFMOs) in specific geographic areas (Rogers and Gianni 2010). Using a combination of satellite tracking data and information on vessel registries, it has been estimated that about 17% of fishing in high seas areas is by potentially unauthorised or unregulated vessels (Park et al. 2023). An example of the scale of this problem is unregulated squid fishing, which has increased 68% from 2017 to 2020, taking place largely in unregulated regions (86%) and amounting to 149,000 to 251,000 fishing days annually. Such fisheries not only have a great potential to damage the targeted stocks but can also cause significant damage to associated biological communities such as VMEs (Brewin et al. 2021). Even where fisheries are regulated, pooling of data on catches or bycatches can obscure the impacts of fisheries on individual species (e.g., pooling all data on shark catch or bycatch; Rogers and Gianni 2010). IUU fishing therefore represents a significant challenge to the conservation and sustainable management of deep-sea ecosystems.

Hydrocarbons

The exploration for and exploitation of hydrocarbon resources (oil and gas) has been extending from the shelf into deeper waters for decades, with technological developments allowing for wells being drilled down to 3000 m water depth (Macreadie et al. 2018). Offshore operations make up 30% of global oil production (Narula 2018) and an increase in offshore exploration has been particularly important in countries such as the United States (Gulf of Mexico), Norway, Brazil and Angola, with 60–80% of maritime oil deeper than 150 m. The main impacts of oil and gas exploitation are mostly local, caused by the physical damage of drilling and the production of drilling muds and sediment plumes (Gray et al. 1990), but can extend to regional scales in the event of large-scale underwater oil spills. The ecosystems associated with regions of interest for hydrocarbon exploitation can be diverse and will depend on the local environmental conditions as well as, in some regions, on variables related to the presence of hydrocarbons (Bernardino et al. 2020) where pore water can be enriched in methane and hard substrata (carbonates), providing habitat for attached fauna, associated with methane seeps (Levin 2005, Sibuet and Olu 1998). Cold seep communities can comprise abundant assemblages of benthic species, including chemoautotrophic symbiont-bearing and habitat-forming species such as vesicomyid clams, bathymodiolid mussels and siboglinid tubeworms that in turn support a rich associated fauna such as polychaetes, molluscs and crustaceans (Sibuet and Olu 1998, Bernardino et al. 2012), as well as cold-water corals in the surrounding areas (Cordes et al. 2010). Drilling muds or increased particle concentrations in the water column can clog respiratory and feeding organs, with lethal to sub-lethal effects on the individuals. Many of the symbiont-bearing species and other biogenic species such as corals and sponges are long-lived, making them particularly vulnerable to physical or indirect damage.

In addition to local effects, large-scale regional environmental impacts can happen in the event of underwater oil spills. The most recent and largest example of an offshore marine oil spill is that of the *Deepwater Horizon* in the Gulf of Mexico, where an explosion of safety valves in April 2010 resulted in large amounts of oil mixed with gas being released directly into the deep sea (Joye et al. 2016). The explosion caused the release of about 5 million barrels (7,806,103 m^3) of crude oil into the water, until the well was finally sealed in September 2010. The oil rose to the euphotic zone, where it adhered to zooplankton creating an oiled marine snow that sank to the seafloor, covering benthic communities (Passsow et al. 2012), including impact on cold-water corals over 20 km from the spill, and at depths between 1000 and 1900 m (Fisher et al. 2014) (Fig. 3C). The spill also created a mid-water plume of oil microdroplets, enhanced by the use of dispersants, affecting the pelagic microbial communities and reducing their capacity to degrade the hydrocarbons (De Leo et al. 2016). Subsequent studies have shown a limited recovery from the spill after 8 years, with lower diversity in the benthic communities (McClain et al. 2019) and continuous degradation of coral colonies (Girard et al. 2019).

Seabed mining

The increase in mineral demand for the development of new technologies, including those for the energy transition, are driving the push from some sectors and governments to explore, and potentially, in the near future, exploit minerals from the deep seabed (Miller et al. 2018). However, the need for sourcing minerals from the deep sea has been challenged by evidence that new technologies and approaches can obtain the necessary minerals or reduce demand for them, including improved efficiency in exploiting current land-based mines, better recycling, urban mining or the development of batteries or other technologies that are not dependent on critical minerals (Simas et al. 2022). Regardless, interest in seabed mining has greatly increased both in national waters and in areas beyond national jurisdiction (Jones et al. 2020). Although commercial-scale seabed mining has not been initiated at the time of writing this chapter, exploration licenses have been granted in international waters and several nations are developing regulations to open their waters for seabed mining. In areas beyond national jurisdiction, the International Seabed Authority (ISA) is the regulatory authority, established under the 1982 United Nations Convention on the Law of the Sea (UNCLOS) and its implementing agreements, which is charged with managing seabed mineral exploitation in The Area (i.e., the seafloor in areas beyond national jurisdiction). By December 2022, the ISA had granted 31 exploration contracts for the 3 main resource types currently being considered for exploitation: 19 for polymetallic nodules, 7 for seafloor massive sulfides and 5 for cobalt crusts (ISA 2022). The proposed exploitation of these three resources will follow a similar overall extraction process, with large mining vehicles on the seafloor, collecting and, in the case of cobalt crusts and seafloor massive sulfides, crushing the materials, which will then be sent to a surface vessel via a riser pipe (Jones et al. 2020). The material will be dewatered in the vessel and the filtered water will be discarded in the water column or close to the seafloor (Muñoz-Royo et al. 2021, Weaver et al. 2022). The main environmental impacts common to the three resources include the destruction of the habitat and elimination of the fauna on the mining site, the creation of dispersing particle plumes in the water column, potential for the release of toxic compounds, and noise and light pollution (Ramirez-Llodra et al. 2011, Gollner et al. 2017, Jones et al. 2020, Williams et al. 2022). These impacts can result in individual behaviour change, biodiversity loss and community changes that may persist over geological time-scales in the mined sites (Gollner et al. 2017, Niner et al. 2018, Williams et al. 2022).

Manganese nodules are polymetallic potato-shaped nodules that occur on abyssal plains between 4000 and 6000 m depth. Nodules grow extremely slowly, with accretion rates of 2–15 mm per million years (Levin et al. 2016a). They are rich in manganese, nickel and cobalt and the main areas of nodule accumulations are in the Clarion Clipperton Zone in the Pacific Ocean, the Central Indian Ocean Basin and in the Peru Basin. To date, the technology to collect nodules is the most advanced for seabed mining, currently based on large vehicles (10 m wide) that would move in a predetermined grid on the seafloor, collecting the nodules in surface sediments down to 50 cm (Jones et al. 2020). Recent studies in the exploration areas of the Clarion Clipperton Zone in the Central Pacific Ocean provide a wealth of information on the potential environmental impacts of nodule mining (e.g., Glover et al. 2016, Amon et al. 2016, Lejzerowicz et al. 2021). Abyssal plain ecosystems are characterized by a high biodiversity composed mostly of small (microbes, meiofauna and macrofauna) organisms (Rex and Etter 2010, Rabone et al. 2023). On the seafloor, the nodules provide substrate for a variety of fauna including, amongst others, sponges, anemones, corals, polychaetes and large foraminifera (Dahlgren et al. 2016, Jones et al. 2020, Rabone et al. 2023). The removal of the nodules will eliminate the substrate and its associated fauna permanently (Boetius and Haeckel 2018, Vanreusel et al. 2016). With the nodules, also the upper 5–10 cm of the sediment will be removed, eliminating the fauna living on or in this bioactive layer. At the same time, several centimeters of sediment from the suspended sediment plume will be redeposited on the seafloor, inside and outside the mined area (Boetius and Haeckel 2018) (Fig. 3D), limiting the re-colonisation potential of fauna living in the sediment (Simon-Lledo et al. 2019). Another major impact to consider from nodule mining is the creation of sediment plumes in the water column, both from the nodule-collecting machines operating on the seafloor (Weaver et al. 2022), but potentially also from particles in the discharge water, which can contain fine sediments from dewatering processes on the vessel (Muñoz-Royo et al. 2021). Sediment plumes will disperse from the mining site and modeling of plume dispersal scenarios suggests particles from these plumes could settle tens to hundreds of kilometers away (Aleynik et al. 2017, Gillard et al. 2019). The transport of particles in the water column will affect midwater organisms (fish, plankton, larvae) that are otherwise adapted to very low-particle concentrations in the water, thus disrupting the pelagic food web (Drazen et al. 2020). In addition, the mid-water microbial community, which regenerates nutrients for the pelagic community, may be affected, with as yet unknown ecological implications (Ocrutt et al. 2020). Subsequently, the deposition of the plume particles on the seabed will impact benthic fauna that is adapted to very low sedimentation rates, especially when exposed for long periods of time (Levin et al. 2016a, Smith et al. 2020).

Seafloor massive sulfide deposits (SMS) are formed at active hydrothermal vents, through the deposition of particles precipitating from the hydrothermal fluids. SMS are rich in copper, zinc, gold and silver (Van Dover 2010). The deposits with potential commercial value are found between 1000 and 3000 m depth. They are associated with both active and inactive vents (i.e., extinct vents that are no longer hydrothermally active) in slow to ultraslow spreading ridges, such as the Mid-Atlantic Ridge or the South West Indian Ridge, where they can take 10s of thousands of years to form. Vent ecosystems have developed specific physiological adaptations resulting in high species endemism, and they are classified as vulnerable marine ecosystems (VMEs) (Van Dover et al. 2018). Vent species may occupy only a limited number of hydrothermal vents with a small geographic area making them vulnerable to extinction should mining take place (e.g., Sigwart et al. 2019). There is also high potential for the discovery of Marine Genetic Resources (MGRs) from vent fauna, with already proven use in biotechnology and biomedicine (Van Dover et al. 2018). Although there is no legal requirement, industry mostly acknowledges that active hydrothermal vents should be excluded from mining exploitation. Industry is focusing on exploration of inactive vents and their associated deposits, where the communities have changed from fauna that relies on chemoautotrophic primary productivity from microbes to heterotrophic deep-sea communities (Van Dover 2019). However, current knowledge on the ecosystem composition and functioning of inactive vents is very limited and further research is needed to assess the potential impact of seabed mining in these communities (Van Dover et al. 2019), as well as on the sphere of influence of active vents into the background communities (Levin et al. 2016b).

Cobalt crusts are thin (20–25 cm) layers of mineral concretions formed by the precipitation of minerals from cold seawater onto rock surfaces of sediment-free seamounts, ridges and plateaus (Hein 2004) at rates of millimeters per million years. Cobalt crusts are rich in cobalt, nickel, platinum and rare earth elements, and crusts of economic interest are found on seamounts between 800 and 2500 m (Hein et al. 2016). Seamounts support a variety of habitats, including soft sediment and rocky outcrops and walls. Seamounts can support rich and abundant fauna, including habitat engineers such as cold-water corals and sponges (Schlacher et al. 2014, Rogers 2018, 2019) that support associated communities. Many of the species are sessile filter feeders and long-lived, slow-growing fauna that are particularly vulnerable to physical impacts (Clark et al. 2010, Rogers 2019, Schlacher et al. 2014) and are indicators of Vulnerable Marine Ecosystems (VMEs; FAO 2009). Mining seamounts would eliminate large areas of such ecosystems, accentuating biodiversity loss and potentially losing important marine genetic resources before we even knew they existed. Mining seamounts could also affect resident fish populations, with commercial value such as the orange roughy, and affect migratory species that forage on seamounts, or which use them as navigational waypoints (Rogers 2019).

Pollution

Marine debris

Surveys of marine debris combining trawl sampling and video data in the North Atlantic from 35 m depth to 4500 m (Pham et al. 2014b) and across the Mediterranean from 900 to 3000 m (Ramirez-Llodra et al. 2013) identified plastic as the most common item (41%), followed by fishing gear (34%) with a variety of other materials such as clinker, glass, metal, wood, paper/cardboard, clothing, pottery and unidentified objects making up the rest (Ramirez-Llodra et al. 2013, Pham et al. 2014b) (Fig. 3E). In some regions, however, plastic debris can form up to 99.9% of the litter (Galgani et al. 2015, Auta et al. 2017, Chassignet et al. 2021). The impact of plastic debris and microplastics in the marine environment has recently erupted into the public consciousness. This is not surprising, given the recent estimated 4.8–12.7 million tonnes of plastic entering the ocean per annum as a result of the generation of 275 million tonnes of plastic waste globally (2010 figures; Jambeck et al. 2015). Less than 5% of this plastic is recovered globally and so it is accumulating in the environment, including the ocean (Auta et al. 2017). Plastic enters the ocean via rivers, from stormwater runoff, wastewater discharges and through the atmosphere, from other land sources (e.g., from beach littering or natural disasters), and directly at sea from shipping, aquaculture and fishing activities (Jambeck et al. 2015, Auta et al. 2017, Lebreton et al. 2017). It is estimated that between 1.15 and 2.41 million tonnes of plastic enter the ocean via rivers each year and the top 20 rivers contribute 67% of this material, mostly in Asia, but also in South America and Africa (Lebreton et al. 2017). Given that plastic production has already reached more than 300 million tonnes a year and continues to increase, it has been estimated that should business as usual continue the amounts entering the ocean may increase by up to an order of magnitude by 2025 (Jambeck et al. 2015).

Most large-scale studies on plastics in the ocean have focused on the accumulation and distribution of plastic litter in the surface layers or the seafloor. However, as plastic decomposes, it is fragmented into smaller pieces. These small plastic particles can be consumed by zooplankton, which in turn are preyed upon by mesopelagic fish (Cózar et al. 2014). Plastic is then transported down to the seafloor through the production of faecal pellets or the sinking of dead mesopelagic organisms. The largest transport of microplastics down the water column is via marine snow, where the microplastics attaches to particles of organic matter that sink into deep waters (Cózar et al. 2014, Porter et al. 2018, Kvale et al. 2020). The role of these different sink processes in delivery of plastic particles to the deep ocean are poorly quantified, as are the impacts on pelagic and benthic food webs (Taylor et al. 2016) and feedbacks on the biological carbon pump, through for example, decreasing the density of marine snow aggregates and faecal pellets and slowing the sinking of carbon into the deep sea (Cole et al. 2016). Oceanographic processes, such as dense shelf water cascading, severe coastal storms, offshore convection and saline subduction can all induce vertical and horizontal transfers, from shallow ocean and coastal regions to deeper layers including sediments and microplastic particles (Pham et al. 2014b). Likewise, human activities, especially deep-sea trawling, can resuspend sediments, causing significant transport of materials from the shelf into the deep sea (Puig et al. 2012). Canyons can also act as a conduit of plastic debris into the deep ocean (Pham et al. 2014b, Galgani et al. 2015). Studies indicate that microplastic concentrations in deep-sea sediments can be significantly higher than in surface waters (Woodall et al. 2014, Avio et al. 2017). Recent meta-analysis of deep-sea imagery from 100 m to 10898 m depth indicates that plastics have reached the deepest part of the ocean in the Marianas Trench (Chiba et al. 2018). Microplastic particles have also been detected in the guts of deep-sea amphipods from a number of ocean trenches from 7000 m depth to 10890 m depth (Jamieson et al. 2019). On deep-sea sedimentary ecosystems, detritivorous species such as holothurians and echinoids will ingest microplastics accumulated in the sediment, and microplastic fibers and particles have been found in deep-sea benthic invertebrates all the way down to the deepest places in the ocean (Taylor et al. 2016, Jamieson et al. 2019). However, the effects of ingestion of microplastics or microfibres on deep-sea organisms is as yet unknown (Ramirez-Llodra et al. 2011).

The effects of large plastic debris on wildlife include ingestion, entanglement and ghost fishing in the case of lost or discarded fishing nets (Fig. 3F). All these can result in serious injury or death of the organisms involved (Ramirez-Llodra et al. 2011, Gall and Thompson 2015). Nearly 700 species of marine animals have been reported to be affected by entanglement and ingestion of marine plastics, with most reports being for sea turtles, marine mammals and seabirds (Gall and Thompson 2015). It is important to note that understanding the levels of mortality in wild animals from entanglement or ingestion of plastic is extremely difficult as many individuals, probably the majority in some cases, will die and sink at sea or are rapidly scavenged (e.g., Stelfox et al. 2016). Entanglement of plastic debris in cold-water corals has been documented (Appah et al. 2022) and *in-situ* observations and laboratory experiments show that both micro- and macro-plastics significantly reduced skeletal growth rates of corals (Appah et al. 2022). Deep-diving whales have also been found to be affected by plastic ingestion. An individual of a Cuvier's beaked whale was found with 378 items of debris weighing 33 kg in its gut (Poncelet et al. 2000). A study of whale strandings on the Canary Islands between 2000 and 2015 identified ingested foreign bodies in 36 out of 475 individuals representing 15 species (Puig-Lozano et al. 2018). Of the 36 individuals, 13 showed evidence of damage arising from ingested materials including ulcerations, gastrointestinal impaction or perforations, which were likely to have caused death (Puig-Lozano et al. 2018). Deep-diving species were found to be more at risk of foreign body ingestion than shallow divers (Puig-Lozano et al. 2018).

There have been relatively few studies evaluating the effects of plastic debris at the level of habitat or ecosystem, and the few available focus mainly on shallow-water systems (Hall et al. 2015, Clemente et al. 2018). Floating plastic debris in the ocean has the potential to reduce light levels and therefore reduce primary productivity in the surface (VishnuRadhan et al. 2019), with consequences on the quantity and quality of food entering the deep ocean. An important process relevant to the ocean carbon cycle is the production of dissolved organic carbon (DOC) by phytoplankton. This pool of reduced carbon is one of the largest on Earth (662 Pg) and is comparable to the amount of CO_2 in the atmosphere (828 Pg; Romera-Castillo et al. 2018). In seawater, plastics leach dissolved organic compounds, which can then be fed upon by heterotrophic bacteria (Romera-Castillo et al. 2018). It has been estimated that between 260 and 23600 tonnes of DOC may be released into the ocean each year from plastics (Romera-Castillo et al. 2018). In areas where there is a high density of debris (e.g., ocean gyres), up to 10% of the surface microlayer DOC may be from plastics. DOC derived from plastics may therefore have a significant impact on microbial productivity in some parts of the ocean (Romera-Castillo et al. 2018). Plastics collected from the ocean have also been found to emit methane and ethylene, both greenhouse gases, which can potentially contribute to overall global warming (Royer et al. 2018). Plastic debris also has the potential to impact marine biodiversity through the introduction of rafting alien invasive species (e.g., Rech et al. 2018). Barnes (2002) identified bryozoans, barnacles, polychaete worms, hydroids and molluscs on ~ 200 items of debris found mainly across the north and south Atlantic but also elsewhere. He estimated that anthropogenic debris may have doubled the opportunities for rafting of organisms at low latitudes and trebled them in high latitudes (Barnes 2002). Microbial communities attaching to plastic debris may also include pathogenic organisms (Zettler et al. 2013).

Chemical and radioactive pollution

Studies on chemical contamination of deep-sea fauna are limited, but the development of analytical techniques has contributed to an increasing wealth of data that provides evidence of deep-sea ecosystems acting as sink for chemical pollutants (Looser et al. 2000). Of all the different potentially toxic chemicals, those of main concern in the marine environment include persistent organic pollutants (POPs), toxic metals such as mercury, cadmium, lead, nickel and isotopic traces, radioelements, pesticides, herbicides and pharmaceuticals (Froescheis et al. 2000). Xenobiotics (i.e., chemicals found in an organism that are not normally produced by the organism) have been shown to bioaccumulate in deep-sea fauna, as a consequence of consumption of enriched organic matter in the sediment, and through the food chain (Looser et al. 2000, Froescheis et al. 2000). Xenobiotics have been detected in a variety of deep-sea fauna, including Atlantic fish (Stegeman 1983, Stegeman et al. 1986), cephalopods (Unger et al. 2008) and crustaceans such as the commercial Mediterranean shrimp *Aristeus antennatus* (Rotllant et al. 2006), amphipods in a canyon in the Gulf of Mexico (Soliman and Wade 2008) and hadal amphipods from the deepest trenches in the ocean (Jamieson et al. 2017). Transport and accumulation of chemicals in the deep sea is highly influenced by hydrodynamics, and regions where local currents are strengthened through natural processes such as hydrodynamics in canyons and on top of seamounts can lead to higher transport and accumulation of chemicals on the recipient sediments (Canals et al. 2006). Although current understanding on the long-term effect of chemical pollution on deep-sea ecosystems is limited, the continuous input of chemicals to the marine environment results in the enrichment of sediments and bioaccumulation in fauna, which could have significant effects, particularly in sink areas such as deep basins and trenches (Ramirez-Llodra et al. 2011, Jamieson et al. 2017). A different type of chemical pollution comes from pharmaceuticals (Choong et al. 2006). In the 1970s, large quantities of pharmaceuticals were dumped in the 6000 m deep trench off Puerto Rico, impacting the microbial communities (Peele et al. 1981, Grimes et al. 1984) and with evidence of acute toxicity in many invertebrates (Nicol et al. 1978). Although dumping pharmaceuticals in the oceans is currently prohibited, compounds such as antibiotics, pain killers, birth control pills, cancer treatments and antidepressants used by humans and for treatment of livestock still enter the ecosystem through water-ways and as a consequence of inadequate disposal and water treatment (Ramirez-Llodra et al. 2011).

Low-level radioactive waste was dumped on slopes and canyons in the north Atlantic between 1949 to 1982 (Thiel 2003), and also off New Zealand, Japan and Korea, as well as waste disposed of by the former Soviet Union (Vartanov and Hollister 1997). In addition to the active dumping of radioactive waste in the deep ocean, sunken nuclear submarines can also provide focused areas of contamination on the seafloor (Juniper et al. 2020). Since 1994, the Parties of the London Convention banned the disposal of low-level radioactive waste. Initial studies indicated that there was no significant impact on deep-sea communities from radioactive waste dumped in drums, but no further studies have been conducted and leakage of degrading drums could be possible (Thiel 2003). Some areas of the ocean have also been subjected to contamination by high levels of radioactive fallout as a result of nuclear weapons testing. An example is the Marshall Islands, which were subject to 67 nuclear weapons tests between 1946 and 1958 (Marcoux 2021). The fallout from these tests was extensive and caused significant health impacts to people in the islands as well as fishermen at sea (Marcoux 2021). Plutonium from these tests has spread throughout the northwestern Pacific via the North Equatorial Current and Kuroshio Current (Wu et al. 2018). Recent measurements suggest that levels of artificial radionuclides are low in reef fish and shellfish but may be immobilised in corals and non-surface sediments (Corcho-Alverado et al. 2021). The significance of such contamination for deep-water ecosystems in the region is not well studied.

Noise pollution

The increased intensity of offshore and seafloor activities in deep water, from shipping, sonar operations, hydrocarbon exploration and exploitation, fishing and mineral exploration, has resulted in an increase in noise pollution in the ocean, including the deep sea (Juniper et al. 2020, Martin et al. 2021, Williams et al. 2022). Underwater noise can affect all marine organisms, from zooplankton to marine mammals (Martin et al. 2021, Williams et al. 2022). Studies show that, for marine mammals, noise levels above 120 dB can cause behaviour changes, while levels of 150 dB can cause severe behaviour disruption to temporary lowering of hearing sensitivity (Temporary Threshold Shift) (Moore et al. 2002). Levels above 170 dB can cause permanent hearing loss and physical damage, and even death in some cases - the Permanent Threshold Shift (Moore et al. 2002). Behavioural change can have impact on foraging, reproduction, the ability to avoid collisions and communication within the population (Kunc et al. 2014, Williams et al. 2021), which can result in impacts not only at the individual level, but also at the population and ecosystem levels (Juniper et al. 2020). Although empiric data on noise levels for deep-sea machinery is not available, a recent study has estimated noise pollution based on similar shallow water machinery (e.g., from dredging) to levels up to 181 dB for seabed mining vehicles, 183 dB for the riser pump, and 188 dB for the surface vessel (Williams et al. 2022). All these levels of noise are above the Permanent Threshold Shift for marine mammals. A noise dispersion model, calculated for polymetallic nodules in the Clarion Clipperton Zone (Pacific Ocean) at 4000 m depth, indicates that noise levels above 120 dB will be distributed throughout the water column and reach a radius of 4 to 6 km per mining operation (Williams et al. 2022). The noise created by military sonars can also lead to mass stranding of whales and other cetaceans, causing large mortality events (Jepsen et al. 2013, Parsons 2017). Research on noise-reduction from vessel and underwater instrumentations and vehicles has mostly been conducted by the military, but it is now increasing from the scientific and industrial communities (Juniper et al. 2020).

Climate change

Temperature rise

Human activities are generating green-house gas (GHG) emissions that, to date, have resulted in a temperature increase of 1°C above pre-industrial levels (IPCC 2022). There are also increasing positive feedbacks to atmospheric GHG concentrations from climate change effects, such as melting of permafrost that produce CO_2 and methane (Zhang et al. 2017, Dean et al. 2018). As on land, habitat destruction or conversion in the ocean can both generate CO_2 emissions and reduce CO_2 sequestration potential (Lovelock et al. 2017). There are also potential positive feedbacks to atmospheric GHG concentrations resulting from changes in the ocean driven by climate change, such as changes in primary production (Moore et al. 2018), or the release of methane from clathrates (Andreassen et al. 2017). With the current increasing trend of 0.2°C per decade, global warming is likely to reach 1.5°C above pre-industrial levels between 2030 and 2052, and it is expected to be 2 to 3 times higher in the Arctic region (IPCC 2018). Because of its physical characteristics, the ocean has absorbed 93% of excess heat generated in the atmosphere as a result of global warming (Levitus et al. 2012).

The absorption of heat and CO_2 exposes vulnerable ecosystems to a combination of climate-change related stressors: ocean warming from the surface to deep waters, increased acidification and decreased oxygenation (Breitberg et al. 2018, Gruber 2011, Gattuso et al. 2015, Levin 2018). Because of the size and complexity of studying the oceans, understanding the impacts of these stressors on the physiology and behaviour of marine species, their biodiversity patterns and the ecosystem functions they sustain, are emerging issues, particularly in the deep sea (Danovaro et al. 2004, Hoegh-Guldberg and Bruno 2010). It is virtually certain that the temperature of the upper 700 m of the global ocean has increased by 0.6°C in the last century (IPCC 2018). This temperature rise is ongoing at 0.1°C per decade (Purkey and Johnson 2010, IPCC 2018) and accelerating (Cheng et al. 2019), with higher rates in regions such as the Arctic and Southern Ocean (Bergmann et al. 2011, Smith et al. 2012). This temperature rise is also detectable in several deep ocean basins in the Arctic Ocean, North Atlantic and Southern Ocean, particularly in areas of high vertical mixing (Wunsh and Heimbach 2014, Le Bris and Levin 2020). Approximately 19% of ocean's heat uptake is accumulated below 2000 m depth, dominated by warming in the Southern Ocean (Talley et al. 2016, Desbruyères et al. 2017).

Although studies on the effects of increased temperature in deep-sea ecosystems are scarce, initial studies suggest that temperature changes can have significant consequences to deep-sea biodiversity (Danovaro et al. 2004, Yasuhara and Danovaro 2016). Many deep-sea species live in relatively stable thermal environments, so small changes in temperature can subject individuals and populations to stress, causing reduced physiological fitness. This will result in bathymetric or latitudinal distribution changes (Danovaro et al. 2004, Levin and Le Bris 2015) and a predicted general deepening of mid-slope communities. Temperature changes can open new habitats for invasive species, such as the spread of the king crab (*Neolithodes yaldwyni*) in Palmer Deep in Antarctica at more than 850 m (Smith et al. 2012), the introduced king crab (*Paralithodes camtschaticus*) in fjords in northern Norway (Pedersen et al. 2018), and the snow crab (*Chionoecetes opilio*) in the Pechora Sea in NW Russia (Gebruk et al. 2020). These invasions and/or expansions of large predators can cause significant changes in the benthic communities resulting from the physical disturbance of the sediments and feeding behaviour.

The stratification of water masses reduces ventilation of the deeper ocean, at the same time that it reduces nutrient supply from deep waters (where organic matter is recycled) to surface waters, where nutrients are used for primary productivity (PP). The reduction of PP by phytoplankton (Chavez et al. 2011, Kwiatkowski et al. 2018, 2020) will result in a decreased flux of Particulate Organic Carbon (POC) and thus in food availability in the deep-water column and seafloor (Ruhl and Smith 2004, Sweetman et al. 2017). The size of nutrient-poor ocean deserts in the Pacific and Atlantic have increased by 15% between 1998–2006 (Polovina et al. 2008) and bathyal areas, particularly, are predicted to undergo significant decline of POC flux in certain regions (Sweetman et al. 2017). However, complex mechanisms of organic matter export at depth over space and time, and the changing drivers of PP at the global scale, have been only recently recognized (Boyd et al. 2019). Current evidence shows that such modifications can alter the composition of benthic communities down to abyssal depths (Smith et al. 2009), modifying structural and functional patterns (e.g., food-web dynamics, reproductive processes) of deep-sea assemblages.

In certain regions like the NW Mediterranean Sea, particularly strong cooling of the coastal and shelf waters driven by cold and dry northerly winds results in the formation of cold and dense shelf water, which overflows the shelf and cascades down to the deep basin, funneled through submarine canyons (Canals et al. 2006). Warming in these regions has the potential to modify the frequency and intensity of these large episodic events and the effects on the faunal communities. In fact, dense shelf water cascading events transport large amounts of organic matter to bathyal and abyssal depths. In the NW Mediterranean Sea, it has been proposed that the significant input of organic matter supports recruitment of the commercial shrimp *Aristeus antennatus*, mitigating the general trend of overexploitation of this natural resource in the region (Company et al. 2008).

A warming ocean has the potential, also, to destabilize gas hydrates on continental margins at bathyal depths (200–2000 m), triggering the release of methane (Ketzer et al. 2019). Several gigatonnes of CH_4 could be released from a shift of the warm Gulf Stream to areas with gas hydrates (Phrampus and Hornbach 2012). Such massive releases of methane could have significant feedback in global climate, as well as impair aerobic methane oxidation from microorganisms (thus affecting the global methane budget). However, there are high levels of uncertainty around such predictions as many aspects of the biogeochemistry of large methane releases in the ocean that are not understood or are not quantified (Ruppel and Kessler 2017).

Ocean acidification

Atmospheric concentrations of CO_2 (carbon dioxide), CH_4 (methane) and N_2O (nitrous oxide) exceed the levels of the last 800000 years, with CO_2 concentrations 40% higher since pre-industrial times, mostly as a result of fossil fuel emissions and land-use emissions (IPCC 2013). Approximately 1 million tonnes of CO_2 are being released per hour into the Earth's atmosphere (Hall-Spencer and Harvey 2019), with about 28% being absorbed by the oceans, increasing partial pressure of CO_2 (PCO_2) and dissolved inorganic carbon, decreasing pH and the saturation state of biologically-important calcium carbonate minerals, mainly aragonite and calcite, thus causing ocean acidification (IPCC 2013, Gattuso et al. 2015, Talley et al. 2015). Ocean pH has decreased at 0.002 units per decade in the last 4 decades, resulting in a 0.1 pH decrease since pre-industrial times that affects upper ocean and intermediate-depth water masses globally (Byrne et al. 2010, Rios et al. 2015). The subsequent sinking of high-CO_2 water in areas of deep-water formation are contributing to the rapid acidification of the deep North Atlantic, which is predicted to experience pH reductions of 0.2 units on 17% of the seafloor areas below 500 m depth by 2100 (Gehlen et al. 2014). The atmospheric CO_2 threshold (tipping point) for major changes was set at 350 ppm (Hoegh-Guldberg and Bruno 2010). Because of the high solubility of CO_2 in cold waters, the strongest changes in water chemistry are expected to take place at high latitudes, and particularly in the Arctic Ocean (Riebesell et al. 2013). This acidification of the upper ocean significantly modifies seawater chemical conditions and biogeochemical cycles of many compounds in the marine environment (Doney et al. 2009).

The greatest impacts of acidification in deep-sea ecosystems have been documented on aragonitic calcifying organisms such as habitat forming cold-water corals, sponges, molluscs, echinoderms and calcareous foraminifera (Doney et al. 2009). Early developmental stages (embryos, larvae, juveniles) can also be affected by acidification with the response depending on the organism (Doney et al. 2009). For example, a reduced fertilization success, reduced developmental rates, and high embryo mortality have been documented in laboratory experiments on sea urchins and bivalves (Kurihara et al. 2007, Kroecker et al. 2010). Acidification can also cause a redistribution of energetic resources towards maintenance of basic metabolism, in detriment to allocations to reproduction, resulting in a decreased reproductive output (Kroeker et al. 2010). Disturbance to these early-life history processes can have a sub-lethal impact on populations by affecting the reproductive success of the individuals, and the long-term effect on the species could be extinction (Tyler and Ramirez-Llodra 2002).

Photosynthesis by marine phytoplankton contributes to about half of the biosphere's net primary production, and is therefore an essential process in carbon recycling between the biotic and abiotic components (Field et al. 1998). The relationship of primary productivity and acidification is complex. To understand and predict the impact of climate change in primary productivity, a combination of many variables that are currently changing (e.g., increases sea-surface temperature, increased ocean acidity, decrease of sea ice, increased storms, increased water stratification) need to be evaluated and their synergies, cumulative or antagonistic reactions understood (Finkel et al. 2010). This is particularly important as modifications in the abundance of calcifying planktonic organisms and plankton community composition will have cascading effects on food-web dynamics and the structure of the marine food web (Doney et al. 2009, Finkel et al. 2010, Riebesell et al. 2013, Richier et al. 2014) affecting the entire marine system, from the surface to the abyssal communities that reply upon the supply of food originating in the euphotic zone (Levin and Le Bris 2015). Jones et al. (2014) modelled long-term (decadal to century scale) global seafloor biomass changes in response to climate change (under RCP8.5) and predict, for the period 2091–2100, a 5.2% decrease in global open ocean benthic biomass compared to the 2006–2015 period. These authors estimated that the benthic faunal biomass of over 80% of deep-sea hotspots, including seamounts, submarine canyons and cold-water coral reefs, will be reduced, with potential major implications in the ecosystem functions they support, and thus the services they provide to society (Jones et al. 2014).

Hypoxia

Oxygen concentrations in the ocean and coastal waters have decreased since the mid-20th century, with an estimated global loss of 2% (Schmidtko et al. 2017, Breitburg et al. 2018) and strongest declines at mid-latitudes and the Northeast Pacific, Southern and Indian Oceans (Helm et al. 2011). Several coupled ocean-atmosphere models predict a decline of oxygen in the global ocean between 1–7% over the next century (Keeling et al. 2010) and current anthropogenic activities could result in widespread ocean hypoxia in the next thousand years (Watson 2016). This decline is caused by several factors: eutrophication of coastal waters, a warming ocean and stratification. Eutrophication of coastal waters since 1950 is mainly caused by riverine runoff with high concentrations of fertilizers that have seen a 10-fold increase. Oxygen is less soluble in warmer waters, so as sea-surface temperature increases, oxygen concentration decreases. This driver of oxygen depletion in the ocean accounts for 15% of the global decrease (Ito et al. 2017). A warming ocean leads to stratification of water masses, reducing vertical mixing, and thus limiting the supply of oxygen to the deeper ocean. This process accounts for 85% of global ocean decline and affects mostly the upper bathyal depths (200–700 m) at mid-latitudes (Breitburg et al. 2018).

Oxygen concentration of 2 mg l^{-1} (61 μmol kg^{-1}) has been used to define hypoxia. However, tolerance and threshold values to low oxygen concentrations (hypoxia) are species specific and vary between developmental stages within species (Vaquer-Sunyer and Duarte 2008, Ekau et al. 2010), so oxygen levels above the 2 mg l^{-1} have been shown to have lethal or sublethal effects for many benthic marine species (Vaquer-sunyer and Duarte 2008). Hypoxia exerts pressure on metabolic processes, particularly when combined with higher temperatures that decrease O_2 solubility and increase respiratory demands (Brown and Thatje 2015). This will affect the habitat range for many species, compressing it for the species with low tolerance to hypoxia and expanding it for low-oxygen tolerant species such as jellyfish

(Stewart et al. 2014), some cephalopods (Stewart et al. 2014) and mysid shrimp (Childress and Seibel 1998). Thresholds for hypoxia vary with species, and modifications in present conditions can affect the relationships between trophic levels. The importance of microorganisms and meiofauna would increase, modifying biogeochemical cycles (Sweetman et al. 2016) and impacting energy flows to upper trophic levels (Sweetman et al. 2016, 2017).

Benthic organisms on sediments of continental margins are responsible for 50% of carbon burial in the ocean. In areas where conditions become sub-oxic, the decline of sediment-burrowing organisms will affect the regulating services of nitrogen removal and carbon sequestration provided by these organisms (Deutsch et al. 2011, Sweetman et al. 2017). Habitat compression into shallower depths is predicted for mobile fauna that can actively escape hypoxic conditions (Sato et al. 2016). If predators have a lower low-oxygen tolerance than their prey, their feeding opportunities will diminish, while other predators may benefit from the expansion of hypoxic regions, thus modifying locally trophic interactions (Breitburg et al. 2018). For example, the shoaling of OMZs can modify the habitat range of large predators adapted to hypoxia, such as the Humboldt squid (*Dosidicus gigas*). This predator has been observed to stay up to 6h in the oxygen minimum zone (OMZ; Gilly et al. 2006) where it feeds on myctophid fishes as they migrate in the water column (Stewart et al. 2014). The increase in squid abundance and its expansion northwards from California to southeastern Alaska have been related to the expansion of the OMZ in the region (Stewart 2012, Gilly et al. 2013). Species that conduct vertical migrations that have to go through the OMZ will also be affected, positively or negatively, depending on their low-oxygen tolerance, thus modifying the trophic interactions inside and outside of the OMZ (Gilly et al. 2013). The shoaling of species pushed towards the sunlight zone by the OMZ may also result in increased predation from species that rely on vision and such impact has been suggested for fish larvae in the California Bight (Koslow et al. 2011). The effects in deep-sea ecosystems are expected to be particularly strong at bathyal depths (500–1000 m) in the Eastern Pacific, where OMZ expansion is predicted to exceed low-oxygen thresholds for marine fauna (Sperling et al. 2016). Observations on modified diving behaviour of large pelagics such as marlin (*Makaira nigircans*), mako shark (*Isurus oxyrinchus*) or sailfish (*Istiophorus albicans*), which often dive from near surface to prey on the deep scattering layer (Vetter et al. 2008, Prince et al. 2010), have led to the concept of a vertical compression of the pelagic habitat in areas that overlie strong OMZs (Gilly et al. 2013). For blue marlin in the eastern Atlantic, the OMZ expansion has resulted in a habitat loss of 1 m year^{-1} (15%) between 1960 and 2010 (Stramma et al. 2011). The combination of lower oxygen concentrations with other climate change stressors, and particularly with warming, will push fish and invertebrates to migrate poleward or to greater depths (Cheung et al. 2011, Deutsch et al. 2015).

On the continental margin, the impact of hypoxia on species that provide habitat (e.g., octocorals) for other invertebrates will result in a reduced diversity of the ecosystem (Roberts et al. 2009, Buhl-Mortensen et al. 2010). Similarly, the reduction of burrowing infauna will impact the bioturbation of the sediments, decreasing habitat heterogeneity and resulting in a lower sediment biodiversity (Levin et al. 2009, Sperling et al. 2016). On cold seeps, lower oxygen levels may inhibit oxidising symbionts, affecting dense host populations (e.g., mussel beds, tubeworm bushes) that provide habitat and nurseries for other species (Cordes et al. 2010, Sweetman et al. 2017).

Cumulative impacts

The full extent of anthropogenic impacts on deep-sea ecosystems is complex to evaluate, as often different stressors will interact in a single location, resulting in an effect which is magnified compared to that of each individual stressor (Ramirez-Llodra et al. 2011). Climate-change related stressors affect the planet as a whole, including the deep ocean, and are the main parameters that can interact with other anthropogenic activities (Levin and Le Bris 2015). Cumulative impacts are particularly important in the upper bathyal slopes (200–400 m), where pressure from anthropogenic activities is currently most intense. Here, physical and ecotoxicological impacts from fishing, hydrocarbon exploitation and pollution can be accentuated by climate-related stressors, resulting in synergistic effects on populations and the ecosystem responses triggered by environmental change (Stramma et al. 2010, Levin and Le Bris 2015, Ramirez-Llodra et al. 2011, 2020). For example, fished populations where individual fitness is weakened because of warmer temperatures may result in population collapse at a faster rate. In addition, increasing ocean temperature, ocean acidification, deoxygenation or changes in primary productivity and thus food input to the seafloor, can interact amongst each other, reducing resilience of species and populations. Under continuous or acute exposure to several stressors, such impacts can result in biodiversity loss, decreased ecosystem functionality and detrimental effects on the services these ecosystems provide to society (Levin and Le Bris 2015, Ramirez-Llodra et al. 2020). Different extractive industries can target the same habitat and ecosystem. For example, benthic fishing and mineral extraction may overlap on seamounts, resulting in additive pressure to Vulnerable Marine Ecosystems such as cold-water corals or sponge fields and their associated fauna. Such activities have great potential for the loss of marine genetic resources (MGRs) from deep-sea ecosystems, where 90% of the species can be new to science (Van Dover et al. 2018, Rabone et al. 2023, Ramirez-Llodra et al. 2020).

Although empiric data of cumulative impacts on deep-sea ecosystems is still scarce, it is essential that synergistic and additive effects of anthropogenic activity and climate change in the deep ocean are taken into account when devising conservation and management measures, including in support of reaching the UN Sustainable Development Goals through the Kunming-Montreal Global Biodiversity Framework (GBF) adopted in 2022. Understanding and including data on cumulative impacts is important both within national waters and in international treaties and agreements for areas beyond national jurisdiction, such as the implementation of the newly adopted UN Treaty on the conservation and sustainable use of marine biodiversity in areas beyond national jurisdiction (BBNJ) (Tiller et al. 2019) and the finalization of the ISA Mining Code that will regulate seabed mining in The Area.

Acknowledgements

We would like to thank REV Ocean and The Nippon Foundation-Nekton Ocean Census programme for funding us to conduct this work, and the Editors for their continuous support during the process of writing this chapter.

References

Agnew, D. J., Pearce, J., Pramod, G., Peatman, T., Watson, R., Beddington, J. R. et al. (2009). Estimating the worldwide extent of illegal fishing. PLoS ONE 4: e4570.

Aleynik, D., Inall, M. E., Dale, A. and Vink, A. (2017). Impact of remotely generated eddies on plume dispersion at abyssal mining sites in the Pacific, Sci. Rep. 7: 16959.

Amon, D., Ziegler, A., Dahlgren, T., Glover A. G., Goineau, A., Gooday, A. J. et al. (2016). Insights into the abundance and diversity of abyssal megafauna in a polymetallic-nodule region in the eastern Clarion-Clipperton Zone. Sci. Rep. 6: 30492.

Anderson, O. R. J., Small, C. J., Croxall, J. P., Dunn, E. K., Sullivan, B. J., Yates, O. et al. (2011). Global seabird bycatch in longline fisheries. Endang. Species Res. 14: 91–106.

Andreassen, K., Hubbard, A., Winsborrow, M., Patton, H., Vadakkepuliyambatta, S., Plaza-Faverola, A. et al. (2017). Massive blow-out craters formed by hydrate-controlled methane expulsion from the Arctic seafloor. Science 356: 948–953.

Appah, J. K. M., Killeen, O., Lim, A., O'Riordan, R., O'Reilly, L. and Wheeler, A. J. (2022). Accumulation of marine litter in cold-water coral habitats: A comparative study of two Irish Special Areas of Conservation, NE Atlantic. Mar. Pollut. Bull. 180: 113764.

Appeltans, W., Ahjong, S. T., Anderson, G., Angel, M. V., Artois, T., Bailly, N. et al. (2012). The magnitude of global marine species diversity. Curr. Biol. 22: 2189–2202.

Armstrong, C. W., Foley, N. S., Tinch, R. and van den Hove, S. (2012). Services from the deep: steps towards valuation of deep sea goods and services. Ecosyst. Services 2: 2–13.

Ashford, O. S., Kenny, A. J., Froján, C. R. S., Bonsall, M. B., Horton, T., Brandt, A. et al. (2018). Phylogenetic and functional evidence suggests that deep-ocean ecosystems are highly sensitive to environmental change and direct human disturbance. Proc. R. Soc. B Biol. Sci. 285: 20180923.

Auster, P. J., Malatesta, R. J., Langton, R. W., Watling, L., Valentine, P. C., Donaldson, C. L. S. et al. (1996). The impacts of mobile fishing gear on seafloor habitats in the Gulf of Maine (Northwest Atlantic): implications for conservation of fish populations. Rev. Fish. Sci. 4: 185–202.

Auta, H. S., Emenike, C. U. and Fauziah, S. H. (2017). Distribution and importance of microplastics in the marine environment: A review of the sources, fate, effects, and potential solutions. Environ. Int. 102: 165–176.

Avio, C. G., Cardelli, L. R., Gorbi, S., Pellegrini, D. and Regoli, F. (2017). Microplastics pollution after the removal of the Costa Concordia wreck: First evidences from a biomonitoring case study. Environ. Pollut. 227: 207–214.

Bailey, D. M., Collins, M. A., Gordon, J. D. M., Zuur, A. F. and Priede, I. G. (2009). Long-term changes in deep-water fish populations in the northeast Atlantic: a deeper reaching effect of fisheries? Proc. R. Soc. B Biol. Sci. 276: 1965–1969.

Baldwin, C., Bradford, L., Carr, M. K., Doig, L. E., Jardine, T. D., Jones, P. D. et al. (2017). Ecological patterns of fish distribution in the Slave River Delta region, Northwest Territories, Canada, as relayed by traditional knowledge and Western science. Int. J. Water Resour. Dev. 34: 305–324.

Barber B. M., Odean, T. and Zhu, N. (2009). Systematic noise. J. Financial Mark. 12: 547–569.

Barnes, D. (2002). Invasions by marine life on plastic debris. Nature 416: 808–809.

Baum, J. K., Myers, R. A., Kehler, D. G., Worm, B., Harley, S. J. and Doherty, P. A. (2003). Collapse and conservation of shark populations in the Northwest Atlantic. Science 299: 389–392.

Beaulieu, S. E. and Szafranski, K. (2020). InterRidge Global Database of Active Submarine Hydrothermal vent Fields, Version 3.4. World Wide Web electronic publication available from: https://vents-data.interridge.org/ Last accessed 18/02/2022.

Bednarsek, N., Tarling, G. A., Bakker, D. C. E., Fielding, S. and Feely, R. A. (2014). Dissolution dominating calcification process in polar pteropods close to the point of aragonite undersaturation. PLoS ONE 9: e109183.

Behrenfeld, M. J., O'Malley, R. T., Siegel, D. A., McClain, C. R., Sarmiento, J. L., Feldman, G. C. et al. (2006). Climate-driven trends in contemporary ocean productivity. Nature 444: 752–755.

Béné, C., Barange, M., Subasinghe, R., Pinstrup-Andersen, P., Merino, G., Hemre, G.-I. et al. (2015). Feeding 9-billion by 2050: Putting fish back on the menu. Food Sec. 7: 261–274.

Bergmann, M., Soltwedel, T. and Klages, M. (2011). The interannual variability of megafaunal assemblages in the Arctic deep sea: Preliminary results from the HAUSGARTEN observatory (79°N). Deep-Sea Res. I 58: 711–723.

Bernardino, A. F., Levin, L. A., Thurber, A. R. and Smith, C. R. (2012). Comparative composition, diversity and trophic ecology of sediment macrofauna at vents, seeps and organic falls. PLOS ONE 7: e33515.

Bernardino, A. F., Cordes, E. E. and Schlacher, T. A. (2020). The natural capital of offshore oil, gas, and methane hydrates in the world ocean. pp. 111–126. *In*: Baker, M., Ramirez-Llodra, E. and Tyler, P. (eds.). Natural Capital and Exploitation of the Deep Ocean. Oxford University Press, Oxford, United Kingdom.

Bett. B. J. (2001). Atlantic margin environmental survey: introduction and overview of bathyal benthic ecology. Cont. Shelf Res. 21: 917–956.

Boetius, A. and Haeckel, M. (2018). Mind the seafloor. Science 359: 34–36.

Bouchet, P., Decock, W., Lonneville, B., Vanhoorne, B. and Vandepitte, L. (2023). Marine biodiversity discovery: The metrics of new species descriptions. Front. Mar. Sci. 10: 929989.

Bourjea, J., Clermont, S., Delgado, A., Murua, H., Ruiz, J., Ciccione, S. et al. (2014). Marine turtle interaction with purse-seine fishery in the Atlantic and Indian oceans: Lessons for management. Biol. Conserv. 178: 74–87.

Boyd, P. W., Claustre, H., Levy, M., Siegel, D. A. and Weber, T. (2019). Multi-faceted particle pumps drive carbon sequestration in the ocean. Nature 568: 327–335.

Breitburg, D., Levin, L. A., Oschlies, A., Grégoire, M., Chavez, F. P., Conley, D. J. et al. (2018). Declining oxygen in the global ocean and coastal waters. Science 359: 46.

Brewin, P. E., Farrugia, T. J., Jenkins, C. and Brickle, P. (2021). Straddling the line: high potential impact on vulnerable marine ecosystems by bottom-set longline fishing in unregulated areas beyond national jurisdiction. ICES J. Mar. Sci. 78: 2132–2145.

Brown, A. and Thatje, S. (2015). The effects of changing climate on faunal depth distributions determine winners and losers. Glob. Change Biol. 21: 173–180.

Buhl-Mortensen, L., Ellingsen, K. E., Buhl-Mortensen, P., Skaar, K. L. and Gonzalez-Mirelis, G. (2016). Trawling disturbance on megabenthos and sediment in the Barents Sea: chronic effects on density, diversity, and composition. ICES J. Mar. Sci. 73: i98–i114.

Buhl-Mortensen, L., Vanreusel, A., Gooday, A. J., Levin, L. A., Priede, I. G., Buhl-Mortensen, P. et al. (2010). Biological structures as a source of habitat heterogeneity and biodiversity on the deep ocean margins. Mar. Ecol. 31: 21–50.

Buhl-Mortensen, P. and Buhl-Mortensen, L. (2018). Impacts of bottom trawling and litter on the seabed in Norwegian waters. Front. Mar. Sci. 5: 42.

Byrne, R. H., Mecking, S., Feely, R. A. and Liu, X. (2010). Direct observations of basin-wide acidification of the North Pacific Ocean. Geophys. Res. Lett. 37: L02601.

Canals, M., Puig, P., de Madron, X., Heussner, S., Palanques, A. and Fabres, J. (2006). Flushing submarine canyons. Nature 444: 354–357.

Canfield, D. E., Stewart, F. J., Thamdrup, B., DeBrabandere, L., Dalsgaard, T., Delong, E. F. et al. (2010). A cryptic sulfur cycle in oxygen-minimum-zone waters off the Chilean coast. Science 330: 1375–78.

Cardinale, M. and Modin, J. (1999). Changes in size-at-maturity of Baltic cod (*Gadus morhua*) during a period of large variations in stock size and environmental conditions. Fish. Res. 41: 285–295.

Chapman, N. R. and Price, A. (2011). Low frequency deep ocean ambient noise trend in the Northeast Pacific Ocean. J. Acoust. Soc. Am. 129: EL161.

Chapron, L., Peru, E., Engler, A., Ghiglione, J. F., Meistertzheim, A. L., Pruski, A. M. et al. (2018). Macro- and microplastics affect cold-water corals growth, feeding and behaviour. Sci. Rep. 8: 15299.

Chassignet, E. P., Xu, X. and Zavala-Romero, O. (2021). Tracking marine litter with a global ocean model: Where does it go? Where does it come from? Front. Mar. Sci. 8: 667591.

Chavez, F. P., Messié, M. and Pennington, J. T. (2011). Marine primary production in relation to climate variability and change. Annu. Rev. Mar. Sci. 3: 227–260.

Cheng, L., Abraham, J., Hausfather, Z. and Trenberth, K. E. (2019). How fast are the oceans warming? Observational records of ocean heat content show that ocean warming is accelerating. Science 363: 128–129.

Cheung, W. W. L., Watson, R., Morato, T., Pitcher, T. J. and Pauly, D. (2007). Intrinsic vulnerability in the global fish catch. Mar. Ecol. Prog. Ser. 333: 1–12.

Cheung, W. W., Dunne, J., Sarmiento, J. L. and Pauly, D. (2011). Integrating ecophysiology and plankton dynamics into projected maximum fisheries catch potential under climate change in the Northeast Atlantic. ICES J. Mar. Sci. 68: 1008–1018.

Childress, J. J. and Seibel, B. A. (1998). Life at stable low oxygen levels: adaptations of animals to oceanic oxygen minimum layers. J. Exp. Biol. 201: 1223–1232.

Chiba, S., Saito, H., Fletcher, R., Yogi, T., Kayo, M., Miyagi, S. et al. (2018). Human footprint in the abyss: 30 year records of deep-sea plastic debris. Mar. Policy 96: 204–212.

Choong, A., Teo, S., Leow, J., Koh, H. and Ho, P. (2006). A Preliminary ecotoxicity study of pharmaceuticals in the marine environment. J. Toxicol. Environ. Health A 69: 1959–1970.

Clark, M. R., Althaus, F., Schlacher, T. A., Williams, A., Bowden, D. A. and Rowden, A. A. (2016). The impacts of deep-sea fisheries on benthic communities: a review. ICES J. Mar. Sci. 73: i51–i69.

Clark, M. R., Anderson, O. F., Francis, R. I. C. C. and Tracey, D. M. (2000). The effects of commercial exploitation on orange roughy (*Hoplostethus atlanticus*) from the continental slope of the Chatham rise, New Zealand, from 1979 to 1997. Fish. Res. 45: 217–238.

Clark, M. R., Bowden, D. A., Rowden, A. A. and Stewart, R. (2019). Little evidence of benthic community resilience to bottom trawling on seamounts after 15 years. Front. Mar. Sci. 6: 63.

Clark, M. R. and Rowden, A. A. (2009). Effect of deep water trawling on the macroinvertebrate assemblages of seamounts on the Chatham Rise, New Zealand. Deep-Sea Res. 56: 1540–1554.

Clark, M. R., Rowden, A. A., Schlacher, T., Williams, A., Consalvey, M., Stocks, K. I. et al. (2010). The ecology of seamounts: structure, function, and human impacts. Annu. Rev. Mar. Sci. 2: 253–278.

Clark, M. R., Vinnichenko, V. I., Gordon, J. D. M., Beck-Bulat, G. Z., Kukharev, N. N. and Kakora, A. F. (2007). Large-scale distant-water trawl fisheries on seamounts. pp. 361–399. *In*: Pitcher, T. J., Morato, T., Hart, P. J. B., Clark, M. R., Haggan, N. and Santos, R. S. (eds.). Seamounts: Ecology, Fisheries & Conservation, Blackwell Publishers, Oxford, United Kingdom.

Clemente, C. C. C., Paresque, K. and Santos, P. J. P. (2018). The effects of plastic bags presence on a macrobenthic community in a polluted estuary. Mar. Pollut. Bull. 135: 630–635.

Cole, M., Lindeque, P. K., Fileman, E., Clark, J., Lewis, C., Halsband, C. et al. (2016). Microplastics alter the properties and sinking rates of zooplankton faecal pellets. Environ. Sci. Technol. 50: 3239–3246.

Collie, J. S., Hall, S. J., Kaiser, M. J. and Poiner, I. R. (2000). A quantitative analysis of fishing impacts on shelf-sea benthos. J. Animal Ecol. 69: 785–798.

Company, J. B., Puig, P., Sardà, F., Palanques, A., Latasa, M. and Scharek, R. (2008). Climate influence on deep sea populations. PLoS ONE 3: e1431.

Corcho-Alvarado, J. A., Guavis, C., McGinnity, P., Röllin, S., Ketedromo, T., Sahli, H. et al. (2021). Assessment of residual radionuclide levels at the Bokak and Bikar Atolls in the northern Marshall Islands. Sci. Total Environ. 801: 149541.

Cordes, E. E., Cunha, M. R., Galéron, J., Mora, C., Olu-Le Roy, K. et al. (2010). The influence of geological, geochemical, and biogenic habitat heterogeneity on seep diversity. Mar. Ecol. 31: 51–65. https:// doi.org/10.1111/ j.1439-0485.2009.00334.x.

Cordier, T., Angeles, I. B., Henry, N., Lejzerowicz, F., Berney, C., Morardet, R. et al. (2022). Patterns of eukaryotic diversity from the surface to the deep-ocean sediment. Sci. Adv. 8: eabj9309.

Costello, M. J., Cheung, A. and De Hauwere, N. (2010). Topography statistics for the surface and seabed area, volume, depth and slope, of the world's seas, oceans and countries. Environ. Sci. Technol. 44: 8821–8828.

Costello, M. J., Wilson, S. P. and Houlding, B. (2012). Predicting total global species richness using rates of species description and estimates of taxonomic effort. Syst. Biol. 61: 871–883.

Cózar, A., Echevarría, F., González-Gordilloa, J. I., Irigoien, X., Úbeda, B., Hernández-Leónd, S. et al. (2014). Plastic debris in the open ocean. Proc. Natl. Acad. Sci. USA 111: 10239–10244.

Dahlgren, T., Wiklund, H., Rabone, M., Amon, D., Ikebe, C., Watling, L. et al. (2016). Abyssal fauna of the UK-1 polymetallic nodule exploration area, Clarion-Clipperton Zone, central Pacific Ocean: Cnidaria. Biodivers. Data J. 4: e9277.

Danovaro, R., Dell'Anno, A. and Pusceddu, A. (2004). Biodiversity response to climate change in a warm deep sea. Ecol. Lett. 7: 821–828.

Davies, R. W. D., Cripps, S. J., Nickson, A. and Porter, G. (2009). Defining and estimating global marine fisheries bycatch. Mar. Policy 33: 661–672.

de Coning, E. and Witbooi, E. (2015). Towards a new 'fisheries crime' paradigm: South Africa as an illustrative example. Mar. Policy 60: 208–215.

Dean, J. F., Middelburg, J. J., Röckmann, T., Aerts, R., Blauw, L. G., Egger, M. et al. (2018). Methane feedbacks to the global climate system in a warmer world. Rev. Geophys. 56: 207–250.

De Leo, D. M., Ruiz-Ramos, D. V., Baums, I. B., Erik, E. and Cordes, E. E. (2016). Response of deep-water corals to oil and chemical dispersant exposure. Deep-Sea Res. II 129: 137–147.

Desbruyères, D., Mcdonagh, E. L. and King, B. A. (2017). Global and full-depth ocean temperature trends during the early Twenty-First Century from Argo and repeat hydrography. J. Clim. 30: 1985–1997.

Deutsch, C., Brix, H., Ito, T., Frenzel, H. and Thompson, L. (2011). Climate-forced variability of ocean hypoxia. Science 333: 336–339.

Deutsch, C., Ferrel, A., Seibel, B., Pörtner, H.-O. and Huey, R. B. (2015). Climate change tightens a metabolic constraint on marine habitats. Science 348: 1132–1135.

Devine, J. A., Baker, K. D. and Haedrich, R. L. (2006). Deep-sea fishes qualify as endangered. Nature 439: 29.

Doney, S. C., Fabry, V. J., Feely, R. A. and Kleypas, J. A. (2009). Ocean acidification: The other CO_2 problem. Annu. Rev. Mar. Sci. 1: 169–192.

Doonan, I. J., Fu, D. and Dunn, M. R. (2015). Harvest control rules for a sustainable orange roughy fishery. Deep-Sea Res. I 98: 53–61.

Doumbouya, A., Camara, O. T., Mamie, J., Intchama, J. F., Jarra, A., Ceesay, S. et al. (2017). Assessing the effectiveness of monitoring control and surveillance of illegal fishing: The case of West Africa. Front. Mar. Sci. 4: 50.

Drazen, J. C., Smith, C. R., Gjerde, K. M., Haddock, S. H. D., Carter, G. S., Choye, C. A. et al. (2020). Midwater ecosystems must be considered when evaluating environmental risks of deep-sea mining. Proc. Natl. Acad. Sci. USA 117: 17455–17460.

Ekau, W., Auel, H., Pörtner, H.-O. and Gilbert, D. (2010). Impacts of hypoxia on the structure and processes in pelagic communities (zooplankton, macro-invertebrates and fish). Biogeosciences 7: 1669–1699.

Fallon, S. J., Thresher, R. E. and Adkins, J. (2014). Age and growth of the cold-water scleractinian *Solenosmilia variabilis* and its reef on SW Pacific seamounts. Coral Reefs 33: 31–38.

FAO. (2009). International Guidelines for the Management of Deep-sea Fisheries in the High Seas. 99 pp. https://www.fao.org/documents/card/en/c/b02fc35e-a0c4-545a-86fb-4fc340e13b52 (last accessed 30/05/2023).

FAO. (2018). The State of World Fisheries and Aquaculture 2018 - Meeting the sustainable development goals. Food and Agricultural Organisation of the United Nations, Rome, Italy, 210pp.

FAO. (2022). The State of World Fisheries and Aquaculture 2022. Towards Blue Transformation. Food and Agricultural Organisation of the United Nations, Rome, Italy, 236pp.

Fernandez-Arcaya, U., Ramirez-Llodra, E., Aguzzi, J., Allcock, A. L., Davies, J. S., Dissanayake, A. et al. (2017). Ecological role of submarine canyons and need for canyon conservation: a review. Front. Mar. Sci. 4: 5.

Field, C. B., Behrenfeld, M. J., Randerson, J. T. and Falkowski, P. (1998). Primary production of the biosphere: integrating terrestrial and oceanic components. Science 281: 237–240.

Finkel, Z. V., Beardall, J., Flynn, K. J., Quigg, A., Alwyn, T., Rees, V. et al. (2010). Phytoplankton in a changing world: cell size and elemental stoichiometry. J. Plank. Res. 32: 119–137.

Fisher, C. R., Demopoulos, A. W. J., Cordes, E. E., Baums, I. B., White, H. K. and Bourque J. R. (2014). Coral communities as indicators of ecosystem-level impacts of the Deepwater Horizon spill. BioScience 64: 796–807.

Friedlander, A. M., Caselle, J. E., Ballesteros, E., Brown, E. K., Turchik, A. and Sala, E. (2014). The real bounty: marine biodiversity in the Pitcairn Islands. PLoS ONE 9: e100142.

Froescheis, O., Looser, R., Cailliet, G. M., Jarman, W. M. and Ballschmiter, K. (2000). The deep sea as a final global sink of semi-volatile persistent pollutants? Part I: PCBs in surface and deep-dwelling fish of the North and South Atlantic and the Monterey Bay Canyon (California). Chemosphere 40: 651–660.

Fosså, J. H., Mortensen, P. B. and Furevik, D. M. (2002). The deep-water coral *Lophelia pertusa* in Norwegian waters: distribution and fishery impacts. Hydrobiologia 471: 1–12.

Gage, J. D. (2003). Food inputs, utilization, carbon flow and energetics. pp. 315–382. *In*: Tyler, P. A. (ed.). Ecosystems of the World 28. Ecosystems of the Deep Oceans. Elsevier Science B.V., Amsterdam, The Netherlands.

Galgani, F., Hanke, G. and Maes, T. (2015). Global distribution, composition and abundance of marine litter. pp. 29–56. *In*: Bergmann, M., Gutow, L. and Klages, M. (eds.). Marine Anthropogenic Litter. Springer, Cham, Switzerland.

Gall, S. C. and Thompson, R. C. (2015). The impact of debris on marine life. Mar. Pollut. Bull. 92: 170–179.

Gattuso, J.-P, Magnan, A., Billé, R., Cheung, W., Howes, E. L., Joos, F. et al. (2015). Risks of warming and acidification for oceans and society. Science 349: aac4722.

Gebruk, A., Zalota, A. K., Dgebuadze, P., Ermilova, Y., Spiridonov, V. A., Shabalin, N. et al. (2021). Trophic niches of benthic crustaceans in the Pechora Sea suggest that the invasive snow crab *Chionoecetes opilio* could be an important competitor. Polar Biol. 44: 57–71.

Gehlen, M., Séférian, R., Jones, D. O. B., Roy, T., Roth, R., Barry, J. et al. (2014). Projected pH reductions by 2100 might put deep North Atlantic biodiversity at risk. Biogeosciences 11: 6955–6967.

Gillard, B., Purkiani, K., Chatzievangelou, D., Vink, A., Iversen, M. H. and Thomsen, L. (2019). Physical and hydrodynamic properties of deep-sea mining-generated, abyssal sediment plumes in the Clarion Clipperton Fracture Zone (eastern-central Pacific). Elem. Sci. Anthr. 7: 5.

Gilly, W. F., Beman, J. M., Litvin, S. Y. and Robison, B. H. (2013). Oceanographic and biological effects of shoaling of the oxygen minimum zone. Annu. Rev. Mar. Sci. 5: 393–420.

Gilly, W. F., Markaida, U., Baxter, C. H., Block, B. A., Boustany, A., Zeidberg, L. et al. (2006). Vertical and horizontal migrations by the jumbo squid *Dosidicus gigas* revealed by electronic tagging. Mar. Ecol. Prog. Ser. 324: 1–17.

Girard, F., Cruz, R., Glickman, O., Harpster, T. and Fisher, C. R. (2019). *In-situ* growth of deep-sea octocorals after the Deepwater Horizon oil spill. Elem. Sci. Anth. 7: 12.

Glover, A. G., Dahlgren, T. G., Wiklund, H., Mohrbeck, I. and Smith, C. R. (2016). An end-to-end DNA taxonomy methodology for benthic biodiversity survey in the Clarion-Clipperton Zone, central Pacific abyss. J. Mar. Sci. Eng. 4: 2.

Gollner, S., Kaiser, S., Menzel, L., Jones, D. O. B., Brown, A., Mestre, N. C. et al. (2017). Resilience of benthic deep-sea fauna to mining activities. Mar. Environ. Res. 129: 76–101.

Gray, J. S., Clarke, K. R., Warwick, R. M. and Hobbs, G. (1990). Detection of initial effects of pollution on marine benthos: an example from the Ekofisk and Eldfisk oilfields, North Sea. Mar. Ecol. Prog. Ser. 66: 285–299.

Grimes, D. J., Singleton, F. L., Stemmler, J., Palmer, L. M., Brayton, P., Colwell, R. R. (1984). Microbiological effects of wastewater effluent discharge into coastal waters of Puerto Rico. Water Res. 18: 613–619.

Gruber, N. (2011). Warming up, turning sour, losing breath: ocean biogeochemistry under global change. Phil. Trans. R. Soc. A 369: 1980–1996.

Hall, N. M., Berry, K. L. E., Rintoul, L. and Hoogenboom, M. O. (2015). Microplastic ingestion by scleractinian corals. Mar. Biol. 162: 725–732.

Hall-Spencer, J., Allain, V. and Fosså J. H. (2002). Trawling damage to Northeast Atlantic ancient coral reefs. Proc. R. Soc. Lond. B 269: 507–511.

Hall-Spencer, J. M. and Harvey, B. P. (2019). Ocean acidification impacts on coastal ecosystem services due to habitat degradation. Emerg. Top. Life Sci. 3: 197–206.

Halpern, B., Frazier, M., Potapenko, J., Casey, K. S., Koenig, K., Longo, C. et al. (2015). Spatial and temporal changes in cumulative human impacts on the world's ocean. Nat. Commun. 6: 7615.

Hartman, S. E. Bett, B. J., Durden, J. M., Henson, Iversen, I., Jeffreys, R. M. et al. (2021). Enduring science: Three decades of observing the Northeast Atlantic from the Porcupine Abyssal Plain Sustained Observatory (PAP-SO). Prog. Oceanogr. 191: 102508.

Hein, J. (2004). Cobalt-rich ferromanganese crusts: global distribution, composition, origin and research activities. pp. 188–256. *In*: Workshop on Minerals Other than Polymetallic Nodules of the International Seabed Area. Prepared by the Office of Resource and Environmental Monitoring, International Seabed Authority, Kingston, Jamaica.

Hein, J. R., Conrad, T., Mizell, K., Banakar, V. K., Frey, F. A., William, W. et al. (2016). Controls on ferromanganese crust composition and reconnaissance resource potential, Ninety East Ridge, Indian Ocean. Deep-Sea Res. I 110: 1–19.

Helm, K. P., Bindoff, N. L. and Church, J. A. (2011). Observed decreases in oxygen content of the global ocean. Geophys. Res. Lett. 38: L23602.

Hiddink, J. G., Jennings, S., Sciberras, M., Bolam, S. G., Cambiè, G., McConnaughey, R. A. et al. (2018). Assessing bottom trawling impacts based on the longevity of benthic invertebrates. J. Appl. Ecol. 56: 1075–1084.

HLPE. (2014). Sustainable Fisheries and Aquaculture for Food Security and Nutrition. A Report by The High Level Panel of Experts on Food Security and Nutrition of the Committee on World Food Security, UN Food and Agricultural Organisation, Rome, Italy, 118pp.

Hoegh-Guldberg, O. and Bruno, J. F. (2010). The impact of climate change on the world's marine ecosystems. Science 328: 1523–1528.

Hourdez, S., Weber, R. E., Green, B. N., Kenney, J. M. and Fisher, C. R. (2002). Respiratory adaptations in a deep-sea orbiniid polychaete from Gulf of Mexico brine pool NR-1: metabolic rates and haemoglobin structure/function relationships. J. Exp. Biol. 205: 1669–1681.

IPCC. (2013). Climate Change 2013: The Physical Science Basis. Contribution of Working Group I to the Fifth Assessment Report of the Intergovernmental Panel on Climate Change. Stocker, T. F., Qin, D., Plattner, G.-K., Tignor, M., Allen, S. K., Boschung, J. et al. (eds.). Cambridge University Press, Cambridge, United Kingdom and New York, NY, USA, 1535 pp.

IPCC. (2018). Global Warming of 1.5°C. An IPCC Special Report on the impacts of global warming of 1.5°C above pre-industrial levels and related global greenhouse gas emission pathways, in the context of strengthening the global response to the threat of climate change, sustainable development, and efforts to eradicate poverty. Masson-Delmotte, V., Zhai, P., Pörtner, H.-O., Roberts, D., Skea, J., Shukla, P.R. et al. (eds.). Cambridge University Press, Cambridge, UK and New York, NY, USA, 562pp.

IPCC. (2022). Climate Change 2022: Impacts, Adaptation and Vulnerability. Contribution of Working Group II to the Sixth Assessment Report of the Intergovernmental Panel on Climate Change. Pörtner, H.-O., Roberts, D. C., Tignor, M., Poloczanska, E. S., Mintenbeck, K., Alegría, A. et al. (eds.). Cambridge University Press, Cambridge, United Kingdom and New York, NY, USA, 3056 pp.

ISA. (2022). International Seabed Authority exploration contracts. https://www.isa.org.jm/exploration-contracts. Last accessed 5 December 2022).

Ito, T., Minobe, S., Long, M. C. and Deutsch, C. (2017). Upper ocean O_2 trends: 1958–2015. Geophys. Res. Lett. 44: 4214–4223.

Jambeck, J. R., Geyer, R., Wilcox, C., Siegler, T., Perryman, M., Andrady. A. et al. (2015). Plastic waste inputs from land into the ocean. Science 347: 768–771.

Jamieson, A. J., Brooks, L. S. R., Reid, W. D. K., Piertney, S. B., Narayanaswamy, B. E. and Linley, T. D. (2019). Microplastics and synthetic particles ingested by deep-sea amphipods in six of the deepest marine ecosystems on Earth. R. Soc. Open Sci. 6: 180667.

Jamieson, A. J., Malkocs, T., Piertney, S., Fujii, T. and Zhang, Z. (2017). Bioaccumulation of persistent organic pollutants in the deepest ocean fauna. Nat. Ecol. Evol. 1: 0051.

Jepson, P. D., Deaville, R., Acevedo-Whitehouse, K., Barnett, J., Brownlow, A., Brownell, R. L. et al. (2013). What caused the UK's largest common dolphin (*Delphinus delphis*) mass stranding event? PLoS ONE 8: e60953.

Jochum, K. P., Wang, X., Vennemann, T. W., Sinha, B. and Müller, W. E. g. (2012). Siliceous deep-sea sponge *Monorhaphis chuni*: A potential paleoclimate archive in ancient animals. Chem. Geol. 300-301: 143–151.

Jones, D. O. B., Amon, D. J. and Chapman, A. S. A. (2018). Mining deep-ocean mineral deposits: What are the ecological risks? Elements 14: 325–330.

Jones, D. O. B., Amon, D. J. and Chapman, A. S. A. (2020). Deep-sea mining: processes and impacts. pp. 91–110. *In*: Baker, M., Ramirez-Llodra, E. and Tyler, P. (eds.). Natural Capital and Exploitation of the Deep Ocean. Oxford University Press, Oxford, United Kingdom.

Jones, D. O. B., Yool, A., Wei, C.-L., Henson, S. A., Ruhl, H. A., Watson, R. A. et al. (2014). Global reductions in seafloor biomass in response to climate change. Glob. Change Biol. 20: 1861–1872.

Joye, S. B., Bracco, A., Özgökmen, T. M., Chanton, J. P., Grosell, M., MacDonald, I. R. et al. (2016). The Gulf of Mexico ecosystem, six years after the Macondo oil well blowout. Deep-Sea Res. II: 129: 4–19.

Juniper, K. S., Thornborough, K., Tyler, P. A. and Randrianarisoa, Y. (2020). Space, the final resource. pp. 183–200. *In*: Baker, M., Ramirez-Llodra, E. and Tyler, P. (eds.). Natural Capital and Exploitation of the Deep Ocean. Oxford University Press, Oxford, United Kingdom.

Keeling, R. F., Körtzinger, A. and Gruber, N. (2010). Ocean deoxygenation in a warming world. Annu. Rev. Mar, Sci. 2: 463–493.

Ketzer, M., Praeg, D., Rodrigues, L. F., Augustin, A., Pivel, M. A. G., Rahmati-Abkenar, M. et al. (2019). Gas hydrate dissociation linked to contemporary ocean warming in the southern hemisphere. Nat. Commun. 11: 3788.

Koslow, J. A., Goericke, R., Lara-Lopez, A. and Watson, W. (2011). Impact of declining intermediate-water oxygen on deepwater fishes in the California Current. Mar. Ecol. Prog. Ser. 436: 207–18.

Kroecker, K. J., Kordas, R. L., Crim, R. N. and Singh, G. G. (2010). Meta-analysis reveals negative yet variable effects of ocean acidification on marine organisms. Ecol. Lett. 13: 1419–1434.

Kunc, H. P., Sigwart, J. D., Lyons, G. N., McLaughlin, K. E. and Houghton, J. D. R. (2014). Anthropogenic noise affects behavior across sensory modalities. Am. Nat. 184: E93–E100.

Kvale, K. F., Prowe, A. E. F. and Oschlies, A. (2020). A critical examination of the role of marine snow and zooplankton fecal pellets in removing ocean surface microplastic. Front. Mar. Sci. 6: 808.

Kwiatkowski, L., Aumont, O. and Bopp, L. (2018). Consistent trophic amplification of marine biomass declines under climate change. Glob. Change Biol. 25: 218–229.

Kwiatkowski, L., Torres, O., Bopp, L., Aumont, O., Chamberlain, M., Christian, J. R. et al. (2020). Twenty-first century ocean warming, acidification, deoxygenation, and upper-ocean nutrient and primary production decline from CMIP6 model projections. Biogeosciences 17: 3439–3470.

Lebreton, L., van der Zwet, J., Damsteeg, J. W., Slat, B., Andrady, A. and Reisser, J. (2017). River plastic emissions to the world's oceans. Nat. Commun. 8: 15611.

Le Bris, N. and Levin, L. A. (2020). Climate change cumulative impacts on deep-sea ecosystems. pp. 161–182. *In*: Baker, M., Ramirez-Llodra, E. and Tyler, P. (eds.). Natural Capital and Exploitation of the Deep Ocean. Oxford University Press, Oxford, United Kingdom.

Lejzerowicz, F., Gooday A. J., Barrenechea, A. I., Cordier, T., Morard, R., Apothéloz-Perret-Gentil, L. et al. (2021). Eukaryotic biodiversity and spatial patterns in the clarion-clipperton zone and other abyssal regions: insights from sediment DNA and RNA metabarcoding. Front. Mar. Sci. 8: 671033.

Levin, L. A. (2005). Ecology of cold seep sediments: Interactions of fauna with flow, chemistry and microbes. Oceanogr. Mar. Biol. Ann. Rev. 43: 1–46.

Levin, L. A. (2018). Manifestation, drivers, and emergence of open ocean deoxygenation. Annu. Rev. Mar. Sci. 10: 229–260.

Levin, L. A., Baco, A. R., Bowden, D. A., Colaco, A., Cordes, E. E., Cunha, M. R. et al. (2016b). Hydrothermal vents and methane seeps: Rethinking the sphere of influence. Front. Mar. Sci. 3: 72.

Levin, L. A. and Le Bris, N. (2015). The deep ocean under climate change. Science 350: 766–768.

Levin, L. A., Mengerink, K., Gjerde, K. M., Rowden, A. A., Van Dover, C. L., Clark, M. R. et al. (2016a). Defining "serious harm" to the marine environment in the context of deep seabed mining. Mar. Policy 74: 245–259.

Levin, L. A., Sibuet, M., Gooday, A. J., Smith, C. R. and Vanreusel, A. (2010). The roles of habitat heterogeneity in generating and maintaining biodiversity on continental margins. Mar. Ecol. 31: 1–5.

Levin, L. A, Whitcraft, C. R., Mendoza, G. F., Gonzalez, J. P. and Cowie, G. (2009). Oxygen and organic matter thresholds for benthic faunal activity on the Pakistan margin oxygen minimum zone (700–1100 m). Deep-Sea Res. II 56: 449–471.

Levitus, S., Antonov, J. I., Boyer, T. P., Baranova, O. K., Garcia, H. E., Locarnini, R. A. et al. (2012). World ocean heat content and thermosteric sea level change (0–2000 m), 1955–2010. Geophys. Res. Lett. 39: L10603.

Looser, R., Froescheis, O., Cailliet, G. M., Jarman, W. M. and Ballschmiter, K. (2000). The deep sea as a final global sink of semi-volatile persistent pollutants? Part II: Organochlorine pesticides in surface and deep-dwelling fish of the North and South Atlantic and the Monterey Bay Canyon (California). Chemosphere 40: 661–670.

Lovelock, C. E., Atwood, T., Baldock, J., Duarte, C. M., Hickey, S., Lavery, P. S. et al. (2017). Assessing the risk of carbon dioxide emissions from blue carbon ecosystems. Front. Ecol. Environ. 15: 257–265.

Macreadie, P., McLean, D., Thomson, P., Partridge, J. C., Jones, D. O. B., Gates, A. et al. (2018). Eyes in the sea: Unlocking the mysteries of the ocean using industrial, remotely operated vehicles (ROVs). Sci. Total Environ. 634: 1077–1091.

Martin, C., Weilgart, L., Amon, D. J. and Müller, J. (2021). Deep-sea Mining: A Noisy Affair. OceanCare Report, OceanCare, Wädenswil, Switzerland 64 pp.

Marcoux, S. (2021). Contamination and a compact: remediless environmental harm in the Marshall Islands. Environ. Claims J. 34: 95–135.

Masson, D. G., Bett, B. J., Billett, D. S. M., Jacobs, C. L., Wheeler, A. J. and Wynn, R. B. (2003). The origin of deep-water, coral-topped mounds in the northern Rockall Trough, Northeast Atlantic. Mar. Geol. 194: 159–180.

McClain C. R., Nunnally, C. and Benfield, M. C. (2019). Persistent and substantial impacts of the Deepwater Horizon oil spill on deep-sea megafauna. R. Soc. Open Sci. 6: 191164.

Menot, L., Sibuet, M., Carney, R. S., Levin, L. A., Rowe, G. T., Billett, D. S. M. et al. (2010). New perceptions of continental margin biodiversity. pp. 79–103. *In*: McIntyre, A. D. (ed.). Life in the World's Oceans: Diversity, Distribution, and Abundance. Wiley-Blackwell, Oxford, United Kingdom.

Miller, K. A., Thompson, K. F., Johnston, P. and Santillo, D. (2018). An overview of seabed mining including the current state of development, environmental impacts, and knowledge gaps. Front. Mar. Sci. 4: 418.

Moore, B. C. J., Alcantara, J. I. and Glasberg, B. P. (2002). Behavioural measurement of level-dependent shifts in the vibration pattern on the basilar membrane at 1 and 2 kHz. Hear. Res. 163: 101–110.

Moore, J. K., Fu, W., Primeau, F., Britten, G. L., Lindsay, K., Long, M. et al. (2018). Sustained climate warming drives declining marine biological productivity. Science 359: 1139–1143.

Mora, C., Tittensor, D. P., Adl, S., Simpson, A. G. and Worm, B. (2011). How many species are there on earth and in the ocean? PLoS Biol. 9: e1001127.

Muñoz-Royo, C., Peacock, T., Alford, M. H., Smith, J. A., Le Boyer, A., Kulkarni, C. S. et al. (2021). Extent of impact of deep-sea nodule mining midwater plumes is influenced by sediment loading, turbulence and thresholds. Commun. Earth Environ. 2: 148.

Myers, R. A., Baum, J. K., Shepherd, T. D., Powers, S. P. and Peterson, C. H. (2007). Cascading Effects of the Loss of apex predatory sharks from a coastal ocean. Science 315: 1846–1850.

Narula, R. (2018). Multinational firms and the extractive sectors in the 21st century: Can they drive development? J. World Bus. 53: 85–91.

Neat, F. C., Burns, F., Jones, E. and Blasdale, T. (2015). The diversity, distribution and status of deep-water elasmobranchs in the Rockall Trough, north-east Atlantic Ocean. J. Fish Biol. 87: 1469–1488.

Nicol, J. A. C., Lee, W. Y. and Hannebaum, N. (1978). Toxicity of Puerto Rican organic waste materials on marine invertebrates. Final Report to National Oceanic and Atmospheric Administration/Ocean Dumping and Monitoring Program, Marine Science Institute, University of Texas, Port Aransas. 37 p.

Niklitschek, E. J., Cornejo-Donoso, J., Oyarzú, C., Hernández, E. and Toledo, P. (2010). Developing seamount fishery produces localized reductions in abundance and changes in species composition of bycatch. Mar. Ecol. 31(Suppl. 1): 168–182.

Niner, H. J., Ardron, J. A., Escobar, E. G., Gianni, M., Jaeckel, A., Jones, D. O. B. et al. (2018). Deep-sea mining with no net loss of biodiversity—an impossible aim. Front. Mar. Sci. 5: 53.

Norse, E. A., Brooke, S., Cheung, W. W. L., Clark, M. R., Ekeland, I., Froese, R. et al. (2012). Sustainability of deep-sea fisheries. Mar. Policy 36: 307–320.

Orcutt, B. N., Bradley, J. A., Brazelton, W. J., Estes, E. R., Goordial, J. M., Huber, J. A. et al. (2020). Impacts of deep-sea mining on microbial ecosystem services. Limnol. Oceanogr. 65: 1489–1510.

Orcutt, B. N., Joye, S. B., Kleindienst, S., Knittel, K., Ramette, A., Reitz, A. et al. (2010). Impact of natural oil and higher hydrocarbons on microbial diversity, distribution, and activity in Gulf of Mexico cold-seep sediments. Deep-Sea Res. II 57: 2008–2021.

Park, J., Van Osdel, J., Turner, J., Farthing, C. M., Miller, N. A., Hannah, L. et al. (2023). Tracking elusive and shifting identities of the global fishing fleet. Sci. Adv. 9: eabp8200.

Parsons, E. C. M. (2017). Impacts of navy sonar on whales and dolphins: Now beyond a smoking gun? Front. Mar. Sci. 4: 295.

Passow, U. and Ziervogel, K. (2016). Marine snow sedimented oil released during the Deepwater Horizon spill. Oceanography 29: 118–125.

Passow, U., Ziervogel, K., Asper, V. and Diercks. A. (2012). Marine snow formation in the aftermath of the Deepwater Horizon oil spill in the Gulf of Mexico. Environ. Res. Lett. 7: 035301.

Paulus, E. (2021). Shedding light on deep-sea biodiversity—a highly vulnerable habitat in the face of anthropogenic change. Front. Mar. Sci. 8: 667048.

Pauly, D. and Zeller, D. (2016). Catch reconstructions reveal that global marine fisheries catches are higher than reported and declining. Nat. Commun. 7: 10244.

Pedersen, T., Fuhrmann, M. M., Lindstrøm, U., Nilssen, E. M., Ivarjord, T., Virginie Ramasco, V. et al. (2018). Effects of the invasive red king crab on food web structure and ecosystem properties in an Atlantic fjord. Mar. Ecol. Prog. Ser. 596: 13–31.

Peele, E. R., Singleton, F. L., Deming, J. W., Cavari, B. and Colwell, R. R. (1981). Effects of pharmaceutical wastes on microbial populations in surface waters at the Puerto Rico dump site in the Atlantic Ocean. Appl. Environ. Microbiol. 41: 873–879.

Pham, C. K., Diogo, H., Menezes, G., Porteiro, F., Braga-Henriques, A., Vandeperre, F. et al. (2014a). Deep-water longline fishing has reduced impact on vulnerable marine ecosystems. Sci. Rep. 4: 4837.

Pham, C. K., Ramirez-Llodra, E., Alt, C. H. S., Amaro, T., Bergmann, M., Canals, M. et al. (2014b). Marine litter distribution and density in European seas, from the shelves to deep basins. PLoS ONE 9: e95839.

Phrampus, B. J. and Hornbach, M. J. (2012). Recent changes to the Gulf Stream causing widespread gas hydrate destabilization. Nature 490: 527–530.

Pitcher, T. J., Clark, M. R., Morato, T. and Watson, R. (2010). Seamount Fisheries: do they have a future? Oceanography 23: 134–144.

Polovina, J. J., Howell, E. A. and Abecassis, M. (2008). The ocean's least productive waters are expanding. Geophys. Res. Lett. 35: L03618.

Poncelet, E., van Canneyt, O. and Boubert, J. J. (2000). Considerable amount of plastic debris in the stomach of a Cuvier's beaked whale (*Ziphius cavirostris*) washed ashore on the French Atlantic coast. pp. 44–47. *In*: Evans, P. G. H., Pitt-Aiken, R. and Rogan, E. (eds.). European Research on Cetaceans, 14. Proceedings of the 14th Annual Conference of the European Cetacean Society, Cork, Ireland, 2–5 April 2000. European Cetacean Society, Rome, Italy.

Porter, A., Lyons, B. P., Galloway, T. S. and Lewis, C. (2018). Role of marine snows in microplastic fate and bioavailability. Environ. Sci. Technol. 52: 7111–7119.

Priede, I. G., Godbold, J. A., Niedzielski, T., Collins, M. A., Bailey, D. M., Gordon, J. D. M. et al. (2011). A review of the spatial extent of fishery effects and species vulnerability of the deep-sea demersal fish assemblage of the Porcupine Seabight, Northeast Atlantic Ocean (ICES Subarea VII). ICES J. Mar. Sci. 68: 281–289.

Prince, E. D. and Goodyear, C. P. (2006). Hypoxia-based habitat compression of tropical pelagic fishes. Fish. Oceanogr. 15: 451–464.

Prince, E. D., Luo, J., Goodyear, C. P., Hoolihan, J. P., Snodgrass, D., Orbesen, E. S. et al. (2010). Ocean scale hypoxia-based habitat compression of Atlantic istiophorid billfishes. Fish. Oceanogr. 19: 448–462.

Puig, P., Canals, M., Company, J. B., Martín, J., Amblas, D., Lastras, G. et al. (2012). Ploughing the deep sea floor. Nature 489: 286–289.

Puig-Lozano, R., Fernández, A., Sierra, E., Saavedra, P., Suárez-Santana, C. M., De la Fuente, J. et al. (2020). Retrospective study of fishery interactions in stranded cetaceans, Canary Islands. Front. Vet. Sci. 7: 567258.

Purkey, S. G. and Johnson, G. C. 2010. Warming of global abyssal and deep Southern Ocean waters between the 1990s and 2000s: Contributions to global heat and sea level rise budgets. J. Clim. 23: 6336–6351.

Pusceddu, A., Bianchelli, S., Martín, J., Puig, P., Palanques, A., Masqué, P. et al. (2014). Chronic and intensive bottom trawling impairs deep-sea biodiversity and ecosystem functioning. Proc. Natl. Acad. Sci. USA 111: 8861–8866.

Rabone, M., Wiethase, J. H., Simon-Lledo, E., Emery, A. M., Jone, D. O. B, Dahlgren, T. G. et al. (2023). How many metazoan species live in the world's largest mineral exploration region? Curr. Biol. 33: 1–14.

Ramirez-Llodra, E., Baker, M. and Tyler, P. A. (2020). A holistic vision for our future deep ocean. pp. 201–205. *In*: Baker, M., Ramirez-Llodra, E. and Tyler, P. (eds.). Natural Capital and Exploitation of the Deep Ocean. Oxford University Press, Oxford, United Kingdom.

Ramirez-Llodra, E., Brandt, A., Danovaro, R., De Mol, B., Escobar, E., German, C. R. et al. (2010). Deep, diverse and definitely different: unique attributes of the world's largest ecosystem. Biogeosciences 7: 2851–2899.

Ramirez-Llodra, E., De Mol, B., Company, J.-B., Coll, M. and Sardà, F. (2013). Effects of natural and anthropogenic processes in the distribution of marine litter in the deep Mediterranean Sea. Prog. Oceanogr. 118: 273–287.

Ramirez-Llodra, E., Tyler, P. A., Baker, M. C., Bergstad, O. A., Clark, M. R., Escobar, E. et al. (2011). Man and the last great wilderness: Human impact on the deep sea. PLoS ONE 6: e22588.

Read, A. J., Drinker, P. and Northridge, S. (2006). Bycatch of marine mammals in U.S. and global fisheries. Conserv. Biol. 20: 163–169.

Rech, S., Salmina, S., Pichs, Y. J. B. and García-Vazquez, E. (2018). Dispersal of alien invasive species on anthropogenic litter from European mariculture areas. Mar. Pollut. Bull. 131: 10–16.

Reed, J. K., Koenig, C. C. and Shepard, A. N. (2007). Impacts of bottom trawling on a deep-water *Oculina* coral ecosystem off Florida. Bull. Mar. Sci., 81: 481–496.

Rex M. A. and Etter R. J. (2010). Deep-sea Biodiversity: Pattern and Scale. Harvard University Press, Cambridge, Massachusetts, USA, 332 pp.

Richardson, K., Gunn, R., Wilcox, C. and Hardesty, B. D. (2018). Understanding causes of gear loss provides a sound basis for fisheries management. Mar. Policy 96: 278–284.

Richier, S., Achterberg, E. P., Dumousseaud, C., Poulton, A. J., Suggett, D. J., Tyrrell, T. et al. (2014). Phytoplankton responses and associated carbon cycling during shipboard carbonate chemistry manipulation experiments conducted around Northwest European shelf seas. Biogeosciences 11: 4733–4752.

Riebesell, U., Gattuso, J.-P., Thingstad, T. F. and Middelburg, J. J. (2013). Arctic ocean acidification: pelagic ecosystem and biogeochemical responses during a mesocosm study. Biogeosciences 10: 5619–5626.

Ríos, A. F., Resplandy, L., García-Ibáñeza, M. I., Fajar, N. M., Velo, A., Padin, X. O. et al. (2015). Decadal acidification in the water masses of the Atlantic Ocean. Proc. Natl. Acad. Sci. USA 112: 9950–9955.

Riskas, K. A., Tobin, R. C., Fuentes, M. M. P. B. and Hamann, M. (2018). Evaluating the threat of IUU fishing to sea turtles in the Indian Ocean and Southeast Asia using expert elicitation. Biol. Conserv. 217: 232–239.

Roark, E. B., Guilderson, T. P., Dunbar, R. B., Fallon, S. J. and Mucciarone, D. A. (2009). Extreme longevity in proteinaceous deep-sea corals. Proc. Natl. Acad. Sci. USA 106: 5204–5208.

Roberts, J. M., Wheeler, A., Freiwald, A. and Cairns, S. (2009). Cold-Water Corals: The Biology and Geology of Deep-Sea Coral Habitats. Cambridge University Press, Cambridge, United Kingdom, 368pp.

Romera-Castillo, C., Pinto, M., Langer, T. M., Álvarez-Salgado, X. A. and Herndl, G. (2018). Dissolved organic carbon leaching from plastics stimulates microbial activity in the ocean. Nat. Commun. 9: 1430.

Rogers, A. D. (1999). The biology of *Lophelia pertusa* (Linnaeus 1758) and other deep-water reef-forming corals and impacts from human activities. Int. Rev. Hydrobiol. 84: 315–406.

Rogers, A. D. (2015). Environmental change in the deep ocean. Ann. Rev. Environ. Resour. 40: 1–38.

Rogers A. D. (2018). The biology of seamounts: 25 years on. Adv. Mar. Biol. 79: 137–224

Rogers, A. D. (2019). Threats to seamount ecosystems and their management. pp. 427–451. *In*: Sheppard, C. (ed.). World Seas: An Environmental Evaluation, Second Edition. Academic Press, London, United Kingdom.

Rogers, A. D. and Gianni, M. (2010). The Implementation of UNGA Resolutions 61/105 and 64/72 in the Management of Deep-Sea Fisheries on the High Seas. Report prepared for the Deep-Sea Conservation Coalition. International Programme on State of the Ocean, London, United Kingdom, 97pp.

Rotllant, G., Abad, E., Sardà, F., Ábalos, M., Company, J. B. and Rivera, J. (2006). Dioxin compounds in the deep-sea rose shrimp *Aristeus antennatus* (Risso, 1816). throughout the Mediterranean Sea. Deep-Sea Res. I 53: 1895–1906.

Royer, S. J., Ferrón, S., Wilson, S. T. and Karl, D. M. (2018). Production of methane and ethylene from plastic in the environment. PLoS ONE 13: e0200574.

Ruhl, H. A. and Smith, K. L. (2004). Shifts in deep-sea community structure linked to climate and food supply. Science 305: 513–515.

Ruppel, C. D. and Kessler, J. D. (2017). The interaction of climate change and methane hydrates. Rev. Geophys. 55: 126–168.

Sato, K. N., Levin, L. A. and Schift, K. (2017). Habitat compression and expansion of sea urchins in response to changing climate conditions on the California continental shelf and slope (1994–2013). Deep-Sea Res. II 137: 377–389.

Seibel, B. A., Thuesen, E. V., Childress, J. J. and Gorodezky, L. A. (1997). Decline in pelagic cephalopod metabolism with habitat depth reflects differences in locomotory efficiency. Biol. Bull. 192: 262–278.

Schlacher, T. A., Rowden, A. A., Dower, J. F. and Consalvey, M. (2010). Seamount science scales undersea mountains: new research and outlook. Mar. Ecol. 31: 1–13.

Schlacher, T. A., Baco, A. R., Rowden, A. A., O'Hara, T. D., Clark, M. R., Kelley, C. et al. (2014). Seamount benthos in a cobalt-rich crust region of the central Pacific: conservation challenges for future seabed mining. Divers. Distrib. 20: 491–502.

Schmidtko, S., Stramma, L. and Visbeck, M. (2017). Decline in global oceanic oxygen content during the past five decades. Nature 542: 335–339.

Sibuet, M. and Olu, K. (1998). Biogeography, biodiversity and fluid dependence of deep-sea cold-seep communities at active and passive margins. Deep-Sea Res. II 45: 517–567.

Sigwart, J. D., Chen, C., Thomas, E. A., Allcock, A. L., Böhm, M. and Seddon, M. (2019). Red Listing can protect deep-sea biodiversity. Nat. Ecol. Evol. 3: 1134.

Simas, M., Aponte, F. and Wiebe, K. (2022). The Future is Circular Circular Economy and Critical Minerals for the Green Transition. SINTEF report 2022:01134. 68 pp. https://sintef.brage.unit.no/sintef-xmlui/bitstream/handle/11250/3032049/CircularEconomyAndCriticalMineralsReport.pdf?sequence=7 (last accessed 30/05/2023).

Simon-Lledó, E., Bett, B. J., Huvenne, V. A. I., Köser, K., Schoening, T., Greinert, J. et al. (2019). Biological effects 26 years after simulated deep-sea mining. Sci. Rep. 9: 8040.

Simpfendorfer, C. A. and Kyne, P. M. (2009). Limited potential to recover from overfishing raises concerns for deep-sea sharks, rays and chimaeras. Environ. Conser. 36: 97–103.

Simpson, D. C., O'Connor, T. P. and Park, P. K. (1981). Deep-ocean dumping of industrial wastes. pp. 379–400. *In*: Geyer, R. A. (ed.). Marine Environmental Pollution, Vol. 2, Dumping and Mining, Elsevier Scientific, New York, NY, USA.

Smith, C. R., De Leo, F. C., Bernardino, A. F., Sweetman, A. K. and Martinez Arbizu, P. (2008). Abyssal food limitation, ecosystem structure and climate change. Trends Ecol. Evol. 23: 518–528.

Smith, C. R., Grange, L. J., Honig, D. L. Naudts, L., Huber, B., Guidi, L. et al. (2012). A large population of king crabs in Palmer Deep on the west Antarctic Peninsula shelf and potential invasive impacts. Proc. R. Soc. B Biol. Sci. 279: 1017–1026.

Smith, C. R., Tunnicliffe, V., Colaço, A., Drazen, J. C., Gollner, S., Levin, L. A. et al. (2020). Deep-sea misconceptions cause underestimation of seabed-mining impacts, Trends Ecol. Evol. 35: 853–857.

Smith, K. L., Jr., Ruhl, H. A., Bett, B. J., Billett, D. S. M., Lampitt, R. S. and Kaufmann, R. S. (2009). Climate, carbon cycling, and deep-ocean ecosystems. Proc. Natl. Acad. Sci. USA 106: 19211–19218.

Snelgrove, P. V. R. and Smith, C. R. (2002). A riot of species in an environmental calm; The paradox of the species-rich deep sea. Oceanogr., Mar. Biol. Ann. Rev. 40: 311–342.

Soliman, Y. and Wade, T. L. (2008). Estimates of PAHs burdens in a population of ampeliscid amphipods at the head of the Mississippi Canyon (N. Gulf of Mexico). Deep-Sea Res. II 55: 2577–2584.

Sperling, E. A., Frieder, C. A. and Levin, L. A. (2016). Biodiversity response to natural gradients of multiple stressors on continental margins. Proc. R. Soc. B Biol. Sci. 283: 20160637.

Stefanoudis, P. V., Gress, E., Pitt, J. M., Smith, S. R., Kincaid, T., Rivers, M. et al. (2019). Depth-dependent structuring of reef fish assemblages from the shallows to the rariphotic zone. Front. Mar. Sci. 6: 307.

Stegeman, J. J. (1983). Hepatic microsomal monooxygenase activity and biotransformation of hydrocarboms in deep benthic fish from the Western North Atlantic. Can. J. Fish. Aqua. Sci. 40: 78–85.

Stegeman, J. J., Kloepper-Sams, P. J. and Farrington, J. W. (1986). Monoxygenase induction and chlorobiphenyls in the deep-sea fish *Coryphenoides amatus*. Science 231: 1287–1289.

Stelfox, M., Hudgins, J. and Sweet, M. (2016). A review of ghost gear entanglement amongst marine mammals, reptiles and elasmobranchs. Mar. Pollut. Bull. 111: 6–17.

Stewart, E. C. D., Bribiesca-Contreras, G., Taboada, S., Wiklund, H., Ravara, A., Pape, E. et al. (2023). Biodiversity, biogeography, and connectivity of polychaetes in the world's largest marine minerals exploration frontier. Divers. Distrib. 29: 727–747.

Stewart, J. S. (2012). Humboldt squid in the northern California Current system. PhD thesis. Stanford Univ., Stanford, CA. 193 pp. http://purl.stanford.edu/pb685tk3754 (Last accessed 25/07/2023).

Stewart, J. S., Hazen, E. L., Bograd, S. J., Byrnes, J. E. K., Foley, D. G., Gilly, W. F. et al. (2014). Combined climate- and prey-mediated range expansion of Humboldt squid (*Dosidicus gigas*), a large marine predator in the California Current System. Glob. Change Biol. 20: 1832–1843.

Stone, R., Stevenson, D. and Brooke, S. (2015). Assessment of a pilot study to collect coral bycatch data from the Alaska commercial fishing fleet. *In*: NOAA technical memorandum NMFS-AFSC-296. Springfield, Virginia: US Department of Commerce, 45pp.

Stramma, L., Prince, E. D., Schmidtko, S., Luo, J., Hoolihan, J. P., Visbeck, M. et al. (2011). Expansion of oxygen minimum zones may reduce available habitat for tropical pelagic fishes. Nat. Clim. Change 2: 33–37.

Stramma, L., Schmidtko, S., Levin, L. A. and Johnson, G. C. (2010). Ocean oxygen minima expansions and their biological impacts. Deep-Sea Res. I: 57: 587–595.

Suess, E. (2014). Marine cold seeps and their manifestations: geological control, biogeo-chemical criteria and environmental conditions. Int. J. Earth Sci. 103: 1889–1916.

Sumaila, U. R., Alder, J. and Keith. H. (2006). Global scope and economics of illegal fishing. Mar. Policy 30: 696–703.

Sweetman, A. K., Chelsky, A., Pitt, K. A., Andrade, H., van Oevelen, D. and Renaud, P. E. (2016). Jellyfish decomposition at the seafloor rapidly alter biogeochemical cycling and carbon flow through benthic food-webs. Limnol. Oceanogr. 61: 1449–1461.

Sweetman, A. K., Andrew, R. Thurber, A. R., Smith, C. R., Levin, L. A., Mora, C. et al. (2017). Major impacts of climate change on deep-sea benthic ecosystems. Elem. Sci. Anth. 5: 4.

Talley, L. D., Feely, R. A., Sloyan, B. M., Wanninkhof, R., Baringer, M. O., Bullister, J. L. et al. (2016). Changes in ocean heat, carbon content, and ventilation: a review of the first decade of GO-SHIP global repeat hydrography. Ann. Rev. Mar. Sci. 8: 185–215.

Taylor, M., Gwinnett, C., Robinson, L. and Woodall, L. C. (2016). Plastic microfibre ingestion by deep-sea organisms. Scientific Reports 6: 33997.

Taylor, M. L. and Roterman, C. N. (2017). Invertebrate population genetics across Earth's largest habitat: The deep-sea floor. Mol. Ecol. 26: 4872–4896.

Thiel, H. (2003). Anthropogenic impacts on the deep sea. pp. 427–471. *In*: Tyler, P. A. (ed.). Ecosystems of the Word. Vol. 10 Ecosystems of the Deep Ocean. Elsevier, Amsterdam, Netherlands.

Thrush, S. E. and Dayton, P. K. (2002). Disturbance to marine benthic habitats by trawling and dredging: implications for marine biodiversity. Annu. Rev. Ecol. Syst. 33: 449–473.

Thurber, A. R., Sweetman, A.K., Narayanaswamy, B. E., Jones, D. O. B., Ingels, J. and Hansman, R. L. (2014). Ecosystem function and services provided by the deep sea. Biogeosciences 11: 3941–3963. doi:10.5194/bg-11-3941-2014.

Tiller, R., De Santo, E., Mendenhall, E. and Nyman, E. (2019). The once and future treaty: Towards a new regime for biodiversity in areas beyond national jurisdiction. Mar. Policy 99: 239–242.

Treude, T., Kiel, S., Linke, P., Peckmann, J. and Goedert, J. L. (2011). Elasmobranch egg capsules associated with modern and ancient cold seeps: A nursery for marine deep-water predators. Mar. Ecol. Prog. Ser. 437: 175–181.

Tyler, P. A. and Ramirez-Llodra, E. (2002). Larval and reproductive strategies on European continental margins. pp. 339–350. *In*: Wefer, G., Billett, D. S. M., Hebbeln, D., Jorgensen, B. B., Schluter, M. and Van Weering. T. C. E. (eds.). Ocean Margin Systems Hanse Conference Report. Springer, Berlin.

Unger, M. A., Harvey, E., Vadas, G. G. and Vecchione, M. (2008). Persistent pollutants in nine species of deep-sea cephalopods. Mar. Pollut. Bull. 56: 1486–1512.

Van Dover, C. L. (2000). The Ecology of Deep-Sea Hydrothermal Vents. Princeton University Press, Princeton, New Jersey, USA. 446 pp.

Van Dover, C. L. (2010). Mining seafloor massive sulphides and biodiversity: what is at risk? ICES J. Mar. Sci. 68: 341–348.

Van Dover C. L., Arnaud-Haond, S., Gianni, M., Helmreich, S., Huber, J. A., Jaeckel, A. L. et al. (2018). Scientific rationale and international obligations for protection of active hydrothermal vent ecosystems from deep-sea mining. Mar. Policy 90: 20–28.

Van Dover, C. L. (2019). Inactive sulfide ecosystems in the deep sea: a review. Front. Mar. Sci. 6: 461.

Vanreusel, A., Hilario, A., Ribeiro, P., Menot, L. and Martinez-Arbizu, P. (2016). Threatened by mining, polymetallic nodules are required to preserve abyssal epifauna. Sci. Rep. 6: 26808.

Vartanov, R. and Hollister, C. D. (1997). Nuclear legacy of the Cold War: Russian policy and ocean disposal. Mar. Policy 21: 1–15.

Vaquer-Sunyer, R. and Duarte, C. M. (2008). Thresholds of hypoxia for marine biodiversity. Proc. Natl. Acad. Sci. USA 105: 15452–15457.

Vetter, R., Kohin, S. and Preti, A. (2008). Predatory interactions and niche overlap between mako shark, *Isurus oxyrinchus*, and jumbo squid, *Dosidicus gigas*, in the California Current. CalCOFI Rep. 49: 142–156.

VishnuRadhan, R., Eldho, T. I. and David, T. D. (2019). Can plastics affect near surface layer ocean processes and climate? Mar. Pollut. Bull. 140: 274–280.

Waller, R., Watling, L., Auster, P. and Shank, T. (2007). Anthropogenic impacts on the corner rise seamounts, north-west Atlantic Ocean. J. Mar. Biol. Assoc. U.K. 87: 1075–1076.

Watson, A. J. 2016. Oceans on the edge of anoxia. Science 354: 1529–1530.

Weaver, P. P. E., Aguzzi, J., Boschen-Rose, R. E., Colaço, A., de Stigter, H., Gollner, S. et al. (2022). Assessing plume impacts caused by polymetallic nodule mining vehicles. Mar. Policy 139: 105011.

Weilgart, L. (2018). The Impact of Ocean Noise Pollution on Fish and Invertebrates. Report by OceanCare, Dalhousie University, Canada. https://tethys.pnnl.gov/publications/impact-ocean-noise-pollution-fish-invertebrates (last accessed 16/12/2022). 36pp.

Wheeler, A. J., Bett, B. J., Billet, D. S. M., Masson, D. G. and Mayor, D. (2005). The impact of demersal trawling on NE Atlantic deep-water coral habitats: the case of the Darwin Mounds, UK. Am. Fish. Soc. Symp. 41: 807–817.

Williams, A., Schlacher, T. A., Rowden, A. A., Althaus, F., Clark, M. R., Bowden, D. A. et al. (2010). Seamount megabenthic assemblages fail to recover from trawling impacts. Mar. Ecol. 31: 183–199.

Williams, R., Ashe, E., Yruretagoyena, L., Mastick, N., Siple, M., Wood, J. et al. (2021). Reducing vessel noise increases foraging in endangered killer whales. Mar. Pollut. Bull. 173: 112976.

Williams, R., Erbe, C., Duncan, A., Nielsen, K., Washburn T. and Smith, C. (2022). Noise from deep-sea mining may span vast ocean areas. Science 377: 6602.

Woodall, L. C., Sanchez-Vidal, A., Canals, M., Paterson G. L. J., Coppock, R., Sleight, V. et al. (2014). The deep sea is a major sink for microplastic debris. R. Soc. Open Sci. 1: 140317.

Worm, B., Davis, B., Kettemer, L., Ward-Paige, C. A., Chapman, D., Heithaus, M. R. et al. (2013). Global catches, exploitation rates, and rebuilding options for sharks. Mar. Policy 40: 194–204.

Wu, J., Dai, M., Xu, Y. and Zheng, J. (2018). Sources and accumulation of plutonium in a large Western Pacific marginal sea: The South China Sea. Sci. Total Environ. 610–611: 200–211.

Wunsch, C. and Heimbach, P. (2014). Bidecadal Thermal changes in the abyssal ocean. J. Phys. Oceanogr. 44: 2013–2030.

Yasuhara, M. and Danovaro, R. (2016). Temperature impacts on deep-sea biodiversity. Biol. Rev. 91: 275–287.

Zettler, E. R., Mincer, T. J. and Amaral-Zettler, L. A. (2013). Life in the "Plastisphere": Microbial communities on plastic marine debris. Environ. Sci. Technol. 47: 7137–7146.

Zhang, Y. X., Chao, Q. C., Zheng, Q.-H. and Huang, L. (2017). The withdrawal of the U.S. from the Paris Agreement and its impact on global climate change governance. Adv. Clim. Change Res. 8: 213–219.

Section III

Tools for Tackling Anthropogenic Impacts

CHAPTER 10

How can DNA Barcoding be Useful in the Conservation of Marine Organisms?

Delpiani, S. M.,[1,*] *Delpiani, G. E.,*[1] *Cruz, V. P.,*[2] *Di Prinzio, C.,*[3] *Mabragaña, E.,*[1] *Oliveira, C.,*[2] *Hanner, R.*[4] and *Díaz de Astarloa, J.M.*[1,5]

Introduction

DNA barcoding is a commonly used protocol that employs a standardized genetic marker for the identification of known species and the discovery of new ones (Fig. 1). For animals, a short DNA sequence of 652 base pairs (bp) from the mitochondrial 5' gene encoding cytochrome c oxidase subunit 1 (COI) has been used to develop a reference sequence library derived from expert-identified reference specimens that can be used as a practical tool for the identification of an unknown sample by correctly matching a specific genetic marker to the reference sequence library (Hebert et al. 2003). However, COI sequences were not as effective as originally proposed in some organisms, including some invertebrates, amphibians, and reptiles (Elias et al. 2007, Evans and Paulay 2012, Vences et al. 2012). In addition, some animals presented faulty identifications due to the presence of ancestral polymorphisms, hybridization and introgression (Fazekas et al. 2009). Despite this, the primary application of DNA barcoding will continue to be species identification of specimens and discovery of new species (Pecnika and Buzan 2014), by ecologists, evolutionary biologists, and conservationists alike (Seberg et al. 2003, Dunn 2003, Díaz de Astarloa et al. 2008, Mabragaña et al. 2011, Azpelicueta et al. 2019).

The use of DNA barcodes has been criticized to some extent to identify novel species or delimit species boundaries, by relying on a single gene to infer relationships between species and for being a form of DNA taxonomy (Taylor and Harris 2012). However, it is evident that the impact on taxonomy has been very important, since it has allowed the description of numerous species, resuscitation of synonymized species (Seberg et al. 2003, Gabbanelli et al. 2018, Azpelicueta et al. 2019), revealing hidden diversity (Rosso et al. 2018) or taxonomic conflicts (León-Romero et al. 2012). Nevertheless, its use in species delimitation requires careful consideration (Hubert and Hanner 2015). The results of DNA barcoding

[1] Laboratorio de Biotaxonomía Morfológica y Molecular de Peces, Instituto de Investigaciones Marinas y Costeras, Consejo Nacional de Investigaciones Científicas y Técnicas (CONICET), Universidad Nacional de Mar del Plata (Argentina).
[2] Laboratório de Biologia e Genética de Peixes, Instituto de Biociências de Botucatu,UNESP, Botucatu, SP, Brazil.
[3] Centro de Investigación Esquel de Montaña y Estepa Patagónica (CONICET-UNPSJB), Roca 780, Esquel, Chubut, Argentina. Facultad de Ciencias Naturales y Ciencias de la Salud, Universidad Nacional de la Patagonia San Juan Bosco, Esquel, Chubut, Argentina.
[4] Department of Integrative Biology, University of Guelph, Guelph, ON N1G 2W1, Canada.
[5] Departamento de Biología, Facultad de Ciencias Exactas y Naturales, Universidad Nacional de Mar del Plata.
* Corresponding author: smdelpiani@untdf.edu.ar

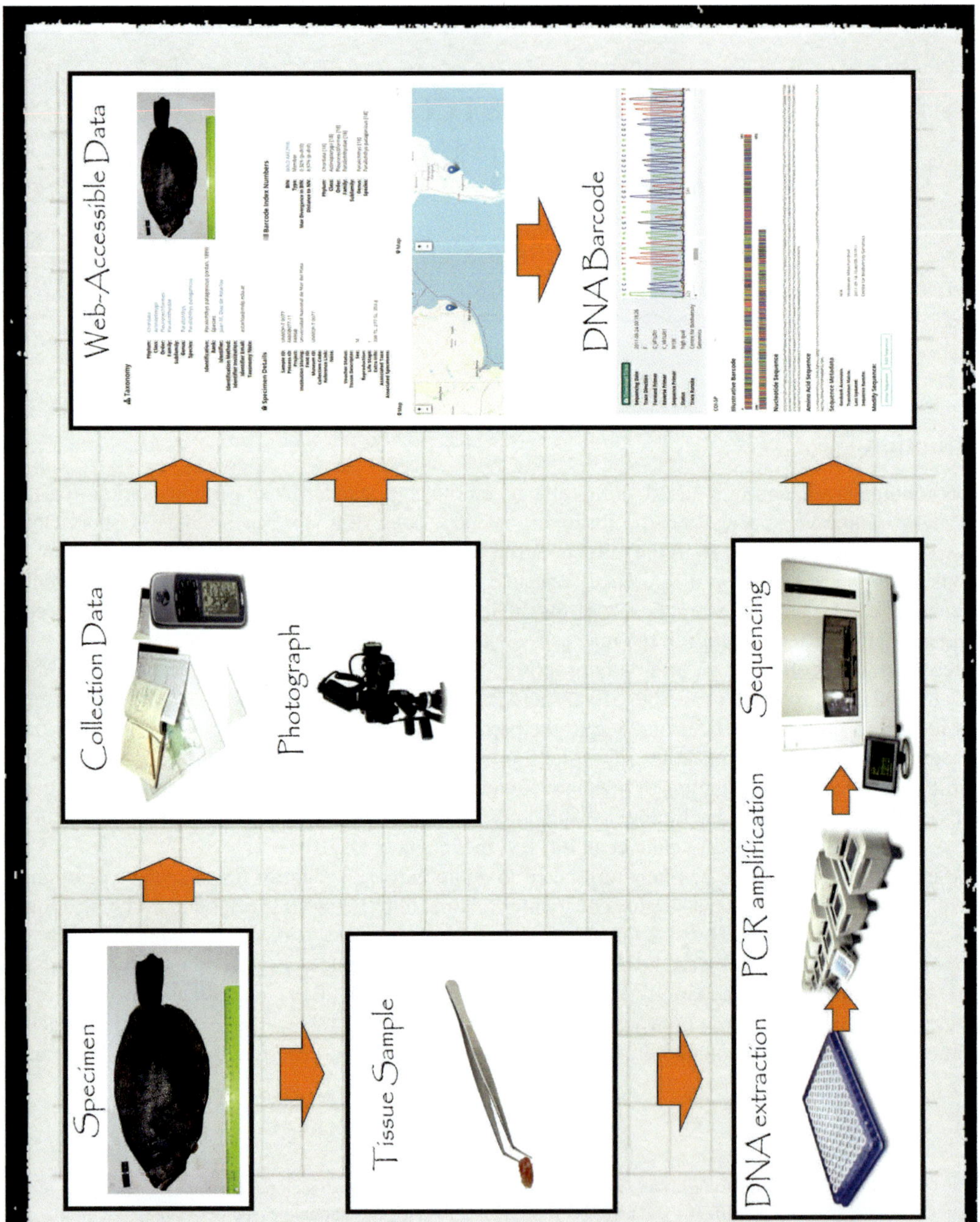

Figure 1. The DNA Barcoding workflow from specimen collection to uploading sequences and web-accessible data.

may highlight taxa that require further investigation. In the case of identifying specimens within well-resolved taxonomic groups, DNA barcoding has been tested to be an excellent tool (Meyer and Paulay 2005, Rach et al. 2008).

The identification of degraded specimens or incomplete remains is often difficult or impossible to resolve by means of traditional taxonomy. The samples such as muscle tissues, eggs, larvae and blood that would usually be difficult or impossible to identify morphologically can be accurately identified by DNA barcoding, with a simple DNA sample. Indeed, DNA barcoding helps successfully in a variety of problems such as wildlife forensics (Dalton and Kotze 2011, Panday et al. 2014, Goncalves et al. 2015), diet analysis (Méheust et al. 2015, Zarzoso-Lacoste et al. 2016), identification of museum specimens (Price et al. 2015, Areta et al. 2016), medicine frauds (Mohammed Abubakar et al. 2017, Mosa et al. 2018), food frauds (Wong and Hanner 2008, Mishra et al. 2015, Delpiani et al. 2020), and conservation biology (Ardura et al. 2010, Krishnamurthy et al. 2012, Miller et al. 2016, Li et al. 2017).

DNA barcoding reveals cryptic diversity in marine organisms

Biological diversity can be quantified within and between ecological communities at different geographic scales (Banks and Paterson 2007). From a conservationist perspective, DNA barcoding can be a premise for the description of new species, it can also be used to assess species delimitation, when looking at species richness in the same communities (Christidis and Boles 2008, Crocker et al. 2010). Traditional methods for identifying, naming, and classifying organisms rely heavily on visible morphology. Limitations to this method exist when attempting to identify organisms during various stages of their development not considered in the original treatments (Hanner et al. 2005).

Our planet surface is more than 70% covered by oceans (Campbell and Taylor 1983), which have higher biodiversity in contrast to freshwater or terrestrial ecosystems. This massive marine ecosystem is the habitat for a large and diverse flora and fauna. Notably, of 35 animal phyla, 14 are exclusively marine animals (Gray 1997). However, this oceanic diversity currently represents only 13% of all described extant species, i.e., approximately between 235,000 and 250,000 species, all groups included (Bouchet 2006).

DNA barcoding aids the identification of known species, as well as the discovery of cryptic diversity (Hebert et al. 2004a, Zemlak et al. 2009, Hubert et al. 2012), clarifying taxonomic uncertainties (Ward et al. 2008, Hubert et al. 2012). The occurrence of cryptic species is more numerous in freshwater organisms, followed by the environment marine and, to a lesser extent, in the terrestrial environment (Poulin and Pérez-Ponce de León 2017). Cryptic species are morphologically similar but genetically different. Here, DNA barcoding can highlight that cryptic species are morphologically indistinguishable (Bickford et al. 2007, Bernardo 2011). Research on cryptic species has increased exponentially over the past decades (Bickford et al. 2007, Jörger and Schrödl 2013, Farooq et al. 2020), fueled in large part by the increasing availability of DNA barcode sequences (Antil et al. 2022). Notably, DNA barcoding has successfully discriminated potential many cryptic species (Kadarusman et al. 2012, Hubert et al. 2012). Cryptic diversity is challenging for management, but once an organism is identified using molecular tools, reliable diagnostic morphological characters may subsequently be recognized (Smith et al. 2011). Barcoding also facilitates numerous related applications in fisheries management and ecology, including identification of ambiguous life-history stages (Taylor et al. 2002, Costa and Carvalho 2007, Fox et al. 2008).

The DNA barcode concept has drawn criticism from traditional taxonomists (Will and Rubinoff 2004, DeSalle et al. 2005, Will et al. 2005, Rubinoff et al. 2006). Limitations such as low resolution for recently diverged species, species complexes, the presence of pseudogenes and mitochondrial introgression have been highlighted as issues for barcoding (Kemppainen et al. 2009). The integration of morphological, physiological, behavioral and ecological data with DNA barcode data can enhance the species discovery and identification process (Waugh 2007, Padial et al. 2010). Therefore, COI has been widely accepted as a marker for molecular identification of many invertebrates and vertebrates, including springtails (Hogg and Hebert 2004), butterflies (Hebert et al. 2004a), crustaceans (Costa et al. 2007), birds (Hebert et al. 2004b) and a large range of fish species (Ward et al. 2005, Pegg et al. 2006, Rock et al. 2008, Ward et al.

2009, Steinke et al. 2009a, b, Hanner et al. 2011, Keskin and Atar 2013, McCusker et al. 2013, Young et al. 2013, Marceniuk et al. 2016, Azpelicueta et al. 2019, Marceniuk et al. 2020).

In addition, DNA barcoding can be useful for the access of more complete information about geographic ranges of fisheries resources and larval recruitment, which can be obtained to increase knowledge on spawning grounds and nursery areas for fisheries management and conservation (Lewis et al. 2016, Antil et al. 2022). The application of molecular techniques in species identification of fish eggs reports that over 60% of the eggs, when phenotypic characters were used, were misidentified (Webb et al. 2006, Valdez-Moreno et al. 2010, Lira et al. 2022). Understanding speciation would be difficult with a misidentification, as well as the diversity, niche partitioning, and many other features of ecosystems. Different authors using DNA barcoding techniques have shown that it is possible to identify fish larvae (Overdyk et al. 2016), but the identification is currently limited by the availability of comparative adult sequences in the DNA sequence database (Webb et al. 2006, Valdez-Moreno et al. 2010, Lira et al. 2022).

DNA barcoding applied to conservation practices

The application of DNA barcoding has included a vast array of fauna, with most of these studies focused on assessing the suitability of DNA barcoding for species assignation of certain animal groups (Gostel and Krass 2022). The identification of specimens from incomplete or damaged samples is often difficult or impossible with traditional morphological classification procedures (Hanner et al. 2005), which makes species identification one of the greatest challenges in conservation practices. The simple online DNA barcoding identification analysis based on the BOLD databases has proven to be efficient and reliable in many studies (Rubinoff 2006, Rubinoff et al. 2006, Dalton and Kotze 2011, Alfonsi et al. 2013). The applications of DNA barcoding to identify unknown specimens in animal conservation practices has been intensively studied (Naro-Maciel et al. 2009, Holmes et al. 2009, Li et al. 2017). For example, the DNA barcoding application to conservation practices to protect endangered sea turtles by assessing turtle meat, carcasses or eggs that are illegally traded (Vargas et al. 2009), many of these species are currently listed as different levels of threat on the IUCN (International Union for Conservation of Nature) red list. It has also supported the identification of endangered sharks in commercial trade (Steinke et al. 2017). There is much evidence concerning the amenability and reliability of this molecular species identification tool in conservation practices (Naro-Maciel et al. 2009, Holmes et al. 2009) because it meets the high accuracy requirement in species assignation of unknown specimens in conservation practice.

Misidentification and mislabeling of reference specimens have been reported in several studies (Ward et al. 2009, Miller and Mariani 2010, Mabragaña et al. 2011) resulting in potential errors and confusion in later research. DNA barcoding provides a robust method to recognize potential cases of morphological misidentifications and mislabelling in inventory samples. For instance, there are species that are prone to misidentification due to subtle morphological differences between sister species, as observed in shark species of the genus *Squalus* (Ariza et al. 2022). In addition, species misidentification can occur if the appropriate genetic marker is not used. For example, Atlantic and Pacific bluefin tunas cannot distinguish based solely on nuclear genetic markers (Ward et al. 2005, Chow et al. 2006).

For the identification of processed biological materials with substantial DNA degradation of protected species (Hajibabaei et al. 2006, Meusnier et al. 2008), a promising alternative is DNA mini-barcodes, which are short DNA sequences; these fragments are usually between 100–350 bp in length (Filonzi et al. 2021). In recent years, different works have shown that the use of DNA mini-barcodes enable accurate identification of processed animal parts and can potentially be used as a powerful tool for combating smuggling and illegal trade in endangered species (Chang et al. 2013, Fields et al. 2015, Hou et al. 2016, Chen et al. 2018, Sultana et al. 2018, Wu et al. 2018, Labrador et al. 2019, Boza et al. 2022).

It is important to recognize different factors that may limit the application of DNA barcoding to conservation practices. Incomplete sampling within species across their range can be a key limitation (Phillips et al. 2015, 2019). Barcode sharing among closely related species is one of the main problems, although it only occurs occasionally. In this case, the use of multiple analytical methods is necessary. Factors that contribute to this problem include incomplete lineage sorting and introgression. For instance, it is recommended to include as many closely related species as possible, or trying another

DNA barcode marker. Particular attention should be paid to sibling species as they tend to be similar in genetic background and morphology (Krück et al. 2013), thus tending to be misidentified either by morphological approaches or DNA barcoding. An important factor is the reference database in DNA barcoding identification. Mainly, involving more barcodes of more species would improve the species assignment performance (Dincă et al. 2011). In the meantime, accurate information from the downloaded barcodes is essential for successful identification, including species name and collection details.

DNA barcoding and its role in food webs

Many management decisions are based on dietary studies, the impact of interactions between species, depending on the importance of food resources, the trophic position of species and their demographic regulation, on both populations and communities (Alonso et al. 2014). Predator-prey interactions explain most of the food web dynamics (Jo et al. 2014), and due to climate change and accelerated anthropogenic impact scenarios, better knowledge is needed to design effective management and conservation plans (Roslin and Majaneva 2016). Conventional identifications are based on undigested portions of prey and considerable taxonomic experience (Sheppard et al. 2005). Furthermore, the morphological identification of such undigested elements depends on the existence of comprehensive regional reference collections of potential prey, added to a current shortage of taxonomists (Wheeler et al. 2004) and is also time-consuming (Packer et al. 2009). The examination of stomach contents presents some deficiencies, such as the significant underestimation of several types of prey (Bestley et al. 2010) or difficulties in the correct identification of prey that are semi-digested (King et al. 2008).

To bridge these gaps, barcoding can be used to analyze the digestive system of predators (Symondson 2002, King et al. 2008) to obtain information from partially digested remains recovered from stomachs (Valentini et al. 2009, Jo et al. 2014). Unlike morphological examination, DNA barcoding can provide taxonomic resolution to species level, allowing a more accurate description of the diversity of prey items (Jeanniard-du-Dot et al. 2017, Granquist et al. 2018). The application of DNA barcoding in fish dietary studies (Sousa et al. 2016) can reveal new and unexpected prey interactions (Bartley et al. 2015), and ultimately how these may change during ontogeny. DNA metabarcoding has become a powerful tool to reconstruct both trophic networks (Roslin and Majaneva 2016) and networks of ecological interactions (Berry et al. 2019, Evans et al. 2016). The metabarcoding application to diet studies could benefit ecosystem management, mainly as species adjustments due to environmental disturbance that could be easily assessed.

Generally, by moving from a traditional analysis of diet (morphological) to applying a DNA barcode, a higher taxonomic resolution is obtained and therefore a greater diversity of prey (Zeale et al. 2011, Berry et al. 2015). Having more detailed information on trophic interactions reveals changes of the species in the trophic networks (Bartley et al. 2015). Remarkable examples of this are the cannibalistic predation of the red lionfish *Pterois volitans* revealed by DNA barcoding and its ecological significance as a predator (Côté et al. 2013, Dahl et al. 2018). Similarly, an expanded view of prey diversity revealed a much more generalist feeding behavior for the Adélie penguin *Pygoscelis adeliae* (Jarman et al. 2013). Modifications of the known diet have also been detected through DNA barcoding analysis. For instance, the absence of salmonids in the diet of the Icelandic harbor seal *Phoca vitulina* demands the interruption of the selective hunting of seals, since the depredation of salmonids promoted the slaughter of the species in north-western Iceland (Granquist et al. 2018). These results have implications for conservation and management decisions for the species in the region.

DNA barcoding and population genetic structure of marine fishes

Marine fishes with a potential dispersal of eggs, larval and adult stages and high migration generally present low genetic variation attributable to differences among populations, thus revealing high connectivity among distant populations (Chiang et al. 2006, Cassista and Hart 2007, Habicht et al. 2010, Nielsen et al. 2012, Hemmer-Hanse et al. 2013). The high levels of gene flow may be constrained by local adaptation (Hauser and Carvalho 2009, Nielsen et al. 2009), leading to a close association between populations at a large spatial scale (Ward et al. 1994). Notwithstanding, several studies have identified population subdivision in marine fishes across small spatial scales (Nielsen et al. 2004, Jørgensen et al. 2005, Fauvelot and Borsa 2011, Limborg et al. 2012, Hemmer-Hansen et al. 2013, Escobar et al. 2019). Population genetic differentiation could be driven by physical barriers and ocean currents (Hare et al. 2005), temperature and salinity (Kelly and Eernisse 2007), and oceanographic fronts (Galarza et al. 2009), resulting in marine population with high dispersal capabilities being at least partially isolated.

Reproductive isolation, an important aspect of the biological species concept, is difficult to investigate in marine ecosystems (Faria et al. 2021). A study involving cosmopolitan marine bryozoans revealed that divergent barcode clusters corresponded to reproductively isolated groups, thereby establishing a link between biological species concept and DNA barcode sequence clusters (Gómez et al. 2007). Different studies have shown that genetic structure is associated with water exchange and between oceans circulation patterns (Froukh and Kochzius 2007, Shui et al. 2008). Species life history traits can also be responsible for genetic differentiation between populations, such as the presence of spawning asynchrony among populations and retention of eggs and larvae (Borsa et al. 1997, Taylor and Hellberg 2003). Also, historical events (Pleistocene glaciations) may explain genetic differentiation within marine species (Lu et al. 2006, Menezes et al. 2006, Crandall et al. 2008, Gaither et al. 2010).

Conservation of fisheries largely depends on the identification of populations (Ovenden et al. 2013, Bernatchez et al. 2017), to keep up exact records to assist correctly with fisheries management. DNA may also enable population-level research, assessing intraspecific genetic structure, population segregation, and phylogeographic patterns (Sales et al. 2019). However, most species have been insufficiently sampled to make inferences about haplotype diversity as noted by Phillips et al. (2015, 2019). In other situations, relatively little population-level variation has been revealed despite extensive sampling (Overdyk et al. 2015). The application of DNA barcoding for estimating within-population genetic diversity of fishes has mainly been applied to species of economic importance (Rasmussen et al. 2009, Ardura et al. 2013). A decline in the genetic variation of a population may diminish its ability to adapt to changing selection pressures (e.g., environmental changes), decreasing its chances of prolonged existence (Wang et al. 2002, Benestan 2019, Manel et al. 2020). Conserving genetic diversity provides more opportunities for evolution, and helps to foster the fitness of populations and their potential to recover from, and adapt to threats ranging from disease to environmental changes (Reed and Frankham 2003). Consequently, regular monitoring of the population distinction of exploited stocks is important for fish management (Benestan 2019) and careful assessment of appropriate DNA markers is warranted.

There are several challenges for the use of DNA barcoding in fisheries management. Incompleteness of reference databases is perhaps the greatest. Although many thousands of fish species are represented by at least one barcode sequence, fewer than half of all described fish have been barcoded. For those species that have been barcoded, most have not been sufficiently sampled to discriminate between marine populations or to estimate effective population sizes and to analyze migration patterns between populations (Waples et al. 2008, Spies and Punt 2015, Bernatchez et al. 2017). For such fine-grained population-level analyses, multiple loci and more sampling may be needed. This is particularly true when distinct populations of migratory fish species intermingle at specific fishing grounds. Ultimately, the application of genetic techniques has been used to address critical conservation and fisheries management issues, such as stock structure and population demography (FAO 2000, 2005). To ensure sustainability, biological processes and management actions must be matched (Reiss et al. 2009, Cadrin 2020). Failure to do so can lead to unintentional overfishing and the risk of fishery collapse due to a mismatch between the management area and the size and limits of biological stocks (Laikre et al. 2005, Reiss et al. 2009, Cadrin 2020).

DNA-barcoding of traceability of products of marine origin

There is a pressing need to address critical management issues, such as the exploitation and conservation of resources, for which genetic techniques such as DNA barcoding have been developed (Dudgeon et al. 2012). The Fish Barcode of Life Initiative (http://www.fishbol.org/) has contributed to BOLD database of sequence information to assist with species identification. This has proven valuable for the identification of seafood mislabeling and myriad other applications, but with some limitations.

Emerging genetic technologies are enabling the development of new markers and greater sequence data production at reduced costs, resulting in a shift away from studies of single marker gene fragments to entire genomes. Many of the limitations of the molecular techniques in current use (e.g., proposed low genetic variation in elasmobranchs and challenges with microsatellite development) are likely to be overcome with these advances in molecular science (Dudgeon et al. 2012). This technique is of value to achieve traceability of fishery products. The traceability systems of sea products allow safe information to be provided to consumers and ensure the use of fishery resources in sustainable economic, social and environmental conditions (CPA 2021). Furthermore, traceability is a key tool to verify the presence or absence of advertised product attributes, and helps to access technical specifications and track the actual history of the product as needed, either to protect the health of citizens or their rights as consumers (CPA 2021). Regarding the food industry, the introduction of supply chain traceability systems has gained momentum in recent decades, especially due to the promotion of food safety regulations.

Measures to prevent fraud are needed for seafood. The increased public awareness of nutritional and environmental issues has resulted in a shift in consumer attitude towards seafood products. With consumer consumption of seafood increasing, a growing number of fish species are being encountered in the market because of increased demand and the globalization of the seafood industry. Subsequent economic deception and food safety concerns are pushing the need for accurately labeled food products and full disclosure of product composition (Wong and Hanner 2008). Countless cases of fraud or fish mislabeling have been reported in many countries around the world (Hanner et al. 2011, Mariani et al. 2015, Chang et al. 2016, Veneza et al. 2018, Delpiani et al. 2020, Dufflocq et al. 2022). Fish names are commonly replaced or adulterated by the fisheries and food industries, and people are often not capable of distinguishing between taxa because the meat is sliced and packaged prior to sale (Jacquet and Pauly 2008, Barbuto et al. 2010, Lamendin et al. 2015).

Of greater concern, is that some traders may deliberately use mislabeling as a means to launder illegally-caught fish into legitimate markets, or simply to defraud consumers for the purpose of accruing greater profits (Ogden 2008). The most alarming cases of replacement are those involving species classified in the highest threat categories. In many countries, chondrichthyans are substituted for other species of bony fishes since the consumer is more reluctant to buy this kind of fish (Ardura et al. 2010, Bornatowski et al. 2015, Delpiani et al. 2020). In Argentina, these cases involve the sharks *Galeorhinus galeus*, which is considered "Critically Endangered" by the IUCN (Walker et al. 2020) and whose population trend is decreasing (Barbini et al. 2015); *Mustelus schmitii*, regarded as "Critically Endangered" (Pollom et al. 2020), and *Squatina guggenheim*, also categorized as "Endangered" (Oddone et al. 2019), as well as the "Critically Endangered" skate *Atlantoraja castelnaui* (Pollom et al. 2020).

In Orange and Los Angeles Counties in California, 35 commercial shark products were purchased online and from restaurants or retail outlets. The list includes winter skate (*Leucoraja ocellata*), shortfin mako (*Isurus oxyrinchus*), thresher (*Alopias vulpinus*), silky shark (*Carcharhinus falciformis*), spot-tail shark (*Carcharhinus sorrah*), blacktip reef shark (*Carcharhinus melanopterus*) and tope shark (*G. galeus*). All of these species are categorized on the IUCN Red List of Threatened Species as near threatened, vulnerable, endangered and critically endangered elasmobranch species (Hellberg et al. 2019). In Brazil, 21 different species were traded under the common name "cação", including hammerhead sharks (*Sphyrna* spp.), the blue shark (*P. glauca*), the tiger shark (*Galeocerdo cuvier*), the bull shark (*Carcharhinus leucas*) and the blacktip shark (*C. limbatus*), all considered as vulnerable and near threatened (Almerón-Souza et al. 2018). A study in north-central Chile indicates that at least 10 pelagic shark species are utilized in the fin trade, with *Alopias superciliosus*, *A. vulpinus*, *Lamna nasus* and *Sphyrna zygaena,* being classified as "Vulnerable" *Carcharhinus obscurus*, *I. oxyrinchus* and *I. paucus* as "Endangered", *G. galeus* and *Sphyrna lewini* as "Critically endangered", and *Prionace glauca* as "Near threatened" (Hernández et al. 2008). The focus is placed on elasmobranchs since several reproductive characteristics, such as long generation times, slow growth rates and low reproductive rates (Cahmi et al. 1998), make them especially susceptible to overexploitation (Hall 1999) and extinction (Dulvy et al. 2014, 2021). There is extensive evidence of population declines for several species worldwide (Heithaus et al. 2008, Worm et al. 2013). In this sense, authenticity and traceability testing can help in the fight against unregulated fishing, collaborating in this way with the conservation, sustainable exploitation and maintenance of fish stocks (CPA 2021).

Traceability systems make it possible to generate added value for the industry, guaranteeing better access to and position in markets. These systems are relevant at a global level to prevent products from illegal or environmentally unsustainable practices from entering their markets (CPA 2021). Fisheries certification is an instrument that recognizes desirable fisheries practices, while eco-labeling provides information to the consumer about the environmental impact caused by the product (Wessells et al. 2001). Enhanced labeling may create market incentives for improved fisheries management (Ward and Phillips 2008). The Marine Stewardship Council (MSC) is a globally recognized standard entity by which marine eco-labeled products are assessed. The basic philosophy underpinning the MSC is similar to that of ecolabeling schemes that operate across other product types, and assumes that the existence of a given 'label' or 'mark' on a product indicates that certain principles and/or practices have been adhered to during its production (Kaiser and Edwards-Jones 2006). However, MSC undertakes limited DNA-based authenticity testing and more development in this area is needed.

DNA-metabarcoding of marine diversity

DNA barcoding simplifies the taxonomic identification of species present in an ecosystem but it requires intense sampling and sequencing effort. However, the development of High-Throughput Sequencing (HTS) offers an opportunity to solve this problem by identifying multiple organisms simultaneously from a single PCR product (Valentini et al. 2009). This technique, called DNA metabarcoding, aims to identify biotic communities from bulk samples (e.g., plankton tows) (Fig. 2). It can also be applied to environmental samples. Environmental DNA (eDNA) is defined as small DNA fragments shed by organisms in the environment (Taberlet et al. 2012b, Pawlowski et al. 2020) or from environmental DNA (eDNA). eDNA metabarcoding extends the identification of a single specimen to the identification of communities of individuals belonging to many species' groups with distinct roles in the ecosystem (Taberlet et al. 2012a, Pawlowski et al. 2020).

This passive sampling method offers advantages over traditional survey methods by reducing cost and working time for sampling, increasing the detection of inconspicuous species, and feasibility to sample in otherwise difficult to survey locations (Lodge et al. 2012, Minamoto et al. 2012). It does not cause harm or stress to animals (Evans et al. 2018), is broadly applicable, fast, efficient, and non-invasive (Ficetola et al. 2008, Blanchet 2012, Thomsen et al. 2012). For those reasons, it emerged as a complementary tool for early detection, for instance, of invasive species, conservation and monitoring of past and present biodiversity (Lacoursière-Roussel et al. 2016, Jo et al. 2019). In recent years, eDNA metabarcoding has been used to study ecological distributions of species, to identify areas used for breeding or spawning, to detect low-density species, to determine areas of critical habitat, and to support more effective allocation of monitoring effort (Jerde et al. 2012, Mächler et al. 2014, Clusa and García-Vázquez 2017, Bracken et al. 2018, Carraro et al. 2020, among others). Ultimately, eDNA metabarcoding has demonstrated significant potential for biomonitoring by expediting the data acquisition and interpretation in biodiversity research (Cristescu 2014, Creer et al. 2016), shedding light on community and/or synecological aspects of biodiversity conservation. Moreover, eDNA metabarcoding can be used in diverse ecosystems to address a variety of questions (Taberlet et al. 2018). It has been used in marine environments to assess the impacts of pollution (Chariton et al. 2015) and fish farming on benthic communities (Pawlowski et al. 2014). eDNA metabarcoding has also been deployed in freshwater ecosystems to survey diatoms, amphibians and fishes (Visco et al. 2015, Zimmermann et al. 2015, Valentini et al. 2016, Apothéloz-Perret-Gentil et al. 2017).

DNA metabarcoding can broaden the spectrum of biomonitoring in different ecosystems and various focal taxa, guiding environmental management decisions not only on taxonomic data, but also on functional capability (Taberlet et al. 2018). However significant hurdles remain. No single DNA marker can reliably detect even fishes (Morey et al. 2020). Conversely, eDNA metabarcoding with the COI marker can yield nonspecific amplification of unwanted taxa (Collins et al. 2019). It can also be prone

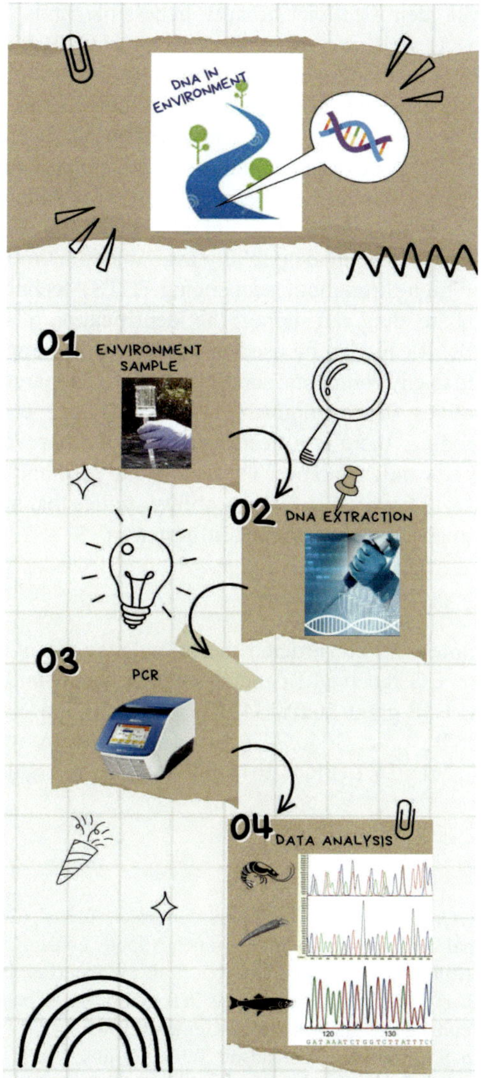

Figure 2. The main steps of an environmental DNA (eDNA) in water study encompassing field and laboratory activities.

to contamination (Sepulveda et al. 2020) and it is also expensive. Moreover, potentially millions of raw sequence reads can be generated from a single environmental sample using eDNA metabarcoding, making the data challenging to analyze and interpret. For these reasons, targeted species detection using a species-specific quantitative polymerase chain reaction (qPCR) assay is a potential alternative when probing environmental samples for a particular taxon of interest and it can also be deployed in the field (Skinner et al. 2020). Ultimately, eDNA-based methods for biomonitoring hold great promise for fisheries management and conservation but the choice of appropriate markers and tools must be carefully considered.

Concluding remarks and future contributions of DNA barcodes

The application of DNA barcoding to species discovery and identification is well-established, including the ability of the methodology to cope with different life stages of animal, provide insights into cryptic species diversity or to address diet-related studies, including market substitution in the seafood supply chain. By making species-level identifications broadly accessible, it is having a substantial impact on basic biodiversity research, as well as conservation monitoring. Beyond biodiversity monitoring, the knowledge provided by DNA barcoding can help mitigate threats to global biodiversity through improved environmental management and informed conservation measures, including the detection of illicit trade in endangered species. However, more work remains if the full value of barcoding is to be achieved. Reference sequence libraries must continue to expand to include more species and more populations within species in order to assess the taxonomic resolution and accuracy that can be achieved with the method. These reference sequences must be linked to expert-identified voucher specimens held in museum collections and data should be made publicly available in a timely manner.

With the development of high-throughput sequencing (HTS) technology (Taberlet et al. 2018, Hebert et al. 2018), DNA metabarcoding can support the identification of stomach contents to enhance our understanding of food webs. It can also be used to identify bulk samples from plankton tows in a highly efficient manner. Applied to environmental samples, eDNA metabarcoding holds great promise for large-scale biomonitoring initiatives. However, given the limitations of specific marker genes and related PCR primer sets for metabarcoding, there is a need to expand the barcode concept to include whole plastid genomes and ultimately to whole genomes. This will greatly enhance population assessments that barcoding might not detect and will have value for traceability studies that need to go beyond just species identification to include more nuanced biogeographic information.

Acknowledgments

We are grateful to all the people who constructed the knowledge about DNA barcode and all those who helped us with this paper. CO received financial support from Fundação de Amparo à Pesquisa do Estado de São Paulo - FAPESP grant 2020/13433-6 and Conselho Nacional de Desenvolvimento Científico e Tecnológico - CNPq proc. 306054/2006-0 (CO). JMDA was supported by CONICET (PIP No. 11220200101475CO), MINCYT (PICT-2018-0790) and Universidad Nacional de Mar del Plata (EXA 970/20).

References

Alfonsi, E., Méheust, E., Fuchs, S., Carpentier, F. G., Quillivic, Y., Viricel, A. et al. (2013). The use of DNA barcoding to monitor the marine mammal biodiversity along the French Atlantic coast. Zookeys 365: 5–24.

Almerón-Souza, F., Sperb, C., Castilho, C. L., Figueiredo, P. I. C. C., Gonçalves, L. T., Machado, R. et al. (2018). Molecular identification of shark meat from local markets in southern Brazil based on DNA barcoding: evidence for mislabeling and trade of endangered species. Front. Genet. 9: 138.

Alonso, H., Granadeiro, J. P., Waap, S., Xavier, J., Symondson, W. O., Ramos, J. A. et al. (2014). An holistic ecological analysis of the diet of Cory's shearwaters using prey morphological characters and DNA barcoding. Mol. Ecol. 23: 3719–3733.

Antil, S., Abraham, J. S., Sripoorna, S., Maurya, S., Dagar, J., Makhija, S. et al. (2022). DNA barcoding, an effective tool for species identification: a review. Mol. Biol. Rep. 1–15.

Apothéloz-Perret-Gentil, L., Cordonier, A., Straub, F., Iseli, J., Esling, P. and Pawlowski, J. (2017). Taxonomy-free molecular diatom index for high-throughput eDNA biomonitoring. Mol. Ecol. Resour. 17(6): 1231–1242.

Ardura, A., Planes, S. and Garcia-Vazquez, E. (2013). Applications of DNA barcoding to fish landings: authentication and diversity assessment. ZooKeys 365: 49.

Ardura, A., Linde, A. R., Moreira, J. C. and García-Vazquez, E. (2010). DNA barcoding for conservation and management of Amazonian commercial fish. Biol. Conserv. 143: 1438–1443.

Areta, J. I., Piacentini, V. D. Q., Haring, E., Gamauf, A., Silveira, L. F., Machado, E. et al. (2016). Tiny bird, huge mystery—the possibly extinct hooded seedeater (*Sporophila melanops*) is a Capuchino with a melanistic cap. PLoS One 11: 1–19.

Ariza, A. A., Adachi, A. M., Roque, P., Hazin, F. H., Vianna, M., Rotundo, M. M. et al. (2022). DNA barcoding and species delimitation for dogfish sharks belonging to the *Squalus* Genus (Squaliformes: Squalidae). Diversity 14(7): 544.

Azpelicueta, M. M., Delpiani, S. M., Cione, A. L., Oliveira, C., Marceniuk, A. P. and Díaz de Astarloa, J. M. (2019). Morphology and molecular evidence support the validity of *Pogonias courbina* (Lacepède, 1803) (Teleostei: Sciaenidae), with a redescription and neotype designation. PLoS One 14: 1–14.

Banks, J. C. and Paterson, A. M. (2007). A preliminary study of the genetic differences in New Zealand oystercatcher species. N. Z. J. Zool. 34: 141–4.

Barbini, S. A., Lucifora, L. O. and Figueroa, D. E. (2015). Using opportunistic records from a recreational fishing magazine to assess population trends of sharks. Can. J. Fish. Aquat. 72(12): 1853–1859.

Barbuto, M., Galimberti, A., Ferri, E., Labra, M., Malandra, R., Galli, P. et al. (2010). DNA barcoding reveals fraudulent substitutions in shark seafood products: the Italian case of "palombo" (*Mustelus* spp.). Food Res. Int. 43: 376–381.

Bartley, T., Braid, H., McCann, K., Lester, N., Shuter, B. and Hanner, R. (2015). DNA barcoding increases resolution and changes structure in Canadian boreal shield lake food webs. DNA Barcodes 3: 30–43.

Benestan, L. (2019). Population genomics applied to fishery management and conservation. pp. 399–421. *In*: Rajora O. P. (ed.). Population Genomics: Marine Organism, Population Genomics. Switzerland: Springer.

Bernardo, J. (2011). A critical appraisal of the meaning and diagnosability of cryptic evolutionary diversity, and its implications for conservation in the face of climate change. pp. 380–438. *In*: Hodkinson, T. R., Jones, M. B., Waldren, S. and Parnell, J. A. N. (eds.). Climate Change, Ecology and Systematics. Cambridge University Press, Cambridge, UK.

Bernatchez, L., Wellenreuther, M., Araneda, C., Ashton, D. T., Barth, J. M. I. et al. (2017). Harnessing the power of genomics to secure the future of seafood. Trends Ecol. Evol. 32: 665–680.

Berry, O., Bulman, C., Bunce, M., Coghlan, M., Murray, D. C. and Ward, R. D. (2015). Comparison of morphological and DNA metabarcoding analyses of diets in exploited marine fishes. Mar. Ecol. Prog. Ser. 540: 167–181.

Berry, T. E., Saunders, B. J., Coghlan, M. L., Stat, M., Jarman, S., Richardson, A. J. et al. (2019). Marine environmental DNA biomonitoring reveals seasonal patterns in biodiversity and identifies ecosystem responses to anomalous climatic events. PLoS Genetics 15(2): e1007943.

Bestley, S., Patterson, T. A., Hindell, M. A. and Gunn, J. S. (2010). Predicting feeding success in a migratory predator: integrating telemetry, environment, and modeling techniques. Ecology 91: 2373–2384.

Bickford, D., Lohman, D. J., Sodhi, N. S., Ng, P. K., Meier, R., Winker, K. et al. (2007). Cryptic species as a window on diversity and conservation. Trends Ecol. Evol. 22(3): 148–155.

Blanchet, S. (2012). The use of molecular tools in invasion biology: An emphasis on freshwater ecosystems. Fish. Manag. Ecol. 19: 120–132.

Bornatowski, H., Braga Rennó, R., Kalinowski, C. and Simões Vitule, J. R. (2015). "Buying a Pig in a Poke": The problem of elasmobranch meat consumption in Southern Brazil. Soc. Ethnobiol. 6: 196–202.

Borsa, P., Planquer, A. and Berrebi, P. (1997). Genetic structure of the flounders *Platichthys flesus* and *P. stellatus* at different geographic scales. Mar. Biol. 129: 233–246.

Bouchet, P. (2006). The magnitude of marine biodiversity. pp. 33–64. *In*: Duarte, C. M. (ed.). The Exploration of Marine Biodiversity: Scientific and Technological Challenges. Fundación BBVA Press, Bilbao, Spain.

Boza, B. R., Cruz, V. P., Stabile, G., Rotundo, M. M., Foresti, F. and Oliveira, C. (2022). Mini DNA barcodes reveal the details of the foraging ecology of the largehead hairtail, *Trichiurus lepturus* (Scombriformes: Trichiuridae), from São Paulo, Brazil. Neotrop. Ichthyol. 20(2): e210166.

Bracken, F. S. A., Rooney, S. M., Kelly-Quinn, M., King, J. J. and Carlsson, J. (2018). Identifying spawning sites and other critical habitat in lotic systems using eDNA "snapshots": A case study using the sea lamprey *Petromyzon marinus* L. Ecol. Evol. 1–15.

Cadrin, S. X. (2020). Defining spatial structure for fishery stock assessment. Fish. Res. 655–221.

Camhi, M., Fowler, S., Musick, J., Bräutigam, A. and Fordham, S. (1998). Sharks and their relatives: ecology and conservation. Occas. Pap. IUCN Species Surviv. Comm. 20: 63.

Campbell, I. H. and Taylor, S. R. (1983). No water, no granites-No oceans, no continents. Geophys. Res. Lett. 10(11): 1061–1064.

Carraro, L., Mächler, E., Wüthrich, R. and Altermatt, F. (2020). Environmental DNA allows upscaling spatial patterns of biodiversity in freshwater ecosystems. Nat. Commun. 11: 3585.

Cassista, M. and Cand Hart, M. W. (2007). Spatial and temporal genetic homogeneity in the Arctic surfclam (*Mactromeris polynyma*). Mar. Biol. 152: 569–579.

Chang, C. H., Lin, H. Y., Ren, Q., Lin, Y. S. and Shao, K. T. (2016). DNA barcode identification of fish products in Taiwan: government-commissioned authentication cases. Food Control. 66: 38–43.

Chang, C. H., Jang-Liaw, N. H., Lin, Y. S., Fang, Y. C. and Shao, K. T. (2013). Authenticating the use of dried seahorses in the traditional Chinese medicine market in Taiwan using molecular forensics. J. Food Drug. Anal. 21: 310–316.

Chariton, A. A., Sun, M., Gibson, J., Webb, J. A., Leung, K. M. Y., Hickey, C. W. et al. (2015). Emergent technologies and analytical approaches for understanding the effects of multiple stressors in aquatic environments. Mar. Freshw. Res. 67: 414–428.

Chen, M. Y., Han, X., Ma, J., Liu, X. Y., Li, M., Lv, X. et al. (2018). Identification of *Plastrum testudinis* used in traditional medicine with DNA mini-barcodes. Revista Brasileira De Farmacognosia. Rev. bras. Farmacogn. 28: 267–272.

Chiang, H. C., Hsu, C. C., Lin, H. D., Ma, G. C., Chiang, T. Y. and Yang, H. Y. (2006). Population structure of bigeye tuna (*Thunnus obesus*) in the South China Sea, Philippine Sea and western Pacific Ocean inferred from mitochondrial DNA. Fish. Res. 79: 219–225.

Chow, S., Nakagawa, T., Suzuki, N., Takeyama, H. and Matsunaga, T. (2006). Phylogenetic relationships among *Thunnus* species inferred from rDNA ITS1 sequence. J. Fish. Biol. 68: 24–35.

Christidis, L. and Boles, W. E. (2008). Systematics and Taxonomy of Australian Birds. Collingwood: CSIRO Publishing.

Clusa, L. and García-Vázquez, E. (2017). A simple, rapid method for detecting seven common invasive fish species in Europe from environmental DNA. Aquatic Conserv: Mar Freshw. Ecosyst. 1–11.

Collins, R., Bakker, J., Wangensteen, O. S., Soto, A. Z., Corrigan, L., Simms, D. W. et al. (2019). Non-specific amplification compromises environmental DNA metabarcoding with COI. Methods Ecol. Evol. 10: 1985–2001.

Costa, F. O. and Carvalho, G. R. (2007). The Barcode of Life Initiative: synopsis and prospective societal impacts of DNA barcoding of Fish. Genomics Soc. Policy 3: 52–56.

Côté, I. M., Green, S. J., Jr Morris, J. A., Akins, J. L. and Steinke, D. (2013). Diet richness of invasive Indo-Pacific lionfish revealed by DNA barcoding. Mar. Ecol. Prog. Ser. 472: 249–256.

CPA, - Circulo de Políticas Ambientales. (2021). https://www.trazabilidaddelapesca.org/.

Crandall, E. D., Frey, M. A., Grosberg, R. K. and Barber, P. H. (2008). Contrasting demographic history and phylogeographical patterns in two Indo-Pacific gastropods. Mol. Ecol. 17(2): 611–626.

Creer, C., Deiner, K., Frey, S., Porazinska, D., Taberlet, P., Thomas, W. K., Potter, C. et al. (2016). The ecologist's field guide to sequence-based identification of biodiversity. Methods Ecol. Evol. 7: 1008–1018.

Cristescu, M. E. (2014). From barcoding single individuals to metabarcoding biological communities: towards an integrative approach to the study of global biodiversity. Trends Ecol. Evol. 29: 566–571.

Crocker, T., Petch, S. and Sagar, P. (2010). Hybridisation by South Island pied oystercatcher (*Haematopus finschi*) and variable oystercatcher (*H. unicolor*) in Canterbury. Notornis 57: 27–32.

Dahl, K. A., Portnoy, D. S., Hogan, J. D., Johnson, J. E., Gold, J. R. and Patterson, W. F. (2018). Genotyping confirms significant cannibalism in northern Gulf of Mexico invasive red lionfish, *Pterois volitans*. Biol. Invasions 20: 3513–3526.

Dalton, D. L. and Kotze, A. (2011). DNA barcoding as a tool for species identification in three forensic wildlife cases in South Africa. Forensic Sci. Int. 207: 51–54.

Delpiani, G., Delpiani, S. M., Deli Antoni, M. Y., Covatti Ale, M., Fischer, L., Lucifora, L. O. et al. (2020). Are we sure we eat what we buy? Fish mislabelling in Buenos Aires province, the largest sea food market in Argentina. Fish. Res. 221.

DeSalle, R. (2006). Species discovery versus species identification in DNA barcoding efforts: response to Rubinoff. Conserv. Biol. 20: 1545–1547.

DeSalle, R., Egan, M. G. and Siddall, M. (2005). The unholy trinity: Taxonomy, species delimitation and DNA barcoding. Philos. Trans. R. Soc. Lond. B Biol. Sci. 360: 1905–1916.

Díaz de Astarloa, J. D., Mabragaña, E., Hanner, R. and Figueroa, D. E. (2008). Morphological and molecular evidence for a new species of longnose skate (Rajiformes: Rajidae*: Dipturus*) from Argentinean waters based on DNA barcoding. Zootaxa. 1921: 35–46.

Dinca, V., Zakharov, E. V., Hebert, P. D. and Vila, R. (2011). Complete DNA barcode reference library for a country's butterfly fauna reveals high performance for temperate Europe. Proc. R. Soc. Lond. B. 278: 347–355.

Dudgeon, C. L., Blower, D. C., Broderick, D., Giles, J. L., Holmes, B. J. and Kashiwagi, T. (2012). A review of the application of molecular genetics for fisheries management and conservation of sharks and rays. J. Fish Biol. 80: 1789–1843.

Dufflocq, P., Larraín, M. A. and Araneda, C. (2022). Species substitution and mislabeling in the swordfish (*Xiphias gladius*) market in Santiago, Chile: Implications in shark conservation. Food Control. 133(9).

Dulvy, N. K., Pacoureau, N., Rigby, C. L., Pollom, R. A., Jabado, R. W., Ebert, D. A. et al. (2021). Overfishing drives over one-third of all sharks and rays toward a global extinction crisis. Curr. Biol. 31: 5118–5119.

Dulvy, N. K., Harisson, L. R., Carlson, J. K., Davidson, L. N. K., Forfham, S. V., Francis, M. P. et al. (2014). Extinction Risk and Conservation of the World's Sharks and Rays. ELife 3: 00590.

Dunn, C. P. (2003). Keeping taxonomy based in morphology. Trends. Ecol. Evol. 18: 269–70.

Elias, M., Hill, R. I., Willmott, K. R., Dasmahapatra, K. K., Brower, A. V. Z. and Mallet, J. (2007). Limited performance of DNA barcoding in a diverse community of tropical butterflies. Proc. R. Soc. B. 274: 2881–2889.

Escobar, R., Luna-Acosta, A. and Caballero, S. (2019). DNA barcoding, fisheries and communities: What do we have? Science and local knowledge to improve resource management in partnership with communities in the Colombian Caribbean. Mar. Policy 99: 407–413.

Evans, D. M., Kitson, J. J. N., Lunt, D. H., Straw, N. A. and Pocock, M. J. O. (2016). Merging DNA metabarcoding and ecological network analysis to understand and build resilient terrestrial ecosystems. Funct. Ecol. 30: 1904–1916.

Evans, N. and Paulay, G. (2012). DNA barcoding methods for invertebrates. pp. 47–78. *In*: Kress, W. J. and Erickson, D. L. (eds.). DNA Barcodes: Methods and Protocols. Methods in Molecular Biology. Humana Press, Springer Science Publishing Media.

Evans, N. T. and Lamberti, G. A. (2018). Freshwater fisheries assessment using environmental DNA: a primer on the method, its potential, and shortcomings as a conservation tool. Fish. Res. 197: 60–66.

FAO. (2000). FAO Technical Guidelines for Responsible Fisheries. Rome: Food and Agriculture Organization of the United Nations.

FAO. (2005). FAO expert consultation on the implementation of the FAO international plan of action for the conservation and management of sharks. In FAO Fisheries Report 295.

Faria, R., Johannesson, K. and Stankowski, S. (2021). Speciation in marine environments: Diving under the surface. J. Evol. Biol. 34: 4–15.

Farooq, Q., Shakir, M., Ejaz, F., Zafar, T., Durrani, K. and Ullah, A. (2020). Role of DNA Barcoding in plant biodiversity conservation. Sch. Int. J. Biochem. 3: 48–52.

Fauvelot, C. and Borsa, P. (2011). Patterns of genetic isolation in a widely distributed pelagic fish, the narrow-barred Spanish mackerel (*Scomberomorus commerson*). Biol. J. Linn. Soc. 104: 886–902.

Fazekas, A. J., Kesanakurti, P. R., Burgess, K. S., Percy, D. M., Graham, S. W., Barrett, S. C. et al. (2009). Are plant species inherently harder to discriminate than animal species using DNA barcoding markers? Mol. Ecol. Resour. 9: 130–139.

Ficetola, G. F., Miaud, C., Pompanon, F. and Taberlet, P. (2008). Species detection using environmental DNA from water samples. Biol. Lett. 4: 423–425.

Fields, A. T., Abercrombie, D. L., Eng, R., Feldheim, K. and Chapman, D. D. (2015). A novel mini-DNA barcoding assay to identify processed fins from internationally protected shark species. PLoS One 10.

Filonzi, L., Vaghi, M., Ardenghi, A., Rontani, P. M., Voccia, A. and Nonnis Marzano, F. (2021). Efficiency of DNA mini-barcoding to assess mislabeling in commercial fish products in Italy: An overview of the last decade. Foods 10: 1449.

Fox, C. J., Taylor, M. I., Dickey-Collas, M., Fossum, P., Kraus, G., Rohlf, N. et al. (2008). Mapping the spawning grounds North Sea cod (*Gadus morhua*) by direct and indirect means. Proc. Royal Soc. B. 275: 1543–1548

Froukh, T. and Kochzius, M. (2007). Genetic population structure of the endemic fourline wrasse (*Larabicus quadrilineatus*) suggests limited larval dispersal distances in the Red Sea. Mol. Ecol. 16: 1359–1367.

Gabbanelli, V., de Astarloa, J. M. D., Gonzalez-Castro, M.,Vazquez, D. M. and Mabragana, E. (2018). Almost a century of oblivion: Integrative taxonomy allows the resurrection of the longnose skate Zearaja brevicaudata (Marini, 1933) (Rajiformes; Rajidae). C R Biol. 341(9–10): 454–470.

Gaither, M. R., Toonen, R. J., Robertson, D. R., Planes, S. and Bowen, B. W. (2010). Genetic evaluation of marine biogeographical barriers: perspectives from two widespread Indo-Pacific snappers (*Lutjanus kasmira* and *Lutjanus fulvus*). J. Biogeogr. 37: 133–147.

Galarza, J. A., Carreras-Carbonell, J., Macpherson, E., Pascual, M., Severine, R., Turner, G. F. et al. (2009). The influence of oceanographic fronts and early-life-history traits on connectivity among littoral fish species. Proc. Natl. Acad. Sci. U.S.A. 106: 1473–1478.

Gómez, A., Wright, P. J., Lunt, D. H., Cancino, J. M., Carvalho, G. R. and Hughes, R. N. (2007). Mating trials validate the use of DNA barcoding to reveal cryptic speciation of a marine bryozoan taxon. Proc. R. Soc. Lond. B. 274: 199–207.

Goncalves, P. F. M., Oliveira-Marques, A. R., Matsumoto, T. E. and Miyaki, C. Y. (2015). DNA barcoding identifies illegal parrot trade. J. Hered. 106: 560–564.

Gostel, M. R. and Kress, W. J. (2022). The expanding role of DNA barcodes: Indispensable tools for ecology, evolution, and conservation. Diversity 14: 213.

Granquist, S. M., Esparza-Salas, R., Hauksson, E., Karlsson, O. and Angerbjörn, A. (2018). Fish consumption of harbour seals (*Phoca vitulina*) in north western Iceland assessed by DNA metabarcoding and morphological analysis. Polar Biol. 41: 2199–2210.

Gray, J. S. (1997). Marine biodiversity: patterns, threats and conservation needs. Biodiv. Conserv. 6: 153–175.

Habicht, C., Seeb, L. W., Myers, K. W., Farley, E. V. and Seeb, J. E. (2010). Summer-fall distribution of stocks of immature sockeye salmon in the Bering Sea as revealed by Single-Nucleotide Polymorphisms. Trans. Am. Fish. Soc. 139: 1171–1191.

Hajibabaei, M., Smith, M. A., Janzen, D. H., Rodriguez, J. J., Whitfield, J. B. and Hebert, P. D. N. (2006). A minimalist barcode can identify a specimen whose DNA is degraded. Mol. Ecol. Notes 6: 959–964.

Hanner, R., Corthals, A. and Dessauer, H. C. (2005). Salvage of genetically valuable tissues following a freezer failure. Mol. Phylogenet. Evol. 34: 452–455.

Hanner, R., Becker, S., Ivanova, N. V. and Steinke, D. (2011). FISH-BOL and seafood identification: geographically dispersed case studies reveal systemic market substitution across Canada. Mitochondrial DNA 22: 106–122.

Hare, M. P., Guenther, C. and Fagan, W. F. (2005). Nonrandom larval dispersal can steepen marine clines. Evolution 59: 2509–2517.

Hauser, L. and Carvalho, G. R. (2008). Paradigm shifts in marine fisheries genetics: ugly hypotheses slain by beautiful facts. Fish Fish. (Oxf). 9: 333–362.

Hebert, P. D. N., Penton, E. H., Burns, J. M., Janzen, D. H. and Hallwachs, W. (2004a). Ten species in one: DNA barcoding reveals cryptic species in the neotropical skipper butterfly Astraptes fulgerator. Proc. Natl. Acad. Sci. U.S.A. 101: 14812–14817.

Hebert, P. D. N., Stoeckle, M. Y., Zemlak, T. S. and Francis, C. M. (2004b). Identification of birds through DNA barcodes. PLoS Biol. 2: 1657–1663.

Hebert, P. D. N., Ratnasingham, S. and DeWaard, J. R. (2003). Barcoding animal life: cytochrome c oxidase subunit 1 divergences among closely related species. Proc. R. Soc. Lond. B. 270: 96–99.

Hebert, P. D. N., Braukmann, T. W., Prosser, S. W., Ratnasingham, S., DeWaard, J. R., Ivanova et al. (2018). A sequel to sanger: amplicon sequencing that scales. BMC Genom. 19: 219.

Heithaus, M. R., Frid, A., Wirsing, A. J. and Worm, B. (2008). Predicting ecological consequences of marine top predator declines. Trends Ecol. Evol. 23: 202–210.

Hellberg, R. S., Isaacs, R. B. and Hernandez, E. L. (2019). Identification of shark species in commercial products using DNA barcoding. Fish. Res. 210: 81–88.

Hemmer-Hansen, J., Nielsen, E. E., Therkildsen, N. O., Taylor, M. I., Ogden, R., Geffen, A. J. et al. (2013). FishPopTrace Consortium, Carvalho GR A genomic island linked to ecotype divergence in Atlantic cod. Mol. Ecol. 22: 2653–2667.

Hernández, S., Haye, P. A. and Shivji, M. S. (2008). Characterization of the pelagic shark-fin trade in north-central Chile by genetic identification and trader surveys. J. Fish Biol. 73: 2293–2304.

Hogg, I. D. and Hebert, P. D. N. (2004). Biological identification of speingtails (Collenbola:Hexapoda) from Canadian arctic, using mitochondrial DNA barcodes. Can. J. Zool. 82: 1–6.

Holmes, B. H., Steinke, D. and Ward, R. D. (2009). Identification of shark and ray fins using DNA barcoding. Fisher. Res. 95: 280–288.

Hou, F., Wen, L., Peng, C. and Guo, J. (2016). Identification of marine traditional Chinese medicine dried seahorses in the traditional Chinese medicine market using DNA barcoding. Mitochondrial DNA Part A: DNA Mapping Sequencing Analysis.

Hubert, N., Meyer, C. P., Bruggemann, H. J., Guerin, F., Komeno, R. J. L., Espiau, B. et al. (2012). Cryptic diversity in Indo-Pacific coral-reef fishes revealed by DNA barcoding provides new support to the Centre-of-Overlap hypothesis. PLoS One 7.

Hubert, N. and Hanner, R. (2015). DNA barcoding, species delineation and taxonomy: a historical perspective. DNA Barcodes 3: 44–58.

Jacquet, J. L. and Pauly, D. (2008). Trade secrets: renaming and mislabeling of seafood. Mar. Policy 32: 309–318.

Jarman, S. N., McInnes, J. C., Faux, C., Polanowski, A. M., Marthick, J., Deagle, B. E. et al. (2013). Adélie penguin population diet monitoring by analysis of food DNA in scats. PLoS One 8: 12.

Jeanniard-du-Dot, T., Thomas, A. C., Cherel, Y., Trites, A. W. and Guinet, C. (2017). Combining hard-part and DNA analyses of scats with biologging and stable isotopes can reveal different diet compositions and feeding strategies within a fur seal population. Mar. Ecol. Prog. Ser. 584: 1–16.

Jerde, C. L., Mahon, A. R., Chadderton, W. L. and Lodge, D. M. (2011). "Sight-unseen" detection of rare species using environmental DNA. Conserv. Lett. 4: 150–157.

Jo, H., Gim, J., Jeong, K., Kim, H. and Joo, G. (2014). Application of DNA barcoding for identification of freshwater carnivorous fish diets: Is number of prey items dependent on size class for *Micropterus salmoides*? Ecol. Evol. 4: 219–229.

Jo, T., Murakami, H., Yamamoto, S., Masuda, R. and Minamoto, T. (2019). Effect of water temperature and fish biomass on environmental DNA shedding, degradation, and size distribution. Ecol. Evol. 9: 1135–1146.

Jørgensen, H. B. H., Hansen, M. M., Bekkevold, D., Ruzzante, D. E. and Loeschcke, V. (2005). Marine landscapes and population genetic structure of herring (*Clupea harengus* L.) in the Baltic Sea. Mar. Ecol. 14: 3219–3234.

Jörger, K. M. and Schrödl, M. (2013). How to describe a cryptic species? Practical challenges of molecular taxonomy. Front. Zool. 10: 1–27.

Kadarusman, Hubert, N., Hadiaty, R. K., Sudarto, Paradis, E. and Pouyaud, L. (2012). Cryptic diversity in Indo-Australian rainbowfishes revealed by DNA barcoding: implications for conservation in a biodiversity hotspot candidate. PLoS One 7.

Kaiser, M. J. and Edwards-Jones, G. (2006). The role of ecolabeling in fisheries management and conservation. Conserv. Biol. 20: 392–398.

Kelly, R. P. and Eernisse, D. J. (2007). Southern hospitality: a latitudinal gradient in gene flow in the marine environment. Evolution 61: 700–707.

Kemppainen, P., Panova, M., Hollander, J. and Johannesson, K. (2009). Complete lack of mitochondrial divergence between two species of NE Atlantic marine intertidal gastropods. J. Evol. Biol. 22: 2000–2011.

Keskin, E. and Atar, H. H. (2013). DNA barcoding commercially important fish species of Turkey. Mol. Ecol. Resour. 13: 788–797.

King, R. A., Read, D. S., Traugott, M. and Symondson, W. O. C. (2008). Molecular analysis of predation: a review of best practice for DNA based approaches. Mol. Ecol. 17: 947–963.

Krishnamurthy, P. K. and Francis, R. A. (2012). A critical review on the utility of DNA barcoding in biodiversity conservation. Biodivers. Conserv. 21: 1901–1919.

Kruck, N. C., Innes, D. I. and Ovenden, J. R. (2013). New SNPs for population genetic analysis reveals possible cryptic speciation of eastern Australian sea mullet (*Mugil cephalus*). Mol. Ecol. Resour. 13: 715–725.

Labrador, K., Agmata, A., Palermo, J. D., Follante, J. and Pante, M. J. (2019). Authentication of processed Philippine sardine products using hotshot DNA extraction and minibarcode amplification. Food Control 98: 150–155.

Lacoursière-Roussel, A., Dubois, Y., Normandeau, E. and Bernatchez, L. (2016). Improving herpetological surveys in eastern North America using the environmental DNA method. Genome 59.

Laikre, L., Palm, S. and Ryman, N. (2005). Genetic population structure of fishes: implications for coastal zone management. AMBIO A J. Hum. Environ. 34: 111–119.

Lamendin, R., Miller, K. and Ward, R. D. (2015). Labelling accuracy in Tasmanian seafood: An investigation using DNA barcoding. Food Control 47: 436–443.

León-Romero, Y., Mejía, O. and Soto-Galera, E. (2012). DNA barcoding reveals taxonomic conflicts in the Herichthys bartoni species group (Pisces: Cichlidae). Mol. Ecol. Resour. 12: 1021–1026.

Lewis, L., Richardson, D., Zakharov, E. and Hanner, R. (2016). Integrating DNA barcoding of fish eggs into ichthyoplankton monitoring programs. Fishery Bulletin 114: 153–165.

Li, J., Cui, Y., Jiang, J., Yu, J., Niu, L., Deng, J. et al. (2017). Applying DNA barcoding to conservation practice: a case study of endangered birds and large mammals in China. Biodivers Conserv. 26: 653–668.

Limborg, M. T., Helyar, S. J., Bruyn, M., Taylor, M. I., Nielsen, E. E., Ogden, R. et al. (2012). Environmental selection on transcriptome-derived SNPs in a high gene flow marine fish, the Atlantic herring (*Clupea harengus*). Mol. Ecol. 21: 3686–3703.

Lira, N. L., Tonello, S., Lui, R. L., Traldi, J. C., Brandão, H., Oliveira, C. et al. (2022). Identifying fish eggs and larvae: from classic methodologies to DNA metabarcoding. Mol. Biol. Rep. 50(2): 1713–1726. https://doi.org/10.1007/s11033-022-08091-9.

Lodge, D. M., Turner, C. R., Jerde, C. L., Barnes, M. A., Chadderton, L., Egan, S. P. et al. (2012). Conservation in a cup of water: estimating biodiversity and population abundance from environmental DNA. Mol. Ecol. 21: 2555–2558.

Lu, C. P., Chen, C. A., Hui, C. F., Tzeng, T. D. and Yeh, S. Y. (2006). Population genetic structure of the swordfish, *Xiphias gladius* (Linnaeus, 1758), in the Indian Ocean and West Pacific inferred from the complete DNA sequence of the mitochondrial control region. Zool. Stud. 45: 269–279.

Mabragaña, E., Díaz De Astarloa, J. M., Hanner, R., Zhang, J. and González Castro, M. (2011). DNA barcoding identifies argentine fishes from marine and brackish waters. PLoS One 6(12).

Mächler, E., Deiner, K., Stenmann, P. and Altermatt, F. (2014). Utility of environmental DNA for monitoring rare and indicator macroinvertebrate species. Freshw. Sci. 33: 1174–1183.

Marceniuk, A. P., Caires, R., Rotundo, M. M., Cerquira, N. N. C. D., Siccha-Ramirez, R., Wosiacki, W. B. et al. (2020). Taxonomic revision of the *Menticirrhus americanus* (Linnaeus, 1758) and *M. littoralis* (Holbrook, 1847) (Percomorphacea: Sciaenidae) species complexes from the western Atlantic. Zootaxa 4822(3): 301–333.

Marceniuk, A. P., Caires, R., Siccha-Ramirez, R. and Oliveira, C. (2016). Review of the harvestfishes, genus *Peprilus* (Perciformes: Stromateidae), of the Atlantic coast of South America. Zootaxa 4098(2): 311–332.

Manel, S., Guerin, P. E., Mouillot, D., Blanchet, S., Velez, L., Albouy, C. et al. (2020). Global determinants of freshwater and marine fish genetic diversity. Nat. Commun. 11: 1–9.

Mariani, S., Griffiths, A. M., Velasco, A., Kappel, K., Jérôme, M., Perez-Martin et al. (2015). Low mislabeling rates indicate marked improvements in European seafood market operations. Front. Ecol. Environ. 13: 536–540.

McCusker, M. R., Denti, D., Guelpen, L. V., Kenchington, E. and Bentzen, P. (2013). Varcoding Atlantic Canada's commonly encountered marine fishes. Mol. Ecol. Resour. 13: 177–188.

Méheust, E., Alfonsi, E., Le Ménec, P., Hassani, S. and Jung, J. L. (2015). DNA barcoding for the identification of soft remains of prey in the stomach contents of grey seals (*Halichoerus grypus*) and harbour porpoises (*Phocoena phocoena*). Mar. Biol. Res. 11: 385–95.

Menezes, M. R., Ikeda, M. and Taniguchi, N. (2006). Genetic variation in skipjack tuna *Katsuwonus pelamis* (L.) using PCR-RFLP analysis of the mitochondrial DNA D-loop region. J. Fish Biol. 68: 156–161.

Meusnier, I., Singer, G. A., Landry, J. F., Hickey, D. A., Hebert, P. D. N. and Hajibabaei, M. (2008). A universal DNA mini-barcode for biodiversity analysis. BMC Genom. 9: 214.

Meyer, C. P. and Paulay, G. (2005). DNA barcoding: error rates based on comprehensive sampling. PLoS Biol. 3: 422.

Miller, S. E., Hausmann, A., Hallwachs, W. and Janzen, D. H. (2016). Advancing taxonomy and bioinventories with DNA barcodes. Philos. Trans. R. Soc. Lond., B, Biol. Sci. 371: 1702.

Miller, D. D. and Mariani, S. (2010). Smoke, mirrors, and mislabeled cod: poor transparency in the European seafood industry. Front. Ecol. Environ. 8(10): 517–521.

Minamoto, T., Yamanaka, H., Takahara, T., Honjo, M. N. and Kawabata, Z. (2012). Surveillance of fish species composition using environmental DNA. Limnol. 13: 193–197.

Mishra, P., Kumar, A., Nagireddy, A., Mani, D. N., Shukla, A. K., Tiwari, R. et al. (2015). DNA barcoding: an efficient tool to overcome authentication challenges in the herbal market. Plant Biotechnol. J. 14: 8–21.

Mohammed Abubakar, B., Mohd Salleh, F., Shamsir Omar, M. S. and Wagiran, A. (2017). Review: DNA barcoding and chromatography fingerprints for the authentication of botanicals in herbal medicinal products. Evid.-based Complement. Altern. Med. 1–28.

Morey, K., Bartley, T. and Hanner, R. (2020). Validating environmental DNA metabarcoding for marine fishes in diverse ecosystems using a public aquarium. Environ. DNA. 2: 330–342.

Mosa, K. A., Soliman, S., El-Keblawy, A., Ali, M. A., Hassan, H. A., Tamim, A. A. B. et al. (2018). Using DNA barcoding to detect adulteration in different herbal plant-based products in the United Arab Emirates: proof of concept and validation. Recent Pat. food, Nutr. Agric. 9: 55–64.

Naro-Maciel, E., Reid, B. and Fitzsimmons, N. N. (2009). DNA barcodes for globally threatened marine turtles: a registry approach to documenting biodiversity. Mol. Ecol. Resour. 10: 252–263.

Nielsen, E. E., Cariani, A., Aoidh, E. M., Maes, G. E., Milano, I., Ogden, R. et al. (2012). Gene-associated markers provide tools for tackling illegal fishing and false eco-certification. Nat. Commun. 3: 851.

Nielsen, E. E., Hemmer-Hansen, J., Larsen, P. F. and Bekkevold, D. (2009). Population genomics of marine fishes: identifying adaptive variation in space and time. Mol. Ecol. 18: 3128–3150.

Nielsen, E. E., Nielsen, P. H., Meldrup, D. and Hansen, M. M. (2004). Genetic population structure of turbot (*Scophthalmus maximus* L.) supports the presence of multiple hybrid zones for marine fishes in the transition zone between the Baltic Sea and the North Sea. Mol. Ecol. 13: 585–595.

Oddone, M., Awruch, C. A., Barreto, R., Charvet, P., Chiaramonte, G. E., Cuevas, J. M. et al. (2019). *Squatina guggenheim*. The IUCN Red List of Threatened Species 2019.

Ogden, R. (2008). Fisheries forensics: The use of DNA tools for improving compliance, traceability and enforcement in the fishing industry. Fish Fish. 9: 462–472.

Ovenden, J. R., Berry, O., Welch, D. J., Buckworth, R. C. and Dichmont, C. M. (2013). Ocean's eleven: a critical evaluation of the role of population, evolutionary and molecular genetics in the management of wild fisheries. Fish Fish. 16: 125–159.

Overdyk, L., Braid, H., Crawford, S. and Hanner, R. (2015). Extending DNA barcoding coverage for lake whitefish (*Coregonus clupeaformis*) across the Three Major Basins of Lake Huron. DNA Barcodes 3: 59–65.

Overdyk, L., Holm, E., Hanner, R. and Crawford, S. (2016). Increased taxonomic resolutions of Laurentian Great Lakes ichthyoplankton through DNA barcoding: case study comparison against visual identification in Stokes Bay, Lake Huron. J. Gt. Lakes Res. 42: 812–818.

Packer, L., Gibbs, J., Sheffield, C. and Hanner, R. (2009). DNA barcoding and the mediocrity of morphology. Mol. Ecol. Resour. 9: 42–50.

Padial, J. M., Miralles, A., De la Riva, I. and Vences, M. (2010). The integrative future of taxonomy. Front Zool. 7: 16–25.

Panday, R., Jha, D. K., Thapa, N., Pokharel, B. R. and Aryal, N. K. (2014). Forensic wildlife parts and their product identification and individualization using DNA barcoding. The Open Forensic Sci. J. 7: 6–13.

Pawlowski, J., Apothéloz-Perret-Gentil, L. and Altermatt, F. (2020). Environmental DNA: What's behind the term? Clarifying the terminology and recommendations for its future use in biomonitoring. Mol. Ecol. 29(22): 4258–4264.

Pawlowski, J., Esling, P., Lejzerowicz, F., Cedhagen, T. and Wilding, T. A. (2014). Environmental monitoring through protist next-generation sequencing metabarcoding: assessing the impact of fish farming on benthic foraminifera communities. Mol. Ecol. Resour. 14: 1129–1140.

Pečnikar, Z. F. and Buzan, E. V. (2014). 20 years since the introduction of DNA barcoding: from theory to application. J. Appl. Genet. 55: 4–52.

Pegg, G. G., Sinclair, B. and Briskey, L. (2006). MtDNA barcode identification of fish larvae in the southern Great Barrier Reef Australia. Sci. Mar. 70(2): 7–12.

Phillips, J., Gillis, D. and Hanner, R. (2019). Incomplete estimates of genetic diversity within species: Implications for DNA barcoding. Ecol. Evol. 9: 2996–3010.

Phillips, J., Gwiazdowski, R., Ashlock, D. and Hanner, R. (2015). An exploration of sufficient sampling effort to describe intraspecific DNA barcode haplotype diversity: examples from the ray-finned fishes (Chordata: Actinopterygii). DNA Barcodes 3: 66–73.

Pollom, R., Ebert, D. A., Leslie, R. and Winker, H. (2020). *Dipturus pullopunctatus*. The IUCN Red List of Threatened Species 2020.

Poulin, R. and Pérez-Ponce de León, G. (2017). Global analysis reveals that cryptic diversity is linked with habitat but not mode of life. J. Evol. Biol. 30(3): 641–649.

Price, B. W., Henry, C. S., Hall, A. C., Mochizuki, A., Duelli, P. and Brooks, S. J. (2015). Singing from the grave: DNA from a 180-year-old type specimen confirms the identity of *Chrysoperla carnea* (Stephens). PLoS One 10(4): 1–11.

Rach, J., DeSalle, R., Sarkar, I., Schierwater, B. and Hadrys, H. (2008). Character-based DNA barcoding allows discrimination of genera, species and populations in Odonata. Proc. R. Soc. B. Biol. Sci. 275: 237–47.

Rasmussen, R. S., Morrissey, M. T. and Hebert, P. D. (2009). DNA barcoding of commercially important salmon and trout species (*Oncorhynchus* and *Salmo*) from North America. J. Agric. Food Chem. 57(18): 8379–8385.

Reed, D. H. and Frankham, R. (2003). Correlation between Fitness and Genetic Diversity. Conserv. Biol. 17: 230–237.

Reiss, H., Hoarau, G., Dickey-Collas, M. and Wolff, W. J. (2009). Genetic population structure of marine fish: mismatch between biological and fisheries management units. Fish Fish. 10: 361–395.

Rock, J., Costa, F. O., Walker, D. I., North, A. W., Hutchinson, W. and Carvalho, G. R. (2008). DNA barcodes of fish of the Scotia Sea, Antarctica indicate priority groups for taxonomic and systematics focus. Antarct. Sci. 20: 253–262.

Roslin, T. and Majaneva, S. (2016). The use of DNA barcodes in food web construction-terrestrial and aquatic ecologists unite! Genome. 59: 603–628.

Rosso, J. J., Rueda, E. C., Sanchez, S., Bruno, M. C., Casciotta, J., Aguilera, G. et al. (2018). Basin-scale distribution and haplotype partitioning in different genetic lineages of the Neotropical migratory fish *Salminus brasiliensis*. Aquat. Conserv. 28(2): 444–456.

Rubinoff, D. (2006). Utility of Mitochondrial DNA barcodes in species conservation. Conserv. Biol. 20: 1026–1033.

Rubinoff, D., Cameron, S. and Will, K. (2006). A genomic perspective on the shortcomings of mitochondrial DNA for 'barcoding' identification. J. Hered. 97: 581–594.

Sales, J. B. L., de Oliveira, C. N., dos Santos, W. C. R., Rotundo, M. M., Ferreira, Y., Ready, J. et al. (2019). Phylogeography of eagle rays of the genus *Aetobatus*: *Aetobatus narinari* is restricted to the continental western Atlantic Ocean. Hydrobiologia 836(1): 169–183.

Seberg, O., Humphries, C. J., Knapp, S., Stevenson, D. W., Petersen, G., Scharff, N. et al. (2003). Shortcuts in systematics? A commentary on DNA-based taxonomy. Trends Ecol Evol. 18: 63–5.

Sepulveda, A., Hutchins, P. R., Forstchen, M., Mckeefry, M. N. and Swigris, A. M. (2020). The Elephant in the Lab (and Field): Contamination in aquatic environmental DNA studies. Front. Ecol. Evol. 8: 1–12.

Sheppard, S. K., Bell, J., Sunderland, K. D., Fenlon, J., Skervin, D. and Symondson, W. O. C. (2005). Detection of secondary predation by PCR analyses of the gut contents of invertebrate generalist predators. Mol. Ecol. 14: 4461–4468.

Shui, B. N., Han, Z. Q., Gao, T. X. and Miao, Z. Q. (2008). Genetic structure of Japanese Spanish mackerel (*Scomberomorus niphonius*) in the East China Sea and Yellow Sea inferred from AFLP data. Afr. J. Biotechnol. 7: 3860–3865.

Skinner, M., Murdoch, M., Loeza-Quintana, T., Crookes, S. and Hanner, R. (2020). A mesocosm comparison of laboratory-based and on-site eDNA solutions for detection and quantification of striped bass (*Morone saxatilis*) in marine ecosystems. Environmental DNA 2: 298–308.

Smith, P. J., Steinke, D., McMillan, P. J., Stewart, A. L., McVeagh, S. M., Diaz de Astarloa, J. M. et al. (2011). DNA barcoding highlights a cryptic species of grenadier (genus *Macrourus*) in the Southern Ocean. J. Fish. Biol. 78: 355–365.

Sousa, L. L., Xavier, R., Costa, V., Humphries, N. E., Trueman, C., Rosa, R. et al. (2016). DNA barcoding identifies a cosmopolitan diet in the ocean sunfish. Sci. Rep. 6: 1–9.

Spies, I. and Punt, A. E. (2015). The utility of genetics in marine fisheries management: a simulation study based on Pacific cod off Alaska. Can. J. Fish. Aquat. Sci. 72(9): 1415–1432.

Steinke, D., Bernard, A., Horn, R., Hilton, P., Hanner, R. and Shivji, M. (2017). DNA analysis of traded shark fins and mobulid gill plates reveals a high proportion of species of conservation concern. Nat. Sci. Rep. 7: 905.

Steinke, D., Zemlak, T. S. and Boutillier, J. A. (2009). DNA barcoding Pacific Canada's fishes. Mar. Biol. 156: 2641–2647.

Steinke, D., Zemlak, T. S. and Hebert, P. D. N. (2009). Barcoding Nemo: DNA-based identifications for the ornamental fish trade. PLoS One 4(7).

Sultana, S., Ali, Hossain, M. E., Asing, M. A. M., Naquiah, N. and Zaidul, I. S. M. (2018). Universal mini COI barcode for the identification of fish species in processed products. Int. Food Res. J. 105: 19–28.

Symondson, W. O. C. (2002). Molecular identification of prey in predator diets. Mol. Ecol. 11: 627–641.

Taberlet, P., Coissac, E., Hajibabaei, M. and Rieseberg, L. H. (2012a). Environmental DNA. Mol. Ecol. 21(8): 1789–1793.

Taberlet, P., Coissac, E., Pompanon, F., Brochmann, C. and Willerslev, E. (2012b). Towards next-generation biodiversity assessment using DNA metabarcoding. Mol. Ecol. 21: 2045–2050.

Taberlet, P., Bonin, A., Zinger, L. and Coissac, E. (2018). Environmental DNA: For Biodiversity Research and Monitoring; Oxford University Press: Oxford, UK.

Taylor, H. R. and Harris, W. E. (2012). An emergent science on the brink of irrelevance: a review of the past 8 years of DNA barcoding. Mol. Ecol. Resour. 12(3): 377–388.

Taylor, M. I., Rico, I., Fox, C. and Rico, C. (2002). Species specific Taqman probes for simultaneous identification of cod (*Gadus morhua* L.), haddock (*Melanogrammus aeglefinus* L.) and whiting (*Merlangius merlangus* L.). Mol. Ecol. Not. 2: 599–601.

Taylor, M. S. and Hellberg, M. E. (2003). Genetic evidence for local retention of pelagic larvae in a Caribbean reef fish. Sci. 299: 107–109.

Thomsen, P. F., Kielgast, J., Iversen, L. L., Møller, P. R., Rasmussen, M. and Willerslev, E. (2012). Detection of a diverse marine fish fauna using environmental DNA from seawater samples. PLoS One 7: 1–9.

Valdez-Moreno, M., Vasquez-Yeomans, L., Elias-Gutierrez, M., Ivanova, N. V. and Hebert, P. D. N. (2010). Using DNA barcodes to connect adults and early life stages of marine fishes from the Yucatan Peninsula, Mexico: potential in fisheries management. Mar. Freshw. Res. 61(6): 665–671.

Valentini, A., Miquel, C., Nawaz, M. A., Bellemain, E., Coissac, E., Pompanon, F. et al. (2009). New perspectives in diet analysis based on DNA barcoding and parallel pyrosequencing: the trnL approach. Mol. Ecol. Resour. 9: 51–60.

Vargas, S. M., Araújo, F. C. F. and Santos, F. R. (2009). DNA barcoding of Brazilian sea turtles (Testudines). Genet. Mol. Biol. 32(3): 608–612.

Vences, M., Nagy, Z. T., Sonet, G. and Verheyen, E. (2012). DNA barcoding amphibians and reptiles. DNA Barcodes. Humana Press, Totowa, NJ. 79–107.

Veneza, I., Da Silva, R., Freitas, L., Silva, S., Martins, K., Sampaio, I. et al. (2018). Molecular authentication of Pargo fillets *Lutjanus purpureus* (Perciformes: Lutjanidae) by DNA barcoding reveals commercial fraud. Neotrop. Ichthyol. 16: 1–6.

Visco, J. A., Apothéloz-Perret-Gentil, L., Cordonier, A., Esling, P., Pillet, L. and Pawlowski, J. (2015). Environmental monitoring: inferring the diatom index from next-generation sequencing data. Environ. Sci. Technol. 49: 7597–7605.

Walker, T. I., Cavanagh, R. D., Stevens, J. D., Carlisle, A. B., Chiaramonte, G. E., Domingo, A. et al. (2006). *Galeorhinus galeus*. The IUCN Red List of Threatened Species 2006.

Wang, S., Hard, J. J. and Utter, F. (2002). Genetic variation and fitness in salmonids. Conserv. Genet. 3(3): 321–333.

Waples, R. S., Punt, A. E. and Cope, J. M. (2008). Integrating genetic data into management of marine resources: how can we do it better? Fish Fish. 9(4): 423–449.

Ward, R. D., Woodwark, M. and Skibinski, D. O. F. (1994). A comparison of genetic diversity levels in marine, freshwater, and anadromous fishes. J. Fish Biol. 44: 213–232.

Ward, R. D., Zemlak, T. S., Innes, B. H., Last, P. R. and Hebert, P. D. N. (2005). DNA barcoding Australia's fish species. Phil. Trans. R Soc B 360: 1847–1857.

Ward, R. D., Hanner, R. and Hebert, P. D. N. (2009). The campaign to DNA barcode all fishes. J. Fish Biol. 74: 329–356.

Ward, T. and Phillips, B. (2008). Ecolabelling of seafood: the basic concepts. pp. 1–37. *In*: Ward, T. and Phillips, B. (eds.). Seafood Ecolabelling. Principles and Practice. Wiley Blackwell. London, UK.

Waugh, J. (2007). DNA barcoding in animal species: progress, potential and pitfalls. BioEssays 29: 188–197.

Webb, K. E., Barnes, D. K. A., Clark, M. S. and Bowden, D. A. (2006). DNA barcoding: a molecular tool to identify Antarctic marine larvae. Deep-Sea Res. II. 53: 1053–1060.

Wessells, C. R., Cochrane, K., Deere, C., Wallis, P. and Willmann, R. (2001). Product Certification and Ecolabelling for Fisheries Sustainability. pp. 1–83. FAO Fish. Tech. Pap 422. Rome, Italy.

Wheeler, Q. D., Raven, P. H. and Wilson, E. O. (2004). Taxonomy: Impediment or expedient? Sci. 303: 285.

Will, K. W., Mishler, B. D. and Wheeler, Q. D. (2005). The perils of DNA barcoding and the need for integrative taxonomy. Syst. Biol. 54: 844–851.

Will, K. W. and Rubinoff, D. (2004). Myth of the molecule: DNA barcodes for species cannot replace morphology for identification and classification. Cladistics 20: 47–55.

Wong, E. H. K. and Hanner, R. H. (2008). DNA barcoding detects market substitution in North American seafood. Int. Food Res. J. 41: 828–837.

Worm, B., Davis, B., Kettemer, L., Ward-Paige, C. A., Chapman, D., Heithaus, M. R. et al. (2013). Global catches, exploitation rates, and rebuilding options for sharks. Mar. Policy 40: 194–204.

Wu, Y., Li, J., Yang, M. G., Jiang, Y. G., Liu, L., Wang, M. C. et al. (2018). Authentication of small berry fruit in fruit products by DNA barcoding method. J. Food Sci. 83(6): 1494–1504.

Young, M. K., McKelvey, K. S., Pilgrim, K. L. and Schwartz, M. K. (2013). DNA barcoding at riverscape scales: assessing biodiversity among fishes of the genus *Cottus* (Teleostei) in northern Rocky Mountain streams. Mol. Ecol. Resour. 13: 583–595.

Zarzoso-Lacoste, D., Bonnaud, E., Corse, E., Gilles, A., Meglecz, E., Costedoat, C. et al. (2016). Improving morphological diet studies with molecular ecology: an application for invasive mammal predation on island birds. Biol Conserv. 193: 134–42.

Zeale, M. R. K., Butlin, R. K., Barker, G. L. A., Lees, D. C. and Jones, G. (2011). Taxon-specific PCR for DNA barcoding arthropod prey in bat faeces. Mol. Ecol. Resour. 11(2): 236–244.

Zemlak, T. S., Ward, R. D. and Connell, A. D. (2009). DNA barcoding reveals overlooked marine fishes. Mol. Ecol. Resour. 9(1): 237–242.

Zimmermann, J., Glöckner, G., Jahn, R., Enke, N. and Gemeinholzer, B. (2015). Metabarcoding vs. morphological identification to assess diatom diversity in environmental studies. Mol. Ecol. Resour. 15: 526–542.

CHAPTER 11

The Tool Box for Mechanistic Understanding of Conservation Problems Facing Marine Wildlife

Steven J. Cooke,[1,*] *Kim Birnie-Gauvin,*[2,3] *Christine Madliger,*[4] *Cory D. Suski,*[5] *Erika J. Eliason,*[3] *Kenneth Jeffries,*[6] *Jeff C. Clements*[7] and *Robert Lennox*[8]

Introduction

Marine ecosystems around the globe are under threat from a diverse array of anthropogenic pressures (Halpern et al. 2007). For centuries, humans have exploited biological resources and removed billions of tonnes of organisms from the oceans (Jackson et al. 2001), while also generating bycatch (Davies et al. 2009) and destroying productive habitats through destructive fishing practices (Dayton et al. 1995). The oceans have also been treated like waste receptacles for everything from plastics (Haward 2018) to sewage (Wood et al. 1993). Runoff from cities (Müller et al. 2020) and agricultural (Beman et al. 2005) landscapes, as well as point source pollution from industry (e.g., mining, oil and gas, manufacturing; Kennish 2017), have all delivered nutrients, silt, and various chemical contaminants to estuaries and oceans. Novel sources of pollution including light (e.g., in coastal habitats; Davies et al. 2014) and noise (e.g., from vessels, drilling; Williams et al. 2015) also threaten marine life. Estuarine and coastal habitats have been modified through infilling and hardening of shorelines (Kennish 2002) and the removal of mangrove forests (Alongi 2002). Invasive species (Molnar et al. 2008) and emerging pathogens (McCallum et al. 2003) further destabilize marine systems. Climate change acts as a threat multiplier bringing with it ocean acidification (Guinotte and Fabry 2008), extreme weather events (including both

[1] Fish Ecology and Conservation Physiology Laboratory, Department of Biology and Institute of Environmental and Interdisciplinary Science, Carleton University, 1125 Colonel By Dr., Ottawa, ON, K1S 5B6, Canada.

[2] Section for Freshwater Fisheries and Ecology, National Institute of Aquatic Resources, Technical University of Denmark, Silkeborg, Denmark.

[3] Department of Ecology, Evolution and Marine Biology, University of California Santa Barbara, California, UCSB Marine Science Institute, Building 520, Santa Barbara, CA, 93106-6150, USA.

[4] Department of Biology, Algoma University, Sault Ste. Marie, ON, P6A 2G4, Canada.

[5] University of Illinois at Urbana-Champaign, Department of Natural Resources and Environmental Sciences, 1102 S Goodwin Ave, Urbana, IL, 61801, USA.

[6] Department of Biological Sciences, University of Manitoba, 50 Sifton Road, Winnipeg, MB, R3T 2N2, Canada.

[7] Fisheries and Oceans Canada, 343 Université Avenue, Moncton, NB, E1C 9B6, Canada.

[8] Department of Biology, Dalhousie University, Halifax, NS, B3H 4R2, Canada.

* Corresponding author: Steven_Cooke@carleton.ca

cold and heat extremes, storms; Ummenhofer and Meehl 2017), and coral bleaching (Hoegh-Guldberg 1999). Rarely do these threats occur singularly; cumulative effects and complex interactions are the norms (Crain et al. 2008, Peterson and Lowe 2009). Quite simply, humans have had an enormous impact on marine ecosystems, and there is little reason to believe that the threat landscape is decreasing (Gray 1997, Luypaert et al. 2020). The aforementioned impacts can be observed at all biological levels and spanning all taxa, yet they are most apparent in marine wildlife (Duarte et al. 2020) given that they often hold higher trophic level positions as predators, scavengers or herbivores and are thus sensitive to indirect effects on ecosystem health (e.g., their nutrition) as well as direct impacts related to exploitation.

Mechanisms matter

Population declines of wildlife in marine ecosystems are widespread. However, understanding the specific mechanisms underpinning such declines is often lacking (McKenzie et al. 2016). We define mechanisms as the suite of physiological and behavioural processes that influence organism function, health, fitness and survival (Cooke et al. 2013, 2023). Without knowledge of the specific mechanisms responsible for population decline, it is difficult to identify the driver of the decline and to therefore take appropriate corrective action (Cooke and O'Connor 2010). Indeed, in many cases multiple stressors and drivers interact in complex and unexpected ways. Mechanistic understanding enables the effective use of management levers to reverse declines and rebuild populations while ensuring limited resources are not wasted (Cooke et al. 2023). Beyond understanding problems, mechanistic research can also be used to identify thresholds that enact regulations (e.g., limiting levels of different pollutants) or, in a predictive sense, to understand how individuals and populations may respond to different threats (e.g., climate change). Because population-level responses (both in terms of decline and recovery) may take decades to observe based on the life-history of a given species, mechanistic tools can also be used as proxies for the state of wildlife populations (Wikelski and Cooke 2006). Finally, mechanistic research can also be used to identify and evaluate different management strategies to determine what works and what doesn't without having to wait for populations to respond.

There is an ever-expanding toolbox that enables researchers to understand the mechanistic basis of conservation problems facing marine wildlife (termed 'the conservation physiology toolbox', but it extends to include behaviour; Madliger et al. 2018). Early work focused on bringing wildlife into the laboratory to conduct controlled experiments and observations. This meant that there would be close access to analytical equipment and scientific apparatus. Such work continues to this day, but is increasingly complemented by mechanistic research conducted in the field, in mesocosms, or the great blue sea (Costa and Sinervo 2004). Field work has been facilitated by new developments including biotelemetry and biologging (Block 2005, Hussey et al. 2015, Clements and Comeau 2019), preservatives for tissue samples (e.g., RNA-later; Camacho-Sanchez et al. 2013), point-of-care devices (Stoot et al. 2014), and mobile respirometers (Byrnes et al. 2020). For work done in the field, or any work involving rare or imperiled species, there has also been interest in the development of tools that are non-lethal yet still provide information on organismal condition (Madliger et al. 2018). Marine wildlife is diverse – from fish in the abyss to gigantic blue whales to intertidal bivalves and barnacles and newly hatched sea turtles undertaking massive trans-oceanic migrations – all of which create their own unique challenges when trying to conduct mechanistic organismal research. From organisms that live under polar ice caps to those that live at deep sea vents, researchers are constantly reminded that marine ecosystems are inherently hostile to humans.

Here, we provide an overview of the toolbox for generating a mechanistic understanding of conservation problems facing marine wildlife using physiology, behaviour, or both (Fig. 1). This is not an exhaustive treatment, given the diversity of taxa with wide ranges of ecologies, behaviours, and physiologies. Indeed, one could easily write about this toolbox for any given taxa and generate a rich and detailed paper. We focus largely on vertebrates but include some examples of macro-invertebrates where relevant. We begin by briefly describing the tools in the toolbox, doing so by generally categorizing them into systems (e.g., sensory physiology, glucocorticoids) and considering the advantages and limitations of said tools. Next, we present several diverse case studies where we demonstrate how mechanistic

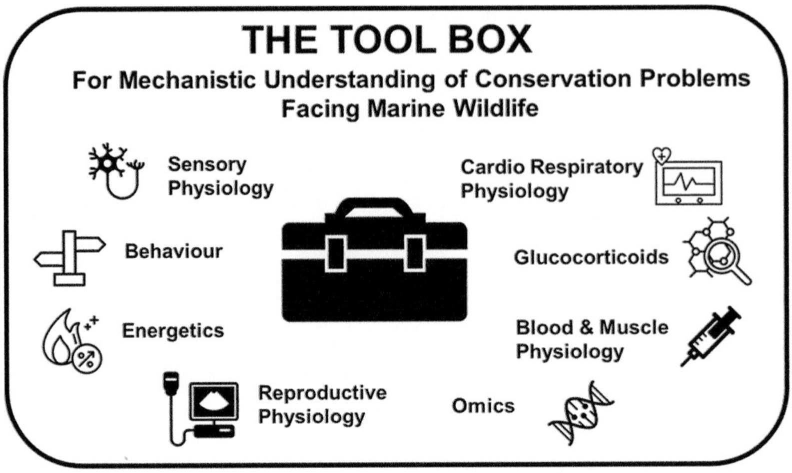

Figure 1. The tool box for the mechanistic understanding of conservation problems facing marine wildlife.

research has helped to understand or solve a pressing marine conservation issue. We conclude with a forward-looking perspective of what is on the horizon for mechanistic research related to the conservation of marine wildlife.

The tool box

Sensory physiology

Animals integrate complex information from their environment through their sensory systems. Seeing, hearing, feeling, smelling, and tasting environmental stimuli in an animals' environment provides immediate feedback to the neuroendocrine system, with cascading effects on physiology and behaviour (Blaxter 1988). Ascertaining the role of different sensory systems in response to abiotic and biotic stimuli is a challenge, but there is broad acceptance that these systems are among the most important to understand animal ecology. Indeed, attracting and repelling species based on their unique sensory modalities is a massive industry with implications for fisheries, aquaculture, and general human-wildlife conflict (Dominoni et al. 2020).

Predicting animal responses to sensory stimuli depends on a fundamental understanding of the function of the system and its responses to stimulation (e.g., Quinn 1984). Once the general mechanisms are understood, laboratory and field investigations can be used to test hypotheses about how the sensory systems integrate information, and whether they can be manipulated for conservation. A salient example is the developing knowledge about the migration biology of fish and their use of a magnetic sensory system for navigation. Ueda et al. (1998) tried to block the magnetoreception of migrating salmon using neodymium-iron magnets around the head to determine whether blocking this sense affected homing compared to controls with copper rings. Later, Putman et al. (2014) not only demonstrated that salmon smolts were guided by a magnetic field, but also that they could be manipulated by simulating different magnetic field strengths to migrate in different vectors. Likewise, marine bivalves can use their sensory system to detect various stimuli from their environment (including chemical and tactile cues) and can use this information to behaviourally close their shells and avoid harmful conditions (Clements et al. 2021). Similar realizations that animals can be manipulated by their ecophysiological responses to environmental stimuli have formed the basis of several management strategies and the development of novel approaches to addressing conflicts and promoting conservation agendas (Dominoni et al. 2020). Acoustic deterrents to dissuade seals from aquaculture farms (Jacobs and Terhune 2002), various sensory-based bycatch reduction approaches in marine fisheries (Jordan et al. 2013), and personal shark deterrents using electromagnetic devices (Huveneers et al. 2018) have been promising developments for conservation efforts aiming to mitigate conflicts using sensory physiology. Exploiting sensory modalities towards fishing gears is an equally prized field, using illuminations (Hazin et al. 2005) or scents (Løkkeborg and Johannessen 1992) to attract fish. Enhancing the precision with which target species are captured and non-target species are avoided using sensory physiology represents a major potential pathway towards marine conservation given the massive scale at which fisheries are impacting marine ecosystems (e.g., Worm et al. 2009).

Behaviour

Behaviour is an ultimate response of an animal to its environment (Sih 2013). Integration of the physiological cues from the environment and internal functioning elicits a response in the animal, for example, to shelter or to flee (Brooker et al. 2016). Individual variation in behaviour is an important aspect of species and population biology and responses to stressors are not necessarily predictable or consistent across space and time (Dingemanse and Dochtermann 2013). Tracking organismal behaviour at a relevant scale for conservation can be challenging in the marine environment (Harcourt et al. 2019). There are myriad stressors operating on animals in the wild, and isolating stressors is really only possible in experimental micro- or mesocosm contexts. Studying stressors individually in the laboratory fails to incorporate the realism inherent in the field where organismal responses are inherently complex.

Tracking behaviour in marine environments can be challenging depending on the scale of the animal's movement and the environment in which it resides. Visual tracking is virtually impossible and mark-recapture designs for resampling individuals are resource intensive and highly variable. Individual tracking using electronic tags is therefore the state-of-the-art for tracking individuals (Hussey et al. 2015). Tracking tools include loggers that take measurements at fixed intervals, such as light-based geolocators that estimate a position at sunrise and sunset (Nielsen et al. 2006) or simple temperature or acceleration loggers (Williams et al. 2020). Loggers must be recovered for accessing the data, either by recapturing the animal or by the tag popping off and sending the data to a satellite (Thorstad et al. 2013). Acoustic transmitters provide data at fixed intervals (usually 60–120 s) to receiver stations that log the data in memory. Receiver stations at known positions can allow analysts to estimate movement paths of tagged animals within the array, usually using network analysis (Whoriskey et al. 2019). Receivers with overlapping detection ranges can be time-synchronized to calculate the exact positions of the tags (e.g., Baktoft et al. 2017).

For an ecophysiological approach, individual tracking should ideally be combined with detailed environmental data within the tracking area to determine what the animals are exposed to and how their behaviour responds. Common parameters to measure are water temperature, dissolved oxygen, current velocity, and salinity, but may be extended to pH, turbidity, or other parameters of interest. Modelling the three dimensional habitat as a morphospace that is constantly changing is challenging and requires a well-designed array of sensors and interpolation among stations. For benthic species, benthoscape maps may be sufficient to provide inferences about how habitats affect behaviour. Sensors on board tags can provide additional detailed information about individual behaviour at finer scales, including the depth the animal resides at, the temperature it experiences, and the activity level based on acceleration data (Williams et al. 2020). Magnetometers may also provide an additional level of detail about the tri-axial orientation of the animal (Williams et al. 2020). While most loggers and transmitters monitor behavioural attributes of vertebrates, biosensors for monitoring invertebrate behaviour exist as well. Perhaps most notably, biosensors to record the valve gaping behaviour of bivalves (in response to sensory stimuli) in both the lab and field have become increasingly common (Andrade et al. 2016, Clements and Comeau 2019). These sensors not only provide information on species biology and ecology, but have been successfully used to monitor responses to various environmental stressors (Andrade et al. 2016). Additionally, regular monitoring of bivalve gaping behaviour serves as a first line of defense in safeguarding human populations from potential negative effects of poor water quality (e.g., marine clams in Poland, The Economist 2021; freshwater mussels in Minneapolis, Lovett 2015).

Energetics

Energy is the currency of life. As such, anthropogenic stressors that have the potential to influence food consumption and energy expenditure (e.g., behaviour, immune response, activity) may lead to less energy available for key functions such as growth and reproduction that are relevant to fitness (Koehn and Bayne 1989, Parsons 1994). Energetics is a relevant mechanistic framework to consider how anthropogenic stressors are influencing organisms and populations (Watson et al. 2020). Energetics models have been used for decades, but the development of dynamic energy budgets (Kooijman and Kooijman 2010) has provided a new tool that has been widely adopted and applied to conservation problems. Unlike many other mechanistic measures that can be taken from marine wildlife, energetics is very directly related to population level processes given the link between energetics and fitness (Nabe-Nielsen et al. 2018).

The most well-established approach to assess the energetic status of an organism involves lethal sampling where tissues are analyzed using bomb calorimetry or through proximate body composition analysis to assess energy density. In larger organisms, it is possible to use tissue biopsies or to assess energy status. A number of proxies have been developed whereby researchers can use simple condition indices (e.g., length-mass relationships are commonly used for fish, while tissue mass-shell mass relationships are often for molluscs) or electronic meters that estimate energy density using microwaves and/or electrical impedance, but those approaches require extensive calibration. However, these aforementioned measures all represent a snapshot in time and are less relevant for estimating field metabolic rates.

While terminology can differ across taxa, fields, and time, the analysis of metabolic rate can generate a number of different metrics that provide valuable insights into how an organism is interacting with its environment. The term 'standard metabolic rate' (SMR), for example, equates to the minimum energetic expenses of a fasting, quiescent, stress-free animal at rest, acting as a fixed energetic requirement on top of which all other energetic costs are added (Rolfe and Brown 1997, Chabot et al. 2016). For freshwater fishes, osmotic regulation is believed to account for 20–50% of resting energy expenditure (Bœuf and Payan 2001), while in mammals, 90% of cellular oxygen consumption takes place in the mitochondria, and is believed to be related to costs with counteracting mitochondrial proton leak (approximately 20%) and ATP production (approximately 80%) (Rolfe and Brown 1997). In contrast, maximum metabolic rate (MMR) corresponds to the maximum rate of aerobic metabolism of an animal, and is a product of the maximal rate at which oxygen can be transported from the environment to mitochondria, often defined when animals are exercising (Rolfe and Brown 1997, Clark et al. 2013, Norin and Clark 2016). The difference between MMR and SMR is called aerobic scope (AS), and represents the scope for metabolic activity for an organism. The concept of AS has received a great deal of attention for fish, particularly in the context of climate change, owing to the hypothesis that a fish's performance, and therefore fitness, can decline at higher temperatures owing to a reduction in the capacity to supply oxygen to tissues as temperature increases and a reduction in aerobic scope (Clark et al. 2013, Farrell 2016). Finally, field metabolic rate refers to metabolism over a period of time when an animal is living in its natural environment and is typically higher than SMR due to the energy requirements of feeding, cold exposure and movement (Rolfe and Brown 1997).

Doubly labelled water (i.e., the isotopic turnover of water, sodium, and energy flux; Nagy 1983, 1989) approaches have been used to assess field metabolic rate in sea turtles (Jones et al. 2009), sea birds (e.g., penguins; Gales and Green 1990; reviewed in Shaffer 2011) and marine mammals (e.g., walrus; Acquarone et al. 2006), but is less commonly used in fish or invertebrates. However, the approach requires recapturing animals for subsequent measurements which may not be feasible with all organisms, particularly in the wild. Over the last few decades, there have been many developments involving biotelemetry and biologging. A variety of sensors have been incorporated into such tools that enable researchers to assess field metabolic rates. For example, electrocardiograms have been measured in fish (Cooke et al. 2016), decapod crustaceans (McGaw et al. 2018), and marine mammals (Green et al. 2009) to assess heart rate in the field, accelerometer sensors have been used to measure swimming activity in penguins (Brisson-Curadeau et al. 2021) and sharks (Cooke et al. 2016), and pressure sensors have been used to measure the jet pressure of squid (Webber and O'Dor 1986). Heartbeat rates have also been measured in benthic invertebrates using non-invasive infrared sensors (Burnett et al. 2013,

Tagliarolo and McQuaid 2016, Gurr et al. 2018, 2021). There are hundreds of studies that use various sensors and tag types (e.g., reviewed by Hussey et al. 2015, Cooke et al. 2016). On their own, data on heart rate or acceleration (which can be looked at in terms of tail beats or wingbeats or flipperbeats) does not provide information on energy use, but using respirometry (see section on cardio respiratory) these sensor outputs can be calibrated allowing researchers to estimate the field energetics of free-ranging animals. The benefit of using these methods is that it allows one to have animals carrying sensors living in the field where anthropogenic stressors can be studied *in situ* either experimentally or observationally.

Cardio respiratory

The cardiorespiratory system is responsible for oxygen delivery and gas exchange that supports animal life (Humphreys 1979). The cardiorespiratory system can be broken down into two major components - the gas exchange apparatus (e.g., gills, lungs, skin) and the gas transportation apparatus (e.g., heart, veins, arteries, blood, hemolymph). These systems are highly variable among marine species - from corals to fish to seabirds. Although many marine organisms solely obtain oxygen from water (e.g., fish, many marine invertebrates), there are also many marine organisms that depend on access to surface air to breathe (e.g., sea turtles and sea snakes, marine mammals, seabirds) or that have the ability to respire in the water and on land (e.g., crabs). Because the cardiorespiratory system is essential to life (i.e., when deprived of oxygen, most marine animals will quickly die), and given that respiration is directly related to metabolism, energetics, and behaviour, it is often studied to understand how different threats and environmental variation influence marine organisms.

Components of the gas exchange apparatus have been shown to respond negatively to a number of natural and anthropogenic stressors, with potential negative consequences to individuals or populations. For example, for fishes, exposure to aquatic toxicants and other irritants (e.g., heavy metals, pesticides, pH, temperature extremes) can result in changes in the structure of gills (Mallatt 1985). The quantification of histological changes in gill tissue, such as atrophy, necrosis, or hyperplasia, can provide insights into the magnitude of disturbances that are present in areas of suspected stressors related to control areas, and help quantify potential declines in performance that can result from impairments to the cardiovascular system (e.g., impairments to uptake of oxygen or removal of wastes) (Bernet et al. 1999). High pressure waves produced by the clearance and underwater disposal of munitions have the potential to result in damage to the lungs of marine vertebrates, as well as possible rupture of blood vessels, inhibiting oxygen supply (Koschinski 2011). Marine anthropogenic noise can lead to altered diving behavior of mammals, which, in turn, can result in decompression injuries to lung tissue, gas-bubble formation, and altered gas dynamics that can negatively impact individuals (Hooker et al. 2012).

The rate of energy consumption can be influenced by a host of factors including physiological state, body size, food intake, anabolism, activity level, and, for ectotherms, temperature (Chabot et al. 2016). Indeed, the metabolic rate in ectotherms can vary up to 10-fold depending on the intersection of these factors (Chabot et al. 2016). There are two main methods to measure the metabolic rate of organisms: direct or indirect calorimetry (Cech 1990, Nelson 2016, Rolfe and Brown 1997). Direct calorimetry involves the measurement of heat production of an animal, which is more commonly used with birds and mammals due to the low heat production of fish and the high heat capacity of water (Gnaiger et al. 1989, Cech 1990, Nelson 2016, Rolfe and Brown 1997). Indirect calorimetry involves quantifying metabolic rates by measuring the production or consumption of components of aerobic respiration (Gnaiger et al. 1989, Cech 1990, Nelson 2016, Rolfe and Brown 1997). The recent development of robust oxygen optodes (Tengberg et al. 2006), capable of accurately measuring dissolved oxygen concentrations at rapid time intervals, has led to the proliferation of oxygen uptake (i.e., the removal of oxygen from water) ($\dot{M}O_2$) as the preferred method of determining the metabolic rate for most terrestrial and aquatic animals, rather than the unit of energy consumed by tissues (Chabot et al. 2016, Nelson 2016). The rate of oxygen uptake is often expressed in terms of oxygen units (i.e., mg O_2/kg/hr), making respiration rate a more appropriate term in this situation (Chabot et al. 2016). Respiration (metabolic) rate data can be generated for animals that use water exclusively to obtain oxygen (e.g., fish, mussels, lobsters) (Shultz et al. 2011, Benítez et al. 2018, Briceño et al. 2018), including larval fishes (Peck and Moyano 2016), animals that breathe air exclusively (e.g., birds (Eichhorn et al. 2019) or bimodal breathers (Lefevre et al. 2016)). Additional details on metabolic rate measures can be found above in the section on *Energetics*.

Research has also focused on the transportation system (i.e., the cardiovascular system) with most effort focused on the heart. The heart can be studied in various ways. For example, hearts can be removed from dead animals and assessed using various morphological and histological (e.g., Santer and Walker 1980, Keen et al. 2016) methods to compare ventricle size, shape, and cellular composition. The vasculature can also be studied for pathologies such as coronary arteriosclerosis (Farrell 2002). Heart tissues can be analyzed for enzymes and metabolites. In some instances, hearts can be harvested and temporarily kept "alive" in various preparations for electrochemistry studies (e.g., Heuer et al. 2019) or studied *in situ* (e.g., Blank et al. 2004). These techniques can identify early life stage cardiac impairment with crude oil exposure in marine fishes (Incardona et al. 2014). Specifically, oil exposure caused pericardial edema, reduced heart rates and impaired survival in early life stages of marine fishes (Incardona et al. 2014, Esbaugh et al. 2016), and this information is being used to help determine oil exposure thresholds and inform best practices for dispersant usage. Similarly, electrochemistry techniques have been used to examine the impact of polycyclic aromatic hydrocarbons (PAHs) on adult marine fish cardiomyocytes (Brette et al. 2017, Heuer et al. 2019). *In situ* perfused heart techniques have been leveraged to examine how marine animals (e.g., crabs, fish, turtles) respond to temperature and hypoxia stressors (e.g., Farrell et al. 1989, Blank et al. 2004, Hanson and Farrell 2007, Overgaard et al. 2004, Farrell et al. 1994, Wilkens 1993). More commonly, researchers are using electrocardiogram (ECGs) techniques to assess the cardiac thermal tolerance of various fish (e.g., Chinook salmon, zebrafish, opaleye) and invertebrate (e.g., snails, crabs, limpets) species in an effort to determine the capacity for species and populations to cope with climate change (e.g., Anttila et al. 2014, Stenseng et al. 2005).

Cardiac function can also be studied in intact, even free-ranging wildlife. Early work in this space occurred on marine mammals and penguins (reviewed in Ponganis 2007). More recently, such work has expanded to other taxa with a focus on anthropogenic impacts. For example, Steell et al. (2020) used heart rate loggers to study Caribbean lobster heart rates in response to light pollution while Vitousek et al. (2010) evaluated the cardiac response (i.e., heart rate) of Galápagos marine iguanas to ecotourism. Heart rate can also be telemetered; Claireaux et al. (1995) used acoustic telemetry tags equipped with heart rate sensors to assess how Atlantic cod respond to thermal variation. However, measuring heart rate (or electrocardiograms) alone may not always be sufficient. Cardiac output (the volume of blood pumped by the heart per unit time) is a product of both heart rate (beats per min) and stroke volume (volume of blood pumped per beat). The strongest predictor of metabolic rate is generally regarded as being cardiac output (Webber et al. 1998). However, stroke volume is much more difficult to measure than heart rate, particularly outside of the laboratory. Nonetheless, Franklin et al. (1964) were able to measure and telemeter the blood velocity in pinnipeds at depth some 60+ years ago while Gräns et al. (2009) were able to successfully measure cardiac blood flow in free-swimming green sturgeon. Nonetheless, most studies of cardiac output where heart rate and stroke volume are measured simultaneously tend to be restricted to lab environments where animals are somewhat restrained (e.g., in a respirometer; Atlantic cod; Webber et al. 1998; sockeye salmon, Eliason et al. 2011; Dungeness crab; Brown and Terwilliger 1999).

Glucocorticoids

Glucocorticoids (GCs; cortisol and corticosterone) are released due to activation of the HPA/HPI (hypothalamic-pituitary-adrenal/interrenal) axis and are an important component of the physiological and behavioural modifications organisms employ to cope with environmental challenges (Wingfield et al. 1998). Glucocorticoids act to maintain homeostasis and promote recovery and energetic balance following constraints such as harsh weather events, predation pressure, or food shortages (Angelier and Wingfield 2013). Human-related activities (e.g., noise, human presence, habitat modification) have been shown to alter GC secretion, and GCs have therefore been assessed extensively as biomarkers of disturbance for wildlife (Dantzer et al. 2014). While blood samples are the most commonly collected sample type for analyzing GC levels, other media are available that can often be acquired less invasively such as faeces, urine, feathers, whale blow (respiratory vapour), blubber, hair, claws, and water samples

(Gormally and Romero 2020). In addition, information about changing GC and other hormone levels over extended time frames has been assessed retrospectively from whale carcasses with sample types like ear wax plugs (Trumble et al. 2013) and baleen (Hunt et al. 2014b).

There are many examples that highlight how the measurement of GCs can lead to better understanding of anthropogenic impacts on marine species. For example, corticosterone levels of nestling brown pelicans (*Pelecanus occidentalis*) predicted fledging success and were related to nutritional condition (as indicated by food delivery rate by parents), thereby making them potentially useful in rapidly assessing the effects of changing food availability (Lamb et al. 2016). Baleen samples from the carcasses of whale calves have allowed researchers to quantify lifetime patterns of GCs and have been shown to correlate positively with levels of wound severity from Kelp Gulls (*Larus dominicanus*), suggesting that parasitism by this species may be contributing to southern right whale (*Eubalaena australis*) chronic stress and calf mortality in Argentina (Ajó et al. 2020). Further, French et al. (2010) found that tourism pressure was reflected in GC levels in marine iguanas (*Amblyrhynchus cristatus*), with tourist-exposed populations having stress-induced elevations in plasma GCs, along with changes in immunological responses, compared to undisturbed populations. However, research by Scheun et al. (2021) illustrated that there are context-specific considerations necessary for GC monitoring in some populations, finding that African penguin (*Spheniscus demersus*) chicks in colonies experiencing low, infrequent presence of tourists had higher urofaecal GC metabolite concentrations than those in colonies with medium and high degrees of human disturbance (Scheun et al. 2021). These results likely indicate that some populations may become habituated to human presence, and managers may need to attempt to limit tourism exposure for breeding colonies not already regularly exposed (Scheun et al. 2021). The measurement of GCs has also been applied to the conservation and well-being in marine invertebrates as well. For example, measuring GCs in response to stocking density under aquaculture conditions can provide information regarding density-related stress and subsequent production for aquaculture and restocking purposes (Lagos et al. 2013). A recent study reported the first successful measurement of GCs in cephalopods, which can aid in the management of these animals in captivity (Chancellor et al. 2021). As the tools available to measure GCs continue to be developed and validated across a variety of systems, there are some broad-scale ways for researchers to improve their application to conservation science. First, because GCs can change in relation to a variety of environmental factors and represent only one component of a full suite of physiological responses (Romero and Beattie 2022), pairing GC measurements with other metrics of physiology and behaviour can allow for clearer interpretation (MacDougall-Shackleton et al. 2019, Gormally and Romero 2020). Indeed, there are many physiological (e.g., heart rate, metabolic rate, telomere length) and behavioural (e.g., neophobia, vigilance) metrics that can be used to measure stress, and choice should be dictated by the biological questions being posed and the timescale to be captured (see extensive review in Gormally and Romero 2020). In addition, it is evident that we cannot equate GC measurements quantified from different types of tissues. Interpretation must be based on knowledge of integration time, which can vary from minutes (e.g., blood) to hours (e.g., faeces) to days (feathers) to months (e.g., claws) (Dantzer et al. 2014, Gormally and Romero 2020). In addition, non-invasive measures, such as those acquired from metabolite levels in faeces, need to be considered in light of diurnal rhythms in GC secretion, handling, assay selection, storage and extraction, and other confounding factors such as diet (Palme 2019). Finally, researchers should note that GCs can show a variety of responses to chronic stress, with increases, suppressions, and lack of change all being recorded (Dickens and Romero 2013). Ideally, each population under study must be validated (MacDougall-Shackleton et al. 2019), and having baseline data or undisturbed sites for comparisons will be important for the interpretation of GC levels in relation to environmental change. Moreover, GCs are increasingly recognized as having manifold roles in organismal function such that they need to be viewed as more than "just" stress hormones (MacDougall-Shackleton et al. 2019).

Reproductive physiology

Reproduction is an important determinant of fitness for individuals, but is one of the most energetically demanding life stage in an organism's life. Its importance within the context of population dynamics and sustainability requires conservation biologists to have an in-depth understanding of reproductive physiology and the factors that affect it (Schreck et al. 2001). The potential impacts of human-induced environmental change on reproductive physiology and other related population-level effects are largely unknown. How often does a species reproduce? What factors affect whether an individual will breed or not in a particular year? Marine animals can be particularly difficult to study within this context because many will reproduce in the depths of the sea, but conservation physiologists have found creative tools to get around this logistical challenge.

Endocrine measures remain one of the most common means of studying the reproductive physiology of marine animals. Hormones such as estradiol, testosterone, 11-keto testosterone and progesterone can be measured from blood samples with relative ease. Reproductive hormones in fishes (Liley and Stacey 1983), sea turtles (Owens 2017), and sea birds (Whittow et al. 2002) are frequently used as biomarkers for reproduction but for some invertebrates (e.g., molluscs), there remains uncertainty about the extent to which hormones are involved in reproduction (Scott 2013). Often, the biggest challenge is collecting such samples from free-ranging wildlife. This has been particularly difficult for large, long-migrating marine animals like whales, which have proven difficult to study using standard techniques (Hunt et al. 2013). Nonetheless, methods to assess reproductive endocrinology in whales now include sampling respiratory vapour ('blow', Hunt et al. 2014a), faecal material (Rolland et al. 2005) and baleen (Hunt et al. 2014b). For example, in baleen (a stratified, cornified epithelial tissue), progesterone could be detected which was indicative of whether female bowhead whales were pregnant or not (Hunt et al. 2014b).

Other non-lethal techniques include tissue biopsies like blood and muscle, or even the collection of carcasses. Blood samples have been used to assess reproductive hormones, but also other metrics such as packed cell volume (PCV). In nesting leatherback sea turtles (Dermochelys coriacea), PCV was found to decrease during reproduction, indicating that females do not feed while breeding (Perrault et al. 2016). Moreover, females with previous breeding experience retained much higher PCV than first-time breeders, suggesting better foraging success prior to breeding. For Pacific salmon species, it is common to dissect female carcasses to assess egg retention to evaluate overall reproductive success (Hruska et al. 2011).

Another method used to study the reproductive condition in marine animals is ultrasonography. This method is common in aquariums where animals are kept in captivity (Daly and Jones 2017), but has also been used in the wild on a wide range of elasmobranchs. In round stingrays (*Urobatis halleri*), ultrasound imaging was combined with telemetry to assess habitat use in restored and natural habitats to show that warmer restored areas may confer a thermal physiological advantage to pregnant female stingrays (Jirik and Lowe 2012). In another study, portable ultrasonography was used to evaluate the reproductive status and fecundity of bonnethead sharks (*Sphyrna tiburo*) and was compared to the traditional method of dissection, showing strong agreements between both techniques (Anderson et al. 2018). An advantage to the ultrasound approach is of course the fact that it is non-lethal. It provides a snapshot of the reproductive state of the animals but it does not provide any particular mechanism understanding unless combined with other methods.

Another commonly used method to assess the reproductive condition in marine organisms is histology. Histological assessment of gonads can provide information on the size of reproductive organs, as well as the number of breeding events by assessing previous mature generations of ova (e.g., de Sylva and Breder 1997). Histology on gonads can be done on lethally sampled animals captured for that purpose, or on fish captured by anglers, where gonads and other organs tend to be discarded but can be used for scientific purposes. This method, along with other direct (yet lethal) measurements of reproductive state (e.g., gonadal weight and condition), remain the most common for marine invertebrates such as bivalves.

Blood and muscle physiology

The collection of blood and tissue samples allows for the quantification of a number of different metrics that can provide a mechanistic understanding of how individual marine organisms respond to different stressors, ultimately influencing populations (Seebacher and Franklin 2012). However, other aspects of organismal condition (e.g., reproductive biology) can also be assessed with these issues. Marine organisms, like all living things, consist of a hierarchical network of components that range from molecules to cells to genes to individuals; individuals can then be grouped into species, populations, and ecosystems (Simon 1962, Dobzhansky 1964, Nielsen 2000). Within this hierarchical framework, alterations or perturbations at one level can lead to changes at the next highest level of organization (coupled with feedback to control responses at lower levels), resulting in vertical integration of responses to stressors (Jørgensen and Nielsen 2013). A number of different stimuli have been shown to result in stressors for marine organisms resulting in a departure from homeostasis. Reductions in pH, for example, can lead to alterations in acid/base balance that can change ion concentrations in blood (Claiborne et al. 2002), while stressors such as high temperatures can lead to the production of hormones involved in the stress response or heat shock proteins intended to restore homeostasis (Sopinka et al. 2016). As such, physiological metrics, quantified through monitoring of blood and muscle physiology of individuals, reflect the response of organisms to a range of different environmental and anthropogenic stressors, integrating the responses to a number of external stimuli. When applied over extended time periods, these stressors, and the physiological changes that they induce, can result in reduced survival, impaired reproduction, and changes at the population level (Calow 1989, Atkinson et al. 2015, Tilbrook et al. 2000). There are an impressive number of physiological responses that can be measured in blood or tissue (reviewed in Mallory et al. 2010, Hunt et al. 2013, Atkinson et al. 2015, Finlayson et al. 2016, Sopinka et al. 2016), and this diversity of potential responses can allow the testing of specific hypotheses related to stressors (e.g., decrease in nutritional status, impairments in reproductive hormones, exposure to pollutants/toxins), along with more general responses that can give indications of animals being 'stressed' (e.g., concentrations of primary stress hormones, loss of ions from plasma). Furthermore, many physiological responses are highly sensitive and respond quickly to external stressors and stimuli providing a short-term snapshot of the condition (Atkinson et al. 2015), while tissues such as hair, feathers, or baleen have the potential to act as a repository for stress markers, allowing a longer-term integration of stress across extended time periods.

A number of different studies can highlight the conservation value for marine organisms in the collection of blood or tissue. For example, projectile biopsies of blubber collected from North Pacific blue whales (*Balaenoptera musculus*) in the Gulf of California have been used to quantify stress condition and reproductive hormones and predict pregnancies (Atkinson et al. 2020). Casini et al. (2018) collected blood, skin, and scute samples from loggerhead turtles (*Caretta caretta*) to quantify contaminants, DNA fragmentation, and indices of oxidative stress that can aid in turtle conservation. Shultz et al. (2014) sampled a number of metabolites in blood, including ion concentrations, in a laboratory setting to quantify the physiological responses of several nearshore tropical fish species in response to anticipated changes in environmental conditions due to climate change, showing that increases in temperature were expected to be more challenging than alterations in salinity or pH. Baleen powder from bowhead whales (*Balaena mysticetus*) was analyzed to provide information on cortisol and progesterone concentrations that can integrate information on an animal's reproductive history or stress response (Hunt et al. 2014b), while Keogh et al. (2020) used lanugo (hair grown in utero) from pups of three different Pacific otariid species (Steller sea lions (*Eumetopias jubatus*), California sea lions (*Zalophus californianus*), and northern fur seals (*Callorhinus ursinus*)) to quantify concentrations of aldosterone, cortisol, corticosterone, and testosterone and highlight spatial differences in maternal stress. Tedeschi et al. (2015) used a laboratory setting to quantify the activity of heat shock proteins (Hsp60, Hsp70, and Hsp90) in heart and brain tissue in embryonic loggerhead sea turtles in an effort to define the response of this species to future heat waves, and heat shock proteins have been used to quantify the impacts of Arctic seawater freshening (due to increasing ice melt) on stress responses during heatwaves in intertidal bivalves (*Mytilus edulis*; Nielsen et al. 2021). Together, the collection of blood and tissues from marine organisms can allow the quantification of a number of different biomarkers that can provide insights into how different stressors can influence individuals and populations.

Omics

Quantifying single or several known physiological endpoints does not factor in the underlying complexity behind how a genotype produces a phenotype in response to environmental factors. Quantifying the comprehensive response using 'omics' approaches, rather than relying on single endpoints, allows researchers to start to examine that complexity as well as discover species-, population- and individual-specific responses. There are several omics technologies that have been developed that are now being applied to wildlife studies, which include genomic, transcriptomic, proteomic, metabolic, and epigenomic approaches. Arguably, the omics tools most commonly applied to wildlife studies are genomics and transcriptomics (e.g., Allendorf et al. 2010, Alvarez et al. 2015, Bernatchez 2016, Oomen and Hutchings 2017), which are studies at the level of DNA and RNA, respectively. Both approaches became increasingly more common in animal research due to the development of "next-generation sequencing", which significantly reduced the cost of DNA and RNA sequencing studies. The general methodology for DNA and RNA sequencing is similar between species, therefore sequencing studies have moved from being focused on model laboratory species to becoming valuable tools for studying marine wildlife and species of conservation concern.

From a mechanistic perspective, transcriptomics approaches (i.e., ultra high-throughput qPCR, microarrays, RNA-sequencing) provide an opportunity to assess the cellular response to environmental stressors in marine organisms by comparing changes in mRNA transcript abundance between individuals. There are too many examples of this approach being applied to marine wildlife to list, but some examples of transcriptomics being used to examine marine animals have focused on the causes of spawning migration mortality in sockeye salmon (Miller et al. 2011), the effects of elevated temperature on eastern oysters (e.g., Johnson et al. 2021) and the marine snail (*Chlorostoma funebralis*; Gleason and Burton 2015), the effects of temperature and CO_2 on larval sea urchins (*Strongylocentrotus purpuratus*; Padilla-Gamiño et al. 2013), effects of pesticides on sea lamprey (*Petromyzon marinus*; Yin et al. 2021), the differences between thermal tolerances of invasive and native fishes in estuaries (e.g., Jeffries et al. 2016, Komoroske et al. 2021), effects of domestication in hatcheries on anadromous steelhead trout (*Oncorhynchus mykiss*; Christie et al. 2016) and the relative salinity tolerances between populations of estuarine fishes (e.g., Whitehead et al. 2011, Jeffries et al. 2019). Some studies have used the RNA-sequencing reads to conduct genomics analyses between populations or individuals as an extension of the transcriptomics approach (e.g., Christie et al. 2016, Jeffries et al. 2019, Johnson et al. 2021, Yin et al. 2021). Furthermore, recent studies have used omics approaches to uncover the roles of an organism's microbiota in modulating the effects of stressors on marine organisms. For example, while eastern oysters (*Crassostrea virginica*) can physiologically cope with prolonged periods of low oxygen, the proliferation of endemic bacteria within the oysters under low oxygen conditions can increase mortality rates in these otherwise hypoxia-tolerant animals (Coffin et al. 2021). There are significant computational challenges to applying omics technology to wildlife; however, as this practice becomes more commonplace, these challenges will be addressed.

A significant benefit of omics approaches is that substantial amounts of data can be generated from relatively small samples, which makes this approach amenable for non-lethal sampling of animals and is becoming relatively common for genomics and transcriptomics studies (i.e., Jeffries et al. 2021). For example, transcriptomics approaches have been used to examine gene expression profiles of wild bottlenose dolphins (*Tursiops truncates*) using skin samples (Trego et al. 2019). Non-lethal blood sampling may be used for transcriptomics work for several species of sea turtle (Banerjee et al. 2011). Further, DNA sequencing has been used to examine migration patterns in tagged wild Arctic Char (*Salvelinus alpinus*; Moore et al. 2017). Although not currently used as frequently in marine wildlife studies, proteomic approaches can be conducted using blood samples to assess the health status of green turtles (*Chelonia mydas*; Marancik et al. 2021). An interesting non-lethal aging alternative to otoliths in fishes is using sequencing approaches to find DNA-methylation patterns that correlate with age in fishes (Mayne et al. 2021, Weber et al. 2021). These are just a few ways that researchers are applying omics technology to non-lethally study mechanisms in marine wildlife. Consequently, as omics tools and approaches continue to be refined or applied in novel ways, they will be valuable tools for studying marine conservation issues in wildlife in the future.

Application of tools to conservation problems

The toolbox described above provides marine scientists with diverse tools needed to reveal the ways in which different threats and stressors influence organismal conditions and behaviour in both the lab and the field. Here we present several case studies as exemplars of how some of these tools have been used to address specific problems and to inform management and conservation.

An integrated perspective on mollusc responses to climate change

Among the multitude of climate change stressors that can affect molluscs within the toolkit framework, ocean warming, acidification, and eutrophication/hypoxia are consequential stressors (Fig. 2). Chief among the effects of warming are impacts on metabolic rates in molluscs. Indeed, metabolic rates associated with warmer climates may serve as a useful mechanistic predictor of mollusc responses to climate change. At macroevolutionary scales, animals with lower whole-body metabolic rates appear to be less susceptible to climate-related extinction events than species with comparatively higher rates (Strotz et al. 2018), which may also explain why larger molluscs appear to be more susceptible to climate-related extinction than smaller ones (Lilliput Effect; Twitchett 2007, Morten and Twitchett 2009). Negative effects of warming on behaviour and energetic physiology can also affect mollusc ecophysiology and potentially translate to population level impacts (increased mortality; Clements et al. 2018, 2021). Ocean acidification can also affect mollusc physiology and effects on metabolic rates may be responsible for size differences in marine snails between sites with low and high seawater CO_2 concentrations (Garilli et al. 2015). Acidification can also affect the shell building capacity of molluscs, which has major implications for species' survival (Gazeau et al. 2013). Effects of acidification on sensory physiology and behaviour have been widely reported (Clements and Hunt 2015); however, recent evidence from both fish (Clark et al. 2020, Clements et al. 2022) and molluscs (Clements et al. 2020, 2021) suggests that behavioural effects may not be as universal or severe as previously thought. Understanding climate change effects on molluscs also requires a recognition of various complexities that can modulate responses. For example, the stressors mentioned above do not always occur in isolation, and the combined effects of multiple stressors can act synergistically, additively, or antagonistically to influence organismal responses (Waldbusser and Salisbury 2014). Likewise, differences in localized environmental variability can result in some populations exhibiting tolerance to climate change conditions and others being susceptible (Vargas et al. 2017, 2022). Along the same vein, the geospatial spread of marine molluscs—from subtidal to intertidal and epifaunal to infaunal—means that microclimatic conditions associated with different habitats can influence mollusc responses to climate change (Waldbusser and Salisbury 2014). Predictions of climate change responses of these animals under the proposed toolkit will assuredly need to account for these (and other) complexities.

Figure 2. Mediterranean mussels (*Mytilus galloprovincialis*) from the northwest coast of Spain that reside in coastal environments subject to climate change impacts. Photo Credit: Jeffrey Clements.

Stress physiology of marine mammals informs spatial planning

Traditionally, the management of marine mammals has focused on "lethal takes" (i.e., situations where animals die due to human interaction), including those associated with vessel collisions, strandings, and bycatch (Wright et al. 2011). However, along with outright mortality, human disturbance can also result in non-lethal impacts that have considerable conservation implications (Nelms et al. 2021). Noise pollution has become increasingly pervasive in marine environments due to commercial shipping, seismic surveys, and naval sonar (Nelms et al. 2021) and the consequences of noise for marine wildlife include disruptions to physiology and behaviour, as well as direct tissue damage (e.g., auditory damage) (Chahouri et al. 2021). For example, a decrease in noise (resulting from a decrease in shipping traffic) in the Bay of Fundy was associated with a decrease in baseline faecal GC levels in North Atlantic right whales (*Eubalaena glacialis*), indicating that low-frequency ship noise may be a source of chronic stress in this critically endangered species (Rolland et al. 2012; Fig. 3). It follows that physiological data can be useful in understanding how stressors impact populations of interest and how to limit exposures. Using an energy budget model that realistically incorporated environmental variability, animal movement, and energetic physiology, Gallagher et al. (2021) showed that harbour porpoises (*Phocoena phocoena*) would be most impacted by seismic surveys in late summer and fall due to lactation costs, water temperature, and levels of body fat. These results indicate that investigations determining how noise pollution will influence marine mammals should consider seasonal changes in energetics and movement. These findings are particularly pertinent for management decisions, such as the designation and extent of Marine Protected Areas (MPAs). Evidence of physiological effects in relation to certain levels of anthropogenic noise will be crucial to limiting disturbance within their boundaries (e.g., through the establishment of buffer zones) (Wright et al. 2011). Using the same strategies, these types of considerations will also help predict and plan for changing distributions of marine mammals under climate change (Silber et al. 2017).

Figure 3. North Atlantic right whale (*Eubalaena glacialis*) resides in areas commonly transited by large vessels leading to potential for collision. Photo Credit: Sean Brilliant, Canadian Wildlife Federation.

Sensory physiology to the rescue of sea turtles

Research on the sensory physiology of sea turtles has been highly effective at addressing two conservation problems - light pollution and bycatch. Coastal light pollution can disrupt nesting behaviour and misdirect adults to inland areas after nesting rather than allowing them to return to the ocean (Witherington 1992; Fig. 4). In addition, light is a confusing cue for newly hatched sea turtles whereby they move inland (towards light) rather than following celestial cues to orient to the ocean, leading to high mortality rates due to predation, dehydration, and starvation (Witherington and Martin 1996). Impacts on nesting and early life survival are highly problematic for sea turtles such that light pollution is recognized as a serious threat (Kamrowski et al. 2012). Extensive sensory physiology research has been used to identify light intensities (Cruz et al. 2018, Price et al. 2018) and colour (Silva et al. 2017), and bulb types (e.g., LED, mercury vapour; Silva et al. 2017) that are least likely to disrupt nesting or alter the behaviour of nesting adults and newly hatched turtles. These experiments have led to changes in ordinances that

Figure 4. Recently hatched sea turtles can be drawn inland rather than heading to sea as a result of light pollution. Photo Credit: Pixaby Licence, Free Public Domain.

restrict lighting, particularly during nesting and hatching periods, and informed outreach and awareness campaigns (Swindall et al. 2019). Sensory physiology can also be exploited for conservation benefits such as in the context of turtle bycatch (i.e., when turtles are captured on hooks or in nets intended to target other taxa). Sea turtle bycatch is regarded as a major conservation issue given that it frequently leads to mortality via injury or drowning (Finkbeiner et al. 2011, Lewison et al. 2014). Efforts to repel sea turtles from fishing gear therefore could be a useful strategy for reducing sea turtle bycatch (Swimmer and Brill 2006). Research has investigated the use of visual deterrents such as LED lights (Wang et al. 2010) and UV lights (Wang et al. 2013), sound deterrents (Southwood et al. 2008, Piniak 2012), and even chemical deterrents (Swimmer et al. 2006, Southwood et al. 2008). Interestingly, although some lights serve as deterrents for some species in some contexts, LED light sticks actually attracted loggerhead turtles to longlines (Wang et al. 2007, Swimmer et al. 2017). Efforts continue to develop effective bycatch reduction strategies informed by sensory physiology but more work is needed (Komoroske and Lewison 2015). This case study highlights how knowledge of sensory physiology can be used to understand threats facing marine wildlife while also identifying and testing mitigation measures (Elmer et al. 2021).

The future

Understanding the ecological, environmental and evolutionary physiology of marine wildlife has led to a novel understanding of how organisms interact with each other and their environment. We now understand the great physiological diversity that exists within and among populations and species (Spicer and Gaston 2009, Deutsch et al. 2020) as well as how organisms perform in diverse environments (Lewontin and Levins 1997). Such fundamental understanding is important and fascinating in its own right, yet given the manifold threats to marine wildlife and ecosystems, such knowledge has become even more timely and relevant. Physiological knowledge and concepts are increasingly recognized as being important to understanding threats and solving conservation problems (Cooke et al. 2013). This has been greatly enabled by the development of a toolbox for studying the physiology of marine wildlife. As outlined above, there now exists a wide range of tools that allow researchers to study various aspects of organism function from a mechanistic perspective that ranges from the gene to the cell to the whole organism, and that span biological systems (e.g., endocrine, cardiorespiratory). There have also been novel developments that allow for the study of animal behaviour at scales that were not imaginable a few decades ago (e.g., across ocean basins and at great depth; Harcourt et al. 2019). Because physiology and behaviour cannot be easily decoupled when studying marine wildlife (Cooke et al. 2014), developments in both fields have been synergistic.

Here we highlighted the suite of tools available for understanding mechanisms associated with threats facing marine wildlife (Cooke et al. 2023; Fig. 1). However, we did so from the position that identifying threats alone is insufficient. There is great urgency to develop and refine conservation and management solutions that are effective (Godet and Devictor 2018). Mechanistic understanding ensures that conservation and management efforts focus on the threats that matter most and in ways that have the potential to mitigate threats (Cooke and O'Connor 2010). This evidence-based approach has great promise. The several case studies that we presented above highlight how diverse tools and conceptual approaches have informed conservation solutions related to very different issues in varied taxa. Given that the threat landscape continues to evolve with the emergence of novel stressors along with the fact that climate change will serve as a threat multiplier (Moe et al. 2013), mechanistic approaches to studying marine wildlife will become even more important.

In the future, we foresee a number of developments related to the mechanistic study of marine wildlife. For example, we anticipate additional efforts to bring realism to studies with more *in situ* work occurring in the field enabled by tools such as biotelemetry (Hussey et al. 2015) and biologging (Chung et al. 2021). We also see a future with more large-scale experimentation that occurs at the scale of reefs or embayments to ensure greater management relevance. Highly controlled laboratory research will remain important but we also anticipate more complex experiments with multiple stressors to better understand the interactive nature of stressors (Folt et al. 1999). We also envision greater reliance on the use of mesocosms as an intermediary between field and lab studies. We anticipate continued efforts to develop sampling and study methods that are less (or non) invasive to benefit animal welfare (Zemanova 2020) and to enable research on rare or threatened species (Pauli et al. 2010). We also anticipate that different tools will be combined in interesting and novel ways to provide a more comprehensive understanding (e.g., combining biotelemetry and omics tools; Jeffries et al. 2021). The toolbox will certainly continue to evolve and grow, supported by developments in other fields such as human medicine and veterinary science. For example, innovations in point-of-care devices for medicine have been of great benefit to those studying marine wildlife (Stoot et al. 2014). Key to all of this is the need for creativity and bold ideas. There will always be challenges with studying the physiology and behaviour of marine wildlife given the diversity of body size, morphology, environmental requirements, and so on. However, as outlined here, the toolbox for doing so is robust and there are examples of not only generating a greater understanding of marine wildlife but developing conservation strategies that work and will help to rebuild marine wildlife populations.

Acknowledgements

Cooke, Madliger and Jeffries are supported by NSERC and GenFish (Genome Canada). Madliger was further supported by FishCAST, an NSERC-CREATE funded program. Michael Dusevic kindly formatted the references. We are thankful to Cynthia Awruch for providing thoughtful feedback on our chapter.

References

Acquarone, M., Born, E. W. and Speakman, J. R. (2006). Field metabolic rates of walrus (Odobenus rosmarus) measured by the doubly labeled water method. Aquatic Mammals 32: 363.

Ajó, A. A. F., Hunt, K. E., Giese, A. C., Sironi, M., Uhart, M., Rowntree, V. J. et al. (2020). Retrospective analysis of the lifetime endocrine response of southern right whale calves to gull wounding and harassment: A baleen hormone approach. General and Comparative Endocrinology 296: 113536.

Allendorf, F.W., Hohenlohe, P. A. and Luikart, G. (2010). Genomics and the future of conservation genetics. Nature Reviews Genetics 11: 697–709.

Alongi, D. M. (2002). Present state and future of the world's mangrove forests. Environmental Conservation 29: 331–349.

Alvarez, M., Schrey, A. W. and Richards, C. L. (2015). Ten years of transcriptomics in wild populations: what have we learned about their ecology and evolution? Molecular Ecology 24: 710–725.

Anderson, B., Belcher, C., Slack, J. and Gelsleichter, J. (2018). Evaluation of the use of portable ultrasonography to determine pregnancy status and fecundity in bonnethead shark Sphyrna tiburo. Journal of Fish Biology 93: 1163–1170.

Andrade. H., Massabuau, J.-C., Cochrane, S., Ciret, P., Tran, D., Sow, M. et al. (2016). High frequency non-invasive (HFNI) bio-sensors as a potential tool for marine monitoring and assessments. Front. Mar. Sci. 3: 187.

Angelier, F. and Wingfield, J. C. (2013). Importance of the glucocorticoid stress response in a changing world: theory, hypotheses and perspectives. General and Comparative Endocrinology 190: 118–128.

Anttila, K., Couturier, C. S., Øverli, Ø., Johnsen, A., Marthinsen, G., Nilsson, G. E. et al. (2014). Atlantic salmon show capability for cardiac acclimation to warm temperatures. Nature Communications 5: 1–6.

Atkinson, S., Crocker, D., Houser, D. and Mashburn, K. (2015). Stress physiology in marine mammals: how well do they fit the terrestrial model?. Journal of Comparative Physiology B 185: 463–486.

Atkinson, S., Gendron, D., Branch, T. A., Mashburn, K. L., Melica, V., Enriquez-Paredes, L. E. et al. (2020). Pregnancy rate and biomarker validations from the blubber of eastern North Pacific blue whales. Marine Mammal Science 36: 6–28.

Baktoft, H., Gjelland, K. Ø., Økland, F. and Thygesen, U. H. (2017). Positioning of aquatic animals based on time-of-arrival and random walk models using YAPS (Yet Another Positioning Solver). Scientific Reports 7(1): 1–10.

Banerjee, S. M., Stoll, J. A., Allen, C. D., Lynch, J. M., Harris, H. S., Kenyon, L. et al. (2011). Species and population specific gene expression in blood transcriptomes of marine turtles. BMC Genomics 22: 346.

Beman, J. M., Arrigo, K. R. and Matson, P. A. (2005). Agricultural runoff fuels large phytoplankton blooms in vulnerable areas of the ocean. Nature 434: 211–214.

Benítez, S., Lagos, N. A., Osores, S., Opitz, T., Duarte, C., Navarro, J. M. and Lardies, M. A. (2018). High pCO_2 levels affect metabolic rate, but not feeding behavior and fitness, of farmed giant mussel Choromytilus chorus. Aquaculture Environment Interactions 10: 267–278.

Bernatchez, L. (2016). On the maintenance of genetic variation and adaptation to environmental change: considerations from population genomics in fishes. J. Fish. Biol. 89: 2519–2556.

Bernet, D., Schmidt, H., Meier, W., Burkhardt-Holm, P. and Wahli, T. (1999). Histopathology in fish: proposal for a protocol to assess aquatic pollution. Journal of Fish Diseases 22: 25–34.

Blank, J. M., Morrissette, J. M., Landeira-Fernandez, A. M., Blackwell, S. B., Williams, T. D. and Block, B. A. (2004). *In situ* cardiac performance of Pacific bluefin tuna hearts in response to acute temperature change. Journal of Experimental Biology 207: 881–890.

Blaxter, J. H. (1988). Sensory performance, behavior, and ecology of fish. pp. 203–232. *In*: Sensory Biology of Aquatic Animals. Springer, New York, NY.

Block, B. A. (2005). Physiological ecology in the 21st century: advancements in biologging science. Integrative and Comparative Biology 45: 305–320.

Bœuf, G. and Payan, P. (2001). How should salinity influence fish growth? Comparative Biochemistry and Physiology Part C: Toxicology & Pharmacology 130: 411–423.

Brette, F., Shiels, H. A., Galli, G. L., Cros, C., Incardona, J. P., Scholz, N. L. et al. (2017). A novel cardiotoxic mechanism for a pervasive global pollutant. Scientific Reports 7: 1–9.

Briceño, F. A., Polymeropoulos, E. T., Fitzgibbon, Q, P., Dambacher, J. M. and Pecl, G. T. (2018). Changes in metabolic rate of spiny lobster under predation risk. Marine Ecology Progress Series 598: 71–84.

Brisson-Curadeau, É., Handrich, Y., Elliott, K. H. and Bost, C. A. (2021). Accelerometry predicts prey-capture rates in the deep-diving king penguin Aptenodytes patagonicus. Marine Biology 168: 1–10.

Brooker, R. M., Feeney, W. E., White, J. R., Manassa, R. P., Johansen, J. L. and Dixson, D. L. (2016). Using insights from animal behaviour and behavioural ecology to inform marine conservation initiatives. Animal Behaviour 120: 211–221.

Brown, A. C. and Terwilliger, N. B. (1999). Developmental changes in oxygen uptake in Cancer magister (Dana) in response to changes in salinity and temperature. Journal of Experimental Marine Biology and Ecology 241: 179–192.

Burnett, N. P., Seabra, R., de Pirro, M., Wethey, D. S., Woodin, S. A., Helmuth, B. et al. (2013). An improved noninvasive method for measuring heartbeat of intertidal animals. Limnology and Oceanography: Methods 11: 91–100.

Byrnes, E. E., Lear, K. O., Morgan, D. L. and Gleiss, A. C. (2020). Respirometer in a box: development and use of a portable field respirometer for estimating oxygen consumption of large-bodied fishes. Journal of Fish Biology 96: 1045–1050.

Calow, P. (1989). Proximate and ultimate responses to stress in biological systems. Biological Journal of the Linnean Society 37(1-2): 173–181.

Camacho-Sanchez, M., Burraco, P., Gomez-Mestre, I. and Leonard, J. A. (2013). Preservation of RNA and DNA from mammal samples under field conditions. Molecular Ecology Resources 13: 663–673.

Casini, S., Caliani, I., Giannetti, M., Marsili, L., Maltese, S., Coppola, D. et al. (2018). First ecotoxicological assessment of Caretta caretta (Linnaeus, 1758) in the Mediterranean Sea using an integrated nondestructive protocol. Science of the Total Environment 631: 1221–1233.

Cech Jr, J. J. (1990). Respirometry. pp. 335–362. *In*: Methods for Fish Biology.

Chabot, D., Steffensen, J. F. and Farrell, A. P. (2016). The determination of standard metabolic rate in fishes. Journal of Fish Biology 88: 81–121.

Chahouri, A., Elouahmani, N. and Ouchene, H. (2021). Recent progress in marine noise pollution: A thorough review. Chemosphere 291: 132983.

Chancellor, S., Abbo, L., Grasse, B., Sakmar, T., Brown, J. S., Scheel, D. et al. (2021). Determining the effectiveness of using dermal swabs to evaluate the stress physiology of laboratory cephalopods: A preliminary investigation. General and Comparative Endocrinology 314: 113903.

Christie, M. R., Marine, M. L., Fox, S. E., French, R. A. and Blouin, M. S. (2016). A single generation of domestication heritably alters the expression of hundreds of genes. Nature Communications 7: 1–6.

Chung, H.,. Lee, J. and Lee, W. Y. (2021). A review: marine bio-logging of animal behaviour and ocean environments. Ocean Science Journal 56: 117–131.

Claiborne, J. B., Edwards, S. L. and Morrison-Shetlar, A. I. (2002). Acid–base regulation in fishes: cellular and molecular mechanisms. Journal of Experimental Zoology 293: 302–319.

Claireaux, G., Webber, D., Kerr, S. and Boutilier, R. (1995). Physiology and behaviour of free-swimming Atlantic cod (Gadus morhua) facing fluctuating temperature conditions. The Journal of Experimental Biology 198: 49–60.

Clark, T. D., Raby, G. D., Roche, D. G., Binning, S. A., Speers-Roesch, B., Jutfelt, F. et al. (2020). Ocean acidification does not impair the behaviour of coral reef fishes. Nature 577: 370–375.

Clark, T. D., Sandblom, E. and Jutfelt, F. (2013). Aerobic scope measurements of fishes in an era of climate change: respirometry, relevance and recommendations. Journal of Experimental Biology. 216: 2771-2782.

Clements, J. C. and Hunt, H. L. (2015). Marine animal behaviour in a high CO_2 ocean. Mar Ecol Prog Ser. 536: 259–279.

Clements, J. C., Hicks, C., Tremblay, R. and Comeau, L. A. (2018). Elevated seawater temperature, not pCO_2, negatively affects post-spawning adult mussels (*Mytilus edulis*) under food limitation. Cons Physiol. 6.

Clements, J. C. and Comeau, L. A. (2019). Use of high-frequency noninvasive electromagnetic biosensors to detect ocean acidification effects on shellfish behavior. J. Shellfish Res. 38: 811–818.

Clements, J. C., Poirier, L. A., Pérez, F. F., Comeau, L. A. and Babarro, J. M. F. (2020). Behavioural responses to predators in Mediterranean mussels (*Mytilus galloprovincialis*) are unaffected by elevated pCO_2. Mar. Environ. Res. 161: 105148.

Clements, J. C., Ramesh, K., Nysveen, J., Dupont, S. and Jutfelt, F. (2021). Animal size and sea water temperature, but not pH, influence a repeatable startle response behaviour in a wide-ranging marine mollusc. Anim Behav. 173: 191–205.

Clements, J. C., Sundin, J., Clark, T. D. and Jutfelt, F. (2022). Meta-analysis reveals an extreme "decline effect" in the impacts of ocean acidification on fish behavior. PLoS Biology 20.

Coffin, M., Clements, J., Comeau, L., Guyondet, T., Maillet, M., Steeves, L. et al. (2021). The killer within: Endogenous bacteria accelerate oyster mortality during sustained anoxia. Limnol. Oceanogr. 66: 2885–2900.

Cooke, S. J. and O'Connor, C. M. (2010). Making conservation physiology relevant to policy makers and conservation practitioners. Conservation Letters 3(3): 159–166.

Cooke, S. J., Blumstein, D. T., Buchholz, R., Caro, T., Fernandez-Juricic, E., Franklin, C. E. et al. (2014). Physiology, behavior, and conservation. Physiological and Biochemical Zoology 87(1): 1–14.

Cooke, S. J., Sack, L., Franklin, C. E., Farrell, A. P., Beardall, J., Wikelski, M. et al. (2013). What is conservation physiology? Perspectives on an increasingly integrated and essential science. Conservation Physiology 1(1).

Cooke, S. J., Brownscombe, J. W., Raby, G. D., Broell, F., Hinch, S. G., Clark, T. D. et al. (2016). Remote bioenergetics measurements in wild fish: opportunities and challenges. Comparative Biochemistry and Physiology Part A: Molecular & Integrative Physiology 202: 23–37.

Cooke, S. J., Madliger, C. L., Lennox, R. J., Olden, J. D., Eliason, E. J., Cramp, R. L. et al. (2023). Biological mechanisms matter in contemporary wildlife conservation. iScience 26: 106192.

Costa, D. P. and Sinervo, B. (2004). Field physiology: physiological insights from animals in nature. Annu. Rev. Physiol. 66: 209–238.

Crain, C. M., Kroeker, K. and Halpern, B. S. (2008). Interactive and cumulative effects of multiple human stressors in marine systems. Ecology Letters 11: 1304–1315.

Cruz, L. M., Shillinger, G. L., Robinson, N. J., Tomillo, P. S. and Paladino, F. V. (2018). Effect of light intensity and wavelength on the in-water orientation of olive ridley turtle hatchlings. Journal of Experimental Marine Biology and Ecology 505: 52–56.

Daly, J. and Jones, R. (2017). The use of reproductive technologies in breeding programs for elasmobranchs in aquaria. The Elasmobranch Husbandry Manual II: Recent Advances in the Care of Sharks, Rays and their Relatives. Columbus, Ohio: Special Publication of the Ohio Biological Survey 363–374.

Dantzer, B., Fletcher, Q. E., Boonstra, R. and Sheriff, M. J. (2014). Measures of physiological stress: a transparent or opaque window into the status, management and conservation of species?. Conservation Physiology 2.

Davies, R. W. D., Cripps, S. J., Nickson, A. and Porter, G. (2009). Defining and estimating global marine fisheries bycatch. Marine Policy 33: 661–672.

Davies, T. W., Duffy, J. P., Bennie, J. and Gaston, K. J. (2014). The nature, extent, and ecological implications of marine light pollution. Frontiers in Ecology and the Environment 12: 347–355.

Dayton, P. K., Thrush, S. F., Agardy, M. T. and Hofman, R. J. (1995). Environmental effects of marine fishing. Aquatic conservation: marine and freshwater ecosystems. 5: 205–232.

de Sylva, D. P. and Breder, P. R. (1997). Reproduction, gonad histology, and spawning cycles of north Atlantic billfishes (Istiophoridae). Bulletin of Marine Science 60: 668–697.

Deutsch, C., Penn, J. L. and Seibel, B. (2020). Metabolic trait diversity shapes marine biogeography. Nature 585: 557–562.

Dickens, M. J. and Romero, L. M. (2013). A consensus endocrine profile for chronically stressed wild animals does not exist. General and Comparative Endocrinology 191: 177–189.

Dingemanse, N. J. and Dochtermann, N. A. (2013). Quantifying individual variation in behaviour: mixed-effect modelling approaches. Journal of Animal Ecology 82: 39–54.

Dobzhansky, T. (1964). Biology, molecular and organismic. American Zoologist 4: 443–452.

Dominoni, D. M., Halfwerk, W., Baird, E., Buxton, R. T., Fernández-Juricic, E., Fristrup, K. M. et al. (2020). Why conservation biology can benefit from sensory ecology. Nature Ecology & Evolution 4: 502–511.

Duarte, C. M., Agusti, S., Barbier, E., Britten, G. L., Castilla, J. C., Gattuso, J. P. et al. (2020). Rebuilding marine life. Nature 580: 39–51.

Eichhorn, G., Enstipp, M. R., Georges, J. Y., Hasselquist, D. and Nolet, B. A. (2019). Resting metabolic rate in migratory and non-migratory geese following range expansion: go south, go low. Oikos 128: 1424–1434.

Eliason, E. J., Clark, T. D., Hague, M. J., Hanson, L. M., Gallagher, Z. S., Jeffries, K. M. et al. (2011). Differences in thermal tolerance among sockeye salmon populations. Science 332(6025): 109–112.

Elmer, L. K., Madliger, C. L., Blumstein, D. T., Elvidge, C. K., Fernández-Juricic, E., Horodysky, A. Z. et al. (2021). Exploiting common senses: sensory ecology meets wildlife conservation and management. Conservation Physiology 9.

Esbaugh, A. J., Mager, E. M., Stieglitz, J. D., Hoenig, R., Brown, T. L., French, B. L. et al. (2016). The effects of weathering and chemical dispersion on Deepwater Horizon crude oil toxicity to mahi-mahi (Coryphaena hippurus) early life stages. Science of the Total Environment 543: 644–651.

Farrell, A. P., Small. S. and Graham, M. S. (1989). Effect of heart rate and hypoxia on the performance of a perfused trout heart. Canadian Journal of Zoology 67: 274–280.

Farrell, A., Franklin, C., Arthur, P., Thorarensen, H. and Cousins, K. (1994). Mechanical performance of an in situ perfused heart from the turtle Chrysemys scripta during normoxia and anoxia at 5 C and 15 C. The Journal of Experimental Biology 191: 207–229.

Farrell, A. P. (2002). Coronary arteriosclerosis in salmon: growing old or growing fast?. Comparative Biochemistry and Physiology Part A: Molecular & Integrative Physiology 132: 723–735.

Farrell, A. P. (2016). Pragmatic perspective on aerobic scope: peaking, plummeting, pejus and apportioning. Journal of Fish Biology 88: 322–343.

Finkbeiner, E. M., Wallace, B. P., Moore, J. E., Lewison, R. L., Crowder, L. B. and Read, A. J. (2011). Cumulative estimates of sea turtle bycatch and mortality in USA fisheries between 1990 and 2007. Biological Conservation 144: 2719–2727.

Finlayson, K. A., Leusch, F. D. and van de Merwe, J. P. (2016). The current state and future directions of marine turtle toxicology research. Environment International 94: 113–123.

Folt, C. L., Chen, C. Y., Moore, M. V. and Burnaford, J. (1999). Synergism and antagonism among multiple stressors. Limnology and Oceanography 44: 864–877.

Franklin, D. L., Watson, N. W. and Van Citters, R. L. (1964). Blood velocity telemetered from untethered animals. pp. 528–530. *In*: Nature, 203.

French, S. S., DeNardo, D. F., Greives, T. J., Strand, C. R. and Demas, G. E. (2010). Human disturbance alters endocrine and immune responses in the Galapagos marine iguana (Amblyrhynchus cristatus). Hormones and Behavior 58: 792–799.

Gales, R. and Green, B. (1990). The annual energetics cycle of little penguins (Eudyptula minor). Ecology 71: 2297–2312.

Gallagher, C. A., Grimm, V., Kyhn, L. A., Kinze, C. C. and Nabe-Nielsen, J. (2021). Movement and seasonal energetics mediate vulnerability to disturbance in marine mammal populations. The American Naturalist 197: 296–311.

Garilli, V., Rodolfo-Metalpa, R., Scuderi, D., Brusca, L., Parrinello, D., Rastrick, S. et al. (2015). Physiological advantages of dwarfing in surviving extinctions in high-CO_2 oceans. Nature Clim. Change 5: 678–682.

Gazeau, F., Parker, L. M., Comeau, S., Gattuso, J-P., O'Connor, W. A., Martin, S. et al. (2013). Impacts of ocean acidification on marine shelled molluscs. Mar Biol. 160: 2207–2245.

Gleason, L. U. and Burton, R. S. (2015). RNA-seq reveals regional differences in transcriptome response to heat stress in the marine snail *Chlorostoma funebralis*. Molecular Ecology 24: 610–627.

Gnaiger, E., Shick, J. M. and Widdows, J. (1989). Metabolic microcalorimetry and respirometry of aquatic animals. pp. 113–135. *In*: Techniques in Comparative Respiratory Physiology: An Experimental Approach. Cambridge University Press, Cambridge.

Godet, L. and Devictor, V. (2018). What conservation does. Trends in Ecology & Evolution 33: 720–730.

Gormally, B. M. and Romero, L. M. (2020). What are you actually measuring? A review of techniques that integrate the stress response on distinct time-scales. Functional Ecology 34: 2030–2044.

Gräns, A., Axelsson, M., Pitsillides, K., Olsson, C., Höjesjö, J., Kaufman, R. C. et al. (2009). A fully implantable multi-channel biotelemetry system for measurement of blood flow and temperature: a first evaluation in the green sturgeon. Hydrobiologia 619: 11–25.

Gray, J. S. (1997). Marine biodiversity: patterns, threats and conservation needs. Biodiversity & Conservation 6: 153–175.

Green, J. A., Haulena, M., Boyd, I. L., Calkins, D., Gulland, F., Woakes, A. J. et al. (2009). Trial implantation of heart rate data loggers in pinnipeds. The Journal of Wildlife Management 73(1): 115–121.

Guinotte, J. M. and Fabry, V. J. (2008). Ocean acidification and its potential effects on marine ecosystems. Annals of the New York Academy of Sciences 1134: 320–342.

Gurr, S. J., Goleski, J., Lima, F. P., Seabra, R., Gobler, C. J. and Volkenborn, N. (2018). Cardiac responses of the bay scallop Argopecten irradians to diel-cycling hypoxia. Journal of Experimental Marine Biology and Ecology 500: 18–29.

Gurr, S. J., Dwyer, I. P., Goleski, J., Lima, F. P., Seabra, R., Gobler, C. J. et al. (2021). Acclimatization in the bay scallop Argopecten irradians along a eutrophication gradient: insights from heartbeat rate measurements during a simulated hypoxic event. Marine and Freshwater Behaviour and Physiology 54: 23–49.

Halpern, B. S., Selkoe, K. A., Micheli, F. and Kappel, C. V. (2007). Evaluating and ranking the vulnerability of global marine ecosystems to anthropogenic threats. Conservation Biology 21: 1301–1315.

Hanson, L. M. and Farrell, A. P. (2007). The hypoxic threshold for maximum cardiac performance in rainbow trout Oncorhynchus mykiss (Walbaum) during simulated exercise conditions at 18 C. Journal of Fish Biology 71: 926–932.

Harcourt, R., Sequeira, A. M., Zhang, X., Roquet, F., Komatsu, K., Heupel, M. et al. (2019). Animal-borne telemetry: an integral component of the ocean observing toolkit. Frontiers in Marine Science 6: 1–21.

Haward, M. (2018). Plastic pollution of the world's seas and oceans as a contemporary challenge in ocean governance. Nature Communications 9: 1–3.

Hazin, H. G., Hazin, F. H., Travassos, P. and Erzini, K. (2005). Effect of light-sticks and electralume attractors on surface-longline catches of swordfish (Xiphias gladius, Linnaeus, 1959) in the southwest equatorial Atlantic. Fisheries Research 72: 271–277.

Heuer, R. M., Galli, G. L., Shiels, H. A., Fieber, L. A., Cox, G. K., Mager, E. M. et al. (2019). Impacts of deepwater horizon crude oil on mahi-mahi (Coryphaena hippurus) heart cell function. Environmental Science & Technology 53: 9895–9904.

Hoegh-Guldberg, O. (1999). Climate change, coral bleaching and the future of the world's coral reefs. Marine and Freshwater Research 50: 839–866.

Hooker, S. K., Fahlman, A., Moore, M. J., Aguilar De Soto, N., Bernaldo de Quirós, Y., Brubakk, A. O. et al. (2012). Deadly diving? Physiological and behavioural management of decompression stress in diving mammals. Proceedings of the Royal Society B: Biological Sciences 279: 1041–1050.

Hruska, K. A., Hinch, S. G., Patterson, D. A. and Healey, M. C. (2011). Egg retention in relation to arrival timing and reproductive longevity in female sockeye salmon (Oncorhynchus nerka). Canadian Journal of Fisheries and Aquatic Sciences 68: 250–259.

Humphreys, W. F. (1979). Production and respiration in animal populations. The Journal of Animal Ecology 48: 427–453.

Hunt, K. E., Moore, M. J., Rolland, R. M., Kellar, N. M., Hall, A. J., Kershaw, J. et al. (2013). Overcoming the challenges of studying conservation physiology in large whales: a review of available methods. Conservation Physiology 1.

Hunt, K. E., Rolland, R. M. and Kraus, S. D. (2014a). Detection of steroid and thyroid hormones via immunoassay of North Atlantic right whale (Eubalaena glacialis) respiratory vapor. Marine Mammal Science 30: 796–809.

Hunt, K. E., Stimmelmayr, R., George, C., Hanns, C., Suydam, R., Brower, H. et al. (2014b). Baleen hormones: a novel tool for retrospective assessment of stress and reproduction in bowhead whales (Balaena mysticetus). Conservation Physiology 2.

Hussey, N. E., Kessel, S. T., Aarestrup, K., Cooke, S. J., Cowley, P. D., Fisk, A. T. et al. (2015). Aquatic animal telemetry: a panoramic window into the underwater world. Science 348.

Huveneers, C., Whitmarsh, S., Thiele, M., Meyer, L., Fox, A. and Bradshaw, C. J. A. (2018). Effectiveness of five personal shark-bite deterrents for surfers. PeerJ 6: e5554–e5554.

Incardona, J. P., Gardner, L. D., Linbo, T. L., Brown, T. L., Esbaugh, A. J., Mager, E. M. et al. (2014). Deepwater Horizon crude oil impacts the developing hearts of large predatory pelagic fish. Proceedings of the National Academy of Sciences 111(15): E1510–E1518.

Jackson, J. B., Kirby, M. X., Berger, W. H., Bjorndal, K. A., Botsford, L. W., Bourque, B. J. et al. (2001). Historical overfishing and the recent collapse of coastal ecosystems. Science 293: 629–637.

Jacobs, S. R. and Terhune, J. M. (2002). The effectiveness of acoustic harassment devices in the Bay of Fundy, Canada: seal reactions and a noise exposure model. Aquatic Mammals 28: 147–158.

Jeffries, K. M., Connon, R. E., Davis, B. E., Komoroske, L. M., Britton, M. T., Sommer, T. et al. (2016). Effects of high temperatures on threatened estuarine fishes during periods of extreme drought. Journal of Experimental Biology 219: 1705–16.

Jeffries, K. M., Connon, R. E., Verhille, C. E., Dabruzzi, T. F., Britton, M. T., Durbin-Johnson, B. P. et al. (2019). Divergent transcriptomic signatures in response to salinity exposure in two populations of an estuarine fish. Evolutionary Applications 12: 1212–1226.

Jeffries, K. M., Teffer, A., Michaleski, S., Bernier, N. J., Heath, D. D. and Miller, K. M. (2021). The use of non-lethal sampling for transcriptomics to assess the physiological status of wild fishes. Comparative Biochemistry and Physiology Part B: Biochemistry and Molecular Biology 110629.

Jirik, K. and Lowe, C. (2012). An elasmobranch maternity ward: female round stingrays Urobatis halleri use warm, restored estuarine habitat during gestation. Journal of Fish Biology 80: 1227–1245.

Johnson, K. M., Jones, H. R., Casas, S. M., La Peyre, J. F. and Kelly, M. W. (2021). Transcriptomic signatures of temperature adaptation in the eastern oyster *Crassostrea virginica*. Journal of Evolutionary Biology 34: 1212–1224.

Jones, T. T., Hastings, M. D., Bostrom, B. L., Andrews, R. D. and Jones, D. R. (2009). Validation of the use of doubly labeled water for estimating metabolic rate in the green turtle (Chelonia mydas L.): a word of caution. Journal of Experimental Biology 212: 2635–2644.

Jordan, L. K., Mandelman, J. W., McComb, D. M., Fordham, S. V., Carlson, J. K. and Werner, T. B. (2013). Linking sensory biology and fisheries bycatch reduction in elasmobranch fishes: a review with new directions for research. Conservation Physiology 1.

Jørgensen, S. E. and Nielsen, S. N. (2013). The properties of the ecological hierarchy and their application as ecological indicators. Ecological Indicators. Elsevier Ltd. 28: 48–53.

Kamrowski, R. L., Limpus, C., Moloney, J. and Hamann, M. (2012). Coastal light pollution and marine turtles: assessing the magnitude of the problem. Endangered Species Research 19: 85–98.

Keen, A. N., Fenna, A. J., McConnell, J. C., Sherratt, M. J., Gardner, P. and Shiels, H. A. (2016). The dynamic nature of hypertrophic and fibrotic remodeling of the fish ventricle. Frontiers in Physiology 6: 427.

Kennish, M. J. (2002). Environmental threats and environmental future of estuaries. Environmental Conservation 29: 78–107.

Kennish, M. J. (2017). Practical Handbook of Estuarine and Marine Pollution. CRC Press.

Koehn, R. K. and Bayne, B. L. (1989). Towards a physiological and genetical understanding of the energetics of the stress response. Biological Journal of the Linnean Society 37: 157–171.

Keogh, M. J., Gastaldi, A., Charapata, P., Melin, S. and Fadely, B. S. (2020). Stress-related and reproductive hormones in hair from three north Pacific otariid species: Steller sea lions, California sea lions and northern fur seals. Conservation Physiology 8.

Komoroske, L. M. and Lewison, R. L. (2015). Addressing fisheries bycatch in a changing world. Frontiers in Marine Science 2: 83.

Komoroske, L. M., Jeffries, K. M., Whitehead, A., Roach, J. L., Britton, M., Connon, R. E. et al. (2021). Transcriptional flexibility during thermal challenge corresponds with expanded thermal tolerance in an invasive compared to native fish. Evolutionary Applications 14: 931–949.

Kooijman, B. and Kooijman, S. A. L. M. (2010). Dynamic Energy Budget Theory for Metabolic Organisation. Cambridge University Press.

Koschinski, S. (2011). Underwater noise pollution from munitions clearance and disposal, possible effects on marine vertebrates, and its mitigation. Marine Technology Society Journal 45: 80–88.

Lagos, L., Herrera, M., Sánchez-Lazo, C. and Martínez-Pita, I. (2013). Effect of larval stocking density on growth, survival and whole body cortisol of the Mediterranean mussel Mytilus galloprovincialis (Lamarck, 1819) larvae reared under laboratory conditions. Aquaculture Research 46: 1648–1656.

Lamb, J. S., O'Reilly, K. M. and Jodice, P. G. (2016). Physical condition and stress levels during early development reflect feeding rates and predict pre-and post-fledging survival in a nearshore seabird. Conservation Physiology 4.

Lefevre, S., Bayley, M. and McKenzie, D. J. (2016). Measuring oxygen uptake in fishes with bimodal respiration. Journal of Fish Biology 88: 206–231.

Lewison, R. L., Crowder, L. B., Wallace, B. P., Moore, J. E., Cox, T., Zydelis, R. et al. (2014). Global patterns of marine mammal, seabird, and sea turtle bycatch reveal taxa-specific and cumulative megafauna hotspots. Proceedings of the National Academy of Sciences 111: 5271–5276.

Lewontin, R. and Levins, R. (1997). Organism and environment. Capitalism Nature Socialism 8: 95–98.

Liley, N. R. and Stacey, N. E. (1983). 1 Hormones, pheromones, and reproductive behavior in fish. Fish Physiology 9: 1–63.

Løkkeborg, S. and Johannessen, T. (1992). The importance of chemical stimuli in bait fishing—fishing trials with presoaked bait. Fisheries Research 14: 21–29.

Lovett. B. (2015). Water-quality canaries: using mussels to detect contaminants. *In*: Treatment Plant Operator Magazine.

Luypaert, T., Hagan, J. G., McCarthy, M. L. and Poti, M. (2020). Status of marine biodiversity in the Anthropocene. pp. 57–82. *In*: YOUMARES 9-The Oceans: Our Research, Our Future. Springer, Cham.

MacDougall-Shackleton, S. A., Bonier, F., Romero, L. M. and Moore, I. T. (2019). Glucocorticoids and "stress" are not synonymous. Integrative Organismal Biology. 1.

Madliger, C. L., Love, O. P., Hultine, K. R. and Cooke, S. J. (2018). The conservation physiology toolbox: status and opportunities. Conservation Physiology 6.

Mallatt, J. (1985). Fish gill structural changes induced by toxicants and other irritants: a statistical review. Canadian Journal of Fisheries and Aquatic Sciences 42: 630–648.

Mallory, M. L., Robinson, S. A., Hebert, C. E. and Forbes, M. R. (2010). Seabirds as indicators of aquatic ecosystem conditions: a case for gathering multiple proxies of seabird health. Marine Pollution Bulletin 60: 7–12.

Marancik, D. P., Perrault, J. R., Komoroske, L. M., Stoll, J. A., Kelley, K. N. and Manire, C. A. (2021). Plasma proteomics of green turtles (*Chelonia mydas*) reveals pathway shifts and potential biomarker candidates associated with health and disease. Conservation Physiology 9.

Mayne, B., Espinoza, T., Roberts, D., Butler, G. L., Brooks, S., Korbie, D. et al. (2021). Nonlethal age estimation of three threatened fish species using DNA methylation: Australian lungfish, Murray cod and Mary River cod. Mol. Ecol. Resour. 21: 2324–2332.

McCallum, H., Harvell, D. and Dobson, A. (2003). Rates of spread of marine pathogens. Ecology Letters 6: 1062–1067.

McGaw, I. J., Steell, S. C., Leeuwen, T. E. V., Eliason, E. J. and Cooke, S. J. (2018). Application of miniature heart rate data loggers for use in large free-moving decapod crustaceans: method development and validation. Physiological and Biochemical Zoology 91: 731–739.

McKenzie, D. J., Axelsson, M., Chabot, D., Claireaux, G., Cooke, S. J., Corner, R. A. et al. (2016). Conservation physiology of marine fishes: state of the art and prospects for policy. Conservation Physiology 4.

Miller, K. M., Li, S., Kaukinen, K. H., Ginther, N., Hammill, E., Curtis, J. M. et al. (2011). Genomic signatures predict migration and spawning failure in wild Canadian salmon. Science 331: 214–217.

Moe, S. J., De Schamphelaere, K., Clements, W. H., Sorensen, M. T., Van den Brink, P. J. and Liess, M. (2013). Combined and interactive effects of global climate change and toxicants on populations and communities. Environmental Toxicology and Chemistry 32: 49–61.

Molnar, J. L., Gamboa, R. L., Revenga, C. and Spalding, M. D. (2008). Assessing the global threat of invasive species to marine biodiversity. Frontiers in Ecology and the Environment 6: 485–492.

Moore, J. S., Harris, L. N., Le Luyer, J., Sutherland, B. J., Rougemont, Q., Tallman, R. F. et al. (2017). Genomics and telemetry suggest a role for migration harshness in determining overwintering habitat choice, but not gene flow, in anadromous Arctic Char. Molecular Ecology 26: 6784–6800.

Morten, S. and Twitchett, R. (2009). Fluctuations in the body size of marine invertebrates through the Pliensbachian–Toarcian extinction event. Palaeogeogr Palaeoclimatol Palaeoecol. 284: 29–38.

Müller, A., Österlund, H., Marsalek J. and Viklander, M. (2020). The pollution conveyed by urban runoff: a review of sources. Science of the Total Environment 709: 136125.

Nabe-Nielsen, J., van Beest, F. M., Grimm, V., Sibly, R. M., Teilmann, J. and Thompson, P. M. (2018). Predicting the impacts of anthropogenic disturbances on marine populations. Conservation Letters 11: e12563.

Nagy, K. A. (1983). The Doubly Labeled Water (3HH18O) Method: A Guide to Its Use. Los Angeles, CA: University of California Publ. 64: 12–1417.

Nagy, K. A. (1989). Doubly-labeled water studies of vertebrate physiological ecology. pp. 270–287. *In*: Stable Isotopes in Ecological Research. Springer, New York, NY.

Nelms, S. E., Alfaro-Shigueto, J., Arnould, J. P., Avila, I. C., Nash, S. B., Campbell, E. et al. (2021). Marine mammal conservation: over the horizon. Endangered Species Research 44: 291–325.

Nelson, J. A. (2016). Oxygen consumption rate v. rate of energy utilization of fishes: a comparison and brief history of the two measurements. Journal of Fish Biology 88: 10–25.

Nielsen, A., Bigelow, K. A., Musyl, M. K. and Sibert, J. R. (2006). Improving light-based geolocation by including sea surface temperature. Fisheries Oceanography 15: 314–325.

Nielsen, M. B., Vogensen, T. K., Thyrring, J., Sørensen, J. G. and Sejr, M. K. (2021). Freshening increases the susceptibility to heat stress in intertidal mussels (Mytilus edulis) from the Arctic. J. Anim Ecol. 90: 1515–1524.

Nielsen, S. N. (2000). Thermodynamics of an ecosystem interpreted as a hierarchy of embedded systems. Ecological Modelling 135: 279–289.

Norin, T. and Clark, T. D. (2016). Measurement and relevance of maximum metabolic rate in fishes. Journal of Fish Biology 88: 122–151.

Oomen, R. A. and Hutchings, J. A. (2017). Transcriptomic responses to environmental change in fishes: Insights from RNA sequencing. Facets. 2: 610–641.

Overgaard, J., Stecyk, J. A., Gesser, H., Wang, T. and Farrell, A. P. (2004). Effects of temperature and anoxia upon the performance of in situ perfused trout hearts. Journal of Experimental Biology 207: 655–665.

Owens, D. W. (2017). Hormones in the life history of sea turtles. pp. 315–341. *In*: The Biology of Sea Turtles. CRC Press.

Padilla-Gamiño, J. L., Kelly, M. W., Evans, T. G. and Hofmann, G. E. (2013). Temperature and CO_2 additively regulate physiology, morphology and genomic responses of larval sea urchins, *Strongylocentrotus purpuratus*. Biological Sciences 280: 20130155.

Palme, R. (2019). Non-invasive measurement of glucocorticoids: advances and problems. Physiology & Behavior 199: 229–243.

Parsons, P. A. (1994). The energetic cost of stress. Can biodiversity be preserved?. Biodiversity Letters 2: 11–15.

Pauli, J. N., Whiteman, J. P., Riley, M. D. and Middleton, A. D. (2010). Defining noninvasive approaches for sampling of vertebrates. Conservation Biology 24: 349–352.

Peck, M. A. and Moyano, M. (2016). Measuring respiration rates in marine fish larvae: challenges and advances. Journal of Fish Biology 88: 173–205.

Perrault, J. R., Page-Karjian, A. and Miller, D. L. (2016). Nesting leatherback sea turtle (Dermochelys coriacea) packed cell volumes indicate decreased foraging during reproduction. Marine Biology 163: 1–10.

Peterson, M. S. and Lowe, M. R. (2009). Implications of cumulative impacts to estuarine and marine habitat quality for fish and invertebrate resources. Reviews in Fisheries Science 17: 505–523.

Piniak, W. E. D. (2012). Acoustic Ecology of Sea Turtles: Implications for Conservation (Doctoral Dissertation, Duke University).

Ponganis, P. J. (2007). Bio-logging of physiological parameters in higher marine vertebrates. Deep Sea Research Part II: Topical Studies in Oceanography 54: 183–192.

Price, J. T., Drye, B., Domangue, R. J. and Paladino, F. V. (2018). Exploring the role of artificial lighting in loggerhead turtle (Caretta caretta) nest-site selection and hatchling disorientation. Herpetological Conservation and Biology 13: 415–422.

Putman, N. F., Scanlan, M. M., Billman, E. J., O'Neil, J. P., Couture, R. B., Quinn, T. P. et al. (2014). An inherited magnetic map guides ocean navigation in juvenile Pacific salmon. Current Biology 24: 446–450.

Quinn, T. P. (1984). An experimental approach to fish compass and map orientation. pp. 113–123. *In*: Mechanisms of Migration in Fishes. Springer, Boston, MA.

Rolfe, D. F. and Brown, G. C. (1997). Cellular energy utilization and molecular origin of standard metabolic rate in mammals. Physiological Reviews 77: 731–758.

Rolland, R. M., Hunt, K. E., Kraus, S. D. and Wasser, S. K. (2005). Assessing reproductive status of right whales (Eubalaena glacialis) using fecal hormone metabolites. General and Comparative Endocrinology 142: 308–317.

Rolland, R. M., Parks, S. E., Hunt, K. E., Castellote, M., Corkeron, P. J., Nowacek, D. P. et al. (2012). Evidence that ship noise increases stress in right whales. Proceedings of the Royal Society B: Biological Sciences 279: 2363–2368.

Romero, L. M. and Beattie, U. K. (2022). Common myths of glucocorticoid function in ecology and conservation. Journal of Experimental Zoology Part A: Ecological and Integrative Physiology 337: 7–14.

Santer, R. M. and Walker, M. G. (1980). Morphological studies on the ventricle of teleost and elasmobranch hearts. Journal of Zoology 190: 259–272.

Scheun, J., Miller, R. J., Ganswindt, A., Waller, L. J., Pichegru, L., Sherley, R. B. et al. (2021). Urofaecal glucocorticoid metabolite concentrations in African penguin (Spheniscus demersus) chick populations experiencing different levels of human disturbance. Conservation Physiology 9.

Schreck, C. B., Contreras-Sanchez, W. and Fitzpatrick, M. S. (2001). Effects of stress on fish reproduction, gamete quality, and progeny. pp. 3–24. *In*: Reproductive Biotechnology in Finfish Aquaculture. Elsevier.

Scott, A. P. (2013). Do mollusks use vertebrate sex steroids as reproductive hormones? II. Critical review of the evidence that steroids have biological effects. Steroids 78: 268–281.

Seebacher, F. and Franklin, C. E. (2012). Determining environmental causes of biological effects: the need for a mechanistic physiological dimension in conservation biology. Philosophical Transactions of the Royal Society B: Biological Sciences 367: 1607–1614.

Shaffer, S. A. (2011). A review of seabird energetics using the doubly labeled water method. Comparative Biochemistry and Physiology Part A: Molecular & Integrative Physiology 158: 315–322.

Shultz, A. D., Murchie, K. J., Griffith, C., Cooke, S. J., Danylchuk, A. J., Goldberg, T. L. et al. (2011). Impacts of dissolved oxygen on the behavior and physiology of bonefish: implications for live release angling tournaments. Journal of Experimental Marine Biology and Ecology 402: 19–26.

Shultz, A. D., Zuckerman, Z., Stewart, H. and Suski, C. D. (2014). Seasonal blood chemistry response of sub-tropical, nearshore fishes to climate change. Conservation Physiology 2.

Sih, A. (2013). Understanding variation in behavioural responses to human-induced rapid environmental change: a conceptual overview. Animal Behaviour 85: 1077–1088.

Silber, G. K., Lettrich, M. D., Thomas, P. O., Baker, J. D., Baumgartner, M., Becker, E. A. et al. (2017). Projecting marine mammal distribution in a changing climate. Frontiers in Marine Science 4: 413.

Silva, E., Marco, A., da Graça, J., Pérez, H., Abella, E., Patino-Martinez, J. et al. (2017). Light pollution affects nesting behavior of loggerhead turtles and predation risk of nests and hatchlings. Journal of Photochemistry and Photobiology B: Biology 173: 240–249.

Simon, H. A. (1962). The architecture of complexity. Proceedings of the American Philosophical Society 106: 467–482.

Sopinka, N. M., Donaldson, M. R., O'Connor, C. M., Suski, C. D. and Cooke, S. J. (2016). Stress indicators in fish. pp. 405–462. *In*: Fish Physiology (Vol. 35). Academic Press.

Southwood, A., Fritsches, K., Brill, R. and Swimmer, Y. (2008). Sound, chemical, and light detection in sea turtles and pelagic fishes: sensory-based approaches to bycatch reduction in longline fisheries. Endangered Species Research 5: 225–238.

Spicer, J. and Gaston, K. (2009). Physiological Diversity: Ecological Implications. John Wiley & Sons.

Steell, S. C., Cooke, S. J. and Eliason, E. J. (2020). Artificial light at night does not alter heart rate or locomotor behaviour in Caribbean spiny lobster (Panulirus argus): insights into light pollution and physiological disturbance using biologgers. Conservation Physiology 8(1): coaa097.

Stenseng, E., Braby, C. E. and Somero, G. N. (2005). Evolutionary and acclimation-induced variation in the thermal limits of heart function in congeneric marine snails (genus Tegula): implications for vertical zonation. The Biological Bulletin 208: 138–144.

Stoot, L. J., Cairns, N. A., Cull, F., Taylor, J. J., Jeffrey, J. D., Morin, F. et al. (2014). Use of portable blood physiology point-of-care devices for basic and applied research on vertebrates: a review. Conservation Physiology 2(1).

Strotz, L. C., Saupe, E. E., Kimmig, J. and Lieberman, B. S. (2018). Metabolic rates, climate and macroevolution: a case study using Neogene molluscs. Proc R Soc B. 285: 20181292.

Swimmer, Y. and Brill, R. (2006). Sea Turtle and pelagic fish sensory biology: developing techniques to reduce sea turtle bycatch in longline fisheries. pp. 107. *In*: NOAA Technical Memorandum NMFS-PIFSC-7.

Swimmer, Y., McNaughton, L., Southwood, A. and Brill, R. (2006). Tests of repellent bait to reduce turtle bycatch in commercial fisheries. Sea Turtle and Pelagic Fish Sensory Biology: Developing Techniques to Reduce Sea Turtle Bycatch in Longline Fisheries 57.

Swimmer, Y., Gutierrez, A., Bigelow, K., Barceló, C., Schroeder, B., Keene, K. et al. (2017). Sea turtle bycatch mitigation in US longline fisheries. Frontiers in Marine Science 4: 260.

Swindall, J. E., Ober, H. K., Lamont, M. M. and Carthy, R. R. (2019). Informing sea turtle outreach efforts to maximize effectiveness. Wildlife Society Bulletin 43: 436–446.

Tagliarolo, M. and McQuaid, C. D. (2016). Field measurements indicate unexpected, serious underestimation of mussel heart rates and thermal tolerance by laboratory studies. PLOS One 11: e0146341.

Tedeschi, J. N., Kennington, W. J., Berry, O., Whiting, S., Meekan, M. and Mitchell, N. J. (2015). Increased expression of Hsp70 and Hsp90 mRNA as biomarkers of thermal stress in loggerhead turtle embryos (*Caretta caretta*). Journal of Thermal Biology 47: 42–50.

Tengberg, A., Hovdenes, J., Andersson, H. J., Brocandel, O., Diaz, R., Hebert, D. et al. (2006). Evaluation of a lifetime-based optode to measure oxygen in aquatic systems. Limnology and Oceanography Methods 4: 7–17.

The Economist. (2021). How Clams keep Polish Water Clean. The Economist Newspaper Limited.

Thorstad, E. B., Rikardsen, A. H., Alp, A. and Økland, F. (2013). The use of electronic tags in fish research–an overview of fish telemetry methods. Turkish Journal of Fisheries and Aquatic Sciences 13: 881–896.

Tilbrook, A. J., Turner, A. I. and Clarke, I. J. (2000). Effects of stress on reproduction in non-rodent mammals: the role of glucocorticoids and sex differences. Reviews of Reproduction 5: 105–113.

Trego, M. L., Whitehead, A., Kellar, N. M., Lauf, M. and Lewison, R. L. (2019). Tracking transcriptomic responses to endogenous and exogenous variation in cetaceans in the Southern California Bight. Conservation Physiology 7.

Trumble, S. J., Robinson, E. M., Berman-Kowalewski, M., Potter, C. W. and Usenko, S. (2013). Blue whale earplug reveals lifetime contaminant exposure and hormone profiles. Proceedings of the National Academy of Sciences 110: 16922–16926.

Twitchett, R. (2007). The Lilliput effect in the aftermath of the end-Permian extinction event. Palaeogeogr. Palaeoclimatol. Palaeoecol. 252: 132–144.

Ueda, H., Kaeriyama, M., Mukasa, K., Urano, A., Kudo, H., Shoji, T. et al. (1998). Lacustrine sockeye salmon return straight to their natal area from open water using both visual and olfactory cues. Chemical Senses 23: 207–212.

Ummenhofer, C. C. and Meehl, G. A. (2017). Extreme weather and climate events with ecological relevance: a review. Philosophical Transactions of the Royal Society B: Biological Sciences 372: 20160135.

Vargas. C. A., Lagos, N. A., Lardies, M. A., Duarte, C., Manríquez, P. H., Aguilera, V. M. et al. (2017). Species-specific responses to ocean acidification should account for local adaptation and adaptive plasticity. Nat. Ecol. Evol. 1: 0084.

Vargas. C. A., Cuevas, L. A., Broitman, B. R., San Martin, V. A., Lagos, N. A., Gaitán-Esoitia, J. D. et al. (2022). Upper environmental pCO_2 drives sensitivity to ocean acidification in marine invertebrates. Nature Clim. Change 12: 200–207.

Vitousek, M. N., Romero, L. M., Tarlow, E., Cyr, N. E. and Wikelski, M. (2010). Island tameness: an altered cardiovascular stress response in Galápagos marine iguanas. Physiology & Behavior 99: 544–548.

Waldbusser. G. G. and Salisbury, J. E. (2014). Ocean acidification in the coastal zone from an organism's perspective: Multiple system parameters, frequency domains, and habitats. Ann. Rev. Mar. Sci. 6: 221–247.

Wang, J. H., Boles, L. C., Higgins, B. and Lohmann, K. J. (2007). Behavioral responses of sea turtles to lightsticks used in longline fisheries. Anim. Conserv. 10: 176–182.

Wang, J. H., Fisler, S. and Swimmer, Y. (2010). Developing visual deterrents to reduce sea turtle bycatch in gill net fisheries. Marine Ecology Progress Series 408: 241–250.

Wang, J., Barkan, J., Fisler, S., Godinez-Reyes, C. and Swimmer, Y. (2013). Developing ultraviolet illumination of gillnets as a method to reduce sea turtle bycatch. Biology Letters 9: 20130383.

Watson, J. W., Hyder, K., Boyd, R., Thorpe, R., Weltersbach, M. S., Ferter, K. et al. (2020). Assessing the sublethal impacts of anthropogenic stressors on fish: An energy-budget approach. Fish and Fisheries 21: 1034–1045.

Webber, D. M. and O'dor, R. K. (1986). Monitoring the metabolic rate and activity of free-swimming squid with telemetered jet pressure. Journal of Experimental Biology 126: 205–224.

Webber, D. M., Boutilier, R. G. and Kerr, S. R. (1998). Cardiac output as a predictor of metabolic rate in cod Gadus morhua. The Journal of Experimental Biology 201: 2779–2789.

Weber, D. N., Fields, A. T., Patterson III, W. F., Barnett, B. K., Hollenbeck, C. M. and Portnoy, D. S. (2021). Novel epigenetic age estimation in wild-caught Gulf of Mexico reef fishes. Can. J. Fish. Aquat. Sci. 79: 1–5.

Whitehead, A., Roach, J. L., Zhang, S. and Galvez, F. (2011). Genomic mechanisms of evolved physiological plasticity in killifish distributed along an environmental salinity gradient. Proc. Natl. Acad. Sci. 108: 6193–6198.

Whittow, G. C., Schreiber, E. A. and Burger, J. (2002). Seabird reproductive physiology and energetics. pp. 409–427. *In*: Biology of Marine Birds. CRC Press, Boca Raton, FL.

Whoriskey, K., Martins, E. G., Auger-Méthé, M., Gutowsky, L. F., Lennox, R. J., Cooke, S. J. et al. (2019). Current and emerging statistical techniques for aquatic telemetry data: A guide to analysing spatially discrete animal detections. Methods in Ecology and Evolution 10: 935–948.

Wikelski, M. and Cooke, S. J. (2006). Conservation physiology. Trends in Ecology & Evolution 21(1): 38–46.

Wilkens, J. L. (1993). Re-evaluation of the stretch sensitivity hypothesis of crustacean hearts: hypoxia, not lack of stretch, causes reduction in heart rate of isolated hearts. Journal of Experimental Biology 176: 223–232.

Williams, R., Wright, A. J., Ashe, E., Blight, L. K., Bruintjes, R. and Canessa, R. (2015). Impacts of anthropogenic noise on marine life: Publication patterns, new discoveries, and future directions in research and management. Ocean & Coastal Management 115: 17–24.

Williams, H. J., Taylor, L. A., Benhamou, S., Bijleveld, A. L., Clay, T. A., de Grissac, S. et al. (2020). Optimizing the use of biologgers for movement ecology research. Journal of Animal Ecology 89: 186–206.

Wingfield, J. C., Maney, D. L., Breuner, C. W., Jacobs, J. D., Lynn, S., Ramenofsky, M. et al. (1998). Ecological bases of hormone—behavior interactions: the "emergency life history stage". American Zoologist 38: 191–206.

Witherington, B. E. (1992). Behavioral responses of nesting sea turtles to artificial lighting. Herpetologica 48: 31–39.

Witherington, B. E. and Martin, R. E. (1996). Understanding, assessing, and resolving light-pollution problems on sea turtle nesting beaches. pp. 1–71. *In*: Florida Mar. Res. Instit. Techn. Rep.

Wood, I. R., Bell, R. G. and Wilkinson, D. L. (1993). Ocean Disposal of Wastewater. World Scientific, Singapore.

Worm, B., Hilborn, R., Baum, J. K., Branch, T. A., Collie, J. S., Costello, C. et al. (2009). Rebuilding global fisheries. Science 325: 578–585.

Wright, A. J., Deak, T. and Parsons, E. C. M. (2011). Size matters: management of stress responses and chronic stress in beaked whales and other marine mammals may require larger exclusion zones. Marine Pollution Bulletin 63: 5–9.

Yin, X., Martinez, A. S., Perkins, A., Sparks, M. M., Harder, A. M., Willoughby, J. R. et al. (2021). Incipient resistance to an effective pesticide results from genetic adaptation and the canalization of gene expression. Evolutionary Applications 14: 847–859.

Zemanova, M. A. (2020). Towards More Compassionate Wildlife Research Through The 3rs Principles: Moving From Invasive to Non-Invasive Methods. Wildlife Biology.

CHAPTER 12

Ecosystem-based Governance for a Sustainable Blue Economy

Insights from Offshore Wind in Australia

*Miguel Frohlich** and *Pedro Fidelman*

Introduction

Humans have long used the ocean for essential resources and services, such as food, trade, transport and recreation. Anthropogenic impacts on marine ecosystems have now reached an unprecedented level that threatens our very existence. According to the Second World Ocean Assessment (UN 2021), examples of anthropogenic pressures on the ocean environment include unsustainable fishing, the introduction of invasive species, excessive inputs of nutrients and hazardous substances (including plastics and microplastics), noise pollution, ill-managed coastal development and extraction of natural resources. The pace of ocean degradation is being accelerated by climate change (Barnett et al. 2001, Halpern et al. 2008, Brander et al. 2010, Halpern et al. 2015). Greenhouse gas emissions from human activities have caused ocean warming and acidification, glacier loss and permafrost degradation, endangering marine ecosystems (IPCC 2019). As a result, coastal communities and livelihoods have been exposed to more frequent and intense storms, as well as the effects of sea level rise (OECD 2019).

Considering this scenario, the United Nations has declared the current decade (2021–2030) as the Decade of Ocean Science for Sustainable Development, also known as the Ocean Decade. This initiative aims to strengthen international cooperation on scientific research and technological innovation for the responsible management of the ocean (IOC-UNESCO 2020). The Ocean Decade promotes what has been called the "blue economy", a term introduced in 2012 at the United Nations Conference on Sustainable Development (Rio+20) to help overcome the ocean governance issues of the 21st century (Silver et al. 2015, Lee et al. 2020, Qi 2022). Although there is still no agreement on its exact meaning and scope (Voyer et al. 2018, Fabinyi et al. 2021, Germond-Duret 2022, Graziano et al. 2022), the blue economy is now generally understood as a framework for boosting economic growth, improving quality of life and increasing social equality while ensuring environmental sustainability (Martínez-Vázquez et al. 2021, European Commission 2021). Its objective is to promote the sustainable industrialization of ocean resources in a way that allows for its benefits to be shared across the global society (Smith-Godfrey 2016).

Centre for Policy Futures, The University of Queensland, St Lucia QLD 4072, Australia.
Email: p.fidelman@uq.edu.au
* Corresponding author: m.frohlich@uq.edu.au

To ensure that the blue economy does not become a framework that legitimates and aggravates existing human pressures on marine ecosystems, ocean governance must promote a sustainable and equitable use of the ocean (Golden et al. 2017, Bennett et al. 2019, Cisneros-Montemayor et al. 2021, Fusco et al. 2022). However, a critical ocean governance challenge is that ocean activities and resources are traditionally managed on a sectoral basis and at different jurisdictional levels (Haward and Vince 2008, Haward 2014, Werle et al. 2019). In this sense, ocean governance is complicated by the existence of multiple (and often competing) actors pertaining to different organizations and/or economic sectors, which are frequently governed by complex, fragmented, and at times overlapping governance arrangements (Fidelman et al. 2012, Keen et al. 2018, Voyer and van Leeuwen 2019).

The adoption of an ecosystem-based approach has been widely advocated for overcoming the issue of ocean governance fragmentation (Ekstrom et al. 2009, Scherer et al. 2016, Smith et al. 2017, Clark et al. 2022). Such an approach emphasises a holistic view of marine ecosystems that consider the complex and dynamic nature of the ocean environment and the cumulative impacts that may be caused by human activities (Zhang et al. 2019, Msomphora et al. 2022). While highlighting that ecosystem-based governance has no formal, universally agreed upon definition, Platjouw (2019) describes the concept as "the governance approach that requires multi-level, polycentric and participatory governance structures to ensure the maintenance of ecosystem integrity and its functioning, while sustainably using the ecosystem's services and enjoying its benefits" (p. 3). Wondolleck and Yaffe (2017) list five key elements that underpin the ecosystem-based approach. First, it gives preference to ecological boundaries instead of political and administrative boundaries. Second, it acknowledges that marine social-ecological systems are complex and considers this complexity when developing management strategies. Third, it pursues a balance between (and the integration of) interests from ocean actors without compromising marine social-ecological systems' health and resilience. Fourth, it promotes cross-scale collaboration between ocean stakeholders and across academic disciplines. Lastly, it adopts adaptive management, emphasising that management actions must be monitored, periodically evaluated and adjusted in response to new information or change.

Ecosystem-based governance depends on the design of appropriate policy and legislative arrangements to achieve its goals (Alexander and Haward 2019). However, whether the approach necessarily requires major policy and legislative reform for its successful implementation remains a valid question (Macpherson et al. 2021). Comprehensive reforms to create perfect ocean regulatory frameworks for ecosystem-based governance are unlikely to occur in the short term due to the complex and challenging political environment (Hoffman and Hildreth 2016). Consequently, scholars have begun exploring potential opportunities in existing policies, legislation and regulations that may enable ecosystem-based governance (Garmestani et al. 2019, Gray et al. 2021). Under this perspective, we must accept and work with existing ocean governance arrangements that are not entirely fit-for-purpose and acknowledge that eliminating regulatory fragmentation is nearly impossible (Macpherson et al. 2021).

Using Australia's blue economy as a case study, with a particular focus on the Australian offshore wind sector, this chapter identifies potential pathways through which existing policy and legislative arrangements may be leveraged to support ecosystem-based governance. After presenting an overview of the evolution of offshore wind in Australia and its policy and legislative framework, we identify three avenues through which ecosystem-based governance may emerge in the Australian offshore wind sector despite existing policy and legal constraints. Our analysis contributes to ongoing efforts towards the development of a sustainable blue economy in Australia and provides insights to other jurisdictions facing significant regulatory complexity and fragmentation in ocean governance.

Overview of the legal framework for offshore wind in Australia

Australia's blue economy is growing at 2–3 times the rate of the rest of Australia's gross domestic growth (DCCEEW 2021). This trend will likely be sustained over the next decades given Australia's enormous potential for offshore wind. Technically accessible offshore wind energy resources in Australian waters are estimated at 2,233 GW – well above the country's projected electricity demand (Briggs et al. 2021). Yet, at the time this chapter was written, no offshore wind project proposed in Australia had moved into the construction phase. While still lagging behind other leading countries on offshore wind development, such as China, the United Kingdom and Germany (GWEC 2022), Australia is now presented with the opportunity to learn from previous experiences and develop its own offshore wind industry in line with the sustainability principles that underpin the blue economy. The offshore wind sector is expected to play a key role in Australia's energy transition towards its target of net zero emissions by 2050, which was established by the Climate Change Act 2022 (Cth). However, the potential socioeconomic and environmental impacts of offshore wind farms must not be ignored (Salvador et al. 2018, Mauricio Hernandez et al. 2021, Lloret et al. 2022, Virtanen et al. 2022). Lessons from onshore wind farms developed in Australia, which are predominantly located in rural areas along the coastline, show concerning levels of public controversy and criticism. These include particularly social and environmental issues of such projects at the local level, despite the overall growing support for renewable energy at the national level (Harvey and Dew 2016, Harvey et al. 2017, Wifa and Hunter 2022). Australia's emerging offshore wind industry should adopt an ecosystem-based governance approach to ensure that the social and environmental impacts of this emerging industry are considered across local, regional, national and global levels.

Until recently, one of the main constraints to investment in offshore wind in Australia was the lack of a specific federal regulatory framework (Briggs et al. 2021, Frohlich et al. 2023). There was a common view among investors and developers that a clear regulatory pathway was needed to support the blooming of the Australian offshore wind industry (O'Kane and Todd 2021). In September 2021, the Commonwealth Government responded to this demand by introducing in the Australian Parliament the Offshore Electricity Infrastructure Bill 2021 (Cth) and the Offshore Electricity Infrastructure (Regulatory Levies) Bill 2021 (Cth). In December 2021, the Bills passed both Houses and received royal assent; then, the Offshore Electricity Infrastructure Act 2021 (Cth) (OEI Act) and the Offshore Electricity Infrastructure (Regulatory Levies) Act 2021 (Cth) (OEI Levies Act) took effect on June 2nd, 2022.

The OEI Act regulates the construction, operation, maintenance and decommissioning of offshore renewable energy infrastructure (such as an offshore wind farm) and offshore electricity transmission infrastructure (such as submarine cables and associated infrastructure). The OEI Levies Act enables the recovery of costs associated with the implementation of the new regulatory framework by the National Offshore Petroleum Safety and Environmental Management Authority (NOPSEMA), the federal agency legally nominated as the Offshore Infrastructure Regulator (OIR), and the Offshore Infrastructure Registrar, who is an appointed person responsible for administering the licensing scheme created by the OEI Act. In addition, the Commonwealth Government has made the Offshore Electricity Infrastructure Regulations 2022 and the Offshore Electricity Infrastructure (Regulatory Levies) Regulations 2022, which set out detailed arrangements for the broad regulatory framework established by the OEI Act and OEI Levies Act.

An analysis of the OEI Act was conducted by Goodman (2022) who highlighted three central components of the new regulatory framework. First, the OEI Act prohibits offshore infrastructure activities to be carried out without a licence. Second, it sets out that the minister administering the OEI Act (the Minister for Climate Change and Energy) may declare specified areas as suitable for offshore renewable energy infrastructure. Third, the OEI Act creates a licensing scheme, allowing the Minister to grant the following types of licences for offshore infrastructure activities: (i) a feasibility licence for assessing the feasibility of an offshore infrastructure project; (ii) a commercial licence, which authorizes the exploitation of renewable energy resources by those who have previously been granted a feasibility licence; (iii) a research and demonstration licence, which authorizes research and demonstration activities related to offshore renewable energy infrastructure and offshore electricity transmission infrastructure; and (iv) a transmission and infrastructure licence, which covers the transmission infrastructure required by an offshore infrastructure project.

One limitation of the OEI Act derives from Australia's Offshore Constitutional Settlement (OCS), which defined the powers and responsibilities of the Commonwealth and the states over Australian waters (Haward 1989, 1996, Goodman 2022).[1] In line with the OCS, the OEI Act only applies to the

[1] The OCS was incorporated into the Australian legislative framework through the enactment of the Coastal Waters (State Powers) Act 1980 (Cth), the Coastal Waters (State Title) Act 1980 (Cth), and equivalent legislation applicable to the Northern Territory.

Commonwealth offshore area, which comprises the waters, seabed and subsoil from three nautical miles to 200 nautical miles from territorial sea baselines. State governments and the Northern Territory have jurisdiction over internal waters, as well as offshore waters and adjacent seabed and subsoil that extend to three nautical miles from territorial sea baselines. These arrangements have significant implications for offshore wind projects proposed in the Commonwealth offshore area because such projects will frequently demand associated infrastructure (e.g., submarine cables) within the first three nautical miles and on land (e.g., port infrastructure, substations and power lines) (Briggs et al. 2021, Frohlich et al. 2023). Thus, to undertake all applicable regulatory and environmental assessments for offshore wind projects, developers are required to deal with multiple government departments and agencies and comply with federal, state and local legislation (Hughes and Longden 2022).

OCS arrangements may be perceived as a major constraint to ecosystem-based governance in the Australian marine space, as jurisdictional boundaries have been defined without taking into account ecological factors. Nevertheless, the OCS did manage to resolve long-standing political disputes between the Commonwealth Government and the states (Vince et al. 2015), and it is unlikely that a comprehensive reform of OCS arrangements will happen any time soon. As such, alternative strategies will need to be designed and implemented to help match jurisdictional and ecosystem boundaries. We present below three avenues through which ecosystem-based governance may still emerge in the context of the Australian offshore wind sector despite existing policy and legislative constraints.

Avenues for ecosystem-based governance

Policy and legislative arrangements for marine spatial planning

While there are many definitions of Marine Spatial Planning (MSP), the Intergovernmental Oceanographic Commission of the United Nations Educational, Scientific and Cultural Organization (IOC-UNESCO) defines the approach as "a public process of analysing and allocating the spatial and temporal distribution of human activities in marine areas to achieve ecological, economic and social objectives that have been specified through a political process" (IOC-UNESCO and European Commission 2021, p. 23). MSP has been widely advocated as a fundamental tool for ecosystem-based governance (Ehler and Douvere 2009, Agardy 2010, Smith et al. 2017, Wondolleck and Yaffe 2017). The legal context in which MSP is undertaken is recognized as a crucial factor for its success (Zuercher et al. 2022). In this sense, policy and legislative arrangements for MSP may influence whether and how the principles of ecosystem-based governance are incorporated into the regulation of marine activities.

MSP has a long history in Australia. The introduction of the zoning plan for the Great Barrier Reef Marine Park (GBRMP) in the early 1980s is often cited as the earliest example of MSP in the world (Jay 2017, Hassan and Alam 2019, Ehler 2021, Kirkfeldt et al. 2022). The country was also at the forefront of MSP in the 1990s, a movement that culminated with the release of Australia's Oceans Policy (AOP) in 1998 (Rodriguez 2017). Despite not meeting all expectations and its limitations, AOP set out the foundation for MSP in marine areas under the jurisdiction of the Commonwealth Government. This involves the development of Marine Bioregional Plans (MBPs)[2] under section 176 of the Environment Protection and Biodiversity Conservation Act 1999 (Cth) (EPBC Act) (Vince 2014). MBPs have been developed for four marine regions in Australia: South-West, North-West, North and Temperate East.

MBPs have made a pioneering contribution to MSP in Australia, particularly in terms of gathering valuable scientific data about Australian marine ecosystems, but they are not without shortcomings. For instance, they have been criticized for the poor involvement of states and territories during their development, their narrow focus on Marine Protected Areas (MPAs) and their deficiencies in cross-sectoral planning (Vince 2013, 2014, Smyth 2017). Moreover, all existing MBPs were completed in 2012, i.e., prior to more detailed assessments of the Australian offshore wind potential (Briggs et al. 2021). While calls for major legislative, institutional and regulatory reform to improve MSP in

[2] For more information about MBPs, see <https://www.dcceew.gov.au/environment/marine/marine-bioregional-plans> (Accessed: October 6th, 2023).

Australia have been made (Smyth 2017), it is unlikely that policy and legislative change will occur at the necessary speed to create MSP arrangements that are fit for the purpose of offshore wind decisions. Therefore, identifying existing policy and legal opportunities that can enable some form of MSP under ecosystem-based governance may be a way forward.

As previously mentioned, one of the central components of the OEI Act is the regulatory process for declaring areas as suitable for offshore renewable energy infrastructure. If the Minister for Climate Change and Energy decides to make a declaration, a notice must be published on the website of the Department of Climate Change, Energy, the Environment and Water (DCCEEW) with the appropriate information about the proposal. The notice must invite submissions from the public for a period of at least 60 days. The OEI Act also mandates the Minister to consult with the Minister for Defence and the Minister administering section 1 of the Navigation Act 2012 (Cth) (the Minister for Infrastructure, Transport, Regional Development and Local Government) prior to making any declaration. Functioning as an MSP tool, the declaration process defined in the OEI Act aims to ensure that offshore wind decisions have regard to the potential impacts of offshore renewable energy infrastructure in the proposed area on other marine users and interests.

In accordance with the process described above, on August 5th, 2022, the Minister for Climate Change and Energy published a notice proposing the creation of Australia's first area for offshore renewable energy infrastructure (DCCEEW 2022a).[3] Located in Commonwealth waters off Gippsland, Victoria, the proposed area was subject to a consultation process conducted by the Commonwealth Government with the involvement of the Victorian Government and the Wellington Shire Council. This process included 765 online submissions, community sessions, and meetings with relevant stakeholders, including First Nations groups, commercial fishing organizations, port authorities and proponents of offshore renewable energy projects (DCCEEW 2022b). The declaration with the final map of the declared area (Fig. 1) was published on December 19th, 2022 (DCCEEW 2022c).

The declared area is broken into three sections to allow for the exclusion of high vessel traffic areas, restricted defence areas and the area of the Beagle Marine Park. The definition of the final location also considered concerns raised by the Gippsland community during the consultation process, including visual impacts resulting from the installation of the wind turbines and on-land transmission, environmental impacts on endangered marine species and migratory bird species and impacts on the tourism and fishing sectors due to the creation of exclusion areas (DCCEEW 2022b).

Because of the OCS, waters under state jurisdiction are not included in the declared area. This jurisdictional fragmentation complicates – but does not prevent – an ecosystem-based MSP. In Victoria, the Department of Energy, Environment, and Climate Action (DEECA) is leading the implementation of the State's MSP Framework, which is part of Victoria's Marine and Coastal Policy 2020 prepared under the Marine and Coastal Act 2018 (Vic) (DEECA 2022a). The implementation of the State's MSP Framework is also an action included in Victoria's Marine and Coastal Strategy 2022 (DEECA 2022b). DEECA is undertaking the first stage of the MSP process, which consists of determining marine planning areas and identifying areas that may benefit from marine spatial planning. On July 12th, 2022, a workshop was held to gather feedback from stakeholders about the proposed marine planning areas and the MSP guidelines that will provide instructions on how the MSP process will be conducted in each of the approved marine planning areas. As with the process for declaring Australia's first area for offshore renewable infrastructure, the Commonwealth and Victorian Governments have worked closely in the

[3] On July 12th, 2023, Australia's second area for offshore renewable energy infrastructure was declared off the coast of the Hunter, New South Wales. At the time of writing this chapter, proposals for declaring two other areas were under assessment. The area proposed in the Southern Ocean, extending offshore from Warrnambool, Victoria, to Port MacDonnell, South Australia, had already undergone public consultation and was awaiting its declaration by the Minister. The area in the Pacific Ocean off the Illawarra, New South Wales, extending offshore from Wombarra to Kiama, remained open for public submissions. The Commonwealth Government's intention is to initiate public consultation processes for other two priority areas in 2023: (i) the Bass Strait region off Northern Tasmania; and the Indian Ocean region off Perth/ Bunbury. For more information, see <https://www.dcceew.gov.au/energy/renewable/establishing-offshore-infrastructure> (Accessed: October 6th, 2023).

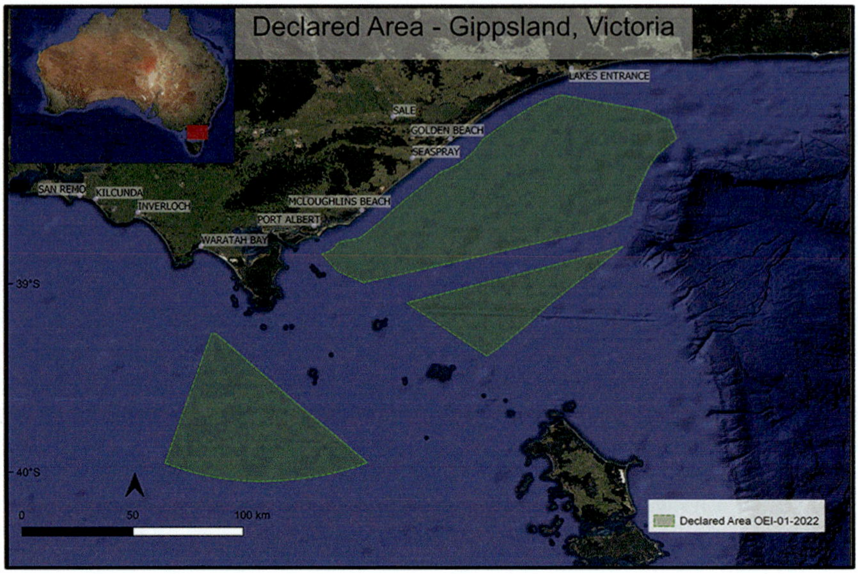

Figure 1. Map of Australia's first area for offshore renewable energy infrastructure (DCCEEW 2022c).

development of Victoria's MSP. Consequently, despite the existing jurisdictional fragmentation resulting from the OCS, ongoing intergovernmental collaboration may lead to the emergence of ecosystem-based governance, ultimately resulting in satisfactory MSP arrangements for offshore wind in Commonwealth waters and adjacent Victorian waters. Notwithstanding these efforts, there are ongoing criticisms regarding the absence of a requirement in the OEI Act for conducting strategic environmental assessments prior to ministerial declarations (Taylor and Taylor 2022), with calls for the supplementation of the federal regulatory frameworks to create binding provisions for MSP (Taylor 2022).

Polycentric development assessment and approval process

Administered by DCCEEW, the EPBC Act is the main piece of federal environmental legislation in Australia. Under its provisions, actions that may have a significant impact on matters of national environmental significance[4] – defined as "controlled actions" – must be approved by the Commonwealth Minister for the Environment and Water. However, in addition to federal environmental approval, offshore wind projects with infrastructure activities within state waters and onshore may also be subject to state environmental permitting legislation. Part 5 of the EPBC Act contains provisions that allow the Commonwealth Government to enter into assessment and approval bilateral agreements with states and territories to delegate its environmental statutory powers. However, approval bilateral agreements have never been executed. Moreover, although assessment bilateral agreements exist with all states and territories since 2015 (Power 2020), they have not contemplated situations where offshore wind projects simultaneously cover areas under federal and state jurisdictions.

At first sight, this jurisdictional fragmentation, stemming from the political and administrative boundaries defined by OCS arrangements, can be viewed as a considerable obstacle to ecosystem-based governance. Current arrangements for development assessment and approval in Australia have long been criticized for creating unnecessary redundancy between federal and state/territory regulatory processes, and recommendations have been made to eliminate regulatory duplication (Samuel 2020). In the context

[4] There are nine matters of national environmental significance: (i) world heritage properties; (ii) national heritage places; (iii) wetlands of international importance; (iv) nationally threatened species and ecological communities; (v) migratory species; (vi) Commonwealth marine areas; (vii) the Great Barrier Reef Marine Park; (viii) nuclear actions; and (ix) water resources (applicable to coal seam gas and large coal mining development).

of offshore wind projects, similar criticisms have led to calls for streamlined assessment and approval processes, aiming to increase regulatory efficiency and achieve both time and cost savings for developers (Hughes and Longden 2022).

However, multiple and overlapping assessment and approval layers may also be perceived as a key checks and balances mechanism that promotes opportunities for discussion among stakeholders about the merits of a proposal (Hollander 2010). Polycentricity, which refers to the distribution of decision-making powers among multiple entities across various governance levels (Ostrom 2010, Chaffin et al. 2014, Frohlich et al. 2022), is considered a critical element of ecosystem-based governance (Heenan et al. 2016, Platjouw 2019, Urlich et al. 2022). Multiple actors with overlapping and redundant responsibilities may enable rapid responses at the level most proximate to the issue being managed, while simultaneously considering the interconnected consequences for other levels (Cosens et al. 2017). Yet, to achieve its full potential, polycentricity necessitates adequate collaboration and coordination across governance levels (Berkes 2010, Fidelman et al. 2013, Tuda et al. 2021).

Australia's first offshore wind project, the Star of the South, illustrates well how collaboration and coordination in a polycentric governance system may contribute to ecosystem-based governance, even in the absence of a fit-for-purpose policy and legal framework. With a capacity of 2.2 GW, enough to power 1.2 million households (Star of the South 2023a), the proposed offshore wind farm is planned to be located within Australia's first declared area for offshore renewable energy infrastructure, in the State of Victoria offshore waters (Fig. 2). While the 400 wind turbines will be mounted on towers founded in the seabed of Commonwealth waters, associated infrastructure will be installed in areas within Victorian

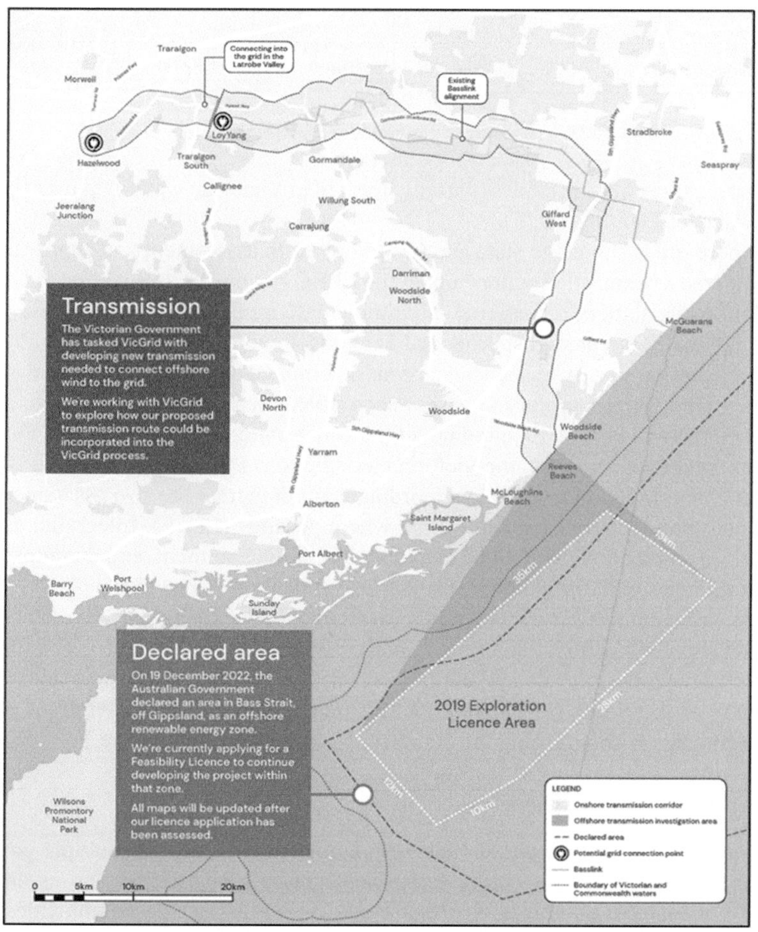

Figure 2. Star of the South – Project map – April 2023 (Star of the South 2023b).

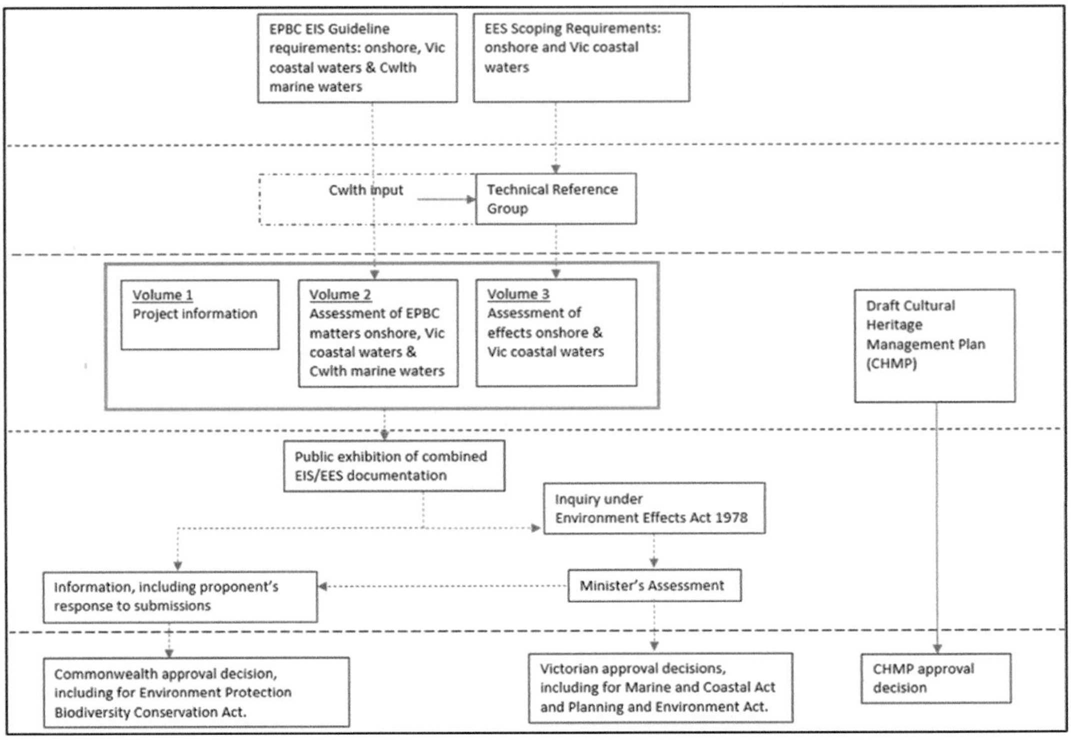

Figure 3. Coordinated assessment process (DTP 2020).

jurisdiction, including transmission cables (partially located in Victorian waters), power lines and onshore substations (Minister for Planning 2020).

The project therefore is subject to state and Commonwealth assessment and approval. On June 20th, 2020, it was considered a "controlled action" under the EPBC Act due to its potential to have a significant impact on the following matters of national environmental significance: (i) wetlands of international importance; (ii) nationally threatened species and ecological communities; (iii) migratory species; and (iv) Commonwealth marine areas. Consequently, it was determined that an environmental impact statement (EIS) must be prepared for the project to assess such potential impacts. In addition, an environment effects statement (EES) will also be required under the Environment Effects Act 1978 (Vic) to assess the project's environmental impacts within the Victorian jurisdiction (Minister for Planning 2020). Given this scenario, DCCEEW and DTP have agreed to coordinate and integrate the two assessment processes for the Star of the South project, reducing duplication where possible (Fig. 3). Integration efforts include a combined EIS/EES document, which will be available for public comment, and coordinated engagement with government agencies with the support of a single technical reference group (TRG) (DTP 2020).

These ad hoc arrangements have attempted to overcome the challenges of regulatory fragmentation between state and Commonwealth jurisdictions resulting from OCS. They will contribute to an integrated assessment of environmental impacts and compliance with both Commonwealth and state legislation. Therefore, the Star of the South development assessment process is an example of how to develop streamlined arrangements without losing the benefits of polycentric governance.

The role of bridging organizations

Bridging organizations—i.e., organizations that help connect actors across multiple governance levels through strategic bridging processes—have been increasingly recognized as an essential element in ecosystem-based governance, playing a significant role in navigating the complexities of fragmented marine conservation settings (Folke et al. 2005, Green et al. 2015, Berdej and Armitage 2016, Berdej et al. 2019). Such organizations facilitate vertical and horizontal collaboration, goal setting, knowledge coproduction and sharing, learning, trust building and conflict resolution (Hahn et al. 2006, Berkes 2009, Crona and Parker 2012, Berdej and Armitage 2016).

When governance is characterized by fragmentation and regulatory complexity, bridging organizations can serve as a vital link to help overcome these challenges. By lowering the transaction costs of collaboration, bridging organizations can improve governance outcomes by spanning the science-policy interface, facilitating interactions among actors operating at different governance levels and promoting collaborative decision-making (Berkes 2009, Crona and Parker 2012, Green et al. 2015, Kowalski and Jenkins 2015, Schultz et al. 2020). These organizations have the potential to create the right environment for various stakeholders, such as government and non-governmental organizations, industry, academia and local communities, to collaborate on solving complex problems (Vignola et al. 2013, Stewart and Tyler 2019), including developing fit-for-purpose policy and legislation.

The role of bridging organizations in enabling ecosystem-based governance can be exemplified by the Blue Economy Cooperative Research Centre (Blue Economy CRC), which was established in 2019 by the Commonwealth Government with a 10-year funding of more than AU$ 329 million (Blue Economy CRC 2021). The initiative is part of the Cooperative Research Centres (CRC) Program, which began in 1990 to support collaborations between industry partners, research institutions and government entities to address major challenges and opportunities in various sectors. Acting as a bridging organization, the Blue Economy CRC brings together 44 partners from 10 countries, including industry, government and research organizations, with expertise in various areas including aquaculture, marine renewable energy, maritime engineering, environmental assessments and policy and regulation (Blue Economy CRC 2023a). Its aim is to conduct high-level, collaborative, industry-oriented research, and training to support the development of the blue economy (Blue Economy CRC 2023a).

The Blue Economy CRC has been instrumental in fostering collaboration and knowledge within the Australian offshore wind sector. In July 2021, it released the report entitled "Offshore Wind Energy in Australia" (Briggs et al. 2021), which concluded that Australia possesses abundant and high-quality offshore wind resources near the existing transmission grid. The associated webinar ("New research on offshore wind energy in Australia")[5] to disseminate the study's findings attracted 744 attendees, the highest number of registrations for a single webinar organised by the Blue Economy CRC in the 2021–22 financial year (Blue Economy CRC 2022). The study, which was developed by the Commonwealth Scientific and Industrial Research Organization (CSIRO) and Saitec Offshore Technologies in collaboration with the University of Technology Sydney, the Maritime Union of Australia, the Electrical Trades Union, the Australian Manufacturing Workers' Union and the Australian Council of Trade Unions, played a significant role in the development of the Australian regulatory framework for offshore wind (Blue Economy CRC 2022). The Blue Economy CRC also provided submissions into the drafting of the Offshore Electricity Infrastructure Bill 2021 (Cth) and the Offshore Electricity Infrastructure (Regulatory Levies) Bill 2021 (Cth), as well as to the Senate's Environment and Communications Legislation Committee's inquiry into such Bills (Blue Economy CRC 2022). Moreover, it gave feedback about the Offshore Electricity Infrastructure Regulations 2022 and the Offshore Electricity Infrastructure (Regulatory Levies) Regulations 2022 (Blue Economy CRC 2022).[6]

Collaboration to support offshore wind development in Australia through the Blue Economy CRC has continued with the announcement of the research project entitled "Pre-conditions for the development of offshore wind energy in Australia" in March 2023 (Blue Economy CRC 2023b). This project involves eight partners (BMT, DNV Australia, Griffith University, SINTEF, the Tasmanian Government, University of Chile, The University of Queensland, and University of Tasmania) working together with five third-party participants (Nexsphere, Flotation Energy, Star of the South, Tasmanian Port Corporation, Copenhagen Energy Australia). The aims of the project are to: (i) identify best practices

[5] The webinar recording is available at: <https://www.youtube.com/watch?v=uGdyC8P-vjQ> (Accessed: October 6th, 2023).

[6] This submission is available at: <https://consult.dcceew.gov.au/oeif-regulations-and-cost-recovery/have-your-say/view/23> (Accessed: October 6th, 2023).

in policy and regulatory processes applicable to offshore wind development that may be considered in the Australian policy and regulatory context; (ii) identify key elements of social acceptability for offshore wind development and assist industry in addressing an integrated integrity system approach to offshore wind projects; and (iii) identify supply chain systems for the development of offshore wind industry and provide strategic direction.

Lastly, it is important to highlight the role the Blue Economy CRC has played in exploring opportunities for offshore wind co-location with other blue economy activities. The focus of the Blue Economy CRC has been on co-location of offshore renewables and offshore aquaculture, which can lead to several benefits, including expanding operational capacities, enhanced management of water quality and waste, decreased likelihood of diseases in aquaculture, decreased operational and maintenance expenses, and a more efficient use of the ocean space (Aryai et al. 2021, Villalba et al. 2022). However, other co-location frontiers, such as combining offshore wind with hydrogen production, have also received attention. Such initiatives show that the role of the Blue Economy CRC as a bridging organization that stimulates the discussion of co-locating activities from different sectors of the blue economy is particularly relevant in the context of institutional complexity and regulatory fragmentation, as discussed in this chapter.

Conclusion

This chapter has explored the emergence of ecosystem-based governance for the blue economy in the context of institutional complexity and regulatory fragmentation. We used the Australian offshore wind sector as a case study to identify ecosystem-based governance pathways amid a policy and legislative framework that is not fit-for-purpose. We specifically focused on three avenues: (i) the use of regulatory process established under the OEI Act for declaring areas as suitable for offshore renewable energy infrastructure as marine spatial planning tool; (ii) the existing polycentric development assessment and approval process for offshore wind projects in Australia; and (iii) the role of bridging organizations in fostering collaboration across governance levels.

Rather than describing the ideal policy and legal conditions for ecosystem-based governance, we adopted in this chapter the perspective that changes in policy and regulatory frameworks are always challenging and may not occur in a timely manner. Yet, ecosystem-based governance may still emerge in difficult policy and legislative environments. While Australia would benefit from improved policy and legislative arrangements for marine spatial planning, consultation requirements for declaring suitable areas for offshore renewable energy infrastructure already allow for consideration of other marine users and interests as part of the offshore wind planning process. The development assessment and approval process established for the Star of the South project also demonstrates that the regulatory fragmentation created by the OCS may be a positive polycentric feature of offshore wind governance in Australia. This does not suggest that efforts to streamline regulations and eliminate unnecessary duplication are not required. When navigating such complex and fragmented policy and legal system, the existence of bridging organizations like the Blue Economy CRC are a crucial element in facilitating ecosystem-based governance. These organizations can act as intermediaries between different stakeholder groups and foster collaboration across governance levels. These examples may provide valuable insights into how to operationalize ecosystem-based governance in other jurisdictions where offshore wind energy and other blue economy activities are developed.

References

Agardy, T. (2010). Ocean Zoning: Making Marine Management More Effective. Earthscan, London, UK.

Alexander, K. A. and Haward, M. (2019). The human side of marine ecosystem-based management (EBM): 'sectoral interplay' as a challenge to implementing EBM. Mar. Policy 101-33-38. https://doi.org/10.1016/j.marpol.2018.12.019.

Aryai, V., Abbassi, R., Abdussamie, N., Salehi, F., Garaniya, V., Asadnia, M. et al. (2021). Reliability of multi-purpose offshore-facilities: present status and future directions in Australia. PSEP 148: 437–461. https://doi.org/10.1016/j.psep.2020.10.016.

Barnett, T. P., Pierce, D. W. and Schnur, R. (2001). Detection of anthropogenic climate change in the world's oceans. Science 292: 270–274. https://doi.org/10.1126/science.1058304.

Bennett, N. J., Cisneros-Montemayor, A. M., Blythe, J., Silver, J. J., Singh, G., Andrews, N. et al. (2019). Towards a sustainable and equitable blue economy. Nat. Sustain. 2: 991993. https://doi.org/10.1038/s41893-019-0404-1.

Berdej, S. M. and Armitage, D. R. (2016). Bridging organizations drive effective governance outcomes for conservation of Indonesia's marine systems. PloS One 11(1): e0147142. https://doi.org/10.1371/journal.pone.0147142.

Berdej, S., Silver, J. and Armitage, D. (2019). A political ecology perspective on bridging organizations and their influence on marine conservation. Soc. Nat. Resour. 32(11): 1258–1275. https://doi.org/10.1080/08941920.2019.1626960.

Berkes, F. (2009). Evolution of co-management: role of knowledge generation, bridging organizations and social learning. J. Environ. Manage. 90(5): 1692–1702. https://10.1016/j.jenvman.2008.12.001.

Berkes, F. (2010). Linkages and multilevel systems for matching governance and ecology: lessons from roving bandits. Bull. Mar. Sci. 86(2): 235–250.

Blue Economy CRC. (2021). Annual Report 2020–2021. [online] Available at: <https://blueeconomycrc.com.au/wpcontent/uploads/2021/11/BECRC_AnnualReport_20202021_A4_e091121.pdf> (Accessed: April 17th, 2023).

Blue Economy CRC. (2022). 2021–22 Annual Report. [online] Available at: < https://blueeconomycrc.com.au/wp-content/uploads/2022/11/BECRC_2021-22-Annual-Report_A4_DPS.pdf> (Accessed: April 17th, 2023).

Blue Economy CRC. (2023a). The Blue Economy CRC – About. [online] Available at: < https://blueeconomycrc.com.au/about/> (Accessed: April 17th, 2023).

Blue Economy CRC. (2023b). Pre-conditions for the Development of Offshore Wind Energy in Australia. [online] Available at: <https://blueeconomycrc.com.au/project/pre-conditions-for-the-development-of-offshore-wind-energy-in-australia/> (Accessed: April 17th 2023).

Brander, K., Botsford, L. W., Ciannelli, L., Fogarty, M. J., Heath, M., Planque, B. et al. (2010). Human impacts on marine systems. Marine ecosystems and global change. pp. 41–71. *In*: Barange, M., Field, J. G., Harris, R. P., Hofmann, E. E., Perry, R. I. and Werner, F. (eds.). Marine Ecosystems and Global Change. Oxford University Press, New York, USA.

Briggs, C., Hemer, M., Howard, P., Langdon, R., Marsh, P., Teske, S. et al. (2021). Offshore Wind Energy in Australia. Blue Economy Cooperative Research Centre, Launceston, Australia. [online] Available at: <https://blueeconomycrc.com.au/project/offshore-wind-potential-for-australia/> (Accessed: September 7th, 2022).

Cisneros-Montemayor, A. M., Moreno-Báez, M., Reygondeau, G., Cheung, W. W. L., Crosman, K. M., González-Espinosa, P. C. et al. (2021). Enabling conditions for an equitable and sustainable blue economy. Nature 591: 396–401. https://doi.org/10.1038/s41586-021-03327-3.

Clark, D. E., Gladstone-Gallagher, R. V., Hewitt, J. E., Stephenson, F. and Ellis, J. I. (2022). Risk assessment for marine ecosystem-based management (EBM). Conserv. Sci. Pract. 4(3): e12636. https://doi.org/10.1111/csp2.12636.

Chaffin, B. C., Gosnell, H. and Cosens, B. A. (2014). A decade of adaptive governance scholarship: synthesis and future directions. Ecol. Soc. 19(3): 56. https://doi.org/10.5751/ES-06824-190356.

Cosens, B. A., Craig, R. K., Hirsch, S. L., Arnold, C. A. T., Benson, M. H., DeCaro, D. A. et al. (2017). The role of law in adaptive governance. Ecol. Soc. 22(1): 30. https://doi.org/10.5751/ES-08731-220130.

Crona, B. I. and Parker, J. N. (2012). Learning in support of governance: theories, methods and a framework to assess how bridging organizations contribute to support adaptive resource governance. Ecol. Soc. 17(1): 32. https://doi.org/10.5751/ES-04534-170132.

DCCEEW (Department of Climate Change, Energy, the Environment and Water). (2021). Australia State of the Environment 2021. Australian government, Canberra. [online] Available at: <https://soe.dcceew.gov.au/> (Accessed: August 25th, 2022).

DCCEEW (Department of Climate Change, Energy, the Environment and Water). (2022a). Establishing offshore renewable energy infrastructure. Australian government, Canberra. [online] Available at: <https://www.dcceew.gov.au/energy/renewable/establishing-offshore-infrastructure> (Accessed: December 20th, 2022).

DCCEEW (Department of Climate Change, Energy, the Environment and Water). (2022b). Fact sheet – Gippsland Assessment and Consultation Process. Australian government, Canberra. [online] Available at: <https://consult.dcceew.gov.au/oeigippsland> (Accessed: December 20th, 2022).

DCCEEW (Department of Climate Change, Energy, the Environment and Water). (2022c). Area in Bass Strait off Gippsland declared suitable for offshore renewable energy. Australian government, Canberra. [online] Available at: <https://www.dcceew.gov.au/energy/renewable/establishing-offshore-infrastructure/gippsland> (Accessed: December 20th, 2022).

DEECA (Department of Energy, Environment and Climate Action). (2022a). Marine Spatial Planning. Victorian government, Melbourne. [online] Available at: <https://www.marineandcoasts.vic.gov.au/marine/marinespatialplanning> (Accessed: October 6th, 2023).

DEECA (Department of Energy, Environment and Climate Action). (2022b). Marine and Coastal Strategy. Victorian government, Melbourne. [online] Available at: <https://www.marineandcoasts.vic.gov.au/coastal-management/marine-and-coastal-strategy> (Accessed: October 6th, 2023).

DTP (Department of Transport and Planning). (2020). Final scoping requirements. [online] Available at: <https://www.planning.vic.gov.au/environmentassessment/browseprojects/projects/star-of-the-south-offshore-wind-farm>.

Ehler, C. (2021). Two decades of progress in marine spatial planning. Mar. Policy 132: 104134. https://doi.org/10.1016/j.marpol.2020.104134.

Ehler, C. and Douvere, F. (2009). Marine Spatial Planning: A Step-by-Step Approach toward Ecosystem-Based Management. Intergovernmental Oceanographic Commission and Man and the Biosphere Programme. UNESCO, Paris, France. [online] Available at: <https://unesdoc.unesco.org/ark:/48223/pf0000186559>. (Accessed: September 19th 2022).

Ekstrom, J. A., Young, O. R., Gaines, S. D., Gordon, M. and McCay, B. (2009). A tool to navigate overlaps in fragmented ocean governance. Mar. Policy 33(3): 532535. https://doi.org/10.1016/j.marpol.2008.11.007.

European Commission. (2021). The EU Blue Economy Report 2021. Publications Office of the European Union, Luxembourg. [online] Available at: <https://op.europa.eu/s/perT> (Accessed: April 27th, 2022).

Fabinyi, M., Wu, A., Lau, S., Mallory, T., Barclay, K., Walsh, K. et al. (2021). China's blue economy: a state project of modernisation. J. Environ. Dev. 30(2): 127148. https://doi.org/10.1177/1070496521995872.

Fidelman, P., Evans, L., Fabinyi, M., Foale, S., Cinner, J. and Rosen, F. (2012). Governing large-scale marine commons: contextual challenges in the Coral Triangle. Mar. Policy 36(1): 42–53. https://doi.org/10.1016/j.marpol.2011.03.007.

Fidelman, P. I. J., Leitch, A. M. and Nelson, D. R. (2013). Unpacking multilevel adaptation to climate change in the Great Barrier Reef, Australia. Glob. Environ. Change 23(1): 800–812. https://doi.org/10.1016/j.gloenvcha.2013.02.016.

Folke, C., Hahn, T., Olsson, P. and Norberg, J. (2005). Adaptive governance of social-ecology systems. Annu. Rev. Environ. Resour. 30: 441473. https://doi.org/10.1146/annurev.energy.30.050504.144511.

Frohlich, M., Smith, T. F., Fidelman, P., Baldwin, C., Jacobson, C. and (Bill) Carter, R. W. (2022). Towards adaptive coastal management law: lessons from Australia and Brazil. Ocean Coast. Manag. 219: 106057. https://doi.org/10.1016/j.ocecoaman.2022.106057.

Frohlich, M., Fidelman, P., Dutton, I., Haward, M., Head, B. W., Maynard, D. et al. (2023). A network approach to analyse Australia's blue economy policy and legislative arrangements. Mar. Policy 151: 105588. https://doi.org/10.1016/j.marpol.2023.105588.

Fusco, L. M., Knott, C., Cisneros-Montemayor, A. M., Singh, G. G. and Spalding, A. K. (2022). Blueing business as usual in the ocean: blue economies, oil, and climate justice. Polit. Geogr. 98: 102670. https://doi.org/10.1016/j.polgeo.2022.102670.

Garmestani, A., Ruhl, J. B., Chaffin, B. C., Craig, R. K., van Rijswick, H. F. M. W., Angeler, D. G. et al. (2019). Untapped capacity for resilience in environmental law. Proc. Natl. Acad. Sci. 116(40): 1989919904. https://doi.org/10.1073/pnas.1906247116.

Germond-Duret, C. (2022). Framing the blue economy: placelessness, development and sustainability. Dev. Change 53(2): 308334. https://doi.org/10.1111/dech.12703.

Golden, J. S., Virdin, J., Nowacek, D., Halpin, P., Bennear, L. and Patil, P. G. (2017). Makin sure the blue economy is green. Nat. Ecol. Evol. 1: 0017. https://doi.org/10.1038/s41559-016-0017.

Goodman, C. (2022). Winds of change in Australian waters: the Offshore Electricity Infrastructure Act 2021. Asia Pacific J. Ocean Law and Policy 7(1): 137150. https://doi.org/10.1163/24519391-0701001.

Gray, B., Harder, J. and Bork, K. (2021). Implementing ecosystem-based management. Duke Environ. Law and Policy F. 31: 215–281.

Graziano, M., Alexander, K. A., McGrane, S. J., Allan, G. J. and Lema, E. (2022). The many sizes and characters of the blue economy. Ecol. Econ. 196: 107419. https://doi.org/10.1016/j.ecolecon.2022.107419.

Green, O. O., Schultz, L., Nekoro, M. and Garmestani, A. S. (2015). The role of bridging organizations in enhancing ecosystem services and facilitating adaptive management of social-ecological systems. pp. 107–122. *In*: Allen, C. R. and Garmestani, A. S. (eds.). Adaptive Management of Social-Ecological Systems. Springer, New York, USA. https://doi.org/10.1007/978-94-017-9682-8.

GWEC (Global Wind Energy Council). (2022). Global Offshore Wind Report 2022. [online] Available at: <https://gwec.net/wp-content/uploads/2022/06/GWEC-Global-Offshore-Wind-Report-2022.pdf> (Accessed: April 11th, 2023).

Halpern, B. S., Walbridge, S., Selkoe, K. A., Kappel, C. V., Micheli, F., D'Agrosa, C. et al. (2008). A global map of human impact on marine ecosystems. Science 319(5865): 948952. https://doi.org/10.1126/science.1149345.

Halpern, B. S., Frazier, M., Potapenko, J., Casey, K. S., Koenig, K., Longo, C. et al. (2015). Spatial and temporal changes in cumulative human impacts on the world's ocean. Nat. Commun. 6: 7615. https://doi.org/10.1038/ncomms8615.

Hahn, T., Olsson, P., Folke, C. and Johansson, K. (2006). Trust-building, knowledge generation and organizational innovations: the role of a bridging organizations for adaptive comanagement of a wetland landscape around Kristianstad, Sweden. Hum Ecol. 34(4): 573–592. https://doi.org/10.1007/s10745-006-9035-z.

Harvey, N. and Dew, R. E. C. (2016). Coastal impact of onshore wind farms in Australia. J. Coast. Res. 75(2): 992–996.

Harvey, N., Dew, R. E. C. and Hender, S. (2017). Rapid land use change by coastal wind farm development: Australian policies, politics and planning. Land Use Policy 61: 368–378. https://doi.org/10/1016/j.landusepol.2016.11.031.

Hassan, D. and Alam, A. (2019). Marine spatial planning and the Great Barrier Reef Marine Park Act 1975: an evaluation. Ocean Coast. Manag. 167: 188196. https://doi.org/10.1016/j.ocecoaman.2018.10.015.

Haward, M. (1989). The Australian offshore constitutional settlement. Mar. Policy 13(4): 334–348. https://doi.org/10.1016/0308-597X(89)90018-3.

Haward, M. (1996). Institutional framework for Australian ocean and coastal management. Ocean Coast. Manag. 33(1-3): 19–39. https://doi.org/10.1016/S0964-5691(96)00043-9.

Haward, M. and Vince, J. (2008). Australian ocean governance–initiatives and challenges. Coast. Manage. 37(1): 1–16. https://doi.org/10.1080/08920750802475368.

Haward, M. (2014). Sectoral approaches to Australian marine and ocean governance. Aust. J. Marit. Ocean Aff. 6(1): 1826. https://doi.org/10.1080/18366503.2014.888134.

Heenan, A., Gorospe, K., Williams, I., Levine, A., Maurin, P., Nadon, M. et al. (2016). Ecosystem monitoring for ecosystem-based management: using a polycentric approach to balance information trade-offs. J. Appl. Ecol. 53(3): 699–704. https://doi.org/10.1111/1365-2664.12633.

Hoffman, A. and Hildreth, R. (2016). Legal aspects of ecosystem-based management (EBM): implementation in Oregon coastal management. Int. J. Environ. Sustain. 12(4): 15–30. https://doi.org/10.18848/2325-1077/CGP/v12i04/15-30.

Hollander, R. (2010). Rethinking overlap and duplication: federalism and environmental assessment in Australia. Publius 40 (1): 136170. https://doi.org/10.1093/publius/pjp028.

Hughes, L. and Longden, T. (2022). Offshore wind will come to Australian waters – as long as we pave the way for this new industry. The Conversation (March 11th, 2022). [online] Available at: <https://theconversation.com/offshorewindwillcometoaustralianwatersaslongaswepavethewayforthisnewindustry178629?utm_medium=Social&utm_source=Facebook#Echobox=1646950595> (Accessed: September 16th, 2022).

IOC-UNESCO (Intergovernmental Oceanographic Commission of the United Nations Educational, Scientific and Cultural Organization). (2020). The Science We Need for the Ocean We Want: The United Nations Decade of Ocean Science for Sustainable Development (20212030). [online] Available at: <https://unesdoc.unesco.org/ark:/48223/pf0000265198 (Accessed: April 27th, 2022).

IOC-UNESCO (Intergovernmental Oceanographic Commission of the United Nations Educational, Scientific and Cultural Organization) and European Commission. (2021). MSPglobal: International Guide on Marine/Maritime Spatial Planning. [online] Available at: <https://unesdoc.unesco.org/ark:/48223/pf0000379196> (Accessed: September 20th, 2022).

IPCC (Intergovernmental Panel on Climate Change). (2019). IPCC Special Report on the Ocean and Cryosphere in a Changing Climate [Pörtner, H.-O., Roberts, D. C., Masson-Delmotte, V., Zhai, P., Tignor, M., Poloczanska, E. et al. (eds.)]. Cambridge University Press, Cambridge, UK and New York, NY, USA, 755 pp. https://doi.org/10.1017/9781009157964.

Jay, S. (2017). Marine Spatial Planning: Assessing Net Benefits and Improving Effectiveness. Issue Paper, 2017 GGSD Forum. OECD, Paris, France. [online] Available at: <https://www.oecd.org/greengrowth/GGSD_2017_Issue%20Paper_Marine%20Spatial%20Planning.pdf> (Accessed: September 20th, 2022).

Keen, M. R., Schwarz, A.-M. and Wini-Simeon, L. (2018). Towards defining the blue economy: practical lessons from pacific ocean governance. Mar. Policy 88: 333–341. https://doi.org/10.1016/j.marpol.2017.03.002.

Kirkfeldt, T. S., van Tatenhove, J. P. M. and Calado, H. M. G. P. (2022). The way forward on ecosystem-based marine spatial planning in the EU. Coast. Manag. 50(1): 29–44. https://doi.org/10.1080/08920753.2022.20068779.

Kowalski, A. A. and Jenkins, L. D. (2015). The role of bridging organizations in environmental management: examining social networks in working groups. Ecol. Soc. 20(2): 16. https://doi.org/10.5751/ES-07541-200216.

Lee, K.-H., Noh, J. and Khim, J. S. (2020). The blue economy and the United Nations' sustainable development goals: challenges and opportunities. Environ. Int. 137: 105528. https://doi.org/10.1016/j.envint.2020.105528.

Lloret, J., Turiel, A., Solé, J., Berdalet, E., Sabatés, A., Olivares, A. et al. (2022). Unravelling the ecological impacts of large scale offshore wind farms in the Mediterranean Sea. Sci. Total Environ. 824: 153803. https://doi.org/10.1016/j.scitotenv.2022.153803.

Macpherson, E., Urlich, S. C., Rennie, H. G., Paul, A., Fisher, K., Braid, L. et al. (2021). 'Hooks' and 'anchors' for relational ecosystem-based marine management. Mar. Policy 130: 104561. https://doi.org/10.1016/j.marpol.2021.104561.

Martínez-Vázquez, R. M., Milán-García, J. and de Pablo Valenciano, J. (2021). Challenges of the Blue Economy: evidence and research trends. Environ. Sci. Eur. 33: 61. https://doi.org/10.1186/s12302-021-00502-1.

Mauricio Hernandez, C. O., Shadman, M., Amiri, M. M., Silva, C., Estefen, S. F. and La Rovere, E. (2021). Environmental impacts of offshore wind installation, operation and maintenance, and decommissioning activities: a case study of Brazil. Renew. Sustain. Energy Rev. 144: 110994. https://doi.org/10.1016/j.rser.2021.110994.

Minister for Planning. (2020). Minister's reasons for decision. [online] Available at: <https://www.planning.vic.gov.au/environmentassessment/browseprojects/projects/star-of-the-south-offshore-wind-farm> (Accessed: April 12th, 2023).

Msomphora, M. R., Njaya, F. and Jentoft, S. (2022). Ecosystem-based governance according to the Malawi principles: a test for the southern Lake Malawi. Marit. Stud. https://doi.org/10.1007/s40152-022-00266-1.

OECD (Organization for Economic Co-operation and Development). (2019). Responding to Rising Seas: OECD Country Approaches to Tackling Coastal Risks. OECD Publishing, Paris. https://doi.org/10.1787/9789264312487-en.

O'Kane, J. and Todd, P. (2021). Proposal for Offshore Renewable Energy Legislation in Australia. Hall & Wilcox. [online] Available at: <https://hallandwilcox.com.au/thinking/proposal-for-offshore-renewable-energy-legislation-in-australia/> (Accessed: September 7th, 2022).

Ostrom, E. (2010). Polycentric systems for coping with collective action and global environmental change. Glob. Environ. Change 20(4): 550557. https://doi.org/10.1016/j.gloenvcha.2010.07.004.

Platjouw, F. M. (2019). Dimensions of transboundary legal coherence to foster ecosystem-based governance in the Artic. Mar. Policy 110: 103666. https://doi.org/10.1016/j.marpol.2019.103666.

Power, S. (2020). Environment Protection and Biodiversity Conservation Amendment (Streamlining Environmental Approvals) Bill 2020. Bill Digest No. 18, 2020–21, Parliament Library, Canberra. [online] Available at: <https://parlinfo.aph.gov.au/parlInfo/download/legislation/billsdgs/7600859/upload_binary/7600859.pdf;fileType=application/pdf> (Accessed: December 21st, 2022).

Qi, X. (2022). Building a bridge between economic complexity and the blue economy. Ocean Coast. Manag. 216: 105987. https://doi.org/10.1016/j.ocecoaman.2021.105987.

Rodriguez, N. J. I. (2017). A comparative analysis of holistic marine management regimes and ecosystem approach in marine spatial planning in developed countries. Ocean Coast. Manag. 137: 185–197. https://doi.org/10.1016/j.ocecoaman.2016.12.023.

Salvador, S., Gimeno, L. and Larruga, F. J. S. (2018). The influence of regulatory framework on environmental impact assessment in the development of offshore wind farms in Spain: issues, challenges and solutions. Ocean Coast. Manag. 161: 165176. https://doi.org/10.1016/j.ocecoaman.2018.05.010.

Samuel, G. (2020). Independent Review of the EPBC Act – Final Report. DAWE, Canberra, Australia. [online] Available at: <https://epbcactreview.environment.gov.au/> (Accessed: April 11th, 2023).

Scherer, M. E. G. and Asmus, M. L. (2016). Ecosystem-based knowledge and management as a tool for integrated coastal and ocean management. J. Coast. Res. 75(1): 690–694.

Schultz, L., West, S. and Florêncio, C. (2020). Nurturing adaptive governance through environmental monitoring: people, practices, politics in the Kruger to Canyons Biosphere Region, South Africa. pp. 293–318. *In*: Glückler, J., Herrigel, G. and Handke, M. (eds.). Knowledge for Governance. Springer, Cham, Switzerland. https://doi.org/10.1007/978-3-030-47150-7.

Silver, J. J., Gray, N. J., Campbell, L. M., Fairbanks, L. W. and Gruby, R. L. (2015). Blue economy and competing discourses in international oceans governance. J. Environ. Dev. 24(2): 135160. https://doi.org/10.1177/1070496515580797.

Smith, D. C., Fulton, E. A., Apfel, P., Cresswell, I. D., Gillanders, B. M., Haward, M. et al. (2017). Implementing marine ecosystem based management: lessons from Australia. ICES J. Mar. Sci. 74(7): 19902003. https://doi.org/10.1093/icesjms/fsx113.

Smith-Godfrey, S. (2016). Defining the blue economy. Marit. Aff.: J. Natl. Marit. Fdn. In. 12(1): 5864. https://doi.org/10.1080/09733159.2016.1175131.

Smyth, C. (2017). Marine spatial planning in Australia. pp. 371–398. *In*: Kitsiou, D. and Karydis, M. (eds.). Marine Spatial Planning: Methodologies, Environmental Issues and Current Trends. Nova Science Publishers, New York, USA.

Star of the South. (2023a). Project overview. [online] Available at: <https://www.starofthesouth.com.au/resources> (Accessed: April 12th, 2023).

Star of the South. (2023b). Project map. [online] Available at: <https://www.starofthesouth.com.au/resources> (Accessed: April 12th, 2023).

Stewart, J. and Tyler, M. E. (2019). Bridging organizations and strategic bridging functions in environmental governance and management. Int. J. Water Resour. Dev. 35(1): 71–94. https://doi.org/10.1080/07900627.2017.1389697.

Taylor, M. (2022). Marine spatial planning and offshore wind: preliminary principles for Australia. RELP 11(2-3): 48–56. https://doi.org/10.4337/relp.2022.02-03.02.

Taylor, M. and Taylor, R. G. (2022). Planning for offshore wind energy: the case for marine spatial planning in Australia. OGEL 5. [online] https://www.ogel.org/article.asp?key=4050 (Accessed: October 6th, 2023).

Tuda, A. O., Kark, S. and Newton, A. (2021). Polycentricity and adaptive governance of transboundary marine socio-ecological systems. Ocean Coast. Manag. 200: 105412. https://doi.org/10.1016/j.ocecoaman.2020.105412.

UN (United Nations). (2021). The Second World Ocean Assessment. [online] Available at: <https://www.un.org/regularprocess/woa2launch> (Accessed: August 10th, 2022).

Urlich, S. C., White, F. R. and Rennie, H. G. (2022). Characterising the regulatory seascape in Aotearoa New Zealand: Bridging local, regional and national scales for marine ecosystembased management. Ocean Coast. Manag. 224: 106193. https://doi.org/10.1016/j.ocecoaman.2022.106193.

Vignola, R., McDaniels, T. L. and Scholz, R. W. (2013). Governance structures for ecosystem-based adaptation: using policy-network analysis to identify key organizations for bridging information across scales and policy areas. Environ. Sci. Policy 31: 7184. https://doi.org/10.1016/j.envsci.2013.03.004.

Villalba, J., Abdussamie, N., Aryai, V., Nikolova, N., Tenekedjiev, K., Wang, C.-M. et al. (2022). Assessment of uncertain alternatives for co-located aquaculture and offshore wind farm in Tasmania. Ocean Eng. 249: 110949. https://doi.org/10.1016/j.oceaneng.2022.110949.

Vince, J. (2013). Marine bioregional plans and implementation issues: Australia's oceans policy process. Mar. Policy 38: 325–329. https://doi.org/10.1016/j.marpol.2012.06.009.

Vince, J. Z. (2014). Oceans governance and marine spatial planning in Australia. Aust. J. Marit. Ocean Aff. 6(1): 5–17. https://doi.org/10.1080/18366503.2014.888137.

Vince, J., Smith, A. D. M., Sainsbury, K. J., Cresswell, I. D., Smith, D. C. and Haward, M. (2015). Australia's Ocean Policy: past, present and future. Mar. Policy 57: 18. https://doi.org/10.1016/j.marpol.2015.02.014.

Virtanen, E. A., Lappalainen, J., Nurmi, M., Viitasalo, M., Tikanma, M., Heinonen, J. et al. (2022). Balancing profitability of energy production, societal impacts and biodiversity in offshore wind farm design. Renew. Sustain. Energy Rev. 158: 112087. https://doi.org/10.1016/j.rser.2022.112087.

Voyer, M., Quirk, G., McIlgorm, A. and Azmi, K. (2018). Shades of blue: What do competing interpretations of the blue economy mean for oceans governance? J. Environ. Policy Plan. 20(5): 595616. https://doi.org/10.1080/1523908X.2018.1473153.

Voyer, M. and van Leeuwen, J. (2019). 'Social license to operate' in the blue economy. Resour. Policy 62: 102–113. https://doi.org/j.resourpol.2019.02.020.

Werle, D., Boudreau, P. R., Brooks, M. R., Butler, M. J., Charles, A., Coffen-Smout, S. et al. (2019). Synthesis: looking ahead: ocean governance challenges in the twenty-first century. pp. 533–541. *In*: Boudreau, P. R., Brooks, M. R., Butler, M. J. A., Charles, A., Coffen-Smout, S., Griffiths, D. et al. (eds.). The Future of Ocean Governance and Capacity Development. Brill | Nijhoff, Leiden, The Netherlands. https://doi.org/10.1163/9789004380271_094.

Wifa, E. and Hunter, T. S. (2022). Mitigating occupational health and safety risks in the proposed Australian offshore wind energy industry: lessons from the safety case regime. J. Energy Nat. Resour. Law 40(1): 83–104. https://doi.org/10.1080/02646811.2021.1879547.

Wondolleck, J. M. and Yaffe, S. L. (2017). Drawing lessons from experience in marine ecosystem-based management. pp. 1–12. *In*: Wondolleck, J. M. and Yaffe, S. L. (eds.). Marine Ecosystem-Based Management in Practice: Different Pathways, Common Lessons. Island Press, Washington DC, USA.

Zhang, Z., Qu, F. and Wang, S. (2019). Sustainable development of the Yellow Sea Large Marine Ecosystem. Deep Sea Res. Part II 163: 102107. https://doi.org/10.1016/j.dsr2.2018.08.009.

Zuercher, R., Ban, N. C., Flannery, W., Guerry, A. D., Halpern, B. S., Magris, R. A. et al. (2022). Enabling conditions for effective marine spatial planning. Mar. Policy 143: 105141. https://doi.org/10.1016/j.marpol.2022.105141.

CHAPTER 13

Marine Ecosystem Restoration

Roberto Danovaro

From the problems to the solutions

The ocean stabilizes climate, stores carbon, produces oxygen, nurtures biodiversity, and directly supports human well-being through food, mineral, and energy resources; it also provides cultural and recreational services (Pendleton et al. 2020, IPBES 2019). The value of the ocean economy (transportation, fishing, tourism, and energy) has been estimated ca $3 trillion USD annually by the Organization for Economic Cooperation and Development (OECD 2016).

Humans, through multiple and synergistic impacts (global development, resource extraction, contamination, climate change) are responsible for the progressive degradation and loss of natural ecosystems, with consequent habitat destruction and progressive loss of biodiversity (UN 2016, Worm et al. 2006, Butchart et al. 2010).

Direct and indirect human pressures on marine ecosystems will likely increase in the next few decades, leading to significant loss of marine habitats and their biodiversity, and impairment of ecosystem functions (Gerovasileiou et al. 2019, Bekkby et al. 2020). These activities are degrading natural capital and marine habitat worldwide with consequent relevant economic losses (Blignaut et al. 2013). Some coastal habitats urgently need protection; seagrass meadows, coral reefs, and kelp forests have declined in spatial extent by 25 to 80% or more, depending on the location. Coral reefs are particularly threatened with projected further declines of 70 to 90% with an ocean temperature increase of 1.5°C and larger losses at 2°C. Studies documented a 35% decline in both marine/coastal and inland natural wetland habitats between 1970 and 2015. The unprecedented declines in marine vegetated ecosystems through anthropogenic impacts have compromised associated essential ecosystem services such as coastal protection and carbon sequestration (Barbier 2017). Despite the development of policies to protect ecosystems from further degradation (see the EU Agenda 2030 that aims to protect 30% of the marine environment by the year 2030), many marine ecosystems have become degraded to the point that effective recovery cannot happen without human intervention, and thus require active restoration projects.

There are three levels of response to contrast and reverse the current trend of habitat and biodiversity loss in the ocean: (1) the so called "zero pollution", which can be translated in the removal of contaminants (including plastics and emerging pollutants) and of the source of contamination which impact marine ecosystems along with the removal of stressors to marine ecosystems with specific actions to mitigate the climate change; (2) the protection of the pristine habitat to promote the recovery of species and populations; (3) the restoration of the degraded habitats (Fig. 1).

Department of Life and Environmental Sciences, Polytechnic University of Marche, Italy.
Email: r.danovaro@univpm.it

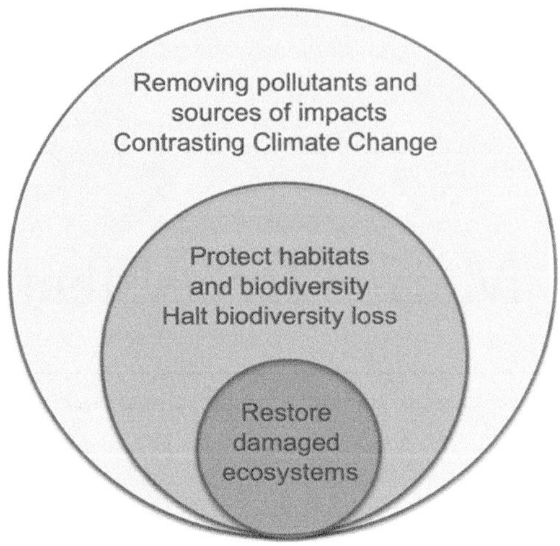

Figure 1. Schematic image of the solutions to reverse current marine ecosystem degradation: concentric approach spanning from the removal of contaminants and mitigation of climate change to the protection of habitats and biodiversity to the restoration of degraded marine habitats.

Environmental managers worldwide, and the EU (through a restoration law) are promoting ecological restoration as a direct response to the degradation and destruction of ecosystems to recover biodiversity, increasing food and water security, delivering goods, services, and economic prosperity, and supporting climate change mitigation, resilience, and adaptation (Gann et al. 2019, Rohr et al. 2018). The ecosystem restoration is defined by the United Nations (UN General Assembly 2019) as *"assisting in the recovery of degraded, damaged, and destroyed ecosystems to regain ecological functionality and provide the goods and services that people value*".

The UN has committed to halting and reversing the decline in the health and productivity of the global ocean and its ecosystems and to protecting and restoring its resilience and ecological integrity. The UN General Assembly of the Conference of the Parties to the Convention on Biological Diversity designated the decade 2021–2030 as the United Nations Decade of Ocean Sciences for Sustainable Development and the United Nations Decade of Ecosystem Restoration (Aronson et al. 2020, CBD 2010).

These designations aim to support and scale up efforts to prevent, halt, and reverse the degradation of ecosystems worldwide and raise awareness of the importance of successful ecosystem restoration. Ecosystem restoration will also contribute to the Paris Agreement adopted under the United Nations Framework Convention on Climate Change, and to achieving the Aichi Biodiversity Targets 10 and the post-2020 global biodiversity framework. These restoration initiatives encourage all countries to:

i) mobilize resources, capacity-building, and scientific research, cooperation, and momentum for ecosystem restoration at global, regional, national, and local levels;
ii) mainstream ecosystem restoration into policies and plans to address current national development priorities and challenges resulting from the degradation of marine and terrestrial ecosystems, biodiversity loss, and climate change vulnerability, thereby creating opportunities for ecosystems to increase their adaptive capacity and enhance opportunities to maintain and improve livelihoods for all;
iii) build on and reinforce existing restoration initiatives in order to scale up good practices;
iv) promote the sharing of experiences and good practices in ecosystem conservation and restoration (Possingham et al. 2015).

The success of the UN Ecosystem Restoration decade will largely depend on our ability to integrate six practical strategies: (1) incorporate ecological actions with physical and chemical metrics as key, explanatory variables, and working at effective spatial and temporal scales; (2) include traditional ecological knowledge to promote eco-cultural relationships; (3) collaborate with allied movements and organizations with similar goals for achieving human and ecological health; (4) advance and apply soil microbiome science and technology in restoration science and applications; (5) provide training and capacity-building opportunities to inspire and educate communities and practitioners; and (6) identify and communicate relationships between ecosystem health and human health. The strategies outlined for terrestrial ecosystems can be transposed to the marine ecosystems and considered priorities for making the ecological restoration in the global ocean effective. Whatever is the approach to be followed, ecosystem restoration is now a global priority (Aronson and Alexander 2013).

Passive and active ecological restoration

The International Standards for the Practice of Ecological Restoration clearly distinguish between "ecological restoration" and other forms of ecosystem repair (see Box 1). Although restoration refers to the process of "assisting the recovery of ecosystems that have been degraded, damaged, or destroyed", the "recovery" of an ecosystem refers to the achievement of a target environment similar to an appropriate local native model or reference ecosystem, in terms of its specific compositional, structural, and functional attributes. Restoration includes an action or multiple actions that jumpstart recovery and place a degraded ecosystem on a trajectory for recovery, regardless of the period required to achieve the recovery outcome (Fig. 2). Passive restoration relies on the spontaneous resilience of an ecosystem to recover simply by removing the environmental stressor(s) that led to degradation, in contrast to active restoration that relies on active interventions to the ecosystem, whether by adding or removing structures, species, or stressors. Ecosystem restoration principally aims to establish a self-supporting habitat similar to the 'original' habitat prior to the impact. The selection of the desired reference system guides ecological restoration and should be both historically inspired and grounded in social processes that include multiparty stakeholders and restoration scientists and practitioners. Application of a group of restoration actions at a site can help in achieving restoration goals, including remediation, reparation, and rehabilitation.

Rehabilitation strives to simply replace structural or functional characteristics damaged by an impact, and to enhance the social, economic, and ecological value of the new ecosystem, thus representing a less ambitious goal than restoration in not necessarily achieving full recovery of processes and returning to "pre-disturbance" conditions. To define the targets of restoration requires identifying "reference ecosystems", referring to "undamaged sites". Alternatively, scientists define reference ecosystems based on multiple sources of information on diverse ecological and biological variables (e.g., biodiversity, life cycles, functional variables, and food webs) supported by abiotic measurements. In any case, restoration planning requires clear definitions of ecosystem baselines. Full recovery requires several steps, each encompassing a range or family of restorative activities that span a continuum of different restorative practices: reduced societal impacts, remediation, rehabilitation, and ecological restoration (O'Connor et al. 2021).

Ecological restoration experts have abandoned the idea of total recovery, recognizing the dynamic nature of ecosystems, and that restoration may not follow the appropriate ecological trajectory to reestablish the original community over time, especially considering the need for ecosystems to adapt and evolve over time in response to climate change.

This solutions-based approach engages communities, scientists, policymakers, and land managers to repair ecological damage and rebuild a healthier relationship between people and nature. When combined with conservation and sustainable use, ecological restoration provides the necessary link to move local, regional, and global environmental conditions from a state of ongoing degradation, to one of net positive improvement.

Historically, research on restoration best practices and methods has mainly focused on terrestrial, rather than on marine ecosystems. Although some of the basic principles developed for terrestrial systems translate to applications in marine environments, the knowledge base regarding factors that enhance or constrain restoration success remains quite limited for marine environments (Wortley et al. 2013).

For example, we now recognize that the complex, expensive, and lengthy process involved in restoration of degraded ecosystem requires considerable time, resources, and knowledge. Restoration actions in the marine environment will require different approaches depending on the specific habitat and restoration activities.

Box 1: Glossary of marine ecosystem restoration

- **Assisted regeneration:** An approach to restoration that focuses on actively triggering any natural regeneration capacity of biota remaining on a site or nearby, as distinct from reintroducing biota to the site or leaving a site to regenerate. Interventions can include removal of pest organisms, reapplying ecological disturbance regimes, and installation of resources to prompt colonization.
- **Baseline conditions:** The condition of a restoration site immediately prior to the initiation of ecological restoration activities.
- **Damage (to ecosystem):** An acute and obvious deleterious impact on an ecosystem.
- **Degradation (of an ecosystem):** A level of deleterious human impact to ecosystems that results in the loss of biodiversity and simplification or disruption in their composition, structure, and functioning, generally leading to a reduction in the flow of ecosystem services.
- **Destruction (of an ecosystem):** When degradation or damage removes all macroscopic life, and commonly ruins the physical environment of an ecosystem.
- **Disturbance regime:** The pattern, frequency, timing, or occurrence of disturbance events that characterize an ecosystem over a given period.
- **Ecological restoration:** The process of assisting the recovery of a degraded, damaged, or destroyed ecosystem. Although some researchers use the terms ecosystem restoration and ecological restoration interchangeably, ecological restoration always addresses biodiversity conservation and ecological integrity, whereas some approaches to restoration may focus solely on ecosystem services.
- **Ecological restoration project:** Any organized effort to achieve substantial recovery of a native ecosystem, from the planning stage through implementation and monitoring. As a long-term program, such a project may require multiple agreements or funding cycles.
- **Ecosystem resilience:** The degree, manner, and pace of recovery of ecosystem properties after natural or human disturbance. In plant and animal communities, this property strongly depends on adaptations by individual species to disturbances or stresses experienced during the species' evolution.
- **Full recovery:** The state whereby all ecosystem attributes closely resemble those of the reference (model) ecosystem. An ecosystem exhibiting self-organization that leads to the full resolution and maturity of ecosystem attributes precedes full recovery. At the point of self-organization, the restoration phase could be considered complete and management shifts to a maintenance phase.
- **Indicators (of recovery):** Characteristics of an ecosystem that scientists can use for measuring progress towards restoration goals or objectives at a particular site (e.g., measures of presence/absence and quality of biotic or abiotic components of the ecosystem).
- **Natural regeneration:** Germination, birth, or other recruitment of biota including plants, animals, and microbiota, that does not involve human intervention.
- **Natural (or spontaneous) regeneration approach:** Ecological restoration that relies only on increases in individuals following removal of causes of degradation (non-assisted regeneration approach).
- **Recovery:** The process by which an ecosystem regains its composition, structure, and function relative to the levels identified for the reference ecosystem. Restoration activities usually assist restoration - and recovery can be partial or full.
- **Reference condition:** See above for "Baseline conditions".
- **Rehabilitation:** Management actions that aim to reinstate a level of ecosystem functioning on degraded sites, where the goal is renewed and ongoing provision of ecosystem services rather than the biodiversity and integrity of a designated native reference ecosystem.

Box 1 contd. ...

...Box 1 contd.

- **Restoration activities:** Any action, intervention, or treatment intended to promote the recovery of an ecosystem or component of an ecosystem, such as substrate amendments, control of invasive species, habitat conditioning, species reintroductions, and population enhancements.
- **Restorative activities:** Activities (including ecological restoration) that reduce degradation or improve conditions for the partial or full recovery of ecosystems.
- **Restorative continuum:** A spectrum of activities that directly or indirectly support or attain at least some recovery of lost or impaired ecosystem attributes.
- **Translocation:** The intentional transportation (by humans) of organisms to a different location.
- **Transplanting:** The translocation of species and/or portions of habitat from one site to another.

Figure 2. Schematic representation of the restoration target: reversing degraded habitats into three-dimensional, highly diverse habitats often dominated by habitat-forming species. Images illustrate: (i) restoration of seagrass meadows (top); (ii) restoration of algal forests and assemblages on hard bottoms (middle); restoration of deep-sea habitats (bottom).

The major constraints to marine restoration

Despite numerous attempts to restore marine ecosystems globally in the last decade, the discipline remains a very young branch of ecosystem restoration, when compared with initiatives carried out in terrestrial ecosystems (Fraschetti et al. 2022, Saunders et al. 2020). The experience acquired from terrestrial ecosystem restoration indicated that solution-based approaches require the engagement of society, scientists, policymakers, and land managers to repair ecological damage (Gann et al. 2019).

Three major logistical constraints hinder marine ecosystem restoration: (1) the long timescales (sometimes spanning several decades or centuries) required to achieve restoration targets (especially for the deep-sea species); (2) the substantial funding and high-technology equipment needed, particularly in the deep sea; and (3) the difficulty in scaling up any restoration intervention to sufficiently large spatial scales to achieve significant impact. The upfront costs of marine ecosystem restoration create a particularly challenging impediment.

Experience accumulated in terrestrial habitats greatly depend upon (i) time; (ii) nature, severity, and extent of degradation, (iii) native biodiversity, (iv) history of the human uses and misuses, (v) edaphic and climatic conditions, (vi) availability of resources, stakeholders support and involvement, and importantly, (vii) effectiveness of the restoration technologies employed (Abhilash 2021). A synthesis of land restoration projects identifies an additional four key elements: (1) the geographic area under restoration; (2) amount of funding invested in restoration; (3) source and limitations of funding, and (4) associated monitoring efforts (Coppus et al. 2019). These basic elements of restoration practice apply equally to terrestrial and marine systems.

In addition, restoration of degraded ecosystems, a complex, expensive, and lengthy process, requires significant time, resources, and knowledge but working 'with' nature and using simple ecological engineering approaches may provide cost-effective solutions (Mitsch 2014). Besides the intrinsic difficulties with working in marine environments, restoration of marine ecosystems has elevated associated costs (Bayraktarov et al. 2016), especially in remote habitats such as the deep sea (Van Dover et al. 2014, Howell et al. 2021). As a result, fundamental gaps in methodologies and scientific knowledge continue to limit ecological restoration in the marine realm (e.g., Elliott et al. 2007, Suding 2011, Duarte et al. 2015). Despite the availability of appropriate technologies, development of effective restoration practices that achieve their potential in ocean environments will require a concerted effort (Abelson et al. 2020, Danovaro et al. 2020).

Ecological restoration of vulnerable marine habitats

Researchers have recently begun to explore the potential for restoration in both shallow soft and hard bottoms (including mesophotic habitats) and in the deep sea. Not surprisingly, restoration actions focus largely on seafloor environments and particularly on ecosystem engineers with specific reference to habitat-forming species.

The most fragile and vulnerable costal and deep-sea habitats include: coral reefs (particularly stony corals), seagrass meadows (*Zostera marina*, *Z. noltii*, *Cymodocea nodosa*, *Posidonia oceanica*), macroalgal forests (including kelp forests), coralligenous outcrops (including *Corallium rubrum*, *Paramuricea clavata*, *Eunicella singularis*, *E. cavolini*), deep or cold-water corals (*Callogorgia verticillata, Paracalyptrophora josephinae, Viminella flagellum, Lophelia pertusa*) inhabiting canyons, seamounts and fjords. Many diverse habitat types could benefit from habitat restoration.

Restoration includes different practices such as species translocation (including transplanting), seedling and grazer removal, the use of artificial (and biodegradable) substrates, or the re-introduction of native species that can facilitate the recovery of specific endangered species and habitats. Each pilot action must consider "restoration efficiency" in order to identify the criteria for the selection of target species and habitats likely to promote successful restoration activities. The experience acquired to date provides evidence that extreme/episodic events (e.g., storms, heat waves) can compromise restoration activities and survival of target species in marine shallow habitats, irrespective of the methodologies utilized for marine ecological restoration. Experience to date suggests that successful marine ecological restoration requires relevant policy decisions, social awareness, adequate financing, and engagement with the private sector.

Recent studies demonstrate that promoting positive interactions between individuals of the same species can increase restoration success, highlighting the importance of facilitative interactions in restoring ecosystem-engineering species. Facilitative interactions among ecosystem engineers may also promote resilience and recovery, but < 3% of restoration projects have integrated interspecific interactions.

The success of restoration actions at large scales varies across ecosystems, and the success of restoration largely depends on the status of anthropogenic pressures. The knowledge developed in pilot actions can help in identifying the environmental conditions necessary to optimize restoration success as well as the best practices (protocols) to apply at larger scales to expedite the recovery of degraded habitats and endangered species in coastal soft and hard bottoms ecosystems.

Coral reef restoration

The complex biological and ecological procedure and the task of emulating natural phenomena of reef biology creates significant challenges in restoring any type of degraded reef habitat. Coral reef restoration has become an increasingly important tool in supporting conservation measures. This type of strategy requires defining the ecological bases and most effective protocols under different environmental constraints appropriate to specific local species assemblages.

During the last decades, coral-reef restoration/rehabilitation efforts have focused on three main approaches: (1) structural restoration; (2) physical restoration and (3) biological restoration.

1) **Structural restoration** involves the construction of habitat structures, in this case artificial reefs, sinking of wrecks, or relocation of rocks/dead coral heads. Structural restoration aims to increase the amount of reef structure and habitat available for corals themselves as well as to create substrate on which other reef organisms can grow. Environments in which disturbance has led to reef loss typically require structural restoration (Fig. 3).
2) **Physical restoration** involves addressing the environmental conditions under which corals grow in order to improve their health, growth rates, or reproductive ability. Some of the techniques used include mid-water coral nurseries placed in locations with superior water quality (such as in the open ocean) but maintaining the ambient light conditions for which the corals and their zooxanthellae are adapted. High survival rates and faster coral growth than similar colonies on the natural reef characterize corals growing on mid-water nurseries because of the decreased stress caused by sedimentation, eutrophication, predation, or pollution on the reef itself.
3) **Biological restoration** usually involves increasing the amount of living corals on the reef in locations with available structure. Scientists generally achieve this objective through methods such as collecting and rehabilitating naturally broken coral fragments, propagating coral colonies, or through direct transplantation of coral fragments or colonies. These techniques align with the concept that the newly established coral colonies in the denuded reefs should originate from an external habitat as similar as possible to the recipient site. However, the techniques used so far for the removal of coral materials, their transportation, and re-attachment (e.g., without an acclimatization period) could compromise the donor sites and the transplanted material. As such, the varying degrees of success in past coral restoration experiments should come as no surprise.

Figure 3. Example of coral restoration conducted through coral gardening on artificial substrates (left) and transplantation of colonies from a donor site to a degraded coral reef (right).

Recently, scientists proposed the "coral gardening" tenet, a strategy associated with the idea of terrestrial forest plantations, as an alternative approach to direct transplantation of corals. This strategy follows a two-step restoration protocol. The first stage involves growing large numbers of farmed nubbins and small coral fragments within specially designed underwater coral nurseries. Deployment of these nurseries in sheltered zones nurture small coral fragments to sizes suitable for transplantation. In the second phase, divers transplant different sizes and species combinations of the nursery-grown coral colonies to degraded reef sites. This approach utilizes the theories and vast experience accrued over many years of terrestrial silviculture. Compared to *ex situ* efforts, comparatively cheap *in situ* coral farming can potentially produce large quantities of coral fragments (thousands of individuals), an important pre-requisite for broad scale reef-restoration.

Furthermore, the establishment of *in situ* nurseries in habitats similar to recipient sites and in sheltered locations may provide cultured corals that have received an appropriate (indeed essential) acclimation period for increasing post-transplantation survivorship and growth. So far, researchers have conducted *in situ* coral "farming" experiments in the Red Sea, East Africa, Southeast Asia, and the Pacific. Varying results from these experiments further emphasize the need to develop protocols specifically adapted to locations characterized by different environmental conditions and coral assemblages.

Seagrass restoration

The choice of restoration techniques and time scales for recovery can vary significantly amongst different seagrass species. For example, *Posidonia oceanica,* a slow growing species, requires longer time scales for recovery than *Cymodocea nodosa* and *Zostera marina*, which exhibit faster clonal growth. Therefore, restoration efforts should align with such specific time frames. However, the extreme vulnerability of seagrass meadows to anthropogenic pressures points to the importance of reducing or removing critical pressures, such as eutrophication (which limits light availability and growth) and habitat destruction, and re-establishing appropriate sediment conditions (Horn et al. 2021). Supportive policies and management should aim to reduce such pressures in combination with restoration projects.

Seagrass restoration generally utilizes two main approaches: (1) transplanting a portion of healthy seagrass into a degraded location (Fig. 4). This approach raises the possibility of transplanting single rhizomes or an entire portion of the habitat (typically in the range of 1 m^2 and including the sediment, the root mass, and the above sediment plants); (2) by implanting *plantulae* (juvenile plants deriving from cultured seagrass seeds) into the degraded habitat. Selection of the receiving system should consider existing pressures, but also refer to historical conditions. For example, restoration efforts should prioritize locations that have hosted seagrass meadows in the past (Da Ros et al. 2021).

Recent studies demonstrate that positive interactions between epifaunal bivalves and plants dominate marine habitats. Plants primarily promote bivalve survival and abundance by providing substrate and shelter, whereas bivalves promote plant growth and survival by stabilizing and fertilizing sediments and reducing water turbidity through filtration. The prevalence of positive interactions increases with water temperature in subtidal habitats. If done properly, co-restoration could increase initial survival, persistence, and resilience of foundation species, and promote the recovery of associated biodiversity and ecosystem services.

A long-term process requires long-term monitoring, and seagrass meadows undergoing restoration efforts require long-term management to monitor response of the ecosystem as a whole. Additional actions might include the introduction of healthy populations of associated species, especially top predators, which can control algal (over)growth through trophic cascades. Researchers have repeatedly identified large-scale planting as an important method for increasing success of seagrass restoration. Greater spatial distribution allows greater opportunity for transplanting donor populations, which increases the probability of recovery success. Therefore, experts strongly encourage active restoration of seagrass meadows at large spatial scales.

Figure 4. Example of restoration of the seagrass Posidonia oceanica in the Mediterranean Sea. The use of a hard substrate elevated from the seafloor might increase the survival of the transplanted seagrass, while ensuring a robust anchorage of the artificial bottom.

Restoration of macroalgal forests

Macroalgal restoration techniques have been considerably developed and refined in recent years (Fig. 5). Kelp-forest restoration requires knowledge of interactions with local predators (such as sea urchins), prevalence of turf algae, as well as an understanding of local and global conditions. Removal of all existing anthropogenic pressures (e.g., from urbanization, eutrophication, and increasing sediment

Figure 5. Example of restoration of macroalgal forests. Reported are: restoration of kelp forests (upper panels) and enhancement techniques (mid and lower panels) in which earthenware substrates collect propagules of Cystoseira, where the tiny propagules grow until they reach a visible size of a few cm prior to transfer to the receiving site, where they can grow further and colonize new surfaces.

loads) should precede restoration because many coastal regions have experienced loss of such forests or transformation into simpler and less productive communities such as barrens and algal turfs as a result of these pressures, which can hinder restoration efforts. Kelp forest restoration might entail transplants from adjacent sites or promotion through assisted restoration where kelp typically show rapid recovery rates following removal of sea urchins and eutrophication. Policies and management practices to support ecosystem recovery should aim to reduce such pressures in combination with restoration efforts. Implementation of restoration actions for kelp forests at large spatial scales in suitable areas typically improves success of restoration efforts and consequently the potential economic returns (Hynes et al. 2021).

Successful planning of restoration of brown algal forests hinges upon understanding growth and health of brown algal forests. Significant loss of such habitat has occurred in several rocky shore habitats of the Mediterranean Sea. Active restoration planting should avoid situations where the status of natural and donor populations reached a critical state; active restoration (planting), and restoration should instead rely on recruitment enhancement and growth of propagules (Fig. 3). In this case, we should expect prolonged recovery for restoration, possibly decades. An optimal restoration practice should combine two approaches: (i) recruitment enhancement techniques, for instance thought the cultivation of propagules on appropriate substrates and transplanting them into receiving habitats, and (ii) sea urchin eradication to control their impact on transplanted algae. These approaches appear to be the most effective approach for *Cystoseira* forestation in degraded shallow water barrens. The maintenance and efficacy of restoration efforts also depends on reducing human pressures, such as eutrophication, chemical pollution, coastal development, and sedimentation. Supportive policies and management interventions should aim to reduce such pressures in combination with restoration efforts.

Oyster reef restoration

Oyster reefs form three-dimensional structures that support diverse and abundant assemblages, and provide important ecosystem services such as coastal protection, water filtration, fisheries productivity and carbon sequestration (Alleway and Connell 2015). Oyster reefs have experienced an 85% decline during the 1700s to early 1900s largely due to overharvest using destructive fishing practices (Gillies et al. 2018). Despite subsequent reductions in fishing pressure, oyster reefs did not recover, thus requiring active restoration interventions. Oyster reefs restoration projects have been focused on the eastern oyster *Crassostera virginica* (Luckenbach et al. 1999), on the Olympia oyster, *Ostrea lurida* (Brumbaugh and Coen 2009, White et al. 2009) and more recently on a number of species, such as: *Saccostrea glomerata, Ostrea edulis, Ostrea angasi, Magallana sikamea, Magallana hongkongensis* (Fitzsimons et al. 2020). Projects conducted at more degraded sites, or sites that are projected to undergo significant environmental change may require greater restoration efforts than those conducted at less degraded sites (McDonald et al. 2016). At the site-scale, three key environmental requirements must be met for reef establishment to occur: (1) site conditions must be suitable for oyster survival, growth and reproduction; (2) hard substrate or other shell reefs should be available; and (3) there must be adequate larval supply to support reef growth. Once the target site is identified, restoration is typically conducted by means of a gardening approach (i.e., oysters are grown in large tanks and then transplanted in target areas). Oyster shells are important for restoration efforts, but they are becoming increasingly scarce. For this reason, oyster shell recycling programs have been developed to recover empty shells from restaurants to be reused in restoration projects. The recycled empty shells are cleaned and placed in huge water tanks containing millions of oyster larvae, which then attach to the shells. Each recycled shell can become home to dozens of those baby oysters, called spat.

Restoring coralligenous habitats

A combination of nutrient enrichment, invasive species, increase of sedimentation, and mechanical impacts, mainly from fishing activities as well as climate change, currently threaten coralligenous assemblages. Managers should prioritize reduction of these pressures prior to starting any restoration

Figure 6. Transplantation of sessile invertebrates: Researchers collect colonies of sessile species (e.g., red corals, sponges and gorgonians) from donor sites and transplant them onto defaunated areas (left panel). Alternatively, the species are reared in aquaria prior to transplanting (right).

actions. Experimental restoration of coralligenous habitats at the local scale has begun in pilot areas characterized by highly fragmented habitats along NW Mediterranean coasts following mass-mortality events. Active restoration on hard bottoms generally focuses on the recovery of structural and/or habitat-forming species that can provide habitat for associated species. Restoration actions can involve gorgonians and red corals, or sponges, which transplant relatively simply with good success rates, resulting in high survival rates and requiring relatively low initial efforts (Fig. 6). However, in the case of transplanting of slow growing, long-lived coralligenous habitat species with limited recruitment, such as sponges (*Petrosia fisciformis*, *Spongia lamella*, *S. officinalis*) and octocorals (*Paramuricea clavata*, *Corallium rubrum*), restoration can take decades. Some other taxa, such as bryozoans (*Pentapora fascialis*), grow at much faster rates and can restore structural complexity within five to ten years.

Depth represents the main limitation for restoration of hard bottom habitats. Work requiring SCUBA divers, long diving times, and considerable expertise can occur within the top 30–40 m depth, but deeper environments create considerable safety challenges.

Restoration of deep-sea habitats

Ecological restoration of deep-sea habitats is particularly challenging because it requires advanced underwater technologies, complex logistics, and significant economic investments compared to shallow-water environments. Future availability of more accessible, cost-effective underwater technology (such as relativity Autonomous Underwater Vehicles or cabled observatories) could enable broader application and upscaling of restoration of vulnerable and critical deep-sea habitats, such as deep-water corals and gorgonian forests, hydrothermal vents, and methane seeps. Restoration of deep-sea ecosystems will surely challenge future decision makers evaluating the costs and benefits of maintaining ecosystem functioning and provisioning of goods against the economic benefits of exploitation. Developing new approaches and methodologies to support and accelerate natural recovery of deep-sea habitats affected by trawling and oil exploitation, and any future mining, remains a major scientific research challenge.

Promoting successful restoration requires the removal of existing stresses and pressures. Cold-water coral habitats are highly sensitive to a range of human activities, including mineral and hydrocarbon

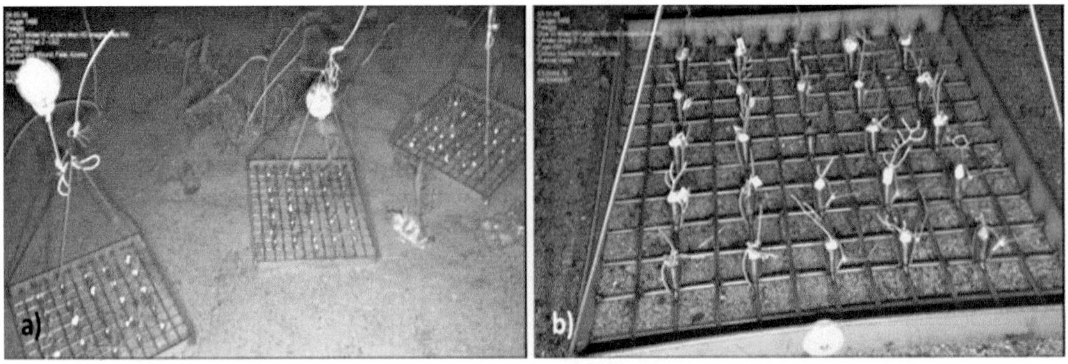

Figure 7. Transplant of sessile invertebrates from a donor site to a degraded deep-sea area.

exploration and extraction, and commercial bottom fisheries. Many experts consider the latter threat as the single greatest pressure, often resulting in the removal of entire communities. The slow recovery potential of cold-water corals points to the need to prioritize the reduction of such pressures prior to and during long-term restoration actions (Da Ros et al. 2019).

The costs required for restoration along with uncertainties of the outcomes raise major concerns, as restoration of deep-sea ecosystems could easily cost 3–4 orders of magnitudes more than for shallow water ecosystems. Pilot actions carried out in the framework of several international projects provide additional evidence in support of the feasibility of deep-sea restoration using transplant techniques (following rearing and acclimatization; Fig. 7) or artificial habitats that enhance larval recruitment of some deep-sea species. A combination of passive and active restoration will optimize restoration success and reduce potential costs. However, this strategy requires time, and full recovery of damaged deep-sea ecosystems within a few years rather than decades or centuries requires active interventions. Moreover, given the life history traits of corals and other deep-sea organisms with great longevity, slow growth, and late reproduction, documentation of fully restored habitats will require long-term monitoring.

The remoteness of deep-sea habitats dictates that restoration actions depend on expensive technologies (e.g., large ships and remotely operated vehicles, ROVs). The considerable cost of deep-sea habitat restoration actions in comparison with shallow water habitats may severely limit the capacity and feasibility for large-scale restoration actions, likely necessitating a combination of restoration approaches with assisted regeneration at small scales and natural regeneration (through fisheries closures, marine protected areas) at large scales. Given the slow growth rate of deep-sea habitats, their recovery will require restoration measures, punctuating the need to carefully consider the wisdom of ongoing and future extraction activities in deep-sea environments.

Challenges in marine ecosystem restoration

Current experience on marine ecosystem restoration suggests the need to consider 5 factors in order to maximize chances of success for restoration: (1) the identification of the best restoration protocols given the recovery target species/habitat; (2) the maintenance and monitoring of the restored habitat; (3) the choice of donor and recipient sites (to ensure physical conditions and biological characteristics in the restoration site as similar as possible to that of the donor site); (4) the use of advanced technologies to promote positive species interactions (some species improve the survival of other species by providing habitat or refuge), which may expedite recovery, and (5) the consideration of the potential climate change effects and regime shifts.

Restoration should prioritize geographic regions in which the target ecosystems for restoration were present prior to the disturbance. One of the main challenges is the identification of the areas where the habitats were historically present and that have completely lost these habitats. The shifts due to changing climate adds further challenge in the lack of knowledge of pre-disturbance baselines. Ultimately, this knowledge gap hinders any effective evaluation of the impact of anthropogenic activities, the actual degree of degradation, and therefore the choice of the restoration goals. The careful selection of an reference ecosystem agreed upon by all stakeholders can greatly facilitate the choice of goals and the establishment of a monitoring and evaluation protocol (Society for Ecological Restoration International Science & Policy Working Group 2004, Clewell and Aronson 2013).

Further challenges in restoration planning arise from complexity of "ownership" of ocean space, multiple stakeholders, jurisdictional considerations (e.g., local, national, international) to the logistical challenges of working in a comparatively vast and "open" system where many major components, including primary producers, constantly move. We must also recognize the conservation and restoration challenges associated with intentional habitat degradation by the emergent deep-sea mining industry that, if and when permitted, will affect immense seabed areas (polymetallic nodule mining), destroy long-lived coral and sponge gardens on seamounts (polymetallic cobalt crust mining), or degrade hydrothermal vent habitats (polymetallic sulfide mining) (Sharma 2019).

Restoration has also a cultural and societal dimension. Restoration actions receiving the support of different stakeholders (e.g., fisher-supplied coral by-catch for coral transplants) can increase efficiency and promote a culture of eco-health. The involvement of key stakeholders, including companies with expertise in offshore and marine industries, is paramount for the implementation phase of restoration projects. Management measures should consider and ensure long-term monitoring which may exceed the typical lifetime of restoration initiatives supported by public funding (e.g., research projects). Short-term monitoring (i.e., a few years) cannot provide a good indication of restoration trajectory or success. Stakeholders can contribute to the monitoring over time of the restoration interventions as well as the maintenance of the restored sites, with the support of experts. Relative success and failure of pilot restoration actions identify some of the best practices/methodologies and can help in developing restoration strategies based on sound ecological, ethical, and economic criteria. Positive restoration actions provide lessons learned, paving the way for steady refinement of future restoration actions and a higher social acceptance of the interventions (Montero-Serra et al. 2019, Tamburello et al. 2019, Verdura et al. 2018, Gagnon et al. 2020, Medrano et al. 2020, Meysick et al. 2020, Temmink et al. 2020).

Future perspectives of marine ecological restoration

Ecosystem restoration efforts have expanded rapidly in different marine environments. All over the world, these efforts have primarily focused on specific shallow-water habitats (e.g., coral reefs, seagrass beds, algal forests, salt marshes, oyster reefs). Availability of new technologies (e.g., ROVs, drones, observatories, satellites) for monitoring the restoration actions (such as satellites and drones for coastal systems and ocean observing networks in the deep sea, Danovaro et al. 2017, Johnston 2019, Yang et al. 2020) creates opportunities to assess the restoration trajectories, but effective restoration science and practice at sea covering a broad range of ecosystems, spanning from coastal habitats to the deep, dark ocean, require a concerted effort. Marine ecosystem restoration also requires upscaling to deliver transformative change. In the context of climate change and urban expansion, management must prioritize actions to ensure effective and resilient ecosystems as well as carbon sequestration and biodiversity benefits. Advancing this approach will also require understanding the socio-economic return and the legal, policy and governance frameworks, which can increase the inclusiveness, appeal, and durability of marine restoration (Wenting et al. 2020).

References

Abelson, A., Reed, D. C., Edgar, G. J., Smith, C. S., Kendrick, G. A., Orth, R. J. et al. (2020). Challenges for restoration of coastal marine ecosystems in the anthropocene. Frontiers in Marine Science 7: 544105.

Abhilash, P. C. (2021). Restoring the unrestored: strategies for restoring global land during the UN Decade on Ecosystem Restoration (UN-DER). Land 10: 201.

Alleway, H. K. and Connell, S. D. (2015). Loss of an ecological baseline through the eradication of oyster reefs from coastal ecosystems and human memory. Conservation Biology 29: 795–804.

Aronson, J. and Alexander, S. (2013). Ecosystem restoration is now a global priority: time to roll up our sleeves. Restoration Ecology 21: 293–296.

Aronson, J., Goodwin, N., Orlando, L., Eisenberg, C. and Cross, A. T. (2020). A world of possibilities: six restoration strategies to support the United Nation's Decade on Ecosystem Restoration. Restoration Ecology 28(4): 730–736.

Barbier, E. B. (2017). Marine ecosystem services. Current Biology 27(11): R507–R510.

Bayraktarov, E., Saunders, M. I., Abdullah, S., Mills, M., Beher, J., Possingham, H. P. et al. (2016). The cost and feasibility of marine coastal restoration. Ecological Applications 26: 1055–1074.

Bekkby, T., Papadopoulou, N., Fiorentino, D., McOwen, C. J., Rinde, E., Boström, C. et al. (2020). Habitat features and their influence on the restoration potential of marine habitats in Europe. Frontiers in Marine Science 7: 184.

Blignaut, J., Esler, K. J., de Wit, M. P., Le Maitre, D., Milton, S. J. and Aronson, J. (2013). Establishing the links between economic development and the restoration of natural capital. Current Opinion in Environmental Sustainability 5: 94–101.

Brumbaugh, R. D. and Coen, L. D. (2009). Contemporary approaches for small-scale oyster reef restoration to address substrate versus recruitment limitation: a review and comments relevant for the Olympia oyster, Ostrea lurida Carpenter 1864 J. Shellfish Res. 28: 147–161.

Butchart, S. H. M., Walpole, M., Collen, B., van Strien, A., Scharlemann, J. P. W., Almond, R. E. A. et al. (2010). Global biodiversity: indicators of recent declines. Science 328(5982): 1164–1168.

CBD. (2010). Secretariat of the Convention on Biological Diversity. Secretariat of the Convention on Biological Diversity, United Nations Environment Programme, Canada, pp. 1–54.

Clewell, A. F. and Aronson, J. (2013). Ecological Restoration: Principles, Values, and Structure of an Emerging Profession. 2nd Edition. Island Press, Washington, D.C.

Coppus, R., Romijn, J. E., Méndez-Toribio, M., Murcia, C., Thomas, E., Guariguata, M. R. et al. (2019). What is out there? a typology of land restoration projects in Latin America and the Caribbean. Environmental Research Communication 1: 041004.

Da Ros, Z., Corinaldesi, C., Dell'Anno, A., Gambi, C., Torsani, F. and Danovaro, R. (2021). Restoration of *Cymodocea nodosa* seagrass meadows: eefficiency and ecological implications. Restoration Ecology. https://doi.org/10.1111/rec.13313.

Da Ros, Z., Dell'Anno, A., Morato, T., Sweetman, A. K., Carreiro-Silva, M., Smith, C. J. et al. (2019). The deep sea: The new frontier for ecological restoration. Marine Policy 108: 103642.

Danovaro, R., Aguzzi, J., Fanelli, E., Billett, D., Gjerde, K., Jamieson, A. et al. (2017). An ecosystem-based deep-ocean strategy. Science 355(6324): 452–454.

Danovaro, R., Fanelli, E., Aguzzi, J., Billett, D., Carugati, L., Corinaldesi, C. et al. (2020). Ecological variables for developing a global deep-ocean monitoring and conservation strategy. Nature Ecology & Evolution 4: 181–192.

Duarte, C. M., Borja, A., Carstensen, J., Elliott, M., Krause-Jensen, D. and Marbà, N. (2015). Paradigms in the recovery of estuarine and coastal ecosystems. Estuaries and Coasts 38: 1202–1212.

Elliott, M., Burdon, D., Hemingway, K. L. and Apitz, S. E. (2007). Estuarine, coastal and marine ecosystem restoration: confusing management and science—A revision of concepts. Estuarine Coastal and Shelf Science 74: 349–366.

Fitzsimons, J. A., Branigan, S., Gillies, C. L., Brumbaugh, R. D., Cheng, J., Deangelis, B. M. et al. (2020). Restoring shellfish reefs: global guidelines for practitioners and scientists. Conserv. Sci. Prac. 2: e198. doi: 10.1111/csp2.198.

Fraschetti, S., McOwen, C., Papa, L., Papadopoulou, N., Bilan, M., Boström, C. et al. (submitted). Where is more important than how: determinants of restoration success in coastal and marine ecosystems. Frontiers in Marine Science.

Gagnon, K., Rinde, E., Bengil, E. G. T., Carugati, L., Christianen, M. J. A., Danovaro, R. et al. (2020). Facilitating foundation species: The potential for plant–bivalve interactions to improve habitat restoration success. Journal of Applied Ecology 57(6): 1161–1179.

Gann, G. D., McDonald, T., Walder, B., Aronson, J., Nelson, C. R., Jonson, J. et al. (2019). International principles and standards for the practice of ecological restoration. Second edition. Restoration Ecology 27(S1): S1–46.

Gerovasileiou, V., Smith, C. J., Sevastou, K., Papadopoulou, K.-N., Dailianis, T., Bekkby, T. et al. (2019). Habitat mapping in the European Seas - is it fit for purpose in the marine restoration agenda? Marine Policy 106: 103521.

Gillies, C. L., McLeod, I. M., Alleway, H. K., Cook, P., Crawford, C., Creighton, C. et al. (2018). Australian shellfish ecosystems: past distribution, current status and future direction. PloS One 13(2): e0190914.

Horn, S., Coll, M., Asmus, H. and Dolch, T. (2021). Food web models reveal potential ecosystem effects of seagrass recovery in the northern Wadden Sea. Restoration Ecology. https://doi.org/10.1111/rec.13328.

Howell, K. L., Hilário, A., Allcock, A. L., Bailey, D., Baker, M., Clark, M. R. et al. (2021). A decade to study deep-sea life. Nature Ecology & Evolution 5: 265–267.

Hynes, S., Chen, W., Vondolia, K., Armstrong, C. and O'Connor, E. (2021). Valuing the ecosystem service benefits from kelp forest restoration: a choice experiment from Norway. Ecological Economics 179: 106833.

IPBES. (2019). Global assessment report on biodiversity and ecosystem services of the Intergovernmental Science-Policy Platform on Biodiversity and Ecosystem Services. Brondizio, E. S., Settele, J., Díaz, S. and Ngo, H. T. (eds.). IPBES Secretariat, Bonn, Germany.

Johnston, D. W. (2019). Unoccupied aircraft systems in marine science and conservation. Annual Review of Marine Science 11: 439–463.

Luckenbach, M., Mann, R. L. and Wesson, J. A. (1999). Oyster Reef Habitat Restoration: a synopsis and synthesis of approaches; proceedings from the symposium, Williamsburg, Virginia, April 1995. Virginia Institute of Marine Science, William & Mary. http://doi.org/10.21220/V5NK51.

McDonald, T., Gann, G. D., Jonson, J. and Dixon, K. W. (2016). International standards for the practice of ecological restoration – including principles and key concepts. Society for Ecological Restoration, Washington, D.C. 48 pp.

Medrano, A., Hereu, B., Cleminson, M., Pagès-Escolà, M., Rovira, G., Solà, J. et al. (2020). From marine deserts to algal beds: *Treptacantha elegans* revegetation to reverse stable degraded ecosystems inside and outside a No-Take marine reserve. Restoration Ecology 28(3): 632–644.

Meysick, L., Norkko, A., Gagnon, K., Gräfnings, M. and Boström, C. (2020). Context-dependency of eelgrass-clam interactions: implications for coastal restoration. Marine Ecology Progress Series 647: 93–108.

Mitsch, W. J., Zhang, L., Waletzko, E. and Bernal, B. (2014). Validation of the ecosystem services of created wetlands: Two decades of plant succession, nutrient retention, and carbon sequestration in experimental riverine marshes. Ecological Engineering 72: 11–24.

Montero-Serra, I., Garrabou, J., Doak, D.F., Ledoux, J.-B. and Linares, C. (2018). Marine protected areas enhance structural complexity but do not buffer the consequences of ocean warming for an overexploited precious coral. Journal of Applied Ecology 56(5): 1063–1074.

O'Connor, E., Hynes, S., Chen, W., Papadopoulou, N. and Smith, C. (2021). Investigating societal attitudes toward marine ecosystem restoration. Restoration Ecology. doi: 10.1111/rec.13239.

OECD. (2016). The Ocean Economy in 2030. (OECD Publishing, Paris, France, 2016). https:/doi.org/10.1787/9789264251724-en.

Pendleton, L., Evans, K, and Visbeck, M. (2020). We need a global movement to transform ocean science for a better world. PNAS 117(18): 9652–9655.

Possingham, H. P., Bode, M. and Klein, C. J. (2015). Optimal conservation outcomes require both restoration and protection. PLoS Biology 13(1): e1002052.

Rohr, J. R., Bernhardt, E., Cadotte, M. W. and Clements, W. (2018). The ecology and economics of restoration: when, what, where, and how to restore ecosystems. Ecology and Society 23(2): 15.

Sala, E., Mayorga, J., Bradley, D., Cabral, R. B., Atwood, T. B., Auber, A. et al. (2021). Protecting the global ocean for biodiversity, food and climate. Nature. https://doi.org/10.1038/s41586-021-03371-z.

Saunders, M. I., Doropoulos, C., Bayraktarov, E., Babcock, R. C., Gorman, D., Eger, A. M. et al. (2020). Bright spots in coastal marine ecosystem restoration. Current Biology 30: R1500–R1510.

Seaman, W. (2007). Artificial habitats and the restoration of degraded marine ecosystems and fisheries. Hydrobiologia 580(1): 143–155.

Sharma, R. (2019). Environmental Issues of Deep-Sea Mining: Impacts, Consequences and Policy Perspectives. Springer, 577 p.

Society for Ecological Restoration International Science & Policy Working Group. (2004). The SER International Primer on Ecological Restoration. www.ser.org & Tucson, AZ: Society for Ecological Restoration International.

Suding, K. N. (2011). Toward an era of restoration in ecology: successes, failures, and opportunities ahead. Annual Review of Ecology, Evolution, and Systematics 42: 465–487.

Tamburello, L., Papa, L., Guarnieri, G., Basconi, L., Zampardi, S., Scipione, M. B. et al. (2019). Are we ready for scaling up restoration actions? An insight from Mediterranean macroalgal canopies. PLoS ONE 14(10): e0224477.

TEEB. (2010). The Economics of Ecosystems and Biodiversity: Mainstreaming the Economics of Nature: A Synthesis of the Approach, Conclusions and Recommendations of TEEB.

Temmink, R. J. M., Christianen, M. J. A., Fivash, G. S., Angelini, C., Boström, C. et al. (2020). Mimicry of emergent traits amplifies coastal restoration success. Nature Communications 11: 3668.

United Nations. (2016). The First Global Integrated Marine Assessment. Cambridge University Press.

Van Dover, C. L., Aronson, J., Pendleton, L., Smith, S., Arnaud-Haond, S., Moreno-Mateos, D. et al. (2014). Ecological restoration in the deep sea: Desiderata. Marine Policy 44: 98–106.

van Tatenhove, J. P. M., Ramírez-Monsalve, P., Carballo-Cárdenas, E., Papadopoulou, N., Smith, C. J., Alferink, L. et al. (2021). The governance of marine restoration: insights from three cases in two European seas. Restoration Ecology. https://doi.org/10.1111/rec.13288.

Verdura, J., Sales, M., Ballesteros, E., Cefalì, M. E. and Cebrian, E. (2018). Restoration of a canopy-forming alga based on recruitment enhancement: methods and long-term success assessment. Frontiers in Plant Sciences 9: 1832.

Vilas, D., Coll, M., Pedersen, T., Corrales, X., Filbee-Dexter, K. and Wernberg, T. (2021). Future trajectories of change for an Arctic deep-sea ecosystem connected to coastal kelp forests. Restoration Ecology. https://doi.org/10.1111/rec.13327.

Wenting, C., Van Assche, K. A. M., Hynes, S., Bekkby, T., Christie, H. C. and Gundersen, H. (2020). Ecosystem accounting's potential to support coastal and marine governance. Marine Policy 112: 103758.

White, J., Ruesink, J. L. and Trimble, A. C. (2009). The nearly forgotten oyster: Ostrea lurida Carpenter 1864 (Olympia oyster) history and management in Washington State J. Shellfish Res. 28(1): 43–50.

Worm, B., Barbier, E. B., Beaumont, N., Duffy, J. E., Folke, C., Halpern, B. S. et al. (2006). Impacts of biodiversity loss on ocean ecosystem services. Science 314: 787–790.

Wortley, L., Hero, J. M. and Howes, M. (2013). Evaluating eco-logical restoration success: A review of the literature. Restoration Ecology 5: 537–543.

Yang, B., Hawthorne, T. L., Hessing-Lewis, M., Duffy, E. J., Reshitnyk, L. Y., Feinman, M. et al. (2020). Developing an introductory UAV/drone mapping training program for seagrass monitoring and research. Drones 4(4): 70.

Index

About the Editors

Juan M. Molina is Assistant Professor of Vertebrate Zoology, Fisheries Resources and Aquaculture at Universidad Nacional del Sur, and also researcher at CONICET (National Research Council) in Argentina. His Ph.D. thesis focused on ecology and biology of fish assemblages. He was awarded an Emerging Leaders of the Americas to conduct a research collaboration in field physiology at Carleton University (Canada) as part of his doctoral studies. As postdoc, he was granted a POGO-SCOR visiting fellow to train in ecological modeling at Hokkaido University (Japan), an Endeavor's Research Fellowship to conduct research on shark metabolism in Monash University (Australia) and an Alexander von Humboldt research fellowship to study climate change implications on fish metabolism at ZMT (Germany) and CCMAR (Portugal). His research revolves around studying fish responses to challenges from their environments and from human activity in order to understand how they deal with these different stressors.

Gabriela E. Blasina is Professor of Fisheries Resources at Universidad Nacional del Sur and Researcher at CONICET. She obtained her Ph.D. in Biological Sciences from Universidad Nacional de Mar del Plata (Argentina) in 2013. Her Ph.D. thesis was based on biology and trophic ecology of estuarine fishes. Blasina's postdoctoral training was completed at Universidad Nacional del Sur. She specialized in Fisheries at Universidad de Cadiz (España). Her professional development focuses on fisheries resources conservation, specifically on topics that cover aspects related to community composition and structure, trophic ecology, ecomorphology, growth and effects of anthropic impact on coastal and estuarine fish species.

For Product Safety Concerns and Information please contact our
EU representative GPSR@taylorandfrancis.com Taylor & Francis
Verlag GmbH, Kaufingerstraße 24, 80331 München, Germany